INTRODUCTION TO NUMERICAL GEODYNAMIC MODELLING

This hands-on introduction to numerical geodynamic modelling provides a solid grounding in the necessary mathematical theory and techniques, including continuum mechanics and partial differential equations, before introducing key numerical modelling methods and applications. Fully updated, this second edition includes four completely new chapters covering the most recent advances in modelling inertial processes, seismic cycles and fluid-solid interactions, and the development of adaptive mesh refinement algorithms. Many well-documented, state-of-the-art visco-elasto-plastic 2D models are presented, which allow robust modelling of key geodynamic processes. Requiring only minimal prerequisite mathematical training, and featuring over 60 practical exercises and 90 MATLAB examples, this user-friendly resource encourages experimentation with geodynamic models. It is an ideal introduction for advanced courses and can be used as a self-study aid for graduates seeking to master geodynamic modelling for their own research projects.

TARAS GERYA is Professor in the Department of Earth Sciences at the Swiss Federal Institute of Technology (ETH-Zürich). He is an expert in numerical geodynamic modelling, with his current research focusing on subduction and collision processes, ridge-transform oceanic spreading patterns, intrusion emplacement into the crust, generation of earthquakes, fluid and melt transport in the lithosphere, Precambrian geodynamics, formation of terrestrial planets and evolution of life. In 2008 he was awarded the Golden Owl Prize by ETH students for his teaching on continuum mechanics and numerical modelling.

INTRODUCTION TO NUMERICAL GEODYNAMIC MODELLING

SECOND EDITION

TARAS GERYA

Swiss Federal Institute of Technology (ETH), Zürich

CAMBRIDGE
UNIVERSITY PRESS

CAMBRIDGE
UNIVERSITY PRESS

University Printing House, Cambridge CB2 8BS, United Kingdom

One Liberty Plaza, 20th Floor, New York, NY 10006, USA

477 Williamstown Road, Port Melbourne, VIC 3207, Australia

314–321, 3rd Floor, Plot 3, Splendor Forum, Jasola District Centre, New Delhi – 110025, India

79 Anson Road, #06–04/06, Singapore 079906

Cambridge University Press is part of the University of Cambridge.

It furthers the University's mission by disseminating knowledge in the pursuit of
education, learning, and research at the highest international levels of excellence.

www.cambridge.org
Information on this title: www.cambridge.org/9781107143142
DOI: 10.1017/9781316534243

© Taras Gerya 2010, 2019

First published 2010
Second edition 2019

Printed in the United Kingdom by TJ International Ltd. Padstow Cornwall

A catalogue record for this publication is available from the British Library.

Library of Congress Cataloging-in-Publication Data
Names: Gerya, Taras, author.
Title: Introduction to numerical geodynamic modelling / Taras Gerya.
Description: Second edition. | Cambridge, United Kingdom ; New York, NY :
Cambridge University Press, 2019. | Includes bibliographical references
and index.
Identifiers: LCCN 2018045345 | ISBN 9781107143142 (hardback)
Subjects: LCSH: Geophysics – Mathematical models. | Geodynamics – Mathematical
models.
Classification: LCC QE501.4.M38 G47 2019 | DDC 550.1/5118–dc23
LC record available at https://lccn.loc.gov/2018045345

ISBN 978-1-107-14314-2 Hardback

Additional resources for this publication at www.cambridge.org/gerya2e.

Contents

Preface to the second edition

The main reason for writing this second edition is the rapid recent progress in the field of numerical geodynamic modelling, which is one of the most dynamic and fast growing fields of the modern Earth sciences. Since the publication of the first edition in 2010 (almost a decade ago ...), several important research directions have become very prominent and advanced in computational geodynamics, such as investigation of coupled solid-fluid processes, coupling of geodynamic evolution to surface processes, modelling of seismic cycles at plate boundaries, development of adaptive grid refinement methods and free surface stabilization approaches, elaboration of more accurate continuity-based Lagrangian advection algorithms, development and broad application of new efficient 3D visco-elasto-plastic highly parallelized numerical modelling tools etc. In order to account for some of these exciting novelties, I both significantly revised some of the previously published chapters (especially numerical modelling of advection processes in Chapter 8 and numerical treatment of visco-elasto-plastic materials in Chapters 12 and 13) and added four new chapters focusing on recent numerical advances in

- modelling of inertial processes (Chapter 14),
- modelling of seismic cycles (Chapter 15),
- modelling of coupled fluid-solid processes (Chapter 16) and
- development of adaptive mesh refinement algorithms (Chapter 17).

As in the first edition, a single relatively simple numerical modelling method (combination of staggered finite differences with marker-in-cell techniques, SFD+MIC) and MATLAB programming are used uniformly throughout this textbook. I hope you will enjoy this new edition!

Acknowledgements

In relation to the first and second editions of this book I would like to acknowledge many people and I will try to do this in chronological order. I am grateful to my wife Irina for inspiration and support. I am grateful to my first supervisor Alexander Nozhkin for giving me a chance to start my scientific career in 1984. I am grateful to my PhD supervisor and good friend Leonid Perchuk (1933–2009) for suggesting that I start with numerical modelling in 1995 (a long time ago, indeed, but I feel like it was yesterday). I am grateful to Alexander Simakin for explaining to me in a few words what numerical modelling is about, when I had just started to learn it and was really puzzled what to do with all these partial differential equations written in textbooks (he told me that I simply have *to compose and solve together as many linear equations as I have unknowns* and this is really the main idea behind numerical modelling). I am grateful to Roberto Weinberg and Harro Schmeling for their excellent paper on polydiapirs, published in 1992, which introduced me to the marker-in-cell techniques when I had just started. I am grateful to Alexey Polyakov for suggesting that I use upwind differences for solving the temperature equation when I was programming my first themomechanical code. I am grateful to Walter Maresch and Bernhard Stöckhert for cooperating with me on modelling of subduction processes, which is a challenging topic and stimulated a lot of my code developments. I am grateful to David Yuen – my co-author in many important numerical modelling papers – for our long-term joint work and friendship (after we met in 2001 at the AGU Fall meeting in San Francisco) and for many great suggestions concerning this book. I am grateful to Paul Tackley for telling me about the fully staggered grid in 2002 (I was using a half-staggered grid before that time) and for introducing me to multigrid in 2005 as well as for many years of our joint studies and good suggestions concerning this book. I am grateful to Jean-Pierre Burg for inviting me to ETH-Zürich and cooperating with me on challenging modelling projects (which again triggered many code developments) and for being a very careful and constructive first reader of the initial version of this book. I am grateful to Yuriy Podladchikov for many stimulating discussions, continuous healthy criticism and challenging suggestions (for example, adding elasticity and plasticity to my codes that 'spoiled' six months of my life in 2004). I am grateful to Boris Kaus for arguing and discussing numerics with me, which we both like, and for great detailed comments and suggestions on the initial version of this book. I am grateful to James Connolly for fruitful work on coupling of thermodynamics and phase petrology with thermomechanical experiments (what I call petrological-thermomechanical numerical modelling). I am very grateful to David May for creative checking of both the first and the second editions of this textbook and for giving many good hints about the content. I am grateful to my colleague and friend Evgenii Burov (1963–2015) for our great cooperation in the field of high-resolution 3D numerical geodynamic modelling, which unfortunately stopped very sadly in 2015. I am very grateful to Viktoriya Yarushina for her invaluable help with the chapter on solid-fluid coupling and to Anton Popov for the generous sharing of his new 3D elastic stress rotation algorithms. I am grateful to Tobias Keller for his detailed and insightful comments on the hydrothermomechanical modelling chapter. I am grateful to Alexey Perchuk for our long-term friendship, exciting discussions and fruitful cooperation in the field of petrology and geodynamic modelling. I am grateful to my son Bogdan for computer and graphic assistance, to my parents Lyudmila and Viktor (1927–2015), my brother Artem and my entire family for moral support. I am grateful to all my students and co-authors for their bright ideas and great work done together. Finally, I am grateful for the generous support of my numerical modelling projects by Alexander von Humboldt foundation research fellowships and by many ETH, SNF and EU research grants.

Introduction

Theory: What is this book? What this book is not. Get started. Seven golden rules for learning the subject. Short history of geodynamics and numerical geodynamic modelling. Few words about programming and visualization. Ten programming rules.
Exercises: Starting with MATLAB. Visualization exercise.

What is this book?

This book is a practical, hands-on introduction to numerical geodynamic modelling for inexperienced people, i.e. for young students and newcomers from other fields. It does not require much background in mathematics or physics and is therefore written with a maximum amount of simple technical details. If you are inexperienced – this book is yours!

What this book is not

This book is not a treatise or a compendium of knowledge for experienced researchers. It does not contain large overviews of existing numerical techniques, and only simple approaches are explained. If you are experienced in numerical methods, look at Chapters 12–21 where some advanced numerical techniques and model examples are discussed. Then you can decide if you wish to read about the technical details presented in these and other chapters.

Get started

Already decided?! Then let us get started! In recent decades numerical modelling has become an essential approach in geosciences in general and in geodynamics in particular. This is a very natural process ('instinctive evolution') since human scales of direct observation are extremely limited in both time and space (depth) and since rapid progress

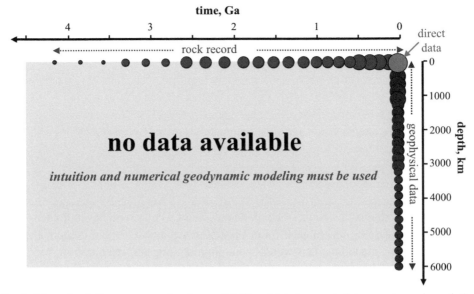

Fig. 1. Time-depth diagram presenting the availability of data for constraining geodynamic evolution of the Earth (Gerya, 2014b). The size of data points reflects the abundance of available data. This is obviously a simplified view since for a spherical Earth such a diagram should be four dimensional.

in computer technology offers every day new and exceptional possibilities to explore sophisticated mathematical models. This is true in every discipline, and even industrial applications.

Numerical geodynamic modelling naturally 'compensates' for the fundamental unavailability of data needed for constraining the evolution of the Earth's interior and surface over time, which is the subject of geodynamics. The following simple exercise explores this subject in the context of the availability of data. Let us imagine an ideally symmetrical Earth with physical properties (density, viscosity, temperature, etc.) as functions of depth and time. A simple two-dimensional time-depth diagram covering the Earth's entire history and its interior will thus be a schematic representation of the subject of geodynamics (Fig. 1). The entire diagram should then be covered by data points characterizing the physical state of the Earth at different depths, ranging from 0 to 6000 km, and for different moments of geological time, ranging from 0 to around 4.5 billion years ago. However, the unfortunate fact for geodynamics is that observations for such systematic coverage are only available along the two axes of the diagram: geophysical data for the present-day Earth structure and the historical record in rocks formed close (typically within a few tens of kilometres) to the Earth's surface. The rest of the diagram is thus fundamentally devoid of observational data, so we have to rely on something else. What else can we use? Scientific intuition based on geological experience and modelling based on fundamental laws of continuum mechanics! However, our intuition cannot always be suitable for geodynamical processes that are completely beyond human scales in time (a few years) and space (a few metres). We have to accept that some of these processes could look completely counterintuitive to us. The ways in which various geodynamic processes interact with each other can also be very difficult to conceive using only scientific intuition. This is why *intuition in geodynamics should be – must be – assisted and calibrated by modelling*. In a way, modelling helps train our intuition for very deep and very slow geological processes that cannot be observed directly. Another role of modelling is the quantification of geodynamic processes based on the sparse array of available observations. Consequently, the systematic use of numerical modelling is crucial to develop, test, and quantify geodynamic hypotheses – and perhaps most questions about the Earth.

At present, numerical modelling in geosciences is widely used for both testing and generating hypotheses, thereby strongly pushing geology from an observational, intuitive to a deductive, predictive natural science. Geo-modelling and geo-visualization play a strong role in relating different branches of geosciences. Therefore, it has become necessary to have some knowledge about numerical techniques before planning and conducting state of the art interdisciplinary research in any branch of geosciences. In this respect, geodynamics is traditionally 'infected' by numerical modelling and pushes the progress of numerical methods in geosciences.

Before starting with numerical modelling we should consider one of the very popular 'myths' among geologists, who often declare (or think) something like:

Numerical modelling is very complicated; it is not affordable for persons with traditional geological background and should be performed by mathematicians.

I was thinking like that before I started. I always remember my feeling when I heard for the first time the expression, 'Navier–Stokes equation'. 'Ok, forget it! This is hopeless.' – did I think at that time, and that was wrong. Therefore, let me formulate the seven 'golden rules' elaborated during my learning experience.

Golden Rule 1. *Numerical modelling is simple and is based on simple mathematics.*
  All you need to know is:

- derivatives and
- linear algebra.

Most of this 'complicated' mathematical knowledge is learned in school before we even start to study at university! I often say to my students that all is needed is:

  strong MOTIVATION,
  algorithmic THINKING (ability to 'translate' generic tasks into code algorithms),
  usual MATH,
  clear EXPLANATIONS,
  regular EXERCISES.

Motivation and algorithmic thinking are most important, indeed …

Golden Rule 2. *When numerical modelling looks complicated see Rule 1.*

Golden Rule 3. *Numerical modelling consists of solving partial differential equations (PDEs).*

There are only a few equations to learn (e.g. Lynch, 2005). They are generally not complicated, but it is essential to learn and understand them gradually and properly. For example, to model the broad spectrum of geodynamic processes discussed in this book, it is necessary to know three principal conservation PDEs only:

- the equation of continuity (conservation of mass),
- the equation of motion (conservation of momentum – *Navier–Stokes equation!*),
- the temperature equation (conservation of energy).

So, only *three* equations have to be understood and not tens or hundreds of them!

Golden Rule 4. *Read books on numerical methods several times.*

There are many excellent books on numerical methods. Many of these books are, however, written for physicists and engineers and need effort to be 'digested' by people with a traditional geological background. The situation has improved recently after several books written by experienced geodynamicists have appeared on the market (Gerya, 2010a; Ismail-Zadeh and Tackley, 2010; Simpson, 2017; Morra, 2018).

Golden Rule 5. *Repeat the transformations of equations involved in numerical modelling.*

These transformations are generally standard and trivial, but repeating them develops a familiarity with the PDEs (maybe you will even start to like them ☺), and allows understanding the structure of the different PDEs. This book, by the way, is full of such trivial detailed transformations – follow them carefully!

Golden Rule 6. *Visualization is important!*

Without proper visualization of results, almost nothing can be done with numerical modelling (Fig. 2). Modellers often spend more time on visualization than on computing and programming.

Golden Rule 7. *Ask!*

This is the most efficient way of learning. In numerical geodynamic modelling, many small numerical know-hows exist. They are extremely important, but rarely discussed in publications (in contrast to this book ☺). Indeed, *do not rely solely on asking* – first try hard to find your own answer to the problem you want to solve numerically.

Short history of geodynamics and numerical geodynamic modelling

The numerical modelling approaches discussed in this book are adopted for solving *thermomechanical* geodynamic problems. Geodynamics is dynamics of the Earth – a

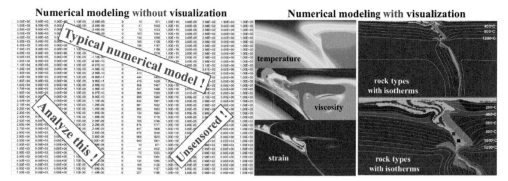

Fig. 2. Rule 6. Visualization is important!

core geological subject that has been very actively progressing during the last century, especially since the establishment of plate tectonics in the 1960s. This was a really great time for geology that 'drifted' strongly and rapidly from a descriptive (qualitative) field, to a predictive (quantitative) physical science. The overall history of the development of geodynamics was not, indeed, very 'dynamic' but rather slow and complicated. A brilliant introduction to this field (which I strongly recommend you to read) is written by Donald L. Turcotte and Gerald Schubert (2002, 2014). According to this introduction and further literature search, the following steps were notable historically in understanding the Earth as a dynamic system.

1620: *Francis Bacon pointed out the similarity in shape between the west coast of Africa and the east coast of South America.*

This was about 400 years ago (!) and several centuries were needed before we could start to interpret this similarity.

1665: *Athanasius Kircher, in his two-volume 'Mundus subterraneus', probably the first printed work on geophysics and volcanology, held that much of the phenomena on Earth were due to the fact that there is 'fire' under the terra firma.*

This was, indeed, very unusual teaching for those days (about 350 years ago!) and very much in line with the thermal origin of mantle convection.

Early part of the seventeenth century: *Gottfried Wilhelm Leibniz proposed that the Earth has a molten core and anticipated the igneous nature of the mantle.*

This began our understanding of the Earth as a hot layered planetary body. One really needed vision to guess this around 300 years ago!

Later part of the nineteenth century: *The fluid-like behaviour of the Earth's mantle was established, based on gravity studies; mountain ranges have low density roots.*

This crucial finding was 'coupled' to Earth dynamics only one hundred years later and was not explored in the continental drift hypothesis.

1895–1915: *The unforeseen discovery of radioactivity.*

That 'killed' the concept of progressive dissipation of the heat of the Earth, and the correlative contraction, as the mechanism for orogenic stresses. It also changed the age of the Earth and stratas by an order of magnitude … All this forced further serious rethinking of geological concepts about dynamic processes shaping the Earth.

1910: Frank B. Taylor formulated the Continental Drift hypothesis.

This was the real beginning of 'drifting' toward plate tectonics, still a long way to go.

1912–1946: Alfred Wegener further developed the continental drift hypothesis, and showed a correspondence of the geological provinces, relict mountain ranges and fossil types. Driving forces – tidal rotation of the Earth. Single protocontinent – Pangaea.

The principal question is considered to be, 'why do continents move?' and 'what are the driving forces?' and not yet, 'how do continents move?' and 'what is the movement mechanism?'

1916: Gustaaf Adolf Frederik Molengraaff proposed that mid-ocean ridges are formed by seafloor spreading as the result of the movement of continents in order to account for the opening of the Atlantic Ocean as well as the East Africa Rift.

The mid-ocean ridges were 're-discovered' for plate tectonics 40 years later …

1924: Harold Jeffreys showed that Wegener's forces were insufficient for moving continents.

Computing forces for testing a geodynamic hypothesis is one of the core principles of modern geodynamics as well! Another point to learn is that opposition to the continental drift hypothesis using physical arguments was always strong and probably strongly delayed the theory of plate tectonics.

1931: Arthur Holmes suggested that thermal convection in the Earth's mantle can drive continental drift.

This crucial idea answered a question about driving forces, but not questions about the movement mechanisms. It was known from seismic studies that the Earth's mantle is in a solid state and that elastic deformation does not allow thousands of kilometres of motion of the continents.

1935: N. A. Haskell established the fluid-like behaviour of the mantle (viscosity 10^{20} Pa s) based on the analysis of beach terraces in Scandinavia and the existence of post-glaciation rebound.

Actually, this had also been established earlier from inferring crustal roots. The question about the physical mechanisms of solid-state mantle deformation remained open.

1937: Alexander du Toit suggested the existence of two protocontinents – Laurasia and Gondwanaland, separated by the Tethys Ocean.

This is a really dramatic story: geologists were continuously developing and supporting the continental drift hypothesis, but the general idea of large lateral displacements of continents was continuously rejected by geophysicists.

1950s: Understanding of the world-wide network of mid-ocean ridges was improved during extensive exploration of the seafloor.

Evidence is critically growing in line with Molengraaf's ideas …

1950s: Mechanisms of solid-state creep of crystalline materials were discoved which were applicable, for example, to the flow of ice in glaciers.

The answer to the second crucial question was finally found in materials science!

Breakthrough! The great 1960s started!

1960s: Paleomagnetic studies led to the finding of regular patterns of magnetic anomalies on the seafloor.

1962: Harry Hess suggested that the seafloor was created at the axis of the ridge.

In fact, this was a refinement of Molengraaf's hypothesis.

1965: B. Gordon proposed the quantitative link between solid-state creep and mantle viscosity.

1968: Jason Morgan formulated the basic hypothesis of plate tectonics (mosaic of rigid plates in relative motion with respect to one another as a natural consequence of mantle convection).

1968: Isacks and co-workers attributed earthquakes, volcanoes, and mountain building to plate boundaries.

1967–1970: This saw the development and broad acceptance of plate tectonics.

Before this time, continental drift was opposed by geophysicists because of the rigidity of the solid elastic mantle and the 'absence' of physical mechanisms allowing horizontal displacements of thousands of kilometres for continents.

The crucial point that was finally understood by the geological community is that both viscous (i.e., fluid-like) and elastic (i.e., solid-like) behaviour is a characteristic of the Earth depending on the time scale of deformation. The Earth's mantle, which is elastic on a human time scale, is viscous on geological time scales (>10 000 years) and can be strongly internally deformed due to solid-state creep. There is an amazing substance demonstrating a similar 'dual' viscous-elastic behaviour. This is silicon putty or 'silly putty' which is frequently used as an analogue of rocks in experimental tectonics. It deforms like clay in the hands, but when dropped on the floor it jumps up like a rubber ball (see animation **Silly_putty.m1v**).

Plate tectonics established both a conceptual and a physical basis of geodynamics. The next rapid development of numerical methods of continuum mechanics in this field is the logical consequence of both theoretical and technological progress. The snapshot-like history of 2D-3D numerical geodynamic modelling (1D models appeared even earlier!) looks as follows (partly subjective literature-web-search-based view, more details on this issue can be found in several overviews of mantle convection modelling: Richter, 1978; Schubert, 1992; Bercovici, 2007).

1970: First 2D numerical models of subduction (Minear and Toksöz, 1970).

Exactly at the time when the 'plate tectonics era' had just started! The first subduction model was thermo-kinematic, with a prescribed velocity field corresponding to a down-going slab inclined at 45°.

1971: First 2D mantle thermal convection models (Torrance and Turcotte, 1971).

This paper discussed possible implications of mantle convection with temperature-dependent viscosity for continental drift. Thermomechanical models based on the stream function formulation for the mechanical part were explored. A rectangular model domain, with a temperature-dependent viscosity and resolution of up to 22×16 nodal points was used.

1972, 1978: *First 2D numerical (finite element) models of salt dome dynamics (Berner et al., 1972; Woidt, 1978).*

Before this, geodynamic modelling studies of crustal diapirism used analytical and analogue modelling approaches. The paper by Woidt (1978) pointed out the inconsistency of the numerical approach used by Berner et al. (1972).

1977–1980: *First 2D thermal-chemical mantle convection models (Keondzhyan and Monin, 1977, 1980).*

A binary stratified medium was used to study the effects of compositional layering on mantle convection.

1978: *First numerical models of continental collision (Bird, 1978; Daignières et al., 1978).*

Mechanical models explored the finite element approach.

1983: *First numerical models of subduction initiation (Matsumoto and Tomoda, 1983).*

Remarkable geodynamic modelling ahead of its time! The numerical solution was based on stream function formulation combined with marker-in-cell technique and free surface implementation based on the 'sticky water' approach, which became widespread *two decades later.*

1985–1986: *First 3D spherical mantle convection models (Baumgardner, 1985; Machetel et al., 1986).*

The first 3D models were spherical and not Cartesian as one would expect. Also, for some reason, the first paper appeared in the *Journal of Statistical Physics*, which is not really a geophysical journal …

1988: *First 3D Cartesian mantle convection models (Cserepes et al., 1988; Houseman, 1988).*

Since the 1980s, numerical geodynamic modelling has been developing very rapidly in terms of both the number of applications and the numerical techniques explored. In the last decade, several textbooks on numerical geodynamic modelling have been published (Gerya, 2010a; Ismail-Zadeh and Tackley, 2010; Simpson, 2017; Morra, 2018), which make it more accessible for geoscientists and help in teaching it to students. At present, geodynamic modelling stands as one of the most dynamic, cross-disciplinary and advanced fields of modern Earth sciences.

Few words about programming, visualization and debugging

In the frame of this book MATLAB is used for the exercises and for visualization. This is a good language of choice for people starting with modelling as it allows both easy computing and visualization. C and FORTRAN are often used for advanced studies that involve usage of supercomputers and computer clusters. In these studies, visualization is mostly done as a post-processing step that allows independent use of specialized visualization packages. In our short course, we are more interested in seeing results instantaneously, during computations. In addition, MATLAB greatly simplifies the solving of systems of linear equations, which is the core of numerical modelling. Another convenient programming language with growing popularity in geodynamics is PYTHON (Morra, 2018).

In this course, we will consider many example programs, since learning *to write programs* (*and not just using them*) is an essential part of numerical geodynamic modelling. There are ten important programming rules (which I call *Bug Rules*), which you may want to follow when writing your own programs.

Bug Rule 1: ***Think before programming!*** Think carefully about the algorithm of your new code and the most efficient way of making modifications to your old code – you will then develop the program faster and more efficiently and will not need too much code re-thinking and re-writing.

Bug Rule 2: ***Comment!*** Making comments in the code is essential to enable the code to be used, debugged and modified correctly. The ratio between comment lines and program lines in a good numerical code is larger then 1:1. Do not be lazy, explain every program line – this will save you a lot of time afterwards!

Bug Rule 3: ***Programming makes bugs!*** We always introduce *bugs* (i.e. programming errors) while writing a code. We typically introduce at least one bug when we modify one single line and we have to test the modified code until we find the bug!

Bug Rule 4: ***Programming means debugging!*** Be prepared that only 1% of the time will be spent on programming and 99% of your time will be spent on debugging.

Bug Rule 5: ***Nice looking codes are often more difficult to debug!*** Do not try to write nice looking codes; try to write codes that are easy to debug! Use the most simple and explicit code logic and structure. Be very pragmatic; do not make changes to previously debugged code sections unless absolutely necessary. Go for important code changes only; do not 'fight' for better looking code structure or minor improvements of computational efficiency.

Bug Rule 6: ***Bugs that are the most difficult to find are trivial ones!*** There are three types of most common bugs:
- errors in index (90% of your bugs!), e.g. $y = x(i,j) + z(12)$ instead of $y = x(j,i) - z(2)$;
- errors in sign, e.g. $y = x + z$ instead of $y = x - z$ or $y = 1e - 19$ instead of $y = 1e + 19$;
- errors in order of magnitude, e.g. $y = 0.0831$ instead of $y = 0.00831$.

Do not be surprised that finding these 'trivial' bugs will sometimes be very difficult (we simply tend to overlook them) and will take a lot of time – this is normal.

Bug Rule 7: ***If you see something strange – there is a bug!*** Be suspicious, do not ignore even small strange things and discrepancies that you see when computing with your code – in 100% of cases you will find that a bug is the cause. Never try to convince yourself (although this is what we typically tend to do) that a single last digit discrepancy in results with the previous version of the code is due to computer accuracy – it is due to either old or new bugs!

Bug Rule 8: *A single bug can ruin a 10 000-line code!* We should really be motivated to carefully debug and test codes. Do not think that one single small error in the code can be ignored – it will spoil the results of months of calculations.

Bug Rule 9: *A wrong model looks beautiful and realistic!* Often erroneous models do not look bad or strange and some of them are really beautiful. Therefore, be prepared that of the numerical modelling results you like, some are actually wrong …

Bug Rule 10: *Creating a good, correct and nicely working code is possible!* This is what should motivate us to follow the nine previous rules!

Many years of correcting students' codes made me convinced that there is only one robust (although not really elegant and efficient) way of finding bugs in a code: write two independent versions (i.e., without copy-pasting) of the same code (preferably by two different people) and compare computational results for well controlled conditions. If the results are different – there is at least one bug in at least one of the two codes. Then, try to copy-paste routines from one code version to the other until the results become identical. This helps to find routines that produce different results and to clarify reasons (bugs) for the discrepancy. Experience shows that it is very unlikely that two independent code versions will have identical bugs (even if both are written by the same person). Adding more code versions (and people) to the 'pool' will further help debugging.

Units

In this book, the metre-kilogram-second (MKS) system is used in all basic equations as a standard, with only occasional specified deviations toward other conventional units widely used in geosciences (kbar, °C etc.).

How to use this book

Once again, this is a textbook, which is primarily aimed at people inexperienced with numerical methods. Therefore, it is organized in a way that, according to my personal learning and teaching experience, provides the easiest path for learning the basics of continuum mechanics and numerical geodynamic modelling. Follow it from one chapter to the next and do all the exercises. Do all the programming by yourself and study code examples ONLY when you are stuck or unsure what to do (all MATLAB codes quoted in the text are provided with this book, see the Appendix). The complexity of the programming exercises gradually increases from one chapter to the next, introducing more and more complex aspects of continuum mechanics and numerical modelling. Just trust this way and *don't give up*!

Programming exercises

Exercise Introduction.1.

Open MATLAB and use it for the first time. Study the following (use MATLAB Help to read about various functions and operations).

(a) Defining variables, vectors and matrices.
(b) Using mathematical functions (+, -, *, .*, /, ./, ^, .^, exp, log, log10, etc.).
(c) Opening/closing text files and loading/writing data from/to them (*fopen, fclose, fscanf, fprintf*).
(d) Plotting of data in 2D and 3D (*figure, plot, pcolor, surf, xlabel, ylabel, shading, light, lighting, axis, colorbar*).
(e) Programming loops (*for, while, end*) and conditions (*if, else, end, switch, case*, &&, ||, ==, ~=, >, <, >=, <= etc.).

Exercise Introduction.2.

Write your first MATLAB code for visualizing the sin and cos functions in 2D (*plot, pcolor, contour*) and 3D (*surf, light, lighting*). An example is in **Visualisation_is_important.m**.

1

The continuity equation

> **Theory:** Definition of a geological medium as a continuum. Field variables used for the representation of a continuum. Methods for definition of the field variables. Eulerian and Lagrangian points of view. Continuity equation in Eulerian and Lagrangian forms and their derivations. Advective transport term. Continuity equation for an incompressible fluid.
> **Exercises:** Computing the divergence of a velocity field in 2D.

1.1 Continuum – what is it?

What we should understand from the very beginning is that geodynamics considers major rock units, such as the Earth's crust and mantle, as *continuous geological media*. *Continuity* of any medium implies that, on a macroscopic scale, the material under consideration does not contain *mass-free voids or gaps* (there can indeed be pores or cavities but they are also filled with some continuous substances). Different physical properties of a continuum may vary at every geometrical point and we thus need a *continuous description*. In *continuum mechanics*, the physical properties of a continuum (*field properties*) are described by *field variables* such as pressure, temperature, density, velocity etc. There are three major types of field variables:

scalars (e.g., pressure, temperature, density),
vectors (e.g., velocity, mass flux, heat flux),
tensors (e.g. stress, strain, strain rate).

Field variables can be represented in a *fully continuous* manner (analytical expressions, Fig. 1.1a) or in a *discrete-continuous* way (by arrays of values which characterize selected *nodal* geometrical points, Fig. 1.1b–d). In the latter case, various linear and non-linear *interpolation* rules are used to calculate values of field variables between the nodal points.

Continuity also implies that displacements of different portions of the medium are not fully independent. These displacements must proceed without creating macroscopic voids and gaps: if some rocks are displaced *from* a certain area (for example due to tectonic

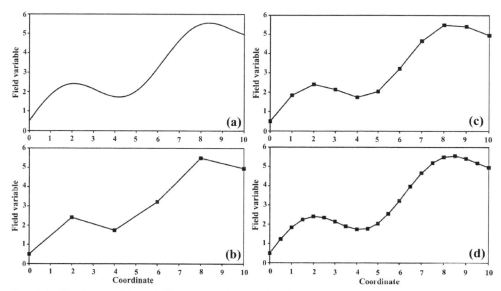

Fig. 1.1. Continuous (a) and discrete-continuous (b)–(d) representations of a field variable as a function of coordinates. Note that in the case of the discrete-continuous representation with linear interpolation between nodal points, the representation accuracy notably increases with increasing number of nodal points (compare a with b, c and d).

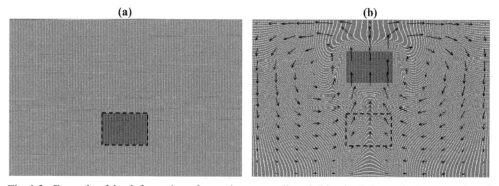

Fig. 1.2. Example of the deformation of a continuous medium (white dots) due to the buoyant rise of a rigid block (black dots): (a) initial stage, (b) final stage with the corresponding velocity field indicated by arrows. Note that no voids are formed where the block was initially located (dashed contour). Individual black and white dots in (a) and (b) correspond to different Lagrangian points displaced by the flow. Diagrams are computed numerically with the code developed by Gerya and Yuen (2003a), and an animation is shown in the file **Continuity.ppt** associated with this chapter.

extrusion), other rocks must come *into* this area and substitute the displaced fragment (Fig. 1.2). In a way, this type of continuous behaviour is very similar to that of water or, generally, any fluid which can be described with *fluid mechanics* (e.g. Landau and Lifshitz, 1987; Kundu, and Cohen, 2002). Since on long time scales rocks behave like slowly creeping fluids, geodynamic processes in the Earth's mantle, as for example mantle convection, are often referred to as processes of *geophysical fluid dynamics*.

1.2 Continuity equation

Our qualitative, intuitive understanding of continuity can, indeed, be transformed into a quantitative mathematical formalism. This formalism is widely used in numerical geodynamic modelling in the form of a *continuity equation*, which describes the *conservation of mass* during the displacement of a continuous medium. Let us write this equation and try to understand its structure in detail.

The first thing that we have to learn is that the form of the mass conservation equation (and many other *time-dependent* conservation equations) can be either *Eulerian* or *Lagrangian* depending on the nature of the geometrical point for which this equation is written.

The Eulerian continuity equation is written for an *immobile* or fixed point in space; it has the form

$$\frac{\partial \rho}{\partial t} + \mathrm{div}(\rho \vec{v}) = 0. \tag{1.1a}$$

Or, in a slightly different symbolic notation often used in continuum mechanics,

$$\frac{\partial \rho}{\partial t} + \nabla \bullet (\rho \vec{v}) = 0, \tag{1.1b}$$

where $\partial/\partial t$ is the Eulerian time derivative; ρ is the local density, which characterizes the amount of mass per unit volume (kg/m^3); $\vec{v}$ is local velocity (m/s) and div() or $\nabla \bullet$ denotes the divergence operator. The divergence is a scalar function of a vector field, and is defined as follows:

$$\text{in one dimension (1D)} \quad \mathrm{div}(\vec{v}) = \frac{\partial v_x}{\partial x}, \tag{1.2a}$$

$$\text{in two dimensions (2D)} \quad \mathrm{div}(\vec{v}) = \frac{\partial v_x}{\partial x} + \frac{\partial v_y}{\partial y}, \tag{1.2b}$$

$$\text{in three dimensions (3D)} \quad \mathrm{div}(\vec{v}) = \frac{\partial v_x}{\partial x} + \frac{\partial v_y}{\partial y} + \frac{\partial v_z}{\partial z}, \tag{1.2c}$$

where x, y, and z are Cartesian coordinates and v_x, v_y and v_z are components parallel to the respective coordinate axes of the velocity vector $\vec{v}$ (Fig. 1.3). In simple words, the divergence of a vector in a given point is positive when the surrounding vector field is directed predominantly outward of the point (divergent flow, Fig. 1.4a) and is negative when this field is directed predominantly toward the point (convergent flow, Fig. 1.4b).

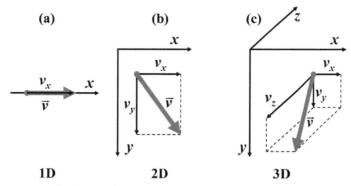

Fig. 1.3. Components of a local velocity vector $\vec{v}$ (grey arrow) in one (a), two (b), and three (c) dimensions.

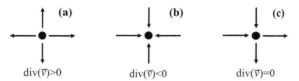

Fig. 1.4. Examples of divergent (a), convergent (b) and neutral (c) 2D velocity fields around a point.

The Lagrangian continuity equation is written for a *moving* point of reference; it has the form

$$\frac{D\rho}{Dt} + \rho \operatorname{div}(\vec{v}) = 0, \tag{1.3a}$$

or

$$\frac{D\rho}{Dt} + \rho \nabla \bullet \vec{v} = 0, \tag{1.3b}$$

where D/Dt is the Lagrangian time derivative and the other parameters were explained before (see Eq. 1.1).

1.3 Eulerian and Lagrangian points – what is the difference?

Understanding the difference between Eulerian and Lagrangian points is *fundamental* for continuum mechanics. A Lagrangian point is strictly connected to a single material point and is moving with this point. Therefore, the same material point is always found in a given Lagrangian point independent of the moment of time. For this reason, the Lagrangian time derivative of density $D\rho/Dt$ (i.e., changes in density with time in the Lagrangian point) is also called the *substantive* or *objective* time derivative. On the other hand, an Eulerian point is an immobile observation point, not related to any specific material point. Therefore, at different moments of time, different Lagrangian material points can be found at the same Eulerian point. In other words, different Lagrangian material points are *passing through* the same Eulerian observation point with time. Good analogies for an Eulerian and a Lagrangian point are respectively a fixed window and a person walking past in front of it. Many equations of continuum mechanics containing time derivatives can be written in both Eulerian and Lagrangian forms which differ from each other (e.g. Eqs. 1.1 and 1.3). Choosing either the Eulerian or Lagrangian form of an equation notably affects the method of representing *advective transport processes* (i.e. the movement of material with the flow) which will be discussed in detail in Chapter 8, together with the advantages and disadvantages of the two approaches.

1.4 Derivation of the Eulerian continuity equation

Let us now analyse the Eulerian continuity equation (Eq. 1.1), which contains both vector (velocity) and scalar (density) variables. This equation establishes the balance of mass in a small fixed observation volume. It implies, in particular, that if mass is leaving (fluxing out of) the volume (i.e., $\text{div}(\rho\vec{v}) > 0$), the local density (i.e., the amount of mass per unit volume) decreases with time (i.e., $\partial\rho/\partial t < 0$).

First we need to understand that $\rho\vec{v}$ is the local *mass flux* vector

$$\rho\vec{v} = \left(\rho v_x,\ \rho v_y,\ \rho v_z\right), \tag{1.4}$$

which has the dimension of unit mass, fluxing through a unit surface, per unit time $(\text{kg}/(\text{m}^2\text{s}))$. This definition follows from the fact that the velocity in a continuous medium can be considered as *material volume flux* (Fig. 1.5), i.e., unit volume fluxing through a unit surface per unit time $(\text{m/s} = \text{m}^3/(\text{m}^2\text{s}))$. Therefore, velocity (i.e. volume flux) multiplied by the density (i.e. mass per unit volume) gives the mass flux.

We can derive the Eulerian continuity equation by analysing material fluxes through a small, immobile, rectangular Eulerian (observation) volume of constant dimensions Δx, Δy and Δz (Fig. 1.6a). Let us assume that the initial mass of fluid in this volume is m_0. Then, the initial average fluid density (ρ_0) within this volume is

$$\rho_0 = \frac{m_0}{\Delta x\,\Delta y\,\Delta z}. \tag{1.5}$$

Mass enters the volume through the boundaries A, C and E and leaves it through the opposite boundaries B, D and F. Material fluxes affect the mass of fluid in the observation volume and after a small period of time Δt (*time increment*), this mass becomes m_1 and the average fluid density changes to the new value (ρ_1)

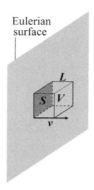

Fig. 1.5. Relationship between the flow velocity v and material volume V, passing through the element S of the immobile Eulerian surface (grey) during the time t. The relations $V = S L$ and $L = v t$ allow one to formulate velocity as the material volume flux $v = V/(St)$.

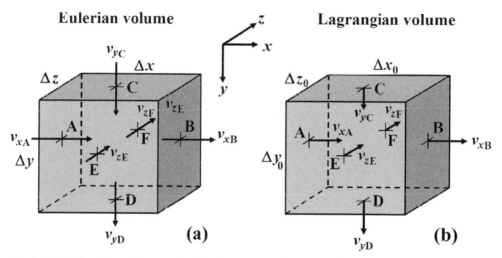

Fig. 1.6. Eulerian (a) and Lagrangian (b) elementary volumes considered for the derivation of the continuity equation. Arrows in (a) show the velocity components which are responsible for *material fluxes through* the respective boundaries (A, B, C, D, E and F). Arrows in (b) show the velocity components responsible for *moving* the respective boundaries.

$$\rho_1 = \frac{m_1}{\Delta x\, \Delta y\, \Delta z}.$$ (1.6)

The balance between the old (m_0) and new (m_1) masses results in the following relations:

$$m_1 = m_0 + m_{in} - m_{out,}$$
$$m_{in} = m_A + m_C + m_E,$$
$$m_{out} = m_B + m_D + m_F,$$
$$m_A = \rho_A v_{xA} \Delta y \Delta z \Delta t,$$
$$m_B = \rho_B v_{xB} \Delta y \Delta z \Delta t,$$
$$m_C = \rho_C v_{yC} \Delta x \Delta z \Delta t,$$
$$m_D = \rho_D v_{yD} \Delta x \Delta z \Delta t,$$
$$m_E = \rho_E v_{zE} \Delta x \Delta y \Delta t,$$
$$m_F = \rho_F v_{zF} \Delta x \Delta y \Delta t,$$

(1.7)

where m_{in} and m_{out} are the incoming and outgoing masses, respectively; m_A to m_F are masses that passed through the respective boundaries during the time Δt; ρ_A to ρ_F are the densities at the respective boundaries; v_{xA} to v_{zF} are the velocity components responsible for material fluxes through the boundaries (Fig. 1.6a). If Δt is small, we can now write an approximate expression for the Eulerian time derivative of the average density in the volume as:

$$\frac{\partial \rho}{\partial t} \approx \frac{\Delta \rho}{\Delta t} = \frac{\rho_1 - \rho_0}{\Delta t} = \frac{m_1 - m_0}{\Delta x \, \Delta y \, \Delta z \, \Delta t}. \tag{1.8a}$$

By using Eq. (1.7) the following expression can be obtained (verify as an exercise)

$$\frac{\Delta \rho}{\Delta t} = -\frac{\rho_B v_{xB} - \rho_A v_{xA}}{\Delta x} - \frac{\rho_D v_{yD} - \rho_C v_{yC}}{\Delta y} - \frac{\rho_F v_{zF} - \rho_E v_{zE}}{\Delta z}, \tag{1.8b}$$

or

$$\frac{\Delta \rho}{\Delta t} = -\frac{\Delta(\rho v_x)}{\Delta x} - \frac{\Delta(\rho v_y)}{\Delta y} - \frac{\Delta(\rho v_z)}{\Delta z}, \tag{1.8c}$$

or

$$\frac{\Delta \rho}{\Delta t} + \frac{\Delta(\rho v_x)}{\Delta x} + \frac{\Delta(\rho v_y)}{\Delta y} + \frac{\Delta(\rho v_z)}{\Delta z} = 0,$$
$$\Delta(\rho v_x) = \rho_B v_{xB} - \rho_A v_{xA},$$
$$\Delta(\rho v_y) = \rho_D v_{yD} - \rho_C v_{yC},$$
$$\Delta(\rho v_z) = \rho_F v_{zF} - \rho_E v_{zE},$$

$(1.8d)$

where $\Delta(\rho v_x)$, $\Delta(\rho v_y)$ and $\Delta(\rho v_z)$ are differences in the mass fluxes in the x, y and z directions respectively (i.e. through the respective pairs of boundaries, Fig. 1.6a). Obviously, in such cases when Δt, Δx, Δy and Δz all tend to zero, the differences can be replaced by derivatives and we obtain the Eulerian continuity equation:

$$\frac{\partial \rho}{\partial t} + \frac{\partial(\rho v_x)}{\partial x} + \frac{\partial(\rho v_y)}{\partial y} + \frac{\partial(\rho v_z)}{\partial z} = 0, \tag{1.9a}$$

or

$$\frac{\partial \rho}{\partial t} + \mathrm{div}(\rho \vec{v}) = 0. \tag{1.9b}$$

1.5 Derivation of the Lagrangian continuity equation

Similarly, we can derive the Lagrangian continuity equation by analysing the motion of a small, mobile, rectangular Lagrangian (material) volume of initial dimensions Δx_0, Δy_0 and Δz_0 (Fig. 1.6b). In contrast to the fixed Eulerian volume, the amount of mass m in the moving Lagrangian volume remains constant (since this volume always contains the same material points), but the dimensions of the volume may change due to *internal* expansion/ contraction processes. The initial average fluid density (ρ_0), within the volume is given by

$$\rho_0 = \frac{m}{\Delta x_0 \, \Delta y_0 \, \Delta z_0}. \tag{1.10}$$

Internal expansion or contraction affects the dimensions of the Lagrangian volume, and after a small period of time Δt, these dimensions become Δx_1, Δy_1 and Δz_1 and the average fluid density (ρ_1) changes to

$$\rho_1 = \frac{m}{\Delta x_1 \, \Delta y_1 \, \Delta z_1}. \tag{1.11}$$

We can establish the relationship between the old (Δx_0, Δy_0, Δz_0) and the new (Δx_1, Δy_1, Δz_1) dimensions of the Lagrangian volume on the basis of the relative movements of the boundaries of the volume (A, B, C, D, E, F, Fig. 1.6b), which leads to the following relations:

$$\begin{aligned} \Delta x_1 &= \Delta x_0 + \Delta t \, \Delta v_x, \\ \Delta v_x &= v_{xB} - v_{xA}, \end{aligned} \tag{1.12}$$

$$\begin{aligned} \Delta y_1 &= \Delta y_0 + \Delta t \, \Delta v_y, \\ \Delta v_y &= v_{yD} - v_{yC}, \end{aligned} \tag{1.13}$$

$$\begin{aligned} \Delta z_1 &= \Delta z_0 + \Delta t \, \Delta v_z, \\ \Delta v_z &= v_{zF} - v_{zE}, \end{aligned} \tag{1.14}$$

where Δv_x, Δv_y and Δv_z are the differences in the velocity components that correspond to the relative movements of respective pairs of boundaries (Fig. 1.6b). Taking Δt to be small, we can now write an approximate expression for the Lagrangian time derivative of the average density in the volume as

$$\frac{D\rho}{Dt} \approx \frac{\Delta\rho}{\Delta t} = \frac{\rho_1 - \rho_0}{\Delta t} = \frac{m}{\Delta x_1\,\Delta y_1\,\Delta z_1\,\Delta t} - \frac{m}{\Delta x_0\,\Delta y_0\,\Delta z_0\,\Delta t}. \qquad (1.15)$$

By using the equations derived for Δx_1, Δy_1 and Δz_1 (Eqs. 1.12–1.14) the following expression can be obtained (verify as an exercise):

$$\frac{\Delta\rho}{\Delta t} = \rho_0 \frac{1 - \left(1 + \Delta t\frac{\Delta v_x}{\Delta x_0}\right)\left(1 + \Delta t\frac{\Delta v_y}{\Delta y_0}\right)\left(1 + \Delta t\frac{\Delta v_z}{\Delta z_0}\right)}{\Delta t\left(1 + \Delta t\frac{\Delta v_x}{\Delta x_0}\right)\left(1 + \Delta t\frac{\Delta v_y}{\Delta y_0}\right)\left(1 + \Delta t\frac{\Delta v_z}{\Delta z_0}\right)}, \qquad (1.16a)$$

or

$$\frac{\Delta\rho}{\Delta t} = \rho_0 \frac{-\frac{\Delta v_x}{\Delta x_0} - \frac{\Delta v_y}{\Delta y_0} - \frac{\Delta v_z}{\Delta z_0} - \Delta t\left(\frac{\Delta v_x}{\Delta x_0}\frac{\Delta v_y}{\Delta y_0} + \frac{\Delta v_x}{\Delta x_0}\frac{\Delta v_z}{\Delta z_0} + \frac{\Delta v_y}{\Delta y_0}\frac{\Delta v_z}{\Delta z_0} + \Delta t\frac{\Delta v_x}{\Delta x_0}\frac{\Delta v_y}{\Delta y_0}\frac{\Delta v_z}{\Delta z_0}\right)}{\left(1 + \Delta t\frac{\Delta v_x}{\Delta x_0}\right)\left(1 + \Delta t\frac{\Delta v_y}{\Delta y_0}\right)\left(1 + \Delta t\frac{\Delta v_z}{\Delta z_0}\right)}, \qquad (1.16b)$$

or

$$\frac{\Delta\rho}{\Delta t} + \rho_0 \frac{\frac{\Delta v_x}{\Delta x_0} + \frac{\Delta v_y}{\Delta y_0} + \frac{\Delta v_z}{\Delta z_0} + K_1}{K_2} = 0, \qquad (1.16c)$$

where

$$K_1 = \Delta t\left(\frac{\Delta v_x}{\Delta x_0}\frac{\Delta v_y}{\Delta y_0} + \frac{\Delta v_x}{\Delta x_0}\frac{\Delta v_z}{\Delta z_0} + \frac{\Delta v_y}{\Delta y_0}\frac{\Delta v_z}{\Delta z_0} + \Delta t\frac{\Delta v_x}{\Delta x_0}\frac{\Delta v_y}{\Delta y_0}\frac{\Delta v_z}{\Delta z_0}\right)$$

$$K_2 = \left(1 + \Delta t\frac{\Delta v_x}{\Delta x_0}\right)\left(1 + \Delta t\frac{\Delta v_y}{\Delta y_0}\right)\left(1 + \Delta t\frac{\Delta v_z}{\Delta z_0}\right).$$

K_1 and K_2 are coefficients which respectively tend to *zero* and *unity* when Δt tends to zero. Obviously, in the case when Δt, Δx_0, Δy_0 and Δz_0 all tend towards zero, the differences in Eq. (1.16c) can be replaced by derivatives and taking $K_1 = 0$ and $K_2 = 1$ we obtain the Lagrangian continuity equation

$$\frac{D\rho}{Dt} + \rho\frac{\partial v_x}{\partial x} + \rho\frac{\partial v_y}{\partial y} + \rho\frac{\partial v_z}{\partial z} = 0, \qquad (1.17a)$$

or

$$\frac{D\rho}{Dt} + \rho\,\mathrm{div}(\vec{v}) = 0. \qquad (1.17b)$$

1.6 Comparing Eulerian and Lagrangian continuity equations: advective transport term

Let us now perform transformations of the Eulerian continuity equation (Eq. 1.1) in order to decipher its structure and to establish a relationship with the Lagrangian continuity equation (Eq. 1.3). It is convenient to decompose div($\rho\vec{v}$) using the standard *product rule* (also called *Leibniz's law*) $(u \cdot v)' = u' \cdot v + v' \cdot u$, or $\partial(uv)/\partial x = (\partial u/\partial x)v + (\partial v/\partial x)u$

$$\text{div}(\rho\vec{v}) = \rho\text{div}(\vec{v}) + \vec{v}\,\text{grad}(\rho), \tag{1.18a}$$

or in a different symbolic notation

$$\nabla\bullet(\rho\vec{v}) = \rho\nabla\bullet\vec{v} + \vec{v}\bullet\nabla\rho, \tag{1.18b}$$

or 'deciphering' what we actually are doing in three dimensions

$$\frac{\partial}{\partial x}(\rho v_x) + \frac{\partial}{\partial y}(\rho v_y) + \frac{\partial}{\partial z}(\rho v_z) = \left(\rho\frac{\partial v_x}{\partial x} + \rho\frac{\partial v_y}{\partial y} + \rho\frac{\partial v_z}{\partial z}\right) + \left(v_x\frac{\partial\rho}{\partial x} + v_y\frac{\partial\rho}{\partial y} + v_z\frac{\partial\rho}{\partial z}\right),$$
$$\tag{1.18c}$$

where grad(ρ) or $\nabla\rho$ is the gradient of the density ρ. The gradient is a vector function of a scalar field defined as follows:

$$\text{in one dimension (1D)} \quad \text{grad}(\rho) = \left(\frac{\partial\rho}{\partial x}\right), \tag{1.19a}$$

$$\text{in two dimensions (2D)} \quad \text{grad}(\rho) = \left(\frac{\partial\rho}{\partial x}, \frac{\partial\rho}{\partial y}\right), \tag{1.19b}$$

$$\text{in three dimensions (3D)} \quad \text{grad}(\rho) = \left(\frac{\partial\rho}{\partial x}, \frac{\partial\rho}{\partial y}, \frac{\partial\rho}{\partial z}\right). \tag{1.19c}$$

Therefore, both $\nabla\rho$ and $\vec{v}$ in Eq. (1.18) are vectors and the *scalar product* (or *dot product*) of these two vectors (1.18c) gives the following result:

$$\text{in one dimension (1D)} \quad \vec{v}\,\text{grad}(\rho) = v_x\frac{\partial\rho}{\partial x}, \tag{1.20a}$$

$$\text{in two dimensions (2D)} \quad \vec{v}\,\text{grad}(\rho) = v_x\frac{\partial\rho}{\partial x} + v_y\frac{\partial\rho}{\partial y}, \tag{1.20b}$$

$$\text{in three dimensions (3D)} \quad \vec{v}\,\text{grad}(\rho) = v_x\frac{\partial\rho}{\partial x} + v_y\frac{\partial\rho}{\partial y} + v_z\frac{\partial\rho}{\partial z}. \tag{1.20c}$$

Now by comparing Eqs. (1.1), (1.3) and (1.18) we can establish the relationship between the Eulerian ($\partial\rho/\partial t$) and Lagrangian ($D\rho/Dt$) time derivatives of density as

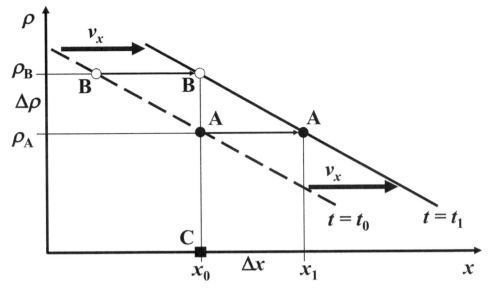

Fig. 1.7. Schematic representation of the advective transport in the case of uniform 1D movement of a fluid with a linear density distribution. The dashed and solid thick lines correspond to the density distribution for the moments of time t_0 and t_1, respectively. Circles A and B denote the density for two different Lagrangian points passing through an Eulerian observation point (solid rectangle C) in the different moments of time (t_0 and t_1, respectively).

$$\frac{D\rho}{Dt} = \frac{\partial \rho}{\partial t} + \vec{v}\,\mathrm{grad}(\rho). \qquad (1.21)$$

The extra term, $\vec{v}\,\mathrm{grad}(\rho)$ in the Eulerian continuity equation is an *advective transport* term that reflects changes of density in an immobile (Eulerian) point, due to the movement of an inhomogeneous medium with some density gradients relative to this point (Fig. 1.7). Obviously, the advective transport terms in the Eulerian continuity equation are only relevant (i.e. non-zero) in situations when both the velocity of the medium and the density gradient are non-zero. On the other hand, *substantive* changes of density ($D\rho/Dt$) in the moving Lagrangian point do not depend on density gradients and the Lagrangian continuity equation (1.3) does not contain advective terms.

When the density in all moving material points does not change with time (i.e. $D\rho/Dt = 0$) the Eulerian continuity equation reduces to the Eulerian *advective transport equation*

$$\frac{\partial \rho}{\partial t} = -\vec{v}\,\mathrm{grad}(\rho). \qquad (1.22)$$

The minus sign in the right hand side of Eq. (1.22) reflects the relation between the density gradient and the direction of motion (Fig. 1.7): if a medium is moving *in the direction* of decreasing density (i.e. $\vec{v}\,\mathrm{grad}(\rho) < 0$), then the density in an immobile observation point *increases* with time (i.e. $\partial\rho/\partial t > 0$).

Let us now derive the advective transport equation (1.22) for the simple 1D case shown in Fig. 1.7. A fluid with a linear density distribution moves at a constant velocity v_x. At the moment of time t_0 the density (ρ_A) in the Eulerian observation point C (solid rectangle) corresponds to the Lagrangian point A (solid circle). Due to fluid motion, the density profile shifts to the right with time. Therefore, at a later moment of time t_1, the density (ρ_B) at the same Eulerian point C will correspond to another Lagrangian point B (open circle). Under the assumption that the difference in time between the two moments ($\Delta t = t_1 - t_0$) is small, we can approximate the time derivative of density at the Eulerian point C as

$$\frac{\partial\rho}{\partial t} \approx \frac{\Delta\rho_t}{\Delta t} = \frac{\rho_B - \rho_A}{t_1 - t_0}, \tag{1.23}$$

where $\Delta\rho_t$ corresponds to changes in density with time at the Eulerian point C. If on the other hand, the displacement of the density profile (i.e. the distance between two Lagrangian points A and B, $\Delta x = x_1 - x_0$) is also small, we can approximate the local derivative of density with respect to x, at the moment of time t_1, as

$$\frac{\partial\rho}{\partial x} \approx \frac{\Delta\rho_x}{\Delta x} = \frac{\rho_A - \rho_B}{x_1 - x_0}, \tag{1.24}$$

where $\Delta\rho_x$ corresponds to a change in density with coordinate at the proximity of the Eulerian point C at the instant when the time equals t_1. Note that $\Delta\rho_t$ and $\Delta\rho_x$ are different quantities (i.e., differences taken in time and in the x direction, respectively), and that in the case considered, they are similar in absolute value but are different in sign (cf. Eqs. 1.23 and 1.24). Taking into account the relationships

$$\Delta x = v_x \Delta t$$

and

$$\Delta\rho_x = -\Delta\rho_t,$$

the following expression can be obtained (derive as an exercise):

$$\frac{\Delta\rho_t}{\Delta t} = -v_x \frac{\Delta\rho_x}{\Delta x}. \tag{1.25}$$

When Δt and respectively Δx tend to zero, the differences in Eq. (1.25) can be replaced by derivatives and we obtain the 1D advection equation

$$\frac{\partial \rho}{\partial t} = -v_x \frac{\partial \rho}{\partial x}. \tag{1.26}$$

The advective transport equation is very important for geodynamic modelling since it describes changes in the distribution of *physical properties* (such as density, temperature, composition etc.) in non-homogeneous, deforming media. We will come back to this issue in Chapter 8.

1.7 Incompressible continuity equation

For many geological media (such as the Earth's crust and mantle), one may assume an incompressibility condition (i.e. density of Lagrangian material points does not change with time)

$$\frac{D\rho}{Dt} = \frac{\partial \rho}{\partial t} + \vec{v} \, \text{grad}(\rho) = 0. \tag{1.27}$$

This is valid in cases when pressure and temperature changes are not very large and no phase transformations leading to volume changes occur in the medium. In this situation, one can assume that $D\rho/Dt = 0$ everywhere in the medium and use an *incompressible continuity equation*, which is the same in both Eulerian and Lagrangian forms

$$\text{div}(\vec{v}) = 0. \tag{1.28}$$

The incompressible continuity equation is widely used in numerical geodynamic modelling, although in many cases it is a rather big simplification (for example in the case of the whole Earth mantle convection, e.g. Tackley, 2008). Typical examples of geodynamic settings where deformations are strongly defined by the incompressibility condition are, for example, corner flow in the mantle wedge above subducting slabs (e.g. Turcotte and Schubert, 2002) and circulation of tectonic melanges in subduction channels (e.g. Cloos, 1982).

Analytical exercise

Exercise 1.1.

In a region of the Earth's mantle, the velocity field is given by

$$v_x = 10^{-10} + x \cdot 10^{-13} + y \cdot 10^{-13} + z \cdot 10^{-13}$$

$$v_y = 10^{-10} - x \cdot 10^{-13} + y \cdot 2 \times 10^{-13} + z \cdot 3 \times 10^{-13}$$

$$v_z = 10^{-10} - x \cdot 10^{-13} - y \cdot 10^{-13} - z \cdot 2 \times 10^{-13}.$$

The mantle density field in the same region is given by

$$\rho = 3300 + x \cdot 0.001 - y \cdot 0.002 + z \cdot 0.001.$$

Calculate ρ, div$(\vec{v})$, $\partial\rho/\partial t$ and $D\rho/Dt$ for the point with coordinates $x = 1000$, $y = 1000$, $z = 1000$.

Programming exercise

Exercise 1.2.

Write a MATLAB code for computing and visualizing a 2D velocity field and its divergence. Model design: an area of the mantle (1000 × 1500 km) is convecting with one central upwelling in the middle of the model box and two downwellings at the sides. The velocity field is given by the following equations

$$v_x = -v_{x0} \sin\left(2\pi\frac{x}{W}\right)\cos\left(\pi\frac{y}{H}\right),$$

$$v_y = v_{y0} \cos\left(2\pi\frac{x}{W}\right)\sin\left(\pi\frac{y}{H}\right),$$

where x and y are respectively horizontal and vertical coordinates inside the box in metres; $W = 1\,000\,000$ m and $H = 1\,500\,000$ m are the width and height of the model, respectively (i.e., 1000 × 1500 km^2); $v_{x0} = 10^{-9}$ m/s and $v_{y0} = 10^{-9}$ m/s are scaling values for respectively horizontal and vertical velocity components (10^{-9} m/s $\approx$ 3 cm/year). Compute (analytically) v_x, v_y, $\partial v_x/\partial x$, $\partial v_y/\partial y$ and div$(\vec{v}) = \partial v_x/\partial x + \partial v_y/\partial y$ on a 2D grid of points (e.g. 31 × 31) which are regularly distributed inside the model, and visualize these parameters separately as colour maps (*pcolor*) in order to see how they are distributed relative to each other. Visualize the velocity as an arrow field (*quiver*). Try to tune the scaling velocity values v_{x0} and v_{y0} such that the divergence of velocity goes to (almost) zero in the entire model (i.e. absolute values of div$(\vec{v})$ should be many orders of magnitude less than those of $\partial v_x/\partial x$ and $\partial v_y/\partial y$). An example is in **Divergence.m**.

2

Density and gravity

Theory: Density of rocks and minerals. Thermal expansion and compressibility. Dependence of density on pressure and temperature. Equations of state. Poisson equation for gravitational potential and its derivation.
Exercises: Computing and visualizing density, thermal expansion and compressibility.

2.1 Density of rocks and minerals: equations of state

Many geodynamic processes are either directly or indirectly driven by the force of gravity due to the spatial variation of rock density inside the Earth. The density of rocks (ρ) depends on the pressure (P), temperature (T), chemical composition (C) and mineralogical composition (M)

$$\rho = f(P, T, C, M). \tag{2.1}$$

These factors are not fully independent. The mineralogical composition of a rock with a constant chemical composition may for example change due to changes in pressure and temperature. *Easy to remember* densities of major rock types used in geodynamics are as follows:

- felsic rocks (typical of the continental crust, e.g., granite, granodiorite) ~2700 kg/m^3,
- mafic rocks (typical of the oceanic crust, e.g., basalt, gabbo) ~3000 kg/m^3,
- ultramafic rocks (typical of the mantle, e.g., peridotite, lherzolite) ~3300 kg/m^3.

Some other relevant rock densities are given in Table 21.2.

Variations in the density of minerals and rocks with T and P are often characterized by the thermal expansion (α) and compressibility (β)

$$\alpha = -\frac{1}{\rho}\frac{\partial \rho}{\partial T}, \tag{2.2}$$

$$\beta = \frac{1}{\rho}\frac{\partial \rho}{\partial P}. \tag{2.3}$$

The thermal expansion of rocks and minerals is typically positive ($\alpha > 0$), i.e. density decreases with increasing temperature (cf. minus in the right hand side of Eq. 2.2). There are, indeed, exceptions, for example, beta-quartz possesses a density which remains almost constant with increasing temperature at room pressure ($\alpha \approx 0$). Typical values of α are on the order of $n \times 10^{-5}$ K^{-1}. Compressibility is always positive ($\beta > 0$) and density always increases with increasing pressure. Typical values of β are on the order of $n \times 10^{-2}$ GPa^{-1} (or $n \times 10^{-11}$ Pa^{-1}). In the case of constant α and β, integration of Eqs. (2.2) and (2.3) versus T and P gives

$$\rho = \rho_r \exp[\beta(P - P_r) - \alpha(T - T_r)], \tag{2.4a}$$

where ρ_r is the density of a given material at *reference* pressure P_r (typically 10^5 Pa = 1 bar) and temperature T_r (typically 298.15 K = 25°C). Since both $\alpha(T - T_r)$ and $\beta(P - P_r)$ are typically very small (much less than unity), Eq. (2.4a) can be simplified to (using the rules that $\exp(a) \approx 1 + a$ and $\exp(-a) \approx 1 - a$ when $a \ll 1$)

$$\rho = \rho_r[1 + \beta(P - P_r)] \times [1 - \alpha(T - T_r)], \tag{2.4b}$$

or to (using the rules that $\exp(a) \approx 1 + a$ and $\exp(-a) = 1/\exp(a) \approx 1/(1 + a)$ when $a \ll 1$)

$$\rho = \rho_r \frac{1 + \beta(P - P_r)}{1 + \alpha(T - T_r)}. \tag{2.4c}$$

Equations (2.4a)–(2.4c), however, do not account for changes in density due to changes in mineralogical composition with changing pressure and temperature. These *equations of state* (*EOS*) are much simplified since they are based on the assumption that α and β remain constant with pressure and temperature. For real rocks and minerals, however, these two parameters strongly depend on both P and T (Fig. 2.1) and therefore a more realistic EOS must be used for describing the density changes (e.g. Anderson, 1995). If compressibility of the mineral depends on pressure, then the following well-known equations can be used.

Murnaghan equation of state (Murnaghan, 1944)

$$\rho = \rho_0 \left(1 + B_0' \frac{P}{B_0}\right)^{1/B_0'}, \tag{2.5}$$

Birch–Murnaghan equation of state (Birch, 1947)

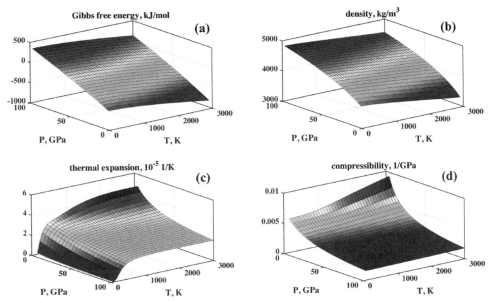

Fig. 2.1. Gibbs free energy (a), density (b), thermal expansion (c) and compressibility (d) of periclase (MgO) computed on the basis of the semi-empirical parameterization of Gerya et al. (2004d) for a broad range of temperature and pressure values. Note that the pressure axis in (c) and (d) is inverted compared to (a) and (b). Results are computed with the code **Periclase_EOS.m**.

$$P(\rho) = \frac{3B_0}{2}\left[\left(\frac{\rho}{\rho_0}\right)^{7/3} - \left(\frac{\rho}{\rho_0}\right)^{5/3}\right]\left\{1 + \frac{3}{4}\left(B_0' - 4\right)\left[\left(\frac{\rho}{\rho_0}\right)^{2/3} - 1\right]\right\}, \qquad (2.6)$$

where $B_0 = 1/\beta_0$ is the bulk modulus (this quantity will also be discussed in relation to elasticity in Chapter 12) at $P = 0$ (β_0 is the compressibility at $P = 0$), $B_0' = (\partial B/\partial P)_{T=\text{const}}$ is the pressure derivative of the bulk modulus at a constant temperature and ρ_0 is the density at $P = 0$. The Murnaghan equation is derived (repeat this derivation as an exercise) with an experimentally based assumption that the pressure derivative of the bulk modulus B_0' is independent of pressure (many substances have a fairly constant B_0' value of about 3.5). The Birch–Murnaghan equation (2.6) is also derived empirically based on measurements of volumes of solid substances, held at a constant temperature. Obviously, this equation has to be solved approximately to obtain a density at a given pressure.

Equations (2.5) and (2.6), however, do not establish the dependence of density upon temperature, and therefore, an even more complicated EOS must be used when the density is described over a wide range of both pressure and temperature values. For example, in mineralogy and petrology, the molar volume of minerals (V) is described in many cases by semi-empirical equations (e.g. the Murnaghan-like EOS) with additional temperature-dependent terms (Holland and Powell, 1998)

$$V = V_r\left[1 + a(T - T_r) + b\left(\sqrt{T} - \sqrt{T_r}\right)\right] \times \left\{1 - \frac{B'_r P}{B_r[1 - c(T - T_r)] + B'_r P}\right\}^{1/B'_r},$$

(2.7)

where V_r, B_r and B'_r are respectively molar volume, bulk modulus and its first derivative at P_r and T_r (i.e. at *reference P – T conditions*) and a, b and c are empirical parameters computed from experimental measurements of molar volume at elevated pressures and temperatures. More recent EOS used in the internally consistent thermodynamic database of Holland and Powell (2011) takes into account variations of the bulk modulus and thermal expansion of minerals in a very broad range of temperature and pressure

$$V = V_r\left\{1 - A_1[1 - [1 + A_2(P - P_{th})]^{-A_3}]\right\},$$

(2.8)

$$A_1 = \frac{1 + B'_r}{1 + B'_r + B_r B''_r},$$

$$A_2 = \frac{B'_r}{B_r} - \frac{B''_r}{1 + B'_r},$$

$$A_3 = \frac{1 + B'_r + B_r B''_r}{B'^2_r + B'_r - B_r B''_r},$$

$$P_{th} = \frac{a_r B_r \theta}{\exp(\theta/T_r)}\left(\frac{\exp(\theta/T_r) - 1}{\theta/T_r}\right)^2\left(\frac{1}{\exp(\theta/T) - 1} - \frac{1}{\exp(\theta/T_r) - 1}\right),$$

where α_r and B''_r are respectively thermal expansion and second derivative of the bulk modulus at P_r and T_r, θ is the Einstein temperature, P_{th} is thermal pressure.

It should also be mentioned that in mineralogy and petrology, self-consistent descriptions of thermodynamic properties (including density) of minerals and fluids are often based on formulating the equations for their molar thermodynamic potentials (typically either for the Gibbs $G_{(P,T)}$ or for the Helmholz $F_{(V,T)}$ potential, e.g., Karpov et al., 1976; Helgeson et al., 1978; Dorogokupets and Karpov, 1984; Berman, 1988; Holland and Powell, 1990, 1998, 2011; Gerya et al., 2004d; Stixrude and Lithgow-Bertelloni, 2005, 2011; Brosh et al., 2008). The density as well as many other physical properties (such as heat capacity, thermal expansion, compressibility etc.) are then computed from the potential (Fig. 2.1) using standard thermodynamic relations

$$V = \left(\frac{\partial G_{(P,T)}}{\partial P}\right)_{T=\text{const}},$$

(2.9)

$$P = -\left(\frac{\partial F_{(V,T)}}{\partial V}\right)_{T=\text{const}},$$

(2.10)

$$V = \frac{m}{\rho},$$ (2.11)

where m is molar mass, kg/mol.

Self-consistent thermodynamic databases (e.g. Karpov et al., 1976; Helgeson et al., 1978; Berman, 1988; Holland and Powell, 1990, 1998, 2011) are often based on the standard formulation of Gibbs free energy in the form:

$$G_{m(P,T)} = \Delta H_r - T \cdot S_r + \int_{T_r}^{T} C_{P(T)} dT - T \cdot \int_{T_r}^{T} \frac{C_{P(T)}}{T} dT + \int_{P_r}^{P} V_{(P,T)} dP,$$ (2.12)

where $G_{m(P,T)}$ is the molar Gibbs free energy (i.e. Gibbs potential) at a given P and T; ΔH_r and S_r are the enthalpy of formation and entropy, respectively, of a substance at standard pressure P_r and temperature T_r; $C_{P(T)}$ is the heat capacity as a function of temperature at a standard pressure P_r; $V_{(P,T)}$ is the molar volume of substance as a function of pressure and temperature defined by a semi-empirical EOS function (such as for example, Eqs. 2.7, 2.8).

Natural rocks typically contain several different minerals and therefore the density of a rock can be calculated from its mineralogical composition as follows

$$\rho_{rock} = \sum_{i=1}^{n} \rho_i X_i,$$ (2.13)

where n is the number of different minerals in the rock, X_i is the *volumetric fraction* of the ith mineral in the rock and ρ_i is the density of the ith mineral as a function of P and T. At any given P, T and chemical composition of the rock, both the amount and the composition of the minerals can be computed using the concept of thermodynamic equilibrium. According to this concept, the Gibbs free energy of the rock in an equilibrium state corresponds to a global minimum. The amount and composition of minerals can then be obtained from internally consistent thermodynamic databases by using the so-called *Gibbs free energy minimization approach* (e.g. Karpov et al., 1976; Dorogokupets and Karpov, 1984; Connolly and Kerrick, 1987; de Capitani and Brown, 1987). In this case, the density of rocks in an equilibrium state (Sobolev and Babeyko, 1994; Gerya et al., 2001; Petrini et al., 2001; Kaus et al., 2005) can be computed from Eqs. (2.9), (2.11)–(2.13) by using the densities and volumetric fractions of minerals composing a computed equilibrium mineral assemblage (Fig. 2.2). Such an approach is used to constrain coupled *petrological-thermomechanical* numerical geodynamic models (e.g. Gerya et al., 2004c, 2006; Mishin et al., 2008; Tackley, 2008), which take into account changes of rock density and thermal properties during the course of various geodynamic processes (we will discuss the details of this approach in Chapter 21).

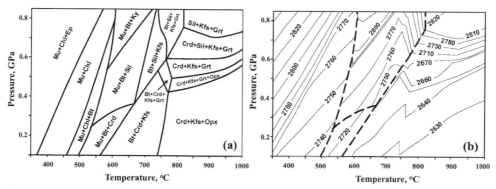

Fig. 2.2. Equilibrium mineral assemblages (a) and the corresponding density (kg/m³) map (b) computed for a typical composition of metamorphosed aluminous sediment (high-grade metapelite) on the basis of Gibbs free energy minimization (Gerya et al., 2001). Quartz, plagioclase and Fe-Ti oxides are present in all mineral assemblages. Other minerals are: Bt biotite, Chl chlorite, Crd cordierite, Ep epidote, Grt garnet, Kfs K-feldspar, Ky kyanite, Mu muscovite, Opx opthopyroxene, Sil sillimanite. Heavy dashed lines in (b) indicate sharp changes in density related to changes in the mineral assemblages.

2.2 Gravity and gravitational potential

The density distribution in a continuous medium is inherently related to the gravitational field in this medium. This can be formulated in the form of the Poisson equation, which describes spatial changes in gravitational potential Φ inside a *self-gravitating* continuum

$$\frac{\partial^2 \Phi}{\partial x^2} + \frac{\partial^2 \Phi}{\partial y^2} + \frac{\partial^2 \Phi}{\partial z^2} = 4\pi G \rho_{(x,y,z)}, \tag{2.14a}$$

or in a different symbolic notation

$$\nabla^2 \Phi = 4\pi G \rho_{(x,y,z)}, \tag{2.14b}$$

where $G = 6.672 \times 10^{-11}$ (Nm²)/kg² is the gravitational constant, $\rho_{(x,y,z)}$ is the spatially variable density and ∇^2 is the Laplace operator or Laplacian (in the literature, *big delta* Δ is also frequently used for the Laplacian but *we will not use it* to avoid confusion with the difference symbol *small delta* Δ often used in this book), which is a *differential operator* (like the divergence and gradient operator from Chapter 1) often used in continuum mechanics for representing the sum of second-order partial derivatives of a variable A:

$$\text{in 1D} \quad \nabla^2 A = \frac{\partial^2 A}{\partial x^2}, \tag{2.15a}$$

$$\text{in 2D} \quad \nabla^2 A = \frac{\partial^2 A}{\partial x^2} + \frac{\partial^2 A}{\partial y^2}, \tag{2.15b}$$

$$\text{in 3D} \quad \nabla^2 A = \frac{\partial^2 A}{\partial x^2} + \frac{\partial^2 A}{\partial y^2} + \frac{\partial^2 A}{\partial z^2}. \tag{2.15c}$$

The gravitational potential Φ (J/kg), characterizes the *amount of potential energy* per unit mass for a given location, related to the interaction of the local mass with all other surrounding masses. Another interpretation of Φ is the *amount of work* needed for (virtually) moving a unit mass (i.e. overcoming its gravitational interactions with surrounding masses) from a given location to infinity (where no interactions with other masses occur). Since such work is always positive, the amount of potential energy and therefore the gravitational potential is maximal at infinity. The relative gravitational potential (i.e. the difference in potential energy relative to infinity) will then always be negative.

The Poisson equation (2.14) can be derived from Newton's law of gravitation that quantifies the gravitational attraction force f_g acting between two bodies with masses m_1 and m_2, separated by a distance r

$$f_g = G \frac{m_1 m_2}{r^2}, \tag{2.16a}$$

or in 3D vector notation

$$f_{x1} = G \frac{m_1 m_2}{r^2} \frac{(x_2 - x_1)}{r}, \tag{2.16b}$$

$$f_{y1} = G \frac{m_1 m_2}{r^2} \frac{(y_2 - y_1)}{r}, \tag{2.16c}$$

$$f_{z1} = G \frac{m_1 m_2}{r^2} \frac{(z_2 - z_1)}{r}, \tag{2.16d}$$

$$g_{x1} = G \frac{m_2}{r^2} \frac{(x_2 - x_1)}{r}, \tag{2.16e}$$

$$g_{y1} = G \frac{m_2}{r^2} \frac{(y_2 - y_1)}{r}, \tag{2.16f}$$

$$g_{z1} = G \frac{m_2}{r^2} \frac{(z_2 - z_1)}{r}, \tag{2.16g}$$

where f_{x1}, f_{y1} and f_{z1} are components of the gravitational force vector $\vec{f}_1$ acting on mass m_1 and g_{x1}, g_{y1} and g_{z1} are components of the gravitational acceleration vector $\vec{g}_1$ felt by mass m_1 (in accordance to Newton's second law of motion, acceleration can be defined as the amount of force per unit mass, $\vec{g}_1 = \vec{f}_1 / m_1$).

The simplified explanation of such a derivation is the following. Firstly, we should realize that local derivatives of gravitational potential by the spatial coordinates are equal to the respective components of the gravitational acceleration vector $\vec{g}$, taken with a negative sign

$$\frac{\partial \Phi}{\partial x} = -g_x, \tag{2.17a}$$

$$\frac{\partial \Phi}{\partial y} = -g_y, \tag{2.17b}$$

$$\frac{\partial \Phi}{\partial z} = -g_z. \tag{2.17c}$$

The minus sign in the right hand side of Eqs. (2.17a)–(2.17c) reflects the fact that the potential energy Φ should increase in the direction opposite to the direction of the local gravity force, i.e. work contributing to the increase in the gravitational potential has to be done by applying a force $\vec{f}$ in the opposite direction compared to the local gravity force

$$d\Phi_x = f_x dx = -g_x dx, \tag{2.18a}$$

$$d\Phi_y = f_y dy = -g_y dy, \tag{2.18b}$$

$$d\Phi_z = f_z dz = -g_z dz, \tag{2.18c}$$

where $d\Phi_x$, $d\Phi_y$ and $d\Phi_z$ are increments in potential energy of a given unit mass due to small changes (dx, dy and dz) in the respective coordinates of this unit mass. According to Newton's law of gravitation (Eq. 2.16), the acceleration (i.e. gravitational force per unit mass) $\vec{g} = (g_x, g_y, g_z)$ felt at any given point within a continuum with coordinates x, y, z due to the gravitational attraction *from surrounding masses* is obtained by summing up (i.e. integrating) the accelerations exerted by each small mass element (δm_i), as follows

$$g_x = \sum_{i=1}^{\infty} G \frac{\delta m_i}{r_i^2} \frac{(x_i - x)}{r_i}, \tag{2.19a}$$

$$g_y = \sum_{i=1}^{\infty} G \frac{\delta m_i}{r_i^2} \frac{(y_i - y)}{r_i}, \tag{2.19b}$$

$$g_z = \sum_{i=1}^{\infty} G \frac{\delta m_i}{r_i^2} \frac{(z_i - z)}{r_i}, \tag{2.19c}$$

$$r_i = \sqrt{(x_i - x)^2 + (y_i - y)^2 + (z_i - z)^2}, \tag{2.20}$$

where x_i, y_i, z_i are the coordinates of the ith mass element and r_i is the distance between this element and the given point. The divergence of the integrated acceleration field at a given point can then be computed as follows (verify as an exercise):

$$-\nabla^2\Phi = \frac{\partial}{\partial x}\left(-\frac{\partial\Phi}{\partial x}\right) + \frac{\partial}{\partial y}\left(-\frac{\partial\Phi}{\partial y}\right) + \frac{\partial}{\partial z}\left(-\frac{\partial\Phi}{\partial z}\right) = \mathrm{div}(\vec{g}) = \frac{\partial g_x}{\partial x} + \frac{\partial g_y}{\partial y} + \frac{\partial g_z}{\partial z},$$

(2.21a)

$$\mathrm{div}(\vec{g}) = -\sum_{i=1}^{\infty} G\delta m_i \left[\frac{\partial}{\partial x}\left(\frac{x-x_i}{r_i^3}\right) + \frac{\partial}{\partial y}\left(\frac{y-y_i}{r_i^3}\right) + \frac{\partial}{\partial z}\left(\frac{z-z_i}{r_i^3}\right)\right],$$ (2.21b)

$$\mathrm{div}(\vec{g}) = -\sum_{i=1}^{\infty} G\delta m_i \left[\left(\frac{3(x-x_i)^2}{r_i^5} - \frac{1}{r_i^3}\right) + \left(\frac{3(y-y_i)^2}{r_i^5} - \frac{1}{r_i^3}\right) + \left(\frac{3(z-z_i)^2}{r_i^5} - \frac{1}{r_i^3}\right)\right],$$

(2.21c)

and finally by using Eq. (2.20)

$$\mathrm{div}(\vec{g}) = -\sum_{i=1}^{\infty} G\delta m_i \left[3\frac{r_i^2}{r_i^5} - 3\frac{1}{r_i^3}\right].$$ (2.21d)

Note that differentiation is done with respect to the coordinates x, y, z of the given point and not by those of the surrounding masses (x_i, y_i, z_i) which are independent of x, y, z.

For all $r_i \neq 0$, Eq. (2.21d) is obviously *zero*. This means that the divergence of gravitational acceleration $\mathrm{div}(\vec{g})$ at a given point is independent of the surrounding masses and must come from the point itself (i.e. for $x = x_i$, $y = y_i$, $z = z_i$ and $r_i = 0$). Then we can restrict the volume of integration to a small sphere with mass δm and radius δr centred on this point (Fig. 2.3). The divergence of acceleration for this small sphere can be approximated by differences with the use of six points (A, B, C, D, E, and F, Fig. 2.3) located on the sphere

$$\mathrm{div}(\vec{g}) \approx \frac{g_{xB} - g_{xA}}{2\delta r} + \frac{g_{yD} - g_{yC}}{2\delta r} + \frac{g_{zF} - g_{zE}}{2\delta r},$$ (2.22)

where g_{xA} to g_{zF} are the respective components of the acceleration vector at different points on the sphere's surface. Density inside the sphere can be considered constant since the mass distribution inside the small sphere is uniform. Therefore, at the considered six points, the respective components obviously coincide with the acceleration vectors directed toward the centre of the sphere and can be computed from the common gravity formula for a sphere (see Turcotte and Schubert, 2002, Eqs. (5–1)–(5–15)) as

$$g_{xA} = g_{yC} = g_{zE} = G\frac{\delta m}{\delta r^2},$$ (2.23a)

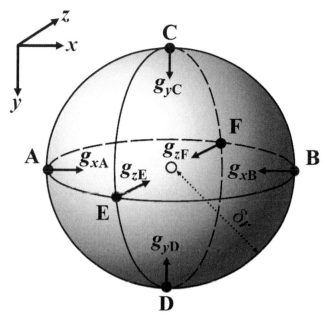

Fig. 2.3. Sphere of radius δr and mass δm considered for the derivation of the Poisson equation. Black arrows show respective components of the gravitational acceleration vector felt at the surface of the sphere in six different points (black dots, A, B, C, D, E and F). The open circle shows the centre of the sphere in which the divergence of the gravitational acceleration is computed.

$$g_{xB} = g_{yD} = g_{zF} - -G\frac{\delta m}{\delta r^2}. \tag{2.23b}$$

Combining Eqs. (2.22) and (2.23) and taking into account that the average density inside the sphere with volume $V = (4\pi\delta r^3)/3$ is given by $\rho = \delta m/V = 3\delta m/(4\pi\delta r^3)$, under the condition that δr tends to zero we obtain the Poisson equation (verify as an exercise)

$$\nabla^2\Phi = -\mathrm{div}(\vec{g}) = 3G\frac{\delta m}{\delta r^3} = 4\pi G\rho. \tag{2.24}$$

It should be noted that our derivation is simplified and uses a standard expression (Eq. 2.23) for the gravitational acceleration at the surface of the sphere, which can in turn be obtained by integrating Eq. (2.19) for a point outside the sphere (see Turcotte and Schubert, 2002, Eqs. (5–1)–(5–15) for details of this derivation). It can also be shown on the basis of *Gauss' theorem* that we can consider not only the small spherical volume, but any arbitrary shape of the local mass, and still obtain the same Poisson equation.

Analytical exercise

Exercise 2.1.
The Molar Gibbs potential of periclase (MgO, molar mass $m = 0.0403044$ kg/mol) is given
by the following equation (Gerya et al., 2004d)

$$G_{m(P,T)} = H_r + V_r\Psi + \sum_{i=1}^{3} c_i[RTln(1 - e_i) - \Delta H_i e_{oi}/(1 - e_{oi})],$$

$$e_i = \exp[-(\Delta H_i + \Delta V_i\Psi)/(RT)],$$

$$e_{oi} = \exp[-\Delta H_i/(RT_r)],$$

$$\Psi = 5/4(P_r + \phi)^{1/5}[(P + \phi)^{4/5} - (P_r + \phi)^{4/5}],$$

where $R = 8.314$ J/mol is the gas constant, $P_r = 100\,000$ Pa and $T_r = 298.15$ K are the reference
pressure and temperature respectively, and $H_r = -601\,500.00$ J, $V_r = 1.12228 \times 10^{-5}$ J/Pa, $\phi =$
$30\,179\,500\,000$ Pa, $c_1 = 1.96612$, $c_2 = 4.12756$, $c_3 = 0.53690$, $\Delta H_1 = 2966.88$ J, $\Delta H_2 =$
5621.69 J, $\Delta H_3 = 27\,787.19$ J, $\Delta V_1 = \Delta V_2 = 3.52971 \times 10^{-8}$ J/Pa, $\Delta V_3 = 1.9849568 \times 10^{-6}$ J/Pa
are empirical parameters.

Derive expressions for the density ρ, thermal expansion α and compressibility β as
functions of pressure and temperature (Fig. 2.1) using the equations given in this chapter.

Programming exercises

Exercise 2.2.
Compute and visualize as colour maps with isolines (*pcolor, contour, contourf*) and surfaces
(*surf, light, lighting*) the density ρ, thermal expansion α and compressibility β for periclase
(Fig. 2.1) based on the equations derived for the analytical exercises. Take a temperature
interval from 100 to 4000 K and a pressure interval from 10^9 to 10^{11} Pa (1 to 100 GPa). Try
also to define the Gibbs free energy equation as an external function $G_{m(P,T)}$ and use
differences instead of derivatives to compute $V_{(P,T)}$, $\rho_{(P,T)}$, $\alpha_{(P,T)}$ and $\beta_{(P,T)}$:

$$\rho_{(P,T)} = \frac{m}{V_{(P,T)}},$$

where

$$V_{(P,T)} = \left(\frac{\partial G_{m(P,T)}}{\partial P}\right)_{T=\text{const}} \approx \frac{\Delta G_{m(P,T)}}{\Delta P} = \frac{G_{m(P + \Delta P,T)} - G_{m(P,T)}}{\Delta P}$$

$$\alpha_{(P,T)} = -\frac{1}{\rho_{(P,T)}} \times \frac{\partial \rho_{(P,T)}}{\partial T} \approx -\frac{1}{\rho_{(P,T)}} \times \frac{\Delta \rho_{(P,T)}}{\Delta T} = -\frac{1}{\rho_{(P,T)}} \times \frac{\rho_{(P,T + \Delta T)} - \rho_{(P,T)}}{\Delta T},$$

$$\beta_{(P,T)} = \frac{1}{\rho_{(P,T)}} \times \frac{\partial \rho_{(P,T)}}{\partial P} \approx \frac{1}{\rho_{(P,T)}} \times \frac{\Delta \rho_{(P,T)}}{\Delta P} = \frac{1}{\rho_{(P,T)}} \times \frac{\rho_{(P+\Delta P,T)} - \rho_{(P,T)}}{\Delta P},$$

where ΔP and ΔT are small increments in pressure and temperature, respectively.

Compare the results based on differences with your analytical solutions. An example is in the code **Periclase_EOS.m** which calls function **G_periclase.m**.

Exercise 2.3.

Load from files (*fopen, fscanf, fclose*) and visualize (*pcolor*) density maps (Mishin et al., 2008) corresponding to the phase diagrams computed for pyrolite (mantle, file **m895_ro**) and MORB (occanic crust, file **morn_ro**) based on a Gibbs free energy minimization approach. Compute and visualize the density difference between these two contrasting types of rocks at various *P-T* parameters. Check at which *P* and *T* this difference is maximal/minimal.

The first nine positions in the data files are as follows:

pl8951 350 350 800.003 10001.5 9.16904 4269.33 T(K) P(bar)

where *pl8951* denotes the rock identification (skip that); *350* and *350* are resolutions for *T* and *P* respectively; *800.003* and *10001.5* are starting values for *T* (K) and *P* (bar, 1 bar = 10^5 Pa), respectively; *9.16904* and *4269.33* are steps for *T* (K) and *P* (bar), respectively; *T(K)* and *P(bar)* are *P* and *T* identifications (to skip). Further data in the files are 350×350 maps of rock density (kg/m³) at variable *T* (inner loop) and *P* (outer loop). An example is in **Density_map.m**.

3

Numerical solutions of partial differential equations

Theory: Analytical and numerical methods for solving partial differential equations. Using finite differences to compute various derivatives. Eulerian and Lagrangian approaches. Transition from partial differential equations to systems of linear equations. Methods of solving large systems of linear equations: iterative methods (Jacobi iteration, Gauss–Seidel iteration), direct methods (Gaussian elimination). Indexing of unknowns in 1D and 2D.
Exercises: Numerical solutions of the Poisson equation in 1D and 2D.

3.1 Finite difference method

Two principal methods are used for solving partial differential equations (PDEs) of continuum mechanics: *analytical* and *numerical*. Analytical methods are restricted to relatively simple problems and cannot be applied to a general case. This caveat is due, in particular, to the fact that it is sometimes impossible to analytically express the distribution of field variables ($T, P, \vec{v}, \eta, \rho$, etc.) in space and time. In effect, analytical methods are very useful for the general understanding of geodynamic processes (e.g. Turcotte and Schubert, 2002). In addition, analytical solutions are widely used for validating (benchmarking) numerical codes and testing their accuracy (Chapter 20).

Numerical methods for solving PDEs are universal and can be applied for both continuous and discontinuous distributions of field variables (e.g. Gustafsson, 2008). The following groups of numerical approaches are most frequently used in geodynamic modelling (e.g. Lynch, 2005; Zhong et al., 2007; Ismail-Zadeh and Tackley, 2010):

(1) finite difference methods (FDM),
(2) finite volume methods (FVM),
(3) finite element methods (FEM),
(4) spectral methods.

In this book, we will concentrate on the finite difference method (e.g. Patankar, 1980), which is the simplest of the four methods both for understanding and from

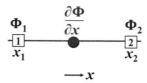

Φ_1 $\quad\dfrac{\partial\Phi}{\partial x}\quad$ Φ_2

Fig. 3.1. 1D numerical grid *stencil* (i.e. pattern of points) used for computing the first-order derivative of the gravity potential $\partial\Phi/\partial x$ using finite differences.

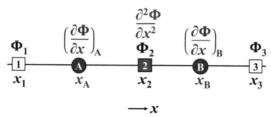

Fig. 3.2. 1D numerical grid stencil used for computing the second derivative of the gravity potential $\partial^2\Phi/\partial x^2$ using finite differences.

a programming point of view. Finite differences are linear mathematical expressions, which are used to represent derivatives to a certain degree of accuracy. For example, the first derivative of gravity potential by the x coordinate (Fig. 3.1) can be computed *within a certain degree of accuracy* (locally) by using finite differences as follows

$$\frac{\partial\Phi}{\partial x}=\frac{\Delta\Phi}{\Delta x}=\frac{\Phi_2-\Phi_1}{x_2-x_1}, \tag{3.1}$$

where $\Delta\Phi=\Phi_2-\Phi_1$ and $\Delta x=x_2-x_1$ are *differences* in respectively gravity potential and x coordinate between points 1 and 2. The smaller the distance Δx between points 1 and 2 becomes, the more accurate the computed derivative is. In fact, the finite difference representation of a derivative naturally follows from its definition

$$\frac{\partial\Phi}{\partial x}=\lim_{\Delta x\to0}\left(\frac{\Delta\Phi}{\Delta x}\right).$$

By analogy, higher order derivatives can be computed using lower order derivatives. For example, the second derivative of the gravity potential (Fig. 3.2) can be computed by repetitive use of finite differences Eq. (3.1) as follows

$$\frac{\partial^2\Phi}{\partial x^2}=\frac{\left(\frac{\partial\Phi}{\partial x}\right)_{\mathrm{B}}-\left(\frac{\partial\Phi}{\partial x}\right)_{\mathrm{A}}}{x_{\mathrm{B}}-x_{\mathrm{A}}}, \tag{3.2}$$

where

$$\left(\frac{\partial\Phi}{\partial x}\right)_A = \frac{\Phi_2 - \Phi_1}{x_2 - x_1}, \quad \left(\frac{\partial\Phi}{\partial x}\right)_B = \frac{\Phi_3 - \Phi_2}{x_3 - x_2}.$$

Using a similar procedure we can formulate third, fourth, fifth and higher order derivatives as well.

As follows from the above examples, we need a *grid* of points representing the distribution of field variables in space (and time) to apply finite differences. This so-called *numerical grid* is also often called a *numerical mesh*. Similarly to the two types of geometrical points (Chapter 1), grid points can be either *Eulerian* or *Lagrangian*. Eulerian points have steady positions and an Eulerian grid does not deform with deformation of the medium. Lagrangian points move according to the local flow and a Lagrangian grid deforms with the deformation of the medium. Time derivatives of field variables for Eulerian and Lagrangian points may differ from each other, for example

$$\frac{D\rho}{Dt} = \frac{\partial\rho}{\partial t} + \vec{v}\,\text{grad}(\rho), \tag{3.3}$$

where $D\rho/Dt$ is the substantive time derivative of density for a moving Lagrangian point and $\partial\rho/\partial t$ is the time derivative of density for an immobile Eulerian point *in the same location*. The main advantage of using an Eulerian grid is the possibility of having a relatively simple grid geometry which does not change during the model deformation; this simplifies the numerical formulation. The main disadvantage is the necessity to account for advective terms in time-dependent PDEs, which often causes numerical problems (e.g., *numerical diffusion*, Chapter 8). For a Lagrangian grid it is the contrary: a deforming grid ultimately produces numerical problems (and requires *re-gridding* or *re-meshing* when it is deformed too strongly) whereas the absence of advective terms in PDEs is an advantage. The use of either an Eulerian or a Lagrangian grid depends on the partial differential equations to be solved as well as on the type of physical processes to be modelled. In geodynamic modelling, combinations of Eulerian and Lagrangian grids for different field variables are often used to explore the advantages of both approaches (see e.g. Zhong et al., 2007; Ismail-Zadeh and Tackley, 2010).

What are we actually gaining by approximating derivatives by finite differences? This is a very important issue at the 'core' of numerical modelling. The use of finite differences allows us to transform partial differential equations, which are applicable to every geometrical point of a continuum (i.e. to an *infinite number of points*), into a *system of a finite amount of linear equations* formulated for a given finite number of grid points. The logical steps of applying finite differences are as follows:

(1) replacing an infinite number of geometrical points of a continuum within the model by a finite number of grid points;
(2) defining physical properties of the continuum in these points;

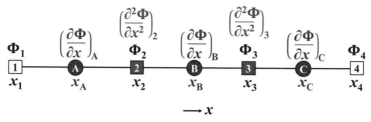

Fig. 3.3. 1D numerical grid for solving the 1D Poisson equation of the form $\partial^2\Phi/\partial x^2 = 1$.

(3) applying partial differential equations (including *boundary condition equations*) to the grid points and substituting them by linear equations expressed via finite differences; these linear equations relate the physical properties defined for different grid points;

(4) solving the resulting system of linear equations and obtaining unknown values of the physical parameters for the grid points.

Let us consider an example of this procedure for solving the 1D Poisson equation of the form $\partial^2\Phi/\partial x^2 = 1$, with the boundary conditions $\Phi(x_1) = R_1$ and $\Phi(x_4) = R_4$, where x_1, x_4 are coordinates of the model boundaries and $\Phi(x_1)$, $\Phi(x_4)$ are the values of the gravity potential at these boundaries.

Step 1: defining a numerical grid. A 1D grid for this problem is shown in Fig. 3.3. Different symbols on this grid show different *grid points* (also called *nodal points* or *nodes*) where different parameters are defined and different equations are applied.

- Nodes [1] and [4] (empty squares) are *basic boundary nodes* where only the gravitational potential Φ is defined and the boundary condition equation is applied

$$\Phi = R_1 \quad \text{when} \quad x = x_1,$$
$$\Phi = R_4 \quad \text{when} \quad x = x_4.$$

- Nodes [2] and [3] (solid squares) are *basic internal nodes* where both the gravitational potential Φ and its second derivative $\partial^2\Phi/\partial x^2$ are defined and the Poisson equation is applied

$$\frac{\partial^2\Phi}{\partial x^2} = 1.$$

- Nodes [A], [B] and [C] (open circles) are *additional nodes* located exactly in the middle of the intervals between the basic nodes. At these additional nodes, the first derivative of the gravitational potential $\partial\Phi/\partial x$ is defined, however no equations are applied. We only need these additional nodes to formulate second derivatives in the basic internal nodes by using finite differences.

Step 2: applying equations for the nodes and converting them to linear equations.
The following equations are applied for different nodes:

$$\text{node } [1] \quad \Phi = R_1$$

$$\text{node } [2] \quad \left(\frac{\partial^2 \Phi}{\partial x^2}\right)_2 = 1$$

$$\text{node } [3] \quad \left(\frac{\partial^2 \Phi}{\partial x^2}\right)_3 = 1$$

$$\text{node } [4] \quad \Phi = R_4.$$

To transform (*discretize*) these equations, we first define a way of computing the derivatives via finite differences for nodes [1] and [2]:

$$\text{node } [2] \quad \left(\frac{\partial^2 \Phi}{\partial x^2}\right)_1 = \frac{\left(\frac{\partial \Phi}{\partial x}\right)_B - \left(\frac{\partial \Phi}{\partial x}\right)_A}{x_B - x_A}$$

where

$$\left(\frac{\partial \Phi}{\partial x}\right)_A = \frac{\Phi_2 - \Phi_1}{x_2 - x_1}, \quad \left(\frac{\partial \Phi}{\partial x}\right)_B = \frac{\Phi_3 - \Phi_2}{x_3 - x_2}$$

$$\text{node } [3] \quad \left(\frac{\partial^2 \Phi}{\partial x^2}\right)_3 = \frac{\left(\frac{\partial \Phi}{\partial x}\right)_C - \left(\frac{\partial \Phi}{\partial x}\right)_B}{x_C - x_B}$$

where

$$\left(\frac{\partial \Phi}{\partial x}\right)_B = \frac{\Phi_3 - \Phi_2}{x_3 - x_2}, \quad \left(\frac{\partial \Phi}{\partial x}\right)_C = \frac{\Phi_4 - \Phi_3}{x_4 - x_3}.$$

Applying these finite differences results in the following system of four equations:

$$\Phi_1 = R_1$$

$$\frac{(\Phi_3 - \Phi_2)/(x_3 - x_2) - (\Phi_2 - \Phi_1)/(x_2 - x_1)}{(x_3 - x_1)/2} = 1$$

$$\frac{(\Phi_4 - \Phi_3)/(x_4 - x_3) - (\Phi_3 - \Phi_2)/(x_3 - x_2)}{(x_4 - x_2)/2} = 1$$

$$\Phi_4 = R_4.$$

Assembling coefficients for each unknown gives a final system of four linear equations:

$$1 \times \Phi_1 = R_1 \tag{3.4}$$

$$\frac{2/(x_2 - x_1)}{(x_3 - x_1)} \times \Phi_1 + \frac{-2/(x_2 - x_1) - 2/(x_3 - x_2)}{(x_3 - x_1)} \times \Phi_2 + \frac{2/(x_3 - x_2)}{(x_3 - x_1)} \times \Phi_3 = 1 \tag{3.5}$$

$$\frac{2/(x_3 - x_2)}{(x_4 - x_2)} \times \Phi_2 + \frac{-2/(x_3 - x_2) - 2/(x_4 - x_3)}{(x_4 - x_2)} \times \Phi_3 + \frac{2/(x_4 - x_3)}{(x_4 - x_2)} \times \Phi_4 = 1 \tag{3.6}$$

$$1 \times \Phi_4 = R_4. \tag{3.7}$$

In order to be able to find a solution, *the number of independent linear equations must be equal to the number of unknown parameters*. In this example, we know the coordinates x_1, x_2, x_3, and x_4 as well as values of R_1 and R_4 for the boundary condition equations. Unknown parameters are the discrete values of gravitational potential Φ in the basic grid points, i.e. Φ_1, Φ_2, Φ_3, and Φ_4. Therefore, the number of equations (four) is equal to the number of unknowns (four) and these equations are *linear with respect to the unknowns* and, thus, can be easily solved.

Step 3: solving the system of linear equations. Solving the above system of equations is trivial since we have only four equations with only four unknowns. As an exercise solve this problem manually for the following values of coordinates and coefficients:

$$x_1 = 0, \quad x_2 = 1, \quad x_3 = 2, \quad x_4 = 3, \quad R_1 = 0, \quad R_4 = 0.$$

This example looks trivial. Why should we spend so much time discussing it in so many details that are apparently so obvious? This is just to be finally convinced that the mathematical basis of numerical modelling *is exactly trivial*. What one needs to know are *derivatives* and *linear algebra* as was postulated in the Introduction (Golden Rule 1).

3.2 Solving linear equations

What is going to change if, instead of four basic nodes, we will have one thousand? Not much ... We will merely have to formulate and solve thousands of linear equations to obtain values of thousands of unknowns. Of course, our numerical solution will be more accurate, and it is not necessary to solve thousands of equations manually. One can obtain much larger systems of equations when trying to numerically solve geodynamic problems in 2D and 3D. Such systems typically contain thousands, millions and even billions of linear equations (such as Eqs. 3.4–3.7) having the following general form

$$L_{1,1}S_1 + L_{1,2}S_2 + L_{1,3}S_3 + \cdots + L_{1,n-1}S_{n-1} + L_{1,n}S_n = R_1$$
$$L_{2,1}S_1 + L_{2,2}S_2 + L_{2,3}S_3 + \cdots + L_{2,n-1}S_{n-1} + L_{2,n}S_n = R_2$$
$$\vdots$$
$$L_{n-1,1}S_1 + L_{n-1,2}S_2 + L_{n-1,3}S_3 + \cdots + L_{n-1,n-1}S_{n-1} + L_{n-1,n}S_n = R_{n-1}$$
$$L_{n,1}S_1 + L_{n,2}S_2 + L_{n,3}S_3 + \cdots + L_{n,n-1}S_{n-1} + L_{n,n}S_n = R_n$$

$$(3.8)$$

where n is the number of equations (and unknowns), S_l are unknowns, which are components of an n-dimensional vector $\{S\}$ (i.e. $n \times 1$ array, line), $L_{k,l}$ are coefficients, which are elements of an $n \times n$ *square matrix* $\{L\}$, R_k are the right hand sides which are components of an n-dimensional vector $\{R\}$ (i.e. $1 \times n$ array, column). Obviously, the system of equations can only be solved when the number of linearly independent equations in the system is equal to the number of unknowns. In the case of numerical modelling with finite differences, most of the coefficients $L_{k,l}$ in the equations are equal to zero, i.e. the $\{L\}$ matrix of coefficients is *sparse*. This is because linear equations formulated with finite differences for a node only contain parameters from the *nearest* nodes and not from all nodes of the grid (e.g., Eq. 3.5). The methods of solving large systems of equations are subdivided into iterative (e.g., *Jacobi iteration, Gauss–Seidel iteration*) and direct (e.g., *Gaussian elimination*) methods.

In order to apply iterative methods, *initial guesses* for the unknown variables are first defined, which represent a current approximation of the solution $S_1^{current}, S_2^{current}, \ldots, S_n^{current}$ (all unknowns can for example initially be set to zero). Then *residuals* (i.e. errors) for each equation can be computed as follows

$$\Delta R_1 = R_1 - L_{1,1}S_1^{current} - L_{1,2}S_2^{current} - L_{1,3}S_3^{current} - \cdots - L_{1,n-1}S_{n-1}^{current} - L_{1,n}S_n^{current}$$
$$\Delta R_2 = R_2 - L_{2,1}S_1^{current} - L_{2,2}S_2^{current} - L_{2,3}S_3^{current} - \cdots - L_{2,n-1}S_{n-1}^{current} - L_{2,n}S_n^{current}$$
$$\vdots$$
$$\Delta R_{n-1} = R_{n-1} - L_{n-1,1}S_1^{current} - L_{n-1,2}S_2^{current} - L_{n-1,3}S_3^{current} - \cdots - L_{n-1,n-1}S_{n-1}^{current} - L_{n-1,n}S_n^{current}$$
$$\Delta R_n = R_n - L_{n,1}S_1^{current} - L_{n,2}S_2^{current} - L_{n,3}S_3^{current} - \cdots - L_{n,n-1}S_{n-1}^{current} - L_{n,n}S_n^{current}.$$

$$(3.9)$$

The values of the residuals can then be used to obtain new, more accurate, values of the unknowns

$$S_1^{new} = S_1^{current} + \theta_1 \frac{\Delta R_1}{L_{1,1}}$$
$$S_2^{new} = S_2^{current} + \theta_2 \frac{\Delta R_2}{L_{2,2}}$$

$$(3.10)$$

$$\vdots$$

$$S_{n-1}^{new} = S_{n-1}^{current} + \theta_{n-1} \frac{\Delta R_{n-1}}{L_{n-1,n-1}}$$
$$S_n^{new} = S_n^{current} + \theta_n \frac{\Delta R_n}{L_{n,n}},$$

where θ_1, θ_2, ..., θ_n are *relaxation parameters* defining how strongly the computed residuals ΔR_1, ΔR_2, ..., ΔR_n will contribute to the changes in the current values of the unknown. Relaxation parameters are typically taken within the range between 0 and 1. In the case when the relaxation parameters are equal to 1, a new solution can be obtained with the use of simplified formulas (please derive as an exercise)

$$S_1^{new} = \frac{R_1 - L_{1,2}S_2^{current} - L_{1,3}S_3^{current} - \cdots - L_{1,n-1}S_{n-1}^{current} - L_{1,n}S_n^{current}}{L_{1,1}}$$

$$S_2^{new} = \frac{R_2 - L_{2,1}S_1^{current} - L_{2,3}S_3^{current} - \cdots - L_{2,n-1}S_{n-1}^{current} - L_{2,n}S_n^{current}}{L_{2,2}}$$

$$\vdots$$

$$S_{n-1}^{new} = \frac{R_{n-1} - L_{n-1,1}S_1^{current} - L_{n-1,2}S_2^{current} - L_{n-1,3}S_3^{current} - \cdots - L_{n-1,n}S_n^{current}}{L_{n-1,n-1}}$$

$$S_n^{new} = \frac{R_n - L_{n,1}S_1^{current} - L_{n,2}S_2^{current} - L_{n,3}S_3^{current} - \cdots - L_{n,n-1}S_{n-1}^{current}}{L_{n,n}}. \quad (3.11)$$

New values of the unknowns are then used for the next iteration to obtain the next (more accurate) solution. Iteration cycles are repeated several times to reach a certain level of accuracy, which is defined by values of residuals.

There are several methods of doing such iterations. In the *Jacobi iteration*, new values are assigned to all unknowns simultaneously *after* finishing one cycle of iterations for all n equations. In *Gauss–Seidel iterations*, new values are assigned to each unknown *during* the cycle of iterations, immediately after obtaining the updated value from the respective equation. Residuals for the following equations in the cycle are then computed using the new values of unknowns obtained from the previous equations.

Computational advantages of iterative methods are (i) a small amount of required computer memory, typically proportional to the number of unknowns, and (ii) a small amount of mathematical operations, also typically proportional to the number of unknowns. Therefore, *iterative solvers* are frequently used in 3D when the number of equations is large. Among the disadvantages are (i) lowered accuracy of the solution and (ii) problems of convergence to an accurate solution, which are especially relevant for solving problems with large variations in material properties (e.g., mechanical problems with strong and sharp variations in viscosity). Indeed, there are various possibilities for notable improvement of the iterative methods discussed above based for example on a *multigrid* approach (Chapter 18).

Direct methods of solution do not require an initial guess and are based on mathematical transformations of the matrix $\{L\}$ and vector $\{R\}$, which allow computation of the vector $\{S\}$ directly. One of the best-known direct methods is *Gaussian elimination*, which produces a diagonal matrix $\{L_{diagonal}\}$ and a corresponding new vector $\{R_{diagonal}\}$ that can then be used to directly compute the vector $\{S\}$. The procedure of Gaussian elimination is as follows:

(a) Divide all Eqs. (3.8) by their respective (non-zero) coefficients at x_1 to obtain a new system with new coefficients at the unknowns and new right hand sides:

$$S_1 + \frac{L_{1,2}}{L_{1,1}}S_2 + \frac{L_{1,3}}{L_{1,1}}S_3 + \cdots + \frac{L_{1,n-1}}{L_{1,1}}S_{n-1} + \frac{L_{1,n}}{L_{1,1}}S_n = \frac{R_1}{L_{1,1}}$$

$$S_1 + \frac{L_{2,2}}{L_{2,1}}S_2 + \frac{L_{2,3}}{L_{2,1}}S_3 + \cdots + \frac{L_{2,n-1}}{L_{2,1}}S_{n-1} + \frac{L_{2,n}}{L_{2,1}}S_n = \frac{R_2}{L_{2,1}}$$

$$\vdots$$

$$S_1 + \frac{L_{n-1,2}}{L_{n-1,1}}S_2 + \frac{L_{n-1,3}}{L_{n-1,1}}S_3 + \cdots + \frac{L_{n-1,n-1}}{L_{n-1,1}}S_{n-1} + \frac{L_{n-1,n}}{L_{n-1,1}}S_n = \frac{R_{n-1}}{L_{n-1,1}}$$

$$S_1 + \frac{L_{n,2}}{L_{n,1}}S_2 + \frac{L_{n,3}}{L_{n,1}}S_3 + \cdots + \frac{L_{n,n-1}}{L_{n,1}}S_{n-1} + \frac{L_{n,n}}{L_{n,1}}S_n = \frac{R_n}{L_{n,1}}.$$

(3.12)

(b) Subtract the first equation from all other equations (starting from the second one) to obtain a subsystem of $n-1$ equations with $n-1$ unknowns (since S_1 will be *eliminated* from all the equations starting from the second one), i.e. new $(n-1) \times (n-1)$ matrix $\{L^1\}$ and new $1 \times (n-1)$ vector $\{R^1\}$

$$\left(\frac{L_{2,2}}{L_{2,1}} - \frac{L_{1,2}}{L_{1,1}}\right)S_2 + \left(\frac{L_{2,3}}{L_{2,1}} - \frac{L_{1,3}}{L_{1,1}}\right)S_3 + \cdots + \left(\frac{L_{2,n-1}}{L_{2,1}} - \frac{L_{1,n-1}}{L_{1,1}}\right)S_{n-1} + \left(\frac{L_{2,n}}{L_{2,1}} - \frac{L_{1,n}}{L_{1,1}}\right)S_n = \frac{R_2}{L_{2,1}} - \frac{R_1}{L_{1,1}}$$

$$\vdots$$

$$\left(\frac{L_{n-1,2}}{L_{n-1,1}} - \frac{L_{1,2}}{L_{1,1}}\right)S_2 + \left(\frac{L_{n-1,3}}{L_{n-1,1}} - \frac{L_{1,3}}{L_{1,1}}\right)S_3 + \cdots + \left(\frac{L_{n-1,n-1}}{L_{n-1,1}} - \frac{L_{1,n-1}}{L_{1,1}}\right)S_{n-1} + \left(\frac{L_{n-1,n}}{L_{n-1,1}} - \frac{L_{1,n}}{L_{1,1}}\right)S_n = \frac{R_{n-1}}{L_{n-1,1}} - \frac{R_1}{L_{1,1}}$$

$$\left(\frac{L_{n,2}}{L_{n,1}} - \frac{L_{1,2}}{L_{1,1}}\right)S_2 + \left(\frac{L_{n,3}}{L_{n,1}} - \frac{L_{1,3}}{L_{1,1}}\right)S_3 + \cdots + \left(\frac{L_{n,n-1}}{L_{n,1}} - \frac{L_{1,n-1}}{L_{1,1}}\right)S_{n-1} + \left(\frac{L_{n,n}}{L_{n,1}} - \frac{L_{1,n}}{L_{1,1}}\right)S_n = \frac{R_n}{L_{n,1}} - \frac{R_1}{L_{1,1}},$$ (3.13)

or by using new notation for the coefficients in the left hand side $L^1_{k,l} = L_{k,l}/L_{k,1} - L_{1,l}/L_{1,1}$ and for the right hand side $R^1_k = R_k/L_{k,1} - R_1/L_{1,1}$ we write

$$L^1_{2,2}S_2 + L^1_{2,3}S_3 + \cdots + L^1_{2,n-1}S_{n-1} + L^1_{2,n}S_n = R^1_2$$

$$\vdots$$

(3.14)

$$L^1_{n-1,2}S_2 + L^1_{n-1,3}S_3 + \cdots + L^1_{n-1,n-1}S_{n-1} + L^1_{n-1,n}S_n = R^1_{n-1}$$

$$L^1_{n,2}S_2 + L^1_{n,3}S_3 + \cdots + L^1_{n,n-1}x_{n-1} + C^1_{n,n}x_n = R^1_n.$$

(c) Repeat (a) and (b) on this new subsystem to obtain a subsystem of $n-2$ equations with $n-2$ unknowns.

(d) Repeat. In total $n-1$ *elimination* cycles are needed.

(e) The last equation in the final system (diagonal matrix) will have the form

$$L_{n,n}^{n-1} S_n = R_n^{n-1},$$ (3.15)

i.e., will contain only one coefficient $L_{n,n}^{n-1}$ for the unknown X_n and right hand side R_n^{n-1}. Then S_n can be directly calculated as

$$S_n = \frac{R_n^{n-1}}{L_{n,n}^{n-1}}.$$ (3.16)

(f) The second last equation will have the form

$$L_{n-1,n-1}^{n-2} S_{n-1} + L_{n-1,n}^{n-2} S_n = R_{n-1}^{n-2},$$ (3.17)

and S_{n-1} can be calculated with the already known S_n as follows

$$S_{n-1} = \frac{R_{n-1}^{n-2} - L_{n-1,n}^{n-2} S_n}{L_{n-1,n-1}^{n-2}}.$$ (3.18)

(g) Repeat for all other unknowns from S_{n-2} to S_1.

The main advantage of direct methods is that the solution can be done to computer accuracy and no iterations are needed. Among their disadvantages are (i) large amounts of consumed memory, typically proportional to the square of the number of unknowns, and (ii) large amounts of operations, typically proportional to the square or even to the cube of the number of unknowns. Due to limitations in computer power, *direct solvers* are more often used in 1D and 2D modelling, particularly for solving numerical problems, where *iterative solvers* are inefficient.

3.3 Geometrical and global indexing of unknowns

In composing the system of Eqs. (3.4)–(3.7), we indexed our unknown parameters Φ_1, Φ_2, Φ_3, and Φ_4 in 1D using the principle of increasing index of the parameter with the x coordinate of the respective geometrical point to which this unknown is assigned (cf. points 1, 2, 3, and 4 in Fig. 3.3). We may then have an impression that a general system of equations (3.8) can only be applicable for 1D problems, since in 2D unknown parameters should have both horizontal and vertical indexes, for example $\Phi_{i,j}$ and respectively $S_{i,j}$. This is not correct and it is a small, but important point to understand. *Geometrical indexing* of unknowns $\Phi_{i,j}$ in a 2D grid is different from overall *global indexing* of these unknowns given by S_l and used in the system of equations (3.8). Global indexing of unknowns (Fig. 3.4) is always needed when direct methods are used for solving the equations, as one has to compose the matrix $\{L\}$ and vector $\{R\}$. In the case of iterative solutions, one can formulate linear equations (such as Eqs. 3.4–3.7) using a geometrical indexing.

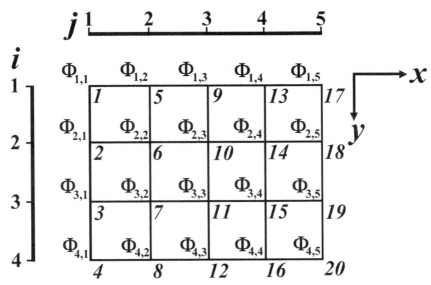

Fig. 3.4. Geometrical indexing of gravity potential values $\Phi_{i,j}$ assigned to the nodes of a 2D grid and global indexing (italic numbers) of these parameters in the vector $\{S\}$. The global indexing is done by columns of nodal points.

Programming exercises

Exercise 3.1.
Solve the 1D Poisson equation, written in the form $\partial^2\Phi/\partial x^2 = 1$, on a regular grid of 101 points with finite differences and visualize the solution. The model length is 1000 km. Use sparse initialization for the matrix of coefficients $\{L\}$, i.e. $L = \text{sparse}(101,101)$. Compose a matrix $\{L\}$ and a right hand side vector $\{R\}$ (use Eqs. 3.4–3.7 as an example) and obtain the solution vector $\{S\}$ with a direct solver (in MATLAB with the command S = L\R). Use $\Phi = 0$ as the boundary condition for the two external nodes of the grid (e.g. Fig. 3.3). An example is in **Poisson1D.m**.

Exercise 3.2.
Solve the 2D Poisson equation with finite differences and visualize the solution. The governing equation is given by

$$\frac{\partial^2\Phi}{\partial x^2} + \frac{\partial^2\Phi}{\partial y^2} = 1, \tag{3.19}$$

on a regular grid of 31×41 points. The model size is $1000 \times 1500 \text{ km}^2$. Use the principle of global indexing in 2D as shown in Fig. 3.4. A finite difference representation of the Poisson equation in 2D can be derived from Eq. (3.19) by analogy with Eq. (3.2), but applied separately for the x and y directions (Fig. 3.5)

$$\frac{\Phi_{i,j-1} - 2\Phi_{i,j} + \Phi_{i,j+1}}{\Delta x^2} + \frac{\Phi_{i-1,j} - 2\Phi_{i,j} + \Phi_{i+1,j}}{\Delta y^2} = 1, \tag{3.20}$$

or by assembling coefficients at each unknown

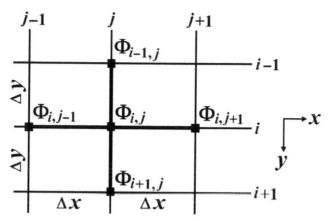

Fig. 3.5. 2D numerical grid stencil (5-point cross) used for formulating the Poisson equation by using finite differences on a regular rectangular grid.

$$\left(\frac{1}{\Delta x^2}\right)\Phi_{i,j-1} + \left(\frac{1}{\Delta x^2}\right)\Phi_{i,j+1} + \left(\frac{-2}{\Delta x^2} + \frac{-2}{\Delta y^2}\right)\Phi_{i,j} + \left(\frac{1}{\Delta y^2}\right)\Phi_{i-1,j} + \left(\frac{1}{\Delta y^2}\right)\Phi_{i+1,j} = 1.$$
(3.21)

Use $\Phi = 0$ as the boundary conditions for all external nodes of the grid. Compute the global index k, based on geometrical indexes i and j (Fig. 3.5) and vertical grid resolution N_y as

$$k = (j - 1) \times N_y + i.$$
(3.22)

An example is in **Poisson2D_direct.m**.

Exercise 3.3.
Solve the same 2D problem using Gauss–Seidel iterations. Use Eq. (3.19) to compute residuals. Use relaxation parameter $\theta = 1.5$ for all points (Eq. 3.10). Plot the gravitational potential and residuals every 10 iterations. An example is in **Poisson2D_Gauss_Seidel.m**.

Exercise 3.4.
Solve the same 2D problem using Jacobi iterations. Use relaxation parameter $\theta = 1.0$ for all points (Eq. 3.10). An example is in **Poisson2D_ Jacobi.m**.

4

Stress and strain

Theory: Deformation and stresses. Definition of stress, strain and strain rate
tensors. Mean normal stress and pressure. Deviatoric stress. Stress and strain
rate invariants.
Exercises: Computing the strain rate tensor components in 2D from the
material velocity fields.

4.1 Stress

Tensors are field variables, which characterize the internal state of a continuum and are,
perhaps, the most difficult quantities to intuitively understand. Indeed, at least three tensors
have to be used in the following and these are the *stress, strain* and *strain rate* tensors.

Stress is the *internal distribution and intensity of force* acting at any point within a continuum
in response to various internal and external loads applied to the continuum. Stress is defined as
a *force per unit area* and we can easily 'apprehend' its effect by pressing two fingers against
each other – *equal force is applied from both sides and therefore nothing moves*, but we have
a *feeling of pressure* between the fingers, which is a sign of the presence of stress. This stress is
directly proportional to the applied force – the stronger we press the stronger the feeling is.
On the other hand, the stress is inversely proportional to the contact surface between the
fingers – if we press one finger with the nail of the other the feeling is much stronger because
the same force is applied to a much smaller area. This is why pricking a finger with a needle is
so painful – the force applied to the needle is not big but the contact surface of the needle with
the finger is very small and the resulting stress is consequently very big.

In order to characterize the stress tensor, let us consider the force $\vec{f}_{(x)}$ acting on a unit
element of a Lagrangian x-surface (i.e. surface orthogonal to the x axis) (Fig. 4.1a). First of
all, we need to understand that the force vector $\vec{f}_{(x)}$ acting on one side of the surface element
is balanced by the counterforce vector $-\vec{f}_{(x)}$ which acts on the other side, and therefore this
stressed surface element does not move. Thus, in order to characterize the *force balance
state* of the stressed surface element, one needs to characterize the magnitude and direction

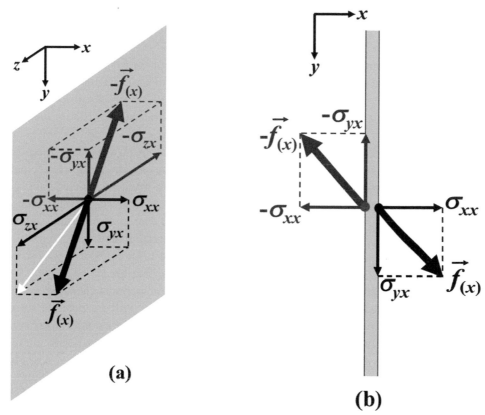

Fig. 4.1. Components of stress tensor defined from the force balance on a surface. (a) Relationship between the stress components (thin arrows σ_{xx}, σ_{yx}, σ_{zx} and $-\sigma_{xx}$, $-\sigma_{yx}$, $-\sigma_{zx}$) and force vectors (thick arrows $\vec{f}_{(x)}$ and $-\vec{f}_{(x)}$) acting on the two sides of the unit element (light blue) of a Lagrangian surface orthogonal to the x-axis (i.e. x-surface). White arrow in (a) shows the direction of shear along the surface. (b) Physical analogy: normal and shear stress components acting on a thin plate (cross-section of the plate in the x-y plane is shown).

of the force (balanced by the counterforce) acting on this element. Let us adopt a convention that the characterization will be based on the force vector $\vec{f}_{(x)}$, applied to the side of the x-surface *from which the x axis is exiting* (i.e. to the side *looking in the direction of the x axis*). Please note that this is a formal choice, which *does not depend on the actual direction of the force vector $\vec{f}_{(x)}$*. According to this convention, *extensional stresses are positive* (force and counterforce go outward from the surface) as is usually assumed in continuum mechanics (e.g. Ranalli, 1995). Notice that this *usual continuum mechanics convention* is opposite to that used in geosciences (e.g. Turcotte and Schubert, 2002), where stresses are taken positive under compression (which geoscientists find more intuitive since pressure is also positive under compression).

The force vector $\vec{f}_{(x)}$ can obviously be decomposed into three components (σ_{xx}, σ_{yx}, σ_{zx}) parallel to each coordinate axis (Fig. 4.1a). These are the components of the stress tensor since force $\vec{f}_{(x)}$ is acting on the unit element. According to the common continuum mechanics convention (e.g. Ranalli, 1995), which is again opposite to that used in geosciences (e.g. Turcotte and Schubert, 2002), the first index (i) of a stress component σ_{ij} denotes the axis along which this stress component is taken (i.e. $i = z$ for the component parallel to the z-axis) and the second index (j) indicates the surface on which force balance is considered (i.e. $j = x$ for the surface orthogonal to the x axis). It should be pointed out that our 'hard choice' of stress definition and notation is, indeed, very convenient for formulating several crucial equations, such as the momentum equation and the rheological constitutive relationships, which is the main reason why we have deviated from the 'geological convention'. On the other hand, our vertical axis y, is always pointing down, thus preserving common 'geological logic' that the vertical coordinate is depth (and not height as in continuum mechanics) and increases downward rather than upward. A stress component that is orthogonal to the surface (cf. σ_{xx} in Fig. 4.1a) is called a *normal stress component* and the components which are parallel to the surface are called *shear stress components* (cf. σ_{yx} and σ_{zx} in Fig. 4.1a). The normal stress component characterizes the magnitude of extension/compression *across the surface* and will be positive under extension and negative under compression in accordance with the adopted *stress convention*. The two shear stress components characterize the magnitude and direction (cf. white arrow in Fig. 4.1a) of shearing applied *along the considered surface*. A useful physical analogy (Fig. 4.1b) – if one imagines that the force and counterforce are applied on two sides of a very thin plate, then the normal component defines how strongly the two opposite sides of the plate are forced to be shifted from/toward each other and the shear stress components define where and how strong these sides are forced to be shifted parallel to each other.

In order to fully characterize the force balance at a point (a small material volume), it is convenient to represent the stress tensor as an $N{\times}N$ matrix where N is the dimension of the problem such that in one, two and three dimensions we will have one, four and nine stress components respectively (Fig. 4.2):

$$\text{1D stress tensor,} \, N = 1 (\text{Fig. 4.2a}) \quad \sigma_{ij} = (\sigma_{xx}),$$

$$\text{2D stress tensor,} \, N = 2 (\text{Fig. 4.2b}) \quad \sigma_{ij} = \begin{pmatrix} \sigma_{xx} & \sigma_{xy} \\ \sigma_{yx} & \sigma_{yy} \end{pmatrix},$$

$$\text{3D stress tensor,} \, N = 3 (\text{Fig. 4.2c}) \quad \sigma_{ij} = \begin{pmatrix} \sigma_{xx} & \sigma_{xy} & \sigma_{xz} \\ \sigma_{yx} & \sigma_{yy} & \sigma_{yz} \\ \sigma_{zx} & \sigma_{zy} & \sigma_{zz} \end{pmatrix},$$

where i and j are *symbolic coordinate indexes* (x, y, z) which vary in vertical and horizontal directions, respectively. In continuum mechanics books a numerical (1, 2, 3) notation for the coordinate indexes i and j and stresses (σ_{11}, σ_{12}, σ_{32} etc.) is commonly used as well (e.g.

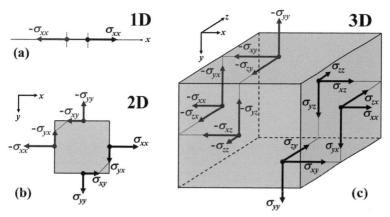

Fig. 4.2. Components of the stress tensor (black and blue arrows) defined in one (a), two (b) and three (c) dimensions on faces of a small interval, square and cube, respectively. The faces are always oriented orthogonal to the main axis. Thin lines in (b) and (c) connect pairs of shear stress components, which should be equal to each other in the absence of internal sources of angular momentum.

Ranalli, 1995). *Note that i and j are symbolic indexes and not physical coordinates of geometric points.*

Normal stresses are always located on the main diagonal of the matrix. Due to the *condition of force balance in the absence of internal sources of angular momentum,* this matrix is symmetric relative to the main diagonal so that

$$\sigma_{ij} = \sigma_{ji},$$

i.e. (Fig. 4.2b,c)

$$\sigma_{xy} = \sigma_{yx},$$

$$\sigma_{xz} = \sigma_{zx},$$

$$\sigma_{yz} = \sigma_{zy}.$$

Like components of a vector, components of the stress tensor at a point depend on the orientation of the coordinate system. We will discuss this in more detail later in relation to elasticity (Chapter 12).

In continuum mechanics, *pressure* is defined as the negative of the mean normal stress

$$P = -(\sigma_{xx} + \sigma_{yy} + \sigma_{zz})/3, \tag{4.1}$$

where the negative sign on the right hand side of Eq. (4.1) reflects a convention that pressure is positive under compression. Pressure is an *invariant* and, thus, *does not change with changing the coordinate system.* In the case of a *hydrostatic stress state* (which is the state of a fluid *at rest*) all shear stresses are zero and all normal stresses are equal to each other

$$\sigma_{xy} = \sigma_{yx} = \sigma_{xz} = \sigma_{zx} = \sigma_{yz} = \sigma_{zy} = 0, \tag{4.2a}$$

$$\sigma_{xx} = \sigma_{yy} = \sigma_{zz} = -P, \tag{4.2b}$$

$$\sigma_{ij(hydrostatic)} = \begin{pmatrix} -P & 0 & 0 \\ 0 & -P & 0 \\ 0 & 0 & -P \end{pmatrix}. \tag{4.2c}$$

In geosciences, pressure is often considered as corresponding to the hydrostatic (litho-static) condition everywhere and it is computed as a function of depth y and vertical density profile $\rho(y)$

$$P(y) = P_0 + g \int_0^y \rho(y)dy, \tag{4.3}$$

where $P_0 = 0.1$ MPa is pressure on the Earth's surface and g is the gravitational acceleration.

This simplification does not hold when deformations of geological media occur and real *dynamic* pressure may notably deviate from the lithostatic value given by Eq. (4.3).

It is often convenient to define the *deviatoric* stress σ_{ij}' (the notation τ_{ij} is also often used), which is *deviation of stress from the hydrostatic stress state* (i.e., deviation from conditions 4.2)

$$\sigma_{ij}' = \sigma_{ij} + P\delta_{ij}, \tag{4.4}$$

where δ_{ij} is the *Kronecker delta*: $\delta_{ij} = 1$ when $i = j$ and $\delta_{ij} = 0$ when $i \neq j$, i and j are symbolic coordinate indexes (x, y, z). The Kronecker delta is a peculiar abbreviation commonly used in continuum mechanics. It only takes a value of either 1 or 0 and is analogous to the logical operator 'if' used in many programming languages. Any equation with δ_{ij} represents a *group of equations*. For example, Eq. (4.4) in 3D represents the following equations:

normal deviatoric stresses

$$\sigma_{xx}' = \sigma_{xx} + P,$$

$$\sigma_{yy}' = \sigma_{yy} + P,$$

$$\sigma_{zz}' = \sigma_{zz} + P,$$

and shear stresses which are entirely deviatoric

$$\sigma_{xy}' = \sigma_{yx}' = \sigma_{xy} = \sigma_{yx},$$

$$\sigma_{xz}' = \sigma_{zx}' = \sigma_{xz} = \sigma_{zx},$$

$$\sigma_{yz}' = \sigma_{zy}' = \sigma_{yz} = \sigma_{zy}.$$

It is worth mentioning that the sum of the normal deviatoric stresses is zero by definition (verify as an exercise on the basis of Eqs. 4.1 and 4.4)

$$\sigma_{xx}' + \sigma_{yy}' + \sigma_{zz}' = 0.$$

The *second invariant* of the deviatoric stress tensor can be calculated as follows:

$$\sigma_{II} = \sqrt{{}^1\!/_2\, \sigma_{ij}'^2}, \qquad (4.5)$$

where the indexes *ij* imply a *summation*! This is another abbreviation that is commonly used in continuum mechanics and makes equations shorter (but, indeed, not easier to understand for inexperienced readers). The spelled-out form of Eq. (4.5) is much longer

$$\sigma_{II} = \sqrt{{}^1\!/_2 \left(\sigma_{xx}'^2 + \sigma_{yy}'^2 + \sigma_{zz}'^2 + \sigma_{xy}^2 + \sigma_{yx}^2 + \sigma_{xz}^2 + \sigma_{zx}^2 + \sigma_{yz}^2 + \sigma_{zy}^2 \right)}, \qquad (4.6a)$$

or, using the condition of force balance $\sigma_{ij} = \sigma_{ji}$

$$\sigma_{II} = \sqrt{{}^1\!/_2 \left(\sigma_{xx}'^2 + \sigma_{yy}'^2 + \sigma_{zz}'^2 \right) + \sigma_{xy}^2 + \sigma_{xz}^2 + \sigma_{yz}^2}. \qquad (4.6b)$$

The second stress invariant σ_{II} does not depend on the coordinate system and characterizes *the magnitude of local deviation of stresses in the medium from the hydrostatic stress state*. Please note that in continuum mechanics the second deviatoric stress invariant is defined differently as $J_2 = {}^1\!/_2\, \sigma_{ij}'^2$ and has units of Pa^2 rather than Pa.

4.2 Strain and strain rate

Another important quantity is the *strain* γ that characterizes the amount of deformation. Strain is dimensionless and is computed as the ratio of displacement ΔL to the initial length of the deforming body L (Fig. 4.3)

$$\gamma = \frac{\Delta L}{L}. \qquad (4.7)$$

By analogy with stress, one can discriminate normal and shear strain corresponding to axial and shear deformation, respectively (Fig. 4.3a,b).

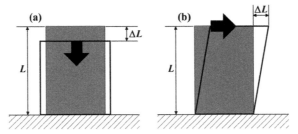

Fig. 4.3. Axial (a) and shear (b) deformation corresponding to normal and shear strain components. The strain in both cases is estimated as $\gamma = \Delta L / L$. Note that in the case of shear deformation, the length L is measured orthogonal to the displacement direction.

The definition of strain given by Eq. (4.7) can only be applied in cases of relatively simple axial and shear deformations. In the case of more complex deformation, the *strain tensor* ε_{ij} is defined as

$$\varepsilon_{ij} = \frac{1}{2}\left(\frac{\partial u_i}{\partial x_j} + \frac{\partial u_j}{\partial x_i}\right), \tag{4.8}$$

where u_i and u_j are components of the *displacement* vector $\vec{u} = (u_x, u_y, u_z)$, which characterizes the displacement of a material point relative to its original position (i.e. before deformation). i and j are *coordinate indexes* (x, y, z) and x_i and x_j *are spatial coordinates* (i.e., x_x, x_y and x_z are x-, y- and z-coordinates respectively). *Note that in contrast to symbolic i and j indexes, x_i and x_j are physical coordinates of geometrical points.* Do not confuse them with each other! In 3D, we can define nine strain tensor components:
three normal strain components

$$\varepsilon_{xx} = \frac{1}{2}\left(\frac{\partial u_x}{\partial x} + \frac{\partial u_x}{\partial x}\right) = \frac{\partial u_x}{\partial x},$$

$$\varepsilon_{yy} = \frac{1}{2}\left(\frac{\partial u_y}{\partial y} + \frac{\partial u_y}{\partial y}\right) = \frac{\partial u_y}{\partial y},$$

$$\varepsilon_{zz} = \frac{1}{2}\left(\frac{\partial u_z}{\partial z} + \frac{\partial u_z}{\partial z}\right) = \frac{\partial u_z}{\partial z},$$

and six shear strain components

$$\varepsilon_{xy} = \varepsilon_{yx} = \frac{1}{2}\left(\frac{\partial u_x}{\partial y} + \frac{\partial u_y}{\partial x}\right),$$

$$\varepsilon_{xz} = \varepsilon_{zx} = \frac{1}{2}\left(\frac{\partial u_x}{\partial z} + \frac{\partial u_z}{\partial x}\right),$$

$$\varepsilon_{yz} = \varepsilon_{zy} = \frac{1}{2}\left(\frac{\partial u_z}{\partial y} + \frac{\partial u_y}{\partial z}\right).$$

Note that stress and strain tensors are very different physical quantities (although they can be strongly correlated in the case of reversible elastic deformation, Chapter 12): stress characterizes the distribution of forces acting in a continuum at a given moment of time, while strain quantifies in an integrated way the entire *deformation history* of the continuum from the initial state, up until this given moment (Fig. 4.3). The symmetric form of the strain tensor *subtracts the rotational component of the velocity field*, which does not contribute to the deformation (rotation of a rigid body has gradients in material displacement, but does not produce any internal deformation). The time derivative of the displacement vector is the velocity vector $\vec{v} = (v_x, v_y, v_z)$ so that

$$v_i = \frac{Du_i}{Dt},\qquad(4.9)$$

and, in 3D deformation

$$v_x = \frac{Du_x}{Dt},$$

$$v_y = \frac{Du_y}{Dt},$$

$$v_z = \frac{Du_z}{Dt}.$$

The strain tensor is widely used when elastic deformation is considered (Chapter 12).

In numerical geodynamic modelling, it is convenient to use the *strain rate*, which characterizes the dynamics of changes in the internal deformation, rather than the strain, which characterizes the total amount of deformation compared to the initial state. The strain rate tensor $\dot{\varepsilon}_{ij}$ is the time derivative (indicated by the dot on top of the strain symbol) of the strain tensor ε_{ij}. Components of the strain rate tensor are defined via spatial gradients of the velocity as follows

$$\dot{\varepsilon}_{ij} = \frac{1}{2}\left(\frac{\partial v_i}{\partial x_j} + \frac{\partial v_j}{\partial x_i}\right),\qquad(4.10)$$

where i and j are coordinate indexes and x_i and x_j are spatial coordinates such that in 3D we can define nine tensor components:

three normal strain rate components

$$\dot{\varepsilon}_{xx} = \frac{1}{2}\left(\frac{\partial v_x}{\partial x} + \frac{\partial v_x}{\partial x}\right) = \frac{\partial v_x}{\partial x},$$

$$\dot{\varepsilon}_{yy} = \frac{1}{2}\left(\frac{\partial v_y}{\partial y} + \frac{\partial v_y}{\partial y}\right) = \frac{\partial v_y}{\partial y},$$

$$\dot{\varepsilon}_{zz} = \frac{1}{2}\left(\frac{\partial v_z}{\partial z} + \frac{\partial v_z}{\partial z}\right) = \frac{\partial v_z}{\partial z},$$

and six shear strain rate components

$$\dot{\varepsilon}_{xy} = \dot{\varepsilon}_{yx} = \frac{1}{2}\left(\frac{\partial v_x}{\partial y} + \frac{\partial v_y}{\partial x}\right),$$

$$\dot{\varepsilon}_{xz} = \dot{\varepsilon}_{zx} = \frac{1}{2}\left(\frac{\partial v_x}{\partial z} + \frac{\partial v_z}{\partial x}\right),$$

$$\dot{\varepsilon}_{yz} = \dot{\varepsilon}_{zy} = \frac{1}{2}\left(\frac{\partial v_y}{\partial z} + \frac{\partial v_z}{\partial y}\right).$$

Similarly to the strain tensor, the symmetric form of the strain rate tensor is obtained by subtracting the rotational component of the velocity field: it is easy to check that rigid body rotation in 2D has gradients in the velocity field which do not produce any internal deformation, i.e.

$$\dot{\varepsilon}_{xy(rotation)} = \frac{1}{2}\left(\frac{\partial v_x}{\partial y} + \frac{\partial v_y}{\partial x}\right) = 0.$$

By analogy to the stress tensor, the strain rate tensor can also be subdivided to isotropic $\dot{\varepsilon}_{kk}$(which is an invariant) and deviatoric $\dot{\varepsilon}'_{ij}$ components

$$\dot{\varepsilon}_{kk} = \dot{\varepsilon}_{xx} + \dot{\varepsilon}_{yy} + \dot{\varepsilon}_{zz} = \frac{\partial v_x}{\partial x} + \frac{\partial v_y}{\partial y} + \frac{\partial v_z}{\partial z} = \mathrm{div}(\vec{v}), \tag{4.11}$$

$$\dot{\varepsilon}'_{ij} = \dot{\varepsilon}_{ij} - \delta_{ij}\frac{1}{3}\dot{\varepsilon}_{kk}, \tag{4.12}$$

where *i, j* and *k* are coordinate indexes.

According to Eqs. (4.11), (4.12) the sum of normal deviatoric strain rate components is zero

$$\dot{\varepsilon}'_{xx} + \dot{\varepsilon}'_{yy} + \dot{\varepsilon}'_{zz} = 0. \tag{4.13}$$

Like the second stress invariant, the second invariant of the deviatoric strain rate tensor is defined as follows

$$\dot{\varepsilon}_{\mathrm{II}} = \sqrt{\tfrac{1}{2}\dot{\varepsilon}'_{ij}{}^{2}}. \tag{4.14}$$

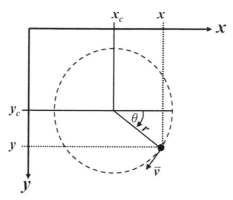

Fig. 4.4. Geometrical relationship in the case of 2D rigid body rotation.

Analytical exercise

Exercise 4.1.

Show that the symmetric form of the strain rate tensor satisfies the condition

$$\dot{\varepsilon}_{xy(rotation)} = \frac{1}{2}\left(\frac{\partial v_x}{\partial y} + \frac{\partial v_y}{\partial x}\right) = 0,$$

in the case of rigid body rotation with constant angular velocity ω. Use the fact that coordinates x, y of a rotating point in 2D are given by (Fig. 4.4)

$$x = x_c + r\cos(\theta),$$

$$y = y_c + r\sin(\theta),$$

$$\theta = \omega t,$$

where r = constant is the distance to the centre of rotation, x_c and y_c are the coordinates of the centre of rotation and θ is the clockwise rotation angle taken from the horizontal axis, t is time.

Programming exercise

Exercise 4.2.

Compute and visualize the deviatoric strain rate tensor components and invariants for the model described in Exercise 1.2. An example is in **Strain_rate.m**.

5

The momentum equation

Theory: Momentum equation. Viscosity and Newtonian law of viscous friction. Navier–Stokes equation. Stokes equation of slow laminar flow of highly viscous incompressible fluid. Simplification of the Stokes equation in the case of constant viscosity and its transformation to the Poisson equation. Analytical example for channel flow. Stream function approach.

Exercises: Solving Stokes and continuity equations for the case of constant viscosity with a stream function approach.

5.1 Momentum equation

The deformation of continuous media always results from the balance of various internal and external forces that act on these media. In order to relate forces and deformation, an equation of motion should be used. This is the so-called *momentum equation*, which describes the *conservation of momentum* for a continuous medium in the gravity field:

$$\text{Eulerian form} \quad \frac{\partial \sigma_{ij}}{\partial x_j} + \rho g_i = \rho \left(\frac{\partial v_i}{\partial t} + v_j \frac{\partial v_i}{\partial x_j} \right), \tag{5.1a}$$

$$\text{Lagrangian form} \quad \frac{\partial \sigma_{ij}}{\partial x_j} + \rho g_i = \rho \frac{D v_i}{Dt}. \tag{5.1b}$$

Please note that in three dimensions Eq. (5.1) *abbreviates three equations*: x-momentum ($i = x$, $j = x, y, z$), y-momentum ($i = y$, $j = x, y, z$) and z-momentum ($i = z$, $j = x, y, z$) equations (write them separately as an exercise). The momentum equation is in fact a continuum mechanics analogue of the famous Newton's second law of motion, which describes changes in velocity of an object with mass m according to

$$f = ma, \tag{5.2}$$

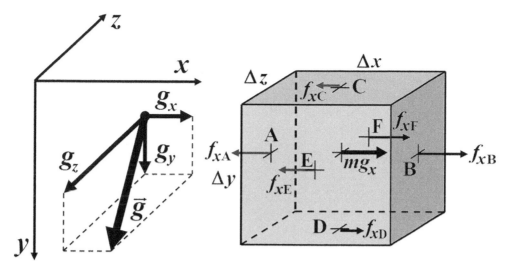

Fig. 5.1. Lagrangian elementary volume considered for the derivation of the respective form of *x*-momentum equation. Thin arrows show *x*-components of stress-related forces acting *from the outside of the volume* on respective boundaries (A, B, C, D, E and F). The thick arrow *inside the volume* shows the *x*-component of the gravity force (mg_x) which is proportional to the mass (m) embedded inside the volume. Orientation and components of gravity vector $\vec{g} = (g_x, g_y, g_z)$ are displayed outside of the volume.

where f is a net force acting on the object and $a = Dv/Dt$ is the acceleration of the object. This law can also be written in vector form as

$$\vec{f} = m\vec{a} \text{ or } f_i = ma_i, \tag{5.3a}$$

which gives three equations for 3D considerations

$$f_x = ma_x, \tag{5.3b}$$

$$f_y = ma_y, \tag{5.3c}$$

$$f_z = ma_z, \tag{5.3d}$$

where the index i denotes the coordinates, f_i are components of the force vector $\vec{f}$ and $a_i = Dv_i/Dt$ are components of the acceleration vector $\vec{a}$. Equation (5.1b) can actually be derived from Eq. (5.3a) while considering each material point of the continuum as a very small Lagrangian volume, for which the net acting force can be computed locally.

Let us make such a derivation on an intuitive basis by considering forces that act on a small Lagrangian volume with dimensions Δx, Δy and Δz (Fig. 5.1). As an example, we will derive the expression for the *x*-momentum equation from Eq. (5.3b). The net force f_x acting in the *x*-direction on the Lagrangian volume in a gravity field can be represented as a sum of seven elementary forces

$$f_x = f_{xA} + f_{xB} + f_{xC} + f_{xD} + f_{xE} + f_{xF} + mg_x, \qquad (5.4)$$

where f_{xA} to f_{xF} are stress-related forces *acting from the outside* of the volume on the respective boundaries (A–F) and mg_x is the gravity force, which is proportional to the mass m embedded in the volume. Stress-related forces are proportional to the surfaces of the respective boundaries and can be computed as follows

$$f_{xA} = -\sigma_{xxA}\Delta y\Delta z, \qquad (5.5a)$$

$$f_{xB} = +\sigma_{xxB}\Delta y\Delta z, \qquad (5.5b)$$

$$f_{xC} = -\sigma_{xyC}\Delta x\Delta z, \qquad (5.5c)$$

$$f_{xD} = +\sigma_{xyD}\Delta x\Delta z, \qquad (5.5d)$$

$$f_{xE} = -\sigma_{xzE}\Delta x\Delta y, \qquad (5.5e)$$

$$f_{xF} = +\sigma_{xzF}\Delta x\Delta y, \qquad (5.5f)$$

where σ_{xxA}, σ_{xxB} and σ_{xyC}, σ_{xyD}, σ_{xzE}, σ_{xzF} are the normal and shear stress components defined at respective boundaries. Note that the sign in front of the stress components on the right hand side of Eqs. (5.5a)–(5.5f) solely depends on whether the force (positive sign, boundaries B, D and F) or the counterforce (negative sign, boundaries A, C and E) is acting on the respective boundary *from outside of the volume* (see Fig. 4.1 for the stress convention). On the other hand, the stress components themselves can also be either positive or negative in sign. By combining Eqs. (5.3b) and (5.4)–(5.5), one can now write Newton's second law of motion for the considered Lagrangian volume (verify as an exercise)

$$f_{xA} + f_{xB} + f_{xC} + f_{xD} + f_{xE} + f_{xF} + mg_x = ma_x, \qquad (5.6a)$$

or

$$(\sigma_{xxB} - \sigma_{xxA})\Delta y\Delta z + (\sigma_{xyD} - \sigma_{xyC})\Delta x\Delta z + (\sigma_{xzF} - \sigma_{xzE})\Delta x\Delta y + mg_x = ma_x. \qquad (5.6b)$$

Normalizing both sides of Eq. (5.6b) by the considered Lagrangian volume V

$$V = \Delta x\Delta y\Delta z, \qquad (5.7)$$

we can now obtain the x-momentum equation in the differences representation (verify as an exercise)

$$\frac{(\sigma_{xxB} - \sigma_{xxA})\Delta y\Delta z}{V} + \frac{(\sigma_{xyD} - \sigma_{xyC})\Delta x\Delta z}{V} + \frac{(\sigma_{xzF} - \sigma_{xzE})\Delta x\Delta y}{V} + \frac{m}{V}g_x = \frac{m}{V}a_x. \qquad (5.8a)$$

or

$$\frac{(\sigma_{xxB} - \sigma_{xxA})}{\Delta x} + \frac{(\sigma_{xyD} - \sigma_{xyC})}{\Delta y} + \frac{(\sigma_{xzF} - \sigma_{xzE})}{\Delta z} + \rho g_x = \rho a_x, \tag{5.8b}$$

or

$$\frac{\Delta \sigma_{xx}}{\Delta x} + \frac{\Delta \sigma_{xy}}{\Delta y} + \frac{\Delta \sigma_{xz}}{\Delta z} + \rho g_x = \rho a_x, \tag{5.8c}$$

$$\rho = \frac{m}{V}, \tag{5.8d}$$

$$\Delta \sigma_{xx} = \sigma_{xxB} - \sigma_{xxA}, \tag{5.8e}$$

$$\Delta \sigma_{xy} = \sigma_{xyD} - \sigma_{xyC}, \tag{5.8f}$$

$$\Delta \sigma_{xz} = \sigma_{xzF} - \sigma_{xzE}, \tag{5.8g}$$

where ρ is the average material density in the Lagrangian volume V, and $\Delta \sigma_{xx}$, $\Delta \sigma_{xy}$, $\Delta \sigma_{xz}$ are differences of the respective stress components taken in x-, y- and z-directions (i.e. between respective pairs of boundaries), respectively.

When Δx, Δy and Δz all tend towards zero, the differences in Eq. (5.8c) can be replaced by derivatives and we obtain the Lagrangian x-momentum equation

$$\frac{\partial \sigma_{xx}}{\partial x} + \frac{\partial \sigma_{xy}}{\partial y} + \frac{\partial \sigma_{xz}}{\partial z} + \rho g_x = \rho a_x, \tag{5.9a}$$

or

$$\frac{\partial \sigma_{xj}}{\partial x_j} + \rho g_x = \rho a_x. \tag{5.9b}$$

In this differential equation, written for an infinitely small Lagrangian volume, the density ρ corresponds to the local density at a point. Obviously, based on similar considerations y- and z-momentum equations can also be derived (verify as an exercise).

5.2 Newtonian law of viscous friction

As we discussed in the Introduction, rocks often behave on geological time scales as highly viscous fluids. For this reason, the viscous *rheological* relationship between stress and strain rate, known as the *Newtonian law of viscous friction*, is widely used in geodynamic modelling. The Newtonian law of viscous friction relates the shear stress τ (Pa) with the shear strain rate $\partial v / \partial x$ (1/s) according to

$$\tau = \eta \frac{\partial v}{\partial x}, \tag{5.10}$$

where η (Pa s), the *viscosity*, characterizes the degree of resistance a material has to shear deformation. Viscosity is generally different for different materials and may also depend on pressure (P), temperature (T), strain rate and some other parameters. The viscosity of rocks is typically greater than 10^{17} Pa s: the viscosity of the asthenospheric upper mantle, for example, is about 10^{21} Pa s (Turcotte and Schubert, 2002). We will discuss this issue in more detail in the next chapter.

In 3D, the law of viscous friction is formulated with the components of both the deviatoric stress (σ'_{ij}) and the deviatoric strain rate ($\dot{\varepsilon}'_{ij}$) tensors in the form of *viscous constitutive relationship* as follows:

$$\sigma'_{ij} = 2\eta \dot{\varepsilon}'_{ij}, \tag{5.11a}$$

or

$$\sigma'_{ij} = 2\eta \left(\dot{\varepsilon}_{ij} - \tfrac{1}{3} \delta_{ij} \dot{\varepsilon}_{kk} \right), \tag{5.11b}$$

where η is the *shear viscosity*; $\dot{\varepsilon}_{kk} = \mathrm{div}(\vec{v})$ is the isotropic strain rate responsible for material volume changes (Eq. 4.11). Note that Eq. (5.11) satisfies the condition that the sum of normal deviatoric stress components should be equal to zero (Chapter 4).

In the absence of mineralogical phase transformations, rocks exhibit relatively small density variations (see Chapter 2). Therefore, the *incompressible fluid approximation* ($\rho =$ const, $D\rho/Dt = 0$, $\mathrm{div}(\vec{v}) = 0$, see Chapter 1) is generally valid. In this case, $\dot{\varepsilon}_{kk} = \mathrm{div}(\vec{v}) = 0$, $\dot{\varepsilon}'_{ij} = \dot{\varepsilon}_{ij}$ and the law of viscous friction can be simplified to

$$\sigma'_{ij} = 2\eta \dot{\varepsilon}_{ij}. \tag{5.12}$$

5.3 Navier–Stokes equation

Using the momentum equation (5.1b) and the relationship between the total (σ_{ij}) and deviatoric (σ'_{ij}) stresses (Eq. 4.3), we can introduce pressure into the momentum equation (5.1a) and obtain the *Navier–Stokes equation of motion*, which describes the conservation of momentum for a fluid in the gravity field:

$$\frac{\partial \sigma'_{ij}}{\partial x_j} - \frac{\partial P}{\partial x_i} + \rho g_i = \rho \frac{Dv_i}{Dt}, \tag{5.13}$$

where i and j are coordinate indexes; x_i and x_j are spatial coordinates; g_i is the ith component of the gravity vector $\vec{g} = (g_x, g_y, g_z)$; Dv_i/Dt is the substantive time derivative of the ith component of the velocity vector (i.e., the acceleration vector component). By analogy to substantive time derivatives of other physical parameters (e.g., density, Chapter 1), it can be related to the Eulerian time derivative as

$$\frac{Dv_i}{Dt} = \frac{\partial v_i}{\partial t} + \vec{v} \cdot \text{grad}(v_i),$$
(5.14)

or in 3D via

$$\frac{Dv_x}{Dt} = \frac{\partial v_x}{\partial t} + v_x \frac{\partial v_x}{\partial x} + v_y \frac{\partial v_x}{\partial y} + v_z \frac{\partial v_x}{\partial z},$$

$$\frac{Dv_y}{Dt} = \frac{\partial v_y}{\partial t} + v_x \frac{\partial v_y}{\partial x} + v_y \frac{\partial v_y}{\partial y} + v_z \frac{\partial v_y}{\partial z},$$

$$\frac{Dv_z}{Dt} = \frac{\partial v_z}{\partial t} + v_x \frac{\partial v_z}{\partial x} + v_y \frac{\partial v_z}{\partial y} + v_z \frac{\partial v_z}{\partial z},$$

which implies that in addition to the acceleration, Eulerian time derivatives also account for advection of respective velocity components with the flow.

In the case of 3D deformation, the Navier–Stokes equation of motion corresponds to a system of three partial differential equations:

$$x \text{ Navier–Stokes equation} \quad \frac{\partial \sigma'_{xx}}{\partial x} + \frac{\partial \sigma'_{xy}}{\partial y} + \frac{\partial \sigma'_{xz}}{\partial z} - \frac{\partial P}{\partial x} + \rho g_x = \rho \frac{Dv_x}{Dt},$$
(5.15)

$$y \text{ Navier–Stokes equation} \quad \frac{\partial \sigma'_{yy}}{\partial y} + \frac{\partial \sigma'_{yx}}{\partial x} + \frac{\partial \sigma'_{yz}}{\partial z} - \frac{\partial P}{\partial y} + \rho g_y = \rho \frac{Dv_y}{Dt},$$
(5.16)

$$z \text{ Navier–Stokes equation} \quad \frac{\partial \sigma'_{zz}}{\partial z} + \frac{\partial \sigma'_{zx}}{\partial x} + \frac{\partial \sigma'_{zy}}{\partial y} - \frac{\partial P}{\partial z} + \rho g_z = \rho \frac{Dv_z}{Dt}.$$
(5.17)

In highly viscous flows, the inertial forces ($\rho \, Dv_i/Dt$) are negligible with respect to viscous resistance and gravitational forces. For example, a typical plate velocity in geodynamics is on the order of several cm/year ($n \times 10^{-9}$ m/s) and it may notably change only within millions of years ($n \times 10^{13}$ s). Consequently, the typical magnitude of mantle flow 'accelerations' will be on the order of

$$\frac{Dv_i}{Dt} \approx \frac{\Delta v}{\Delta t} = \frac{n \times 10^{-9}}{n \times 10^{13}} = n \times 10^{-22} \quad \text{m/s}^2.$$

This makes the right hand side of the Navier–Stokes equation $\rho \, Dv_i/Dt$ negligible compared to the ρg_i term in the left hand side of the equation, since the magnitude of the gravitational acceleration g_i is on the order of 10 m/s^2, i.e. 10^{23} times bigger than the mantle flow accelerations. Under such circumstances, deformation of highly viscous flows can be accurately described by the *Stokes equation of slow flow*:

$$\frac{\partial \sigma'_{ij}}{\partial x_j} - \frac{\partial P}{\partial x_i} + \rho g_i = 0, \tag{5.18}$$

or

$$x \text{ Stokes equation} \quad \frac{\partial \sigma'_{xx}}{\partial x} + \frac{\partial \sigma'_{xy}}{\partial y} + \frac{\partial \sigma'_{xz}}{\partial z} - \frac{\partial P}{\partial x} + \rho g_x = 0, \tag{5.19}$$

$$y \text{ Stokes equation} \quad \frac{\partial \sigma'_{yy}}{\partial y} + \frac{\partial \sigma'_{yx}}{\partial x} + \frac{\partial \sigma'_{yz}}{\partial z} - \frac{\partial P}{\partial y} + \rho g_y = 0, \tag{5.20}$$

$$z \text{ Stokes equation} \quad \frac{\partial \sigma'_{zz}}{\partial z} + \frac{\partial \sigma'_{zx}}{\partial x} + \frac{\partial \sigma'_{zy}}{\partial y} - \frac{\partial P}{\partial z} + \rho g_z = 0 . \tag{5.21}$$

The equations can be further simplified if the viscosity is constant and the fluid is incompressible. In this case, the Stokes equations simplify to:

$$\eta \frac{\partial^2 v_i}{\partial x_j^2} - \frac{\partial P}{\partial x_i} + \rho g_i = 0. \tag{5.22}$$

Let us go through the logic of this simplification for the x Stokes equation (5.19). First, applying Eq. (5.12) we obtain

$$\frac{\partial}{\partial x}(2\eta \dot{\varepsilon}_{xx}) + \frac{\partial}{\partial y}(2\eta \dot{\varepsilon}_{xy}) + \frac{\partial}{\partial z}(2\eta \dot{\varepsilon}_{xz}) - \frac{\partial P}{\partial x} + \rho g_x = 0. \tag{5.23}$$

Then using Eq. (4.10) for strain rate components, we get

$$\eta \left[\frac{\partial}{\partial x}\left(\frac{\partial v_x}{\partial x} + \frac{\partial v_x}{\partial x}\right) + \frac{\partial}{\partial y}\left(\frac{\partial v_x}{\partial y} + \frac{\partial v_y}{\partial x}\right) + \frac{\partial}{\partial z}\left(\frac{\partial v_x}{\partial z} + \frac{\partial v_z}{\partial x}\right) \right] - \frac{\partial P}{\partial x} + \rho g_x = 0. \tag{5.24a}$$

Further regrouping of the above equation gives

$$\eta \left[\left(\frac{\partial^2 v_x}{\partial x^2} + \frac{\partial^2 v_x}{\partial y^2} + \frac{\partial^2 v_x}{\partial z^2}\right) + \left(\frac{\partial^2 v_x}{\partial x \partial x} + \frac{\partial^2 v_y}{\partial x \partial y} + \frac{\partial^2 v_z}{\partial x \partial z}\right) \right] - \frac{\partial P}{\partial x} + \rho g_x = 0 \tag{5.24b}$$

$$\eta \left[\left(\frac{\partial^2 v_x}{\partial x^2} + \frac{\partial^2 v_x}{\partial y^2} + \frac{\partial^2 v_x}{\partial z^2}\right) + \frac{\partial}{\partial x}\left(\frac{\partial v_x}{\partial x} + \frac{\partial v_y}{\partial y} + \frac{\partial v_z}{\partial z}\right) \right] - \frac{\partial P}{\partial x} + \rho g_x = 0. \tag{5.24c}$$

And finally using the fact that for the incompressible fluid

$$\frac{\partial v_x}{\partial x}+\frac{\partial v_y}{\partial y}+\frac{\partial v_z}{\partial z}=div(\vec{v})=0,$$

we obtain

$$\eta\left(\frac{\partial^2 v_x}{\partial x^2}+\frac{\partial^2 v_x}{\partial y^2}+\frac{\partial^2 v_x}{\partial z^2}\right)-\frac{\partial P}{\partial x}+\rho g_x=0, \qquad (5.24d)$$

which is analogous to Eq. (5.22). Obviously, one can get similar expressions for both the y Stokes and z Stokes equations (derive as an exercise).

As usual, for the case of 3D deformation we have three equations:

$$\eta\left(\frac{\partial^2 v_x}{\partial x^2}+\frac{\partial^2 v_x}{\partial y^2}+\frac{\partial^2 v_x}{\partial z^2}\right)-\frac{\partial P}{\partial x}+\rho g_x=0 \quad \text{or} \quad \eta\nabla^2 v_x=\frac{\partial P}{\partial x}-\rho g_x,$$

$$\eta\left(\frac{\partial^2 v_y}{\partial x^2}+\frac{\partial^2 v_y}{\partial y^2}+\frac{\partial^2 v_y}{\partial z^2}\right)-\frac{\partial P}{\partial y}+\rho g_y=0 \quad \text{or} \quad \eta\nabla^2 v_y=\frac{\partial P}{\partial y}-\rho g_y,$$

$$\eta\left(\frac{\partial^2 v_z}{\partial x^2}+\frac{\partial^2 v_z}{\partial y^2}+\frac{\partial^2 v_z}{\partial z^2}\right)-\frac{\partial P}{\partial z}+\rho g_z=0 \quad \text{or} \quad \eta\nabla^2 v_z=\frac{\partial P}{\partial z}-\rho g_z,$$

where

$$\nabla^2=\frac{\partial^2}{\partial x^2}+\frac{\partial^2}{\partial y^2}+\frac{\partial^2}{\partial z^2}$$

is the *Laplace operator*.

5.4 Poisson equation

If the pressure gradients are constant, Eq. (5.22) can be written in the form of the *Poisson equation*

$$\nabla^2 v_i=\text{const}_i \text{ where const}_i=\frac{1}{\eta}\left(\frac{\partial P}{\partial x_i}-\rho g_i\right). \qquad (5.25)$$

In 3D, it is the system of three equations that are always worth writing out explicitly:

$$x \text{ Poisson equation } \nabla^2 v_x=\text{const}_x, \text{ where const}_x=\frac{1}{\eta}\left(\frac{\partial P}{\partial x}-\rho g_x\right),$$

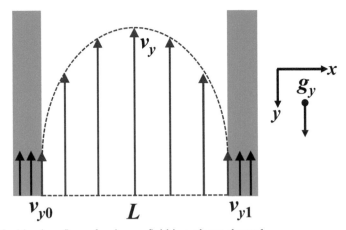

Fig. 5.2. Vertical laminar flow of a viscous fluid in a planar channel.

$$y \text{ Poisson equation } \nabla^2 v_y = \text{const}_y, \text{ where const}_y = \frac{1}{\eta}\left(\frac{\partial P}{\partial y} - \rho g_y\right),$$

$$z \text{ Poisson equation } \nabla^2 v_z = \text{const}_z, \text{ where const}_z = \frac{1}{\eta}\left(\frac{\partial P}{\partial z} - \rho g_z\right).$$

Despite its simplicity, the Poisson equation is valid for several important geodynamic problems, for example in the case of unidirectional (e.g. purely vertical) flow of an incompressible fluid in a planar channel, for example magma flow in the magmatic channel or rock flow in the subduction channel.

In a vertical planar channel of width L (Fig. 5.2), the vertical velocity depends on the x-coordinate only, $v_x = 0$, $v_z = 0$, $\partial v_y/\partial y = 0$, $\partial v_y/\partial z = 0$ and the system of three Poisson equations (5.25) reduces to

$$\frac{\partial^2 v_y}{\partial x^2} = \frac{1}{\eta}\left(\frac{\partial P}{\partial y} - \rho g_y\right), \tag{5.26}$$

which can be solved analytically by a two-fold integration:

$$\frac{\partial v_y}{\partial x} = \int \frac{\partial^2 v_y}{\partial x^2}\,dx = \int \frac{1}{\eta}\left(\frac{\partial P}{\partial y} - \rho g_y\right)dx = \frac{1}{\eta}\left(\frac{\partial P}{\partial y} - \rho g_y\right)x + C_1, \tag{5.27}$$

$$v_y = \int \frac{\partial v_y}{\partial x}\,dx = \int\left[\frac{1}{\eta}\left(\frac{\partial P}{\partial y} - \rho g_y\right)x + C_1\right]dx = \frac{1}{2\eta}\left(\frac{\partial P}{\partial y} - \rho g_y\right)x^2 + C_1 x + C_2, \tag{5.28}$$

where C_1 and C_2 are integration constants which can be defined from the boundary conditions:

left boundary, $v_y = v_{y0}$ when $x = 0$ and then,

$$v_{y0} = \frac{1}{2\eta}\left(\frac{\partial P}{\partial y} - \rho g_y\right)x^2 + C_1 x + C_2 = C_2$$

so that

$$C_2 = v_{y0};$$

right boundary, $v_y = v_{y1}$ when $x = L$ and then,

$$v_{y1} = \frac{1}{2\eta}\left(\frac{\partial P}{\partial y} - \rho g_y\right)x^2 + C_1 x + C_2 = \frac{1}{2\eta}\left(\frac{\partial P}{\partial y} - \rho g_y\right)L^2 + C_1 L + C_2$$

so that

$$C_1 = \frac{v_{y1} - v_{y0}}{L} - \frac{1}{2\eta}\left(\frac{\partial P}{\partial y} - \rho g_y\right)L,$$

where v_{y0} and v_{y1} are the vertical velocities of the left and right walls, respectively. The final expression for the vertical velocity profile is then

$$v_y = \frac{1}{2\eta}\left(\frac{\partial P}{\partial y} - \rho g_y\right)\left(x^2 - Lx\right) + v_{y0} + \left(v_{y1} - v_{y0}\right)\frac{x}{L}. \tag{5.29}$$

If the walls are immobile, $v_{y0} = v_{y1} = 0$ and Eq. (5.29) simplifies to

$$v_y = \frac{1}{2\eta}\left(\frac{\partial P}{\partial y} - \rho g_y\right)\left(x^2 - Lx\right). \tag{5.30}$$

Analytical solutions of the Poisson equation are very useful and are often used for *benchmarking* (i.e., testing the accuracy of) numerical solutions (Chapter 20).

5.5 Stream function approach

The stream function approach is a way to formulate and solve the coupled momentum and continuity equations for 2D incompressible flow with a single scalar variable – *the stream function* (Ψ) – whose derivatives are given by

$$\frac{\partial \Psi}{\partial y} = v_x, \tag{5.31}$$

$$\frac{\partial \Psi}{\partial x} = -v_y. \tag{5.32}$$

Consequently, the 2D incompressibility condition is automatically satisfied

$$\text{div}(\vec{v}) = \frac{\partial v_x}{\partial x} + \frac{\partial v_y}{\partial y} = \frac{\partial}{\partial x}\left(\frac{\partial \Psi}{\partial y}\right) + \frac{\partial}{\partial y}\left(-\frac{\partial \Psi}{\partial x}\right) = \frac{\partial^2 \Psi}{\partial x \partial y} - \frac{\partial^2 \Psi}{\partial y \partial x} = 0.$$

Also, the 2D Stokes equations for incompressible fluid are first differentiated as

$$x \text{ Stokes equation}\quad \frac{\partial}{\partial y}\left(\frac{\partial \sigma'_{xx}}{\partial x} + \frac{\partial \sigma'_{xy}}{\partial y} - \frac{\partial P}{\partial x} + \rho g_x\right) = 0, \tag{5.33}$$

$$y \text{ Stokes equation}\quad \frac{\partial}{\partial x}\left(\frac{\partial \sigma'_{yy}}{\partial y} + \frac{\partial \sigma'_{yx}}{\partial x} - \frac{\partial P}{\partial y} + \rho g_y\right) = 0, \tag{5.34}$$

and then Eq. (5.34) is subtracted from Eq. (5.33) in order to eliminate pressure

$$\left(\frac{\partial^2 \sigma'_{xx}}{\partial x \partial y} + \frac{\partial^2 \sigma'_{xy}}{\partial y^2}\right) - \left(\frac{\partial^2 \sigma'_{yy}}{\partial x \partial y} + \frac{\partial^2 \sigma'_{yx}}{\partial x^2}\right) = \frac{\partial \rho}{\partial x}g_y - \frac{\partial \rho}{\partial y}g_x. \tag{5.35}$$

Using the viscous *constitutive relation* (5.12) and Eqs. (5.31), (5.32) we can now reformulate Eq. (5.35) using the stream function only

$$\frac{\partial^2}{\partial x \partial y}(2\eta \dot{\varepsilon}_{xx}) + \frac{\partial^2}{\partial y^2}(2\eta \dot{\varepsilon}_{xy}) - \frac{\partial^2}{\partial x \partial y}(2\eta \dot{\varepsilon}_{yy}) - \frac{\partial^2}{\partial x^2}(2\eta \dot{\varepsilon}_{xy}) = \frac{\partial \rho}{\partial x}g_y - \frac{\partial \rho}{\partial y}g_x, \tag{5.36a}$$

$$\frac{\partial^2}{\partial x \partial y}\left(2\eta \frac{\partial v_x}{\partial x} - 2\eta \frac{\partial v_y}{\partial y}\right) + \frac{\partial^2}{\partial y^2}\left(\eta \frac{\partial v_x}{\partial y} + \eta \frac{\partial v_y}{\partial x}\right) - \frac{\partial^2}{\partial x^2}\left(\eta \frac{\partial v_x}{\partial y} + \eta \frac{\partial v_y}{\partial x}\right) = \frac{\partial \rho}{\partial x}g_y - \frac{\partial \rho}{\partial y}g_x, \tag{5.36b}$$

$$\frac{\partial^2}{\partial x \partial y}\left(4\eta \frac{\partial^2 \Psi}{\partial x \partial y}\right) + \frac{\partial^2}{\partial y^2}\left(\eta \frac{\partial^2 \Psi}{\partial y^2} - \eta \frac{\partial^2 \Psi}{\partial x^2}\right) - \frac{\partial^2}{\partial x^2}\left(\eta \frac{\partial^2 \Psi}{\partial y^2} - \eta \frac{\partial^2 \Psi}{\partial x^2}\right) = \frac{\partial \rho}{\partial x}g_y - \frac{\partial \rho}{\partial y}g_x. \tag{5.36c}$$

In the case of constant viscosity η, Eq. (5.36c) can be further simplified to

$$\eta\left[4\frac{\partial^4 \Psi}{\partial x^2 \partial y^2} + \frac{\partial^4 \Psi}{\partial y^2 \partial y^2} - \frac{\partial^4 \Psi}{\partial x^2 \partial y^2} - \frac{\partial^4 \Psi}{\partial x^2 \partial y^2} + \frac{\partial^4 \Psi}{\partial x^2 \partial x^2}\right] = \frac{\partial \rho}{\partial x}g_y - \frac{\partial \rho}{\partial y}g_x, \tag{5.37a}$$

$$\eta\left[\frac{\partial^2}{\partial y^2}\left(\frac{\partial^2 \Psi}{\partial x^2} + \frac{\partial^2 \Psi}{\partial y^2}\right) + \frac{\partial^2}{\partial x^2}\left(\frac{\partial^2 \Psi}{\partial x^2} + \frac{\partial^2 \Psi}{\partial y^2}\right)\right] = \frac{\partial \rho}{\partial x}g_y - \frac{\partial \rho}{\partial y}g_x, \tag{5.37b}$$

to obtain the so-called *vorticity formulation*

$$\frac{\partial^2 \omega}{\partial y^2} + \frac{\partial^2 \omega}{\partial x^2} = \frac{1}{\eta}\left(\frac{\partial \rho}{\partial x}g_y - \frac{\partial \rho}{\partial y}g_x\right), \tag{5.38}$$

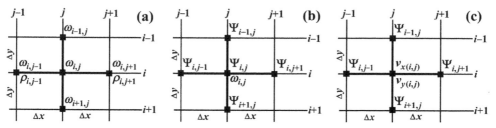

Fig. 5.3. Numerical grid stencils used for solving the coupled momentum and continuity equations in 2D with the stream function-vorticity formulation via finite differences, on a regular rectangular grid, in the case of constant viscosity and purely vertical gravity. (a), (b) Stencils for formulating the Poisson equation for vorticity (Eq. 5.39) (a) and stream function (Eq. 5.38) (b). (c) Stencil for computing the velocity components from the stream function.

where ω is vorticity defined as

$$\omega = \frac{\partial^2 \Psi}{\partial x^2} + \frac{\partial^2 \Psi}{\partial y^2} = \frac{\partial v_y}{\partial x} - \frac{\partial v_x}{\partial y}, \qquad (5.39)$$

which quantifies the rotational component of the velocity field.

As we can see, Eqs. (5.39) and (5.38) are both Poisson equations which can be solved subsequently, such that the solution of Eq. (5.38) for vorticity goes into the left hand side of Eq. (5.39). Obviously, boundary conditions should be formulated for both the vorticity and the stream function by using their relations to velocity.

Analytical exercise

Exercise 5.1.
Calculate the vertical velocity of a magma flow (v_y) in the middle of the vertical planar channel (Fig. 5.2) of width $L = 100$ m, for magma with a viscosity $\eta = 10^{16}$ Pa s and a density $\rho = 2800$ kg/m^3. The pressure gradient along the channel is $\partial P/\partial y = 31000$ Pa/m. Gravitational acceleration is $g_y = 10$ m/s^2, directed downward. The velocity on the channel boundaries is zero. Also calculate the solution when the right boundary is moving upward at a velocity $v_{y1} = 10^{-9}$ m/s.

Programming exercise

Exercise 5.2.
Solve the momentum and continuity equations with finite differences using the stream function-vorticity formulation (Eqs. 5.38, 5.39) on a regular grid (Fig. 5.3) of 51 × 41 points. Program a numerical model for buoyancy-driven flow in a vertical gravity field ($g_x = 0$, $g_y = 10$ m/s^2 in Eq. 5.38) for a density structure with two vertical layers (3200 kg/m^3 and 3300 kg/m^3 for the left and right layers, respectively). The model size is 1000 × 1500 km^2

(i.e. $1\,000\,000 \times 1\,500\,000$ m^2). Use a constant viscosity $\eta = 10^{21}$ Pa s for the entire model. Compose a matrix of coefficients $\{L\}$ and a right hand side vector $\{R\}$ and obtain the solution vector $\{S\}$ with direct solver (S = L\R) for the two Poisson equations (first for Eq. 5.38 and then for Eq. 5.39). A finite difference representation of the Poisson equations in 2D can be formulated by analogy with Eq. (3.20) as follows (Fig. 5.3a,b)

$$\frac{\omega_{i,j-1} - 2\omega_{i,j} + \omega_{i,j+1}}{\Delta x^2} + \frac{\omega_{i-1,j} - 2\omega_{i,j} + \omega_{i+1,j}}{\Delta y^2} = g_y \frac{p_{i,j+1} - p_{i,j-1}}{\eta 2\Delta x}, \tag{5.40}$$

$$\frac{\Psi_{i,j-1} - 2\Psi_{i,j} + \Psi_{i,j+1}}{\Delta x^2} + \frac{\Psi_{i-1,j} - 2\Psi_{i,j} + \Psi_{i+1,j}}{\Delta y^2} = \omega_{i,j}. \tag{5.41}$$

Use $\omega = 0$ and $\Psi = 0$ as boundary conditions for all external nodes of the grid. Demonstrate analytically that these boundary conditions imply $\partial v_x/\partial y = 0$, $v_y = 0$ and $\partial v_y/\partial x = 0$, $v_x = 0$ (i.e. *free slip*) along the horizontal and vertical model boundaries, respectively. Use the principle of global indexing of ω and Ψ in 2D shown in Fig. 3.4. Compute the global index k based on geometrical indexes i and j (Fig. 5.3 a,b) according to Eq. (3.22).

After obtaining the solution for the stream function, convert it into the velocity field using finite differences (Fig. 5.3c) based on Eqs. (5.31), (5.32)

$$v_{x(i,j)} = \frac{\Psi_{i+1,j} - \Psi_{i-1,j}}{2\Delta y}, \tag{5.42}$$

$$v_{y(i,j)} = -\frac{\Psi_{i,j+1} - \Psi_{i,j-1}}{2\Delta x}. \tag{5.43}$$

Note that velocity components should only be computed for the *internal* nodes of the grid (i.e. for 49×39 points).

Plot the results for vorticity (*pcolor*), stream function (*contour*) and velocity (*quiver*) on the same diagram with the vertical axis directed downward (*axis ij*).

An example is in **Streamfunction2D.m**.

6

Viscous rheology of rocks

Theory: Rock rheology. Solid-state creep of minerals and rocks as the major mechanism of deformation of the Earth's interior. Dislocation and diffusion creep mechanisms. Empirical flow laws for minerals and rocks. Effective viscosity and its dependence on temperature, pressure, and strain rate. Formulation of the effective viscosity from empirical flow laws.

Exercises: Programming viscous rheological equations for computing effective viscosities from empirical flow laws.

6.1 Rock rheology

Rheology is the composite physical property characterizing *deformation behaviour of a material*, which can be either solid or liquid. Rheology of rocks includes several different deformation mechanisms and is in general *visco-elasto-plastic*. Here, we will discuss in more detail the physical meaning of viscosity and, generally, the *viscous rheology* of rocks reflecting physics of *solid-state creep*, which is the major mechanism of rock deformation at high temperature. Solid-state creep is the ability of crystalline substances to deform *irreversibly* under applied stresses. Solid-state creep is characteristic of the Earth's middle-lower crust and asthenospheric mantle. Two major types of creep are known: *diffusion creep* and *dislocation creep*.

Diffusion creep is typically dominant at relatively low stresses and results from the diffusion of atoms through the interior (Herring–Nabarro creep) and along the boundaries (Coble creep) of crystalline grains subjected to stresses. As a result of this diffusion, grain deformation leads to bulk rock deformation. Diffusion creep is characterized by a linear (*Newtonian*) relationship between the strain rate $\dot{\gamma}$ and an applied shear stress τ

$$\dot{\gamma} = A_{diff}\tau, \tag{6.1}$$

where A_{diff} is a proportionality coefficient which is independent of stress, but depends on grain size, pressure, temperature, oxygen and water fugacity.

Dislocation creep is dominant at higher stresses and results from migration of dislocations (imperfections in the crystalline lattice structure). Dislocation density increases with increasing stress, and therefore dislocation creep results in a non-linear (*non-Newtonian*) relationship between the strain rate and deviatoric stress

$$\dot{\gamma} = A_{disl}\tau^n, \tag{6.2}$$

where A_{disl} is a proportionality coefficient which is independent of stress and grain size, but depends on pressure, temperature, oxygen and water fugacity, and $n > 1$ is the stress exponent.

Both diffusion and dislocation creep rheologies are often calibrated from experimental data using a simple empirical parameterized relationship (also called *flow law*) between the applied *differential stress* σ_d (the difference between maximal and minimal applied stress) and the resulting *ordinary strain rate* $\dot{\gamma}$ (which is taken positive)

$$\dot{\gamma} = A_D h^m (\sigma_d)^n \exp\left(-\frac{E_a + V_a P}{RT}\right), \tag{6.3}$$

where P is pressure (Pa), T is temperature (K), R is the gas constant (8.314 J/(K mol)), h is grain size (m) and A_D, n, m, E_a and V_a are experimentally determined *rheological parameters*: A_D is a material constant ($Pa^{-n}s^{-1}$), n is the stress exponent ($n = 1$ in the case of diffusion creep and $n > 1$ in the case of dislocation creep), m is the grain size exponent, E_a is the activation energy (J/mol) and V_a is the activation volume (J/Pa). Dislocation creep is grain size independent and therefore $m = 0$ and $h^m = 1$. In contrast, diffusion creep notably depends on grain size and the grain size exponent m is negative (i.e. strain rate increases with decreasing grain size).

6.2 Effective viscosity

In order to use the experimentally parameterized equation (6.3) in numerical modelling, one needs to reformulate it in terms of an effective viscosity (η_{eff}), written as a function of the second invariant of either the deviatoric stress (σ_{II}) or strain rate ($\dot{\varepsilon}_{II}$) and the following general relation that is valid for isotropic materials

$$\sigma_{II} = 2\eta_{eff}\dot{\varepsilon}_{II}, \tag{6.4a}$$

or

$$\eta_{eff} = \frac{\sigma_{II}}{2\dot{\varepsilon}_{II}}. \tag{6.4b}$$

The reformulation should also take into account the type of rheological experiments performed in order to establish the proper relations between σ_{II} and σ_d, as well as between $\dot{\varepsilon}_{II}$ and $\dot{\gamma}$. The resulting expressions for the effective viscosity are as follows

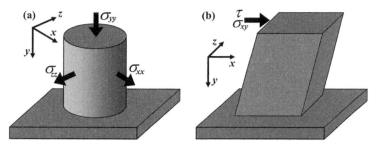

Fig. 6.1. Schematic geometrical relations for axial compression (a) and simple shear (b) experiments.

$$\eta_{eff} = F_1 \frac{1}{A_D h^m (\sigma_{II})^{(n-1)}} \exp\left(\frac{E_a + V_a P}{RT}\right),$$ (6.5a)

or

$$\eta_{eff} = F_2 \frac{1}{(A_D)^{1/n} h^{m/n} (\dot{\varepsilon}_{II})^{(n-1)/n}} \exp\left(\frac{E_a + V_a P}{nRT}\right),$$ (6.5b)

where dimensionless coefficients F_1 and F_2 depend on the type of experiments used for calibration of Eq. (6.3). A strain rate based formulation (6.5b) is more suitable for numerical modelling with viscous (visco-plastic) rheology, whereas a stress-based formulation (6.5a) is appropriate for visco-elastic (visco-elasto-plastic) problems (Chapters 12, 13). Below, two important types of rheological experiments are considered in order to derive F_1 and F_2: axial compression (Fig. 6.1a) and simple shear (Fig. 6.1b). Note that in our consideration we will always use the incompressibility assumption $\mathrm{div}(\vec{v}) = 0$.

In the case of an axial compression experiment (Fig. 6.1a) the F_1 and F_2 coefficients are obtained as follows (under compression oriented along the y axis).

(1) Establish the relationship between $\dot{\varepsilon}_{II}$ and $\dot{\gamma}$:

$$\dot{\varepsilon}_{xy} = \dot{\varepsilon}_{xz} = \dot{\varepsilon}_{yz} = 0,$$

$$\dot{\gamma} = \left|\frac{\partial v_y}{\partial y}\right| = |\dot{\varepsilon}_{yy}|,$$

$$\dot{\varepsilon}_{xx} = \dot{\varepsilon}_{zz} = -\frac{1}{2}\dot{\varepsilon}_{yy},$$ (6.6)

$$\dot{\varepsilon}_{II} = \sqrt{\frac{1}{2}\dot{\varepsilon}_{xx}^2 + \frac{1}{2}\dot{\varepsilon}_{yy}^2 + \frac{1}{2}\dot{\varepsilon}_{zz}^2} = \sqrt{\frac{3}{4}\dot{\varepsilon}_{yy}^2} = \frac{\sqrt{3}}{2}|\dot{\varepsilon}_{yy}| = \frac{\sqrt{3}}{2}\dot{\gamma},$$

$$\dot{\gamma} = \frac{2}{\sqrt{3}}\dot{\varepsilon}_{II}.$$

(2) Establish the relationship between σ_{II} and σ_d:

$$\sigma_{xx} = \sigma_{zz},$$

$$\sigma'_{xx} = \sigma'_{zz} = -\frac{1}{2}\sigma'_{yy},$$

$$\sigma_d = \sigma_{max} - \sigma_{min} = \sigma_{xx} - \sigma_{yy} = \left(\sigma'_{xx} - P\right) - \left(\sigma'_{yy} - P\right) = \sigma'_{xx} - \sigma'_{yy} = \frac{3}{2}\left|\sigma'_{yy}\right|, \quad (6.7)$$

$$\sigma_{II} = \sqrt{\frac{1}{2}\sigma'^2_{xx} + \frac{1}{2}\sigma'^2_{yy} + \frac{1}{2}\sigma'^2_{zz}} = \sqrt{\frac{3}{4}\sigma'^2_{yy}} = \frac{\sqrt{3}}{2}\left|\sigma'_{yy}\right| = \frac{1}{\sqrt{3}}\sigma_d,$$

$$\sigma_d = \sqrt{3}\sigma_{II}.$$

(3) Rewrite Eq. (6.3) in terms of $\dot{\varepsilon}_{II}$ and σ_{II} using Eqs. (6.6) and (6.7)

$$\frac{2}{\sqrt{3}}\dot{\varepsilon}_{II} = A_D h^m \left(\sqrt{3}\sigma_{II}\right)^n \exp\left(-\frac{E_a + V_a P}{RT}\right), \quad (6.8a)$$

or

$$\dot{\varepsilon}_{II} = \frac{3^{(n+1)/2}}{2} A_D h^m (\sigma_{II})^n \exp\left(-\frac{E_a + V_a P}{RT}\right), \quad (6.8b)$$

or

$$\sigma_{II} = \frac{2^{1/n}}{3^{(n+1)/(2n)}} (\dot{\varepsilon}_{II})^{1/n} \frac{1}{(A_D)^{1/n} h^{m/n}} \exp\left(\frac{E_a + V_a P}{nRT}\right). \quad (6.8c)$$

(4) Write an expression for η_{eff} as a function of the second invariant of deviatoric stress σ_{II} using Eqs. (6.4b) and (6.8b). Define F_1 by comparing with Eq. (6.5a)

$$\eta_{eff} = \frac{\sigma_{II}}{2\dot{\varepsilon}_{II}} = \frac{\sigma_{II}}{2\left[\dfrac{3^{(n+1)/2}}{2} A_D h^m (\sigma_{II})^n \exp\left(-\dfrac{E_a + V_a P}{RT}\right)\right]},$$

$$\eta_{eff} = \frac{1}{3^{(n+1)/2}} \times \frac{1}{A_D h^m (\sigma_{II})^{(n-1)}} \exp\left(\frac{E_a + V_a P}{RT}\right), \quad (6.9)$$

$$F_1 = \frac{1}{3^{(n+1)/2}}.$$

(5) Write an expression for η_{eff} as a function of the second strain rate invariant $\dot{\varepsilon}_{II}$ using Eqs. (6.4b) and (6.8c). Define F_2 by comparing with Eq. (6.5b)

$$\eta_{eff} = \frac{\sigma_{\mathrm{II}}}{2\dot{\varepsilon}_{\mathrm{II}}} = \frac{\left[\dfrac{2^{1/n}}{3^{(n+1)/(2n)}} \left(\dot{\varepsilon}_{\mathrm{II}}\right)^{1/n} \dfrac{1}{(A_D)^{1/n}\, h^{m/n}} \exp\left(\dfrac{E_a + V_a P}{nRT}\right)\right]}{2\dot{\varepsilon}_{\mathrm{II}}},$$

$$\eta_{eff} = \frac{1}{2^{(n-1)/n} 3^{(n+1)/(2n)}} \times \frac{1}{(A_D)^{1/n}\, h^{m/n} (\dot{\varepsilon}_{\mathrm{II}})^{(n-1)/n}} \exp\left(\frac{E_a + V_a P}{nRT}\right), \qquad (6.10)$$

$$F_2 = \frac{1}{2^{(n-1)/n} 3^{(n+1)/(2n)}}.$$

In the case of simple shear experiments (Fig. 6.1b), under applied shear stress $\tau = \frac{1}{2}\,\sigma_d$ (by the way, τ may also be sometimes used in Eq. (6.3) instead of σ_d for calibrating experiments). In the case of an xy shear $\tau = |\sigma_{xy}|$, the coefficients F_1 and F_2 are obtained as follows.

(1) Establish a relationship between $\dot{\varepsilon}_{\mathrm{II}}$ and $\dot{\gamma}$

$$\dot{\varepsilon}_{xx} = \dot{\varepsilon}_{yy} = \dot{\varepsilon}_{zz} = \dot{\varepsilon}_{xz} = \dot{\varepsilon}_{yz} = 0,$$
$$\dot{\gamma} = \left|\frac{\partial v_x}{\partial y}\right| = 2|\dot{\varepsilon}_{xy}|,$$
$$\dot{\varepsilon}_{\mathrm{II}} = \sqrt{\dot{\varepsilon}_{xy}{}^2} = |\dot{\varepsilon}_{xy}| = \frac{1}{2}\dot{\gamma}, \qquad (6.11)$$
$$\dot{\gamma} = 2\dot{\varepsilon}_{\mathrm{II}}.$$

(2) Establish a relationship between σ_{II} and σ_d

$$\sigma'_{xx} = \sigma'_{yy} = \sigma'_{zz} = \sigma_{xz} = \sigma_{yz} = 0,$$
$$\sigma_d = 2\tau = 2|\sigma_{xy}|,$$
$$\sigma_{\mathrm{II}} = \sqrt{\sigma_{xy}{}^2} = |\sigma_{xy}| = \frac{1}{2}\sigma_d, \qquad (6.12)$$
$$\sigma_d - 2\sigma_{\mathrm{II}}.$$

(3) Rewrite Eq. (6.3) in terms of $\dot{\varepsilon}_{\mathrm{II}}$ and σ_{II} using Eqs. (6.11) and (6.12)

$$2\dot{\varepsilon}_{\mathrm{II}} = A_D h^m (2\sigma_{\mathrm{II}})^n \exp\left(-\frac{E_a + V_a P}{RT}\right), \qquad (6.13a)$$

or

$$\dot{\varepsilon}_{\mathrm{II}} = 2^{n-1} A_D h^m (\sigma_{\mathrm{II}})^n \exp\left(-\frac{E_a + V_a P}{RT}\right), \qquad (6.13b)$$

or

$$\sigma_{\mathrm{II}} = \frac{1}{2^{(n-1)/n}} \left(\dot{\varepsilon}_{\mathrm{II}}\right)^{1/n} \frac{1}{(A_D)^{1/n}\, h^{m/n}} \exp\left(\frac{E_a + V_a P}{nRT}\right). \qquad (6.13c)$$

(4) Write an expression for η_{eff} as a function of the second invariant of deviatoric stress σ_{II} by using Eqs. (6.4b) and (6.13b). Define F_1 by comparing with Eq. (6.5a)

$$\eta_{eff} = \frac{\sigma_{II}}{2\dot{\varepsilon}_{II}} = \frac{\sigma_{II}}{2\left[2^{(n-1)}A_D h^m (\sigma_{II})^n \exp\left(-\frac{E_a + V_a P}{RT}\right)\right]},$$

$$\eta_{eff} = \frac{1}{2^n} \times \frac{1}{A_D h^m (\sigma_{II})^{(n-1)}} \exp\left(\frac{E_a + V_a P}{RT}\right), \tag{6.14}$$

$$F_1 = \frac{1}{2^n}.$$

(5) Write an expression for η_{eff} as a function of second strain rate invariant $\dot{\varepsilon}_{II}$ using Eqs. (6.4b) and (6.13c). Define F_2 by comparing with Eq. (6.5b)

$$\eta_{eff} = \frac{\sigma_{II}}{2\dot{\varepsilon}_{II}} = \frac{\left[\frac{1}{2^{(n-1)/n}}(\dot{\varepsilon}_{II})^{1/n}\frac{1}{(A_D)^{1/n} h^{m/n}}\exp\left(\frac{E_a + V_a P}{nRT}\right)\right]}{2\dot{\varepsilon}_{II}},$$

$$\eta_{eff} = \frac{1}{2^{(2n-1)/n}} \times \frac{1}{(A_D)^{1/n} h^{m/n}(\dot{\varepsilon}_{II})^{(n-1)/n}}\exp\left(\frac{E_a + V_a P}{nRT}\right), \tag{6.15}$$

$$F_2 = \frac{1}{2^{(2n-1)/n}}.$$

It is also important to mention that in rocks and mineral aggregates, both dislocation and diffusion creep occur simultaneously under applied stress, which can be expressed in the following relation for the effective viscosity

$$\frac{1}{\eta_{eff}} = \frac{1}{\eta_{diff}} + \frac{1}{\eta_{disl}}, \tag{6.16}$$

where η_{diff} and η_{disl} are the viscosities for diffusion and dislocation creep respectively, which are defined separately by Eq. (6.5) with different empirical parameters. This relation follows from the assumption that under the condition of constant applied deviatoric stress σ'_{ij}, the deviatoric strain rate $\dot{\varepsilon}'_{ij}$ can be decomposed to the contribution of dislocation ($\dot{\varepsilon}'_{ij(disl)}$) and diffusion ($\dot{\varepsilon}'_{ij(diff)}$) creep

$$\dot{\varepsilon}'_{ij} = \dot{\varepsilon}'_{ij(disl)} + \dot{\varepsilon}'_{ij(diff)} \tag{6.17}$$

where

$$\dot{\varepsilon}'_{ij} = \frac{\sigma'_{ij}}{2\eta_{eff}}, \quad \dot{\varepsilon}'_{ij(disl)} = \frac{\sigma'_{ij}}{2\eta_{disl}}, \quad \dot{\varepsilon}'_{ij(diff)} = \frac{\sigma'_{ij}}{2\eta_{diff}}. \tag{6.18}$$

Substituting relations (6.18) in Eq. (6.17) gives formula (6.16) (verify as an exercise).

6.3 Non-Newtonian channel flow

One important consequence of having a fluid with a *non-Newtonian rheology* is that the Poisson equation is *no longer valid* for channel flow (see Chapter 5) and the Stokes equation should be applied instead. Let us analyse such a non-Newtonian channel flow. In the case of a planar channel (Fig. 5.2), the Stokes equation reduces to

$$\frac{\partial \sigma_{xy}}{\partial x} = \frac{\partial P}{\partial y} - \rho g_y. \tag{6.19}$$

Under the assumption of a non-linear relationship between the stress and strain rate of the form $\dot{\varepsilon}_{xy} = A\sigma_{xy}^{n}$, the Stokes equation can be solved as follows. First we integrate Eq. (6.19) to obtain a horizontal stress profile

$$\sigma_{xy} = \int \left(\frac{\partial \sigma_{xy}}{\partial x}\right) dx = \left(\frac{\partial P}{\partial y} - \rho g_y\right) x + C_1, \tag{6.19}$$

where C_1 is our first integration constant. Then we represent the stress as a function of strain rate

$$\sigma_{xy} = \left(\frac{\dot{\varepsilon}_{xy}}{A}\right)^{1/n}, \tag{6.20}$$

and modify Eq. (6.19) using Eq. (6.20) and the relation $\dot{\varepsilon}_{xy} = \frac{1}{2}\partial v_y/\partial x$ to give

$$\left(\frac{\dot{\varepsilon}_{xy}}{A}\right)^{1/n} = \left(\frac{\partial P}{\partial y} - \rho g_y\right) x + C_1,$$
$$\dot{\varepsilon}_{xy} = A\left[\left(\frac{\partial P}{\partial y} - \rho g_y\right) x + C_1\right]^n, \tag{6.21}$$
$$\frac{\partial v_y}{\partial x} = 2A\left[\left(\frac{\partial P}{\partial y} - \rho g_y\right) x + C_1\right]^n.$$

Integrating Eq. (6.21) we obtain an expression for the horizontal profile of v_y velocity

$$v_y = \int \left(\frac{\partial v_y}{\partial x}\right) dx = \frac{2A/(n+1)}{\partial P/\partial y - \rho g_y}\left[\left(\frac{\partial P}{\partial y} - \rho g_y\right) x + C_1\right]^{n+1} + C_2, \tag{6.22}$$

where C_1 and C_2 are integration constants which can be defined from the boundary conditions.

For example, let us assume the following boundary conditions:

$$v_y = v_{y0} \quad \text{when} \quad x = 0,$$

$$\sigma_{xy} = 0 \quad \text{when} \quad x = L/2.$$

The latter condition is in fact an equivalent of the condition of symmetric velocity profile

$$v_y = v_{y0} = v_{y1} \quad \text{when } x = L.$$

Then our integration constants become as follows (derive as an exercise based on Eqs. 6.19, 6.22)

$$C_1 = -\left(\frac{\partial P}{\partial y} - \rho g_y\right)\frac{L}{2}, \tag{6.23}$$

$$C_2 = v_{y0} - \frac{2A/(n+1)}{\partial P/\partial y - \rho g_y} C_1{}^{n+1}. \tag{6.24}$$

Programming exercises

Exercise 6.1.
Calculate and visualize a logarithmic viscosity profile across a 100 000 m (100 km) thick mantle lithosphere with a temperature that linearly increases from 400°C at the top, to 1200 °C at the bottom. Apply the conditions of constant strain rate $\dot{\varepsilon}_{II} = 10^{-14}\text{s}^{-1}$. Take the following mantle rheological parameters: $A_D = 2.5 \times 10^{-17}$ 1/(Pa^n s), $n = 3.5$, $E_a = 532000$ J/mol, $m = 0$, $V_a = 0$ (representative for dry olivine under upper mantle conditions). Take into account that all parameters are based on axial compression experiments (Fig. 6.1a). An example is in **Viscosity_profile.m**.

Exercise 6.2.
Use the flow law from the previous exercise to compute and visualize a logarithmic viscosity map for the mantle in coordinates of temperature (400–1400°C) and $\log_{10}$ stress (10^3–10^9 Pa, second stress invariant). An example is in **Viscosity_map.m**.

Exercise 6.3.
Repeat the previous exercise for an effective mantle viscosity based on Eqs. (6.4) and (6.16) using the flow laws for diffusion and dislocation creep written in the following form (Karato and Wu, 1993):

$$\dot{\varepsilon}_{II} = A\left(\frac{h}{b}\right)^m \left(\frac{\sigma_{II}}{\mu}\right)^n \exp\left(-\frac{E_a + V_a P}{RT}\right), \tag{6.25}$$

Table 6.1 *Flow parameters for olivine (Karato and Wu, 1993)*

Mechanism	Dry	Wet
Dislocation creep		
A, 1/s	3.5×10^{22}	2.0×10^{18}
n	3.5	3
m	0	0
E_a, J/mol	540000	430000
V_a, m³/mol	15×10^{-6} to 25×10^{-6}	10×10^{-6} to 20×10^{-6}
Diffusion creep		
A, 1/s	8.7×10^{15}	5.3×10^{15}
n	1	1
m	−2.5	−2.5
E_a, J/mol	300000	240000
V_a, m³/mol	6×10^{-6}	5×10^{-6}

where $b = 5 \times 10^{-10}$ m is the Burgers vector, $\mu = 8 \times 10^{10}$ Pa, $R = 8.314$ J/(K mol). The other flow law parameters are given in Table 6.1. In your calculations, assume that the grain size $h = 0.001$ m (1 mm) and $P = 0$. Produce and compare logarithmic viscosity plots characteristic for dry and wet conditions. An example is in **Viscosity_comparison.m**.

7

Numerical solutions of the momentum and continuity equations

Theory: Types of numerical grids and their applicability for different differential equations. Staggered, half-staggered and non-staggered grids in one and two dimensions. Discretization of the continuity and Stokes equations on a rectangular grid. Conservative and non-conservative discretization schemes for Stokes equations. Velocity boundary conditions and their numerical implementation. No slip and free slip boundary conditions.

Exercises: Programming different boundary conditions. Solving momentum and continuity equations for the case of constant and variable viscosity Stokes flows.

7.1 Grids

As we have already discussed, the numerical solution of partial differential equations (PDEs) requires the definition of a grid of nodal points within the numerical model. The choice of this grid depends strongly on the type of equations to be solved. *Discretization schemes* for these equations will also change with the changing types of numerical grids. The following types of numerical grid exist in numerical geodynamic modelling.

- Depending on the dimension of the problem, the numerical grid can be one, two or three dimensional (1D, 2D, 3D) (Fig. 7.1).
- Depending on the shape of the basic elements, the grid can be rectangular or triangular (Fig. 7.2).
- Depending on the distribution of nodal points, the grid can be regular or non-regular (irregular) (Fig. 7.3).
- Depending on the distribution of different variables within the grid, it can be non-staggered (collocated) or staggered (Fig. 7.4).

The simplest grids are *non-staggered* (*collocated*). All variables are defined at the same nodal points (Fig. 7.5). When using finite differences (FD) with such a grid, all equations

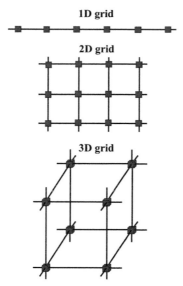

Fig. 7.1. Examples of 1D, 2D and 3D numerical grids.

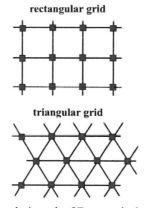

Fig. 7.2. Examples of rectangular and triangular 2D numerical grids.

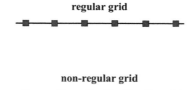

Fig. 7.3. Examples of regular and non-regular 1D numerical grids.

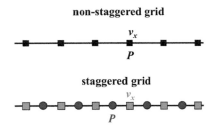

Fig. 7.4. Examples of non-staggered and staggered 1D numerical grids.

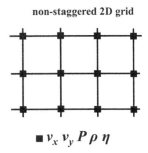

Fig. 7.5. Example of a non-staggered 2D numerical grid.

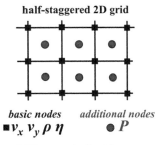

Fig. 7.6. Example of a half-staggered 2D numerical grid.

are formulated at the same nodal points. Non-staggered grids are a natural choice for solving the Poisson, heat conservation and advective transport equations.

In the *staggered grid*, several types of nodal points exist (Fig. 7.6) at which different variables are defined. A *half-staggered* grid in two dimensions (Fig. 7.6) is a combination of a basic non-staggered grid, with an additional set of points defined at the centres of cells formed by the basic grid. Some variables are then defined at these additional nodes and not at the basic nodal points. A half-staggered grid is suitable for solving the combination of Stokes and continuity equations in the case of *constant viscosity* when unknown parameters are components of velocity (v_x, v_y) defined at the basic nodes, and pressure (P) defined at the additional nodes (Fig. 7.6). Accordingly, x and y Stokes equations (Eq. 5.24) are formulated at basic nodes, whereas the continuity equation is formulated at the additional nodes.

fully staggered 2D grid

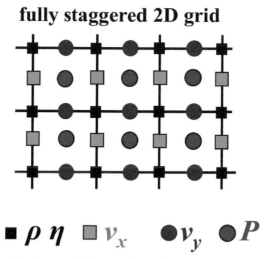

Fig. 7.7. Example of a fully staggered 2D numerical grid.

However, the half-staggered grid is less convenient for mechanical and thermomechanical problems with *variable viscosity.*

Fully staggered grids are applied in one, two and three dimensions and consist of a combination of several types of nodal points having different geometrical positions (Fig. 7.7). Different variables are then defined at different nodal points. Different equations are also formulated at different nodal points. Despite the apparent geometrical complexity, fully staggered grids are the most convenient choice for thermomechanical numerical problems with variable viscosity when finite differences are used for solving the continuity, Stokes and temperature equations. Discretization of thermomechanical equations on the fully staggered grid is very natural and gives simple FD formulas. Also, the accuracy of a numerical solution obtained on a fully staggered grid is notably (up to four times, Fornberg, 1995) higher then that obtained on a non-staggered grid. Therefore: 'Think in a fully staggered way!'

7.2 Discretization of the equations

The discretization of the equations depends on the type of numerical grid that is employed. The following discretization schemes can for example be used to represent the 2D incompressible continuity equation $\partial v_x/\partial x + \partial v_y/\partial y = 0$ in the case of two different 2D grids: non-staggered (half-staggered) grid (Fig. 7.8)

$$\frac{v_{x3} - v_{x1} + v_{x4} - v_{x2}}{2\Delta x} + \frac{v_{y2} - v_{y1} + v_{y4} - v_{y3}}{2\Delta y} = 0, \tag{7.1}$$

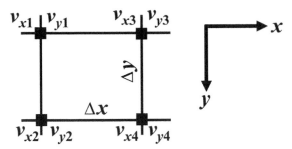

Fig. 7.8. Stencil used for the discretization of the continuity equation for a 2D non-staggered grid.

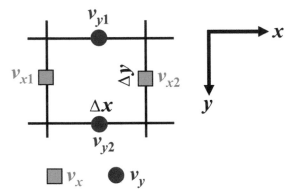

Fig. 7.9. Stencil used for the discretization of the continuity equation for a 2D fully staggered grid.

fully staggered grid (Fig. 7.9)

$$\frac{v_{x2} - v_{x1}}{\Delta x} + \frac{v_{y2} - v_{y1}}{\Delta y} = 0. \tag{7.2}$$

In both examples, the continuity equation is formulated at the centre of a cell that forms an elementary volume of the numerical grid. However, in the case of a fully staggered grid, the *stencil* (i.e. pattern of points) used to represent the equation on the grid is the most natural for the continuity equation (cf. Figs. 7.8 and 7.9) and therefore the resulting finite difference formula involves a minimum number of unknowns (cf. Eqs. 7.1 and 7.2).

7.3 Conservative finite differences

In order to discretize the Stokes equations in the case of variable viscosity, *conservative finite differences* should be used. Such finite differences provide *conservation of stresses* and thus allow a correct numerical solution. Below, examples of non-conservative and conservative finite differences for the 1D incompressible Stokes equation

$$\frac{\partial \sigma'_{xx}}{\partial x} - \frac{\partial P}{\partial x} = 0, \quad \text{where} \quad \sigma'_{xx} = 2\eta \frac{\partial v_x}{\partial x},$$

are compared for a 1D staggered grid (Fig. 7.10).

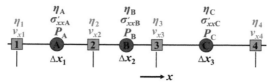

Fig. 7.10. Example of a 1D staggered grid used for the discretization of the Stokes equation with a variable viscosity. 1, 2, 3, 4 are basic nodes of the grid where Stokes equations are formulated. A, B, C are additional (stress) nodes of the grid where the stresses are formulated.

An ***erroneous non-conservative*** FD formulation of the Stokes equation for two basic nodes 2 and 3 can be constructed, for example (we can 'arrive' at this formula assuming erroneously that all we need to do is to apply Eq. (5.24) and use a different viscosity for each of the different nodes)

$$\text{node 2}\quad 2\eta_2 \frac{(v_{x3}-v_{x2})/\Delta x_2 - (v_{x2}-v_{x1})/\Delta x_1}{(\Delta x_1 + \Delta x_2)/2} - \frac{P_B - P_A}{(\Delta x_1 + \Delta x_2)/2} = 0, \qquad (7.3a)$$

$$\text{node 3}\quad 2\eta_3 \frac{(v_{x4}-v_{x3})/\Delta x_3 - (v_{x3}-v_{x2})/\Delta x_2}{(\Delta x_2 + \Delta x_3)/2} - \frac{P_C - P_B}{(\Delta x_2 + \Delta x_3)/2} = 0, \qquad (7.3b)$$

which implicitly means that formulations of the deviatoric stress σ'_{xxB} in the Stokes equation written for nodes 2 and 3 are different due to different viscosities η_2 and η_3:

$$\text{node 2}\quad \sigma'_{xxB} = 2\eta_2 \frac{v_{x3}-v_{x2}}{\Delta x_2},$$

$$\text{node 3}\quad \sigma'_{xxB} = 2\eta_3 \frac{v_{x3}-v_{x2}}{\Delta x_2}.$$

This implies that stress is *not conserved* and artificially 'jumps' between the two nodes in response to the changing viscosity.

On the other hand, a ***proper conservative*** FD formulation of the Stokes equation is

$$\text{node 2}\quad \frac{2\eta_B(v_{x3}-v_{x2})/\Delta x_2 - 2\eta_A(v_{x2}-v_{x1})/\Delta x_1}{(\Delta x_1 + \Delta x_2)/2} - \frac{P_B - P_A}{(\Delta x_1 + \Delta x_2)/2} = 0, \qquad (7.4a)$$

$$\text{node 3}\quad \frac{2\eta_C(v_{x4}-v_{x3})/\Delta x_3 - 2\eta_B(v_{x3}-v_{x2})/\Delta x_2}{(\Delta x_2 + \Delta x_3)/2} - \frac{P_C - P_B}{(\Delta x_2 + \Delta x_3)/2} = 0, \qquad (7.4b)$$

which means that formulations of the deviatoric stress σ'_{xxB} used in the Stokes equation written for nodes 2 and 3 are the same:

$$\sigma'_{xxB} = 2\eta_B \frac{v_{x3}-v_{x2}}{\Delta x_2},$$

implying that stress is *conserved* between the two nodes.

Thus, the conservative FD formulation is based on the following three formal rules.

(1) The Stokes equation is initially discretized in terms of stress components for the *basic (velocity) nodes* of the grid (cf. nodes 2, 3, Fig. 7.10),

$$\text{node 2} \qquad \frac{\sigma'_{xxB} - \sigma'_{xxA}}{(\Delta x_1 + \Delta x_2)/2} - \frac{P_B - P_A}{(\Delta x_1 + \Delta x_2)/2} = 0,$$

$$\text{node 3} \qquad \frac{\sigma'_{xxC} - \sigma'_{xxB}}{(\Delta x_2 + \Delta x_3)/2} - \frac{P_C - P_B}{(\Delta x_2 + \Delta x_3)/2} = 0.$$

(2) These stress components are formulated at the *additional (stress) nodes* of the grid (cf. nodes A, B, C, Fig. 4.2)

$$\text{node A} \qquad \sigma'_{xxA} = 2\eta_A \frac{v_{x2} - v_{x1}}{\Delta x_1},$$

$$\text{node B} \qquad \sigma'_{xxB} = 2\eta_B \frac{v_{x3} - v_{x2}}{\Delta x_2},$$

$$\text{node C} \qquad \sigma'_{xxC} = 2\eta_C \frac{v_{x4} - v_{x3}}{\Delta x_3}.$$

Note that we have to use viscosity values η_A, η_B and η_C for the additional nodes (A, B, C) where the stress components are defined. If these values are not directly available at these locations, they can be computed by for example *harmonic averaging* of the known viscosity values from the basic nodes (1, 2, 3, 4)

$$\eta_A = \frac{2}{1/\eta_1 + 1/\eta_2},$$

$$\eta_B = \frac{2}{1/\eta_2 + 1/\eta_3},$$

$$\eta_C = \frac{2}{1/\eta_3 + 1/\eta_4}.$$

(3) Identical formulations of the stress components are used for the Stokes equation on the different basic nodes.

Applying these rules in 2D, the following conservative formulations can be derived in 2D for x and y Stokes equations (derive them as an exercise based on the above principles), x Stokes equation (Fig. 7.11)

$$2\frac{\sigma'_{xxB} - \sigma'_{xxA}}{\Delta x_1 + \Delta x_2} + \frac{\sigma'_{xy2} - \sigma'_{xy1}}{\Delta y_2} - 2\frac{P_B - P_A}{\Delta x_1 + \Delta x_2} = -\frac{\rho_1 + \rho_2}{2}g_x, \qquad (7.5)$$

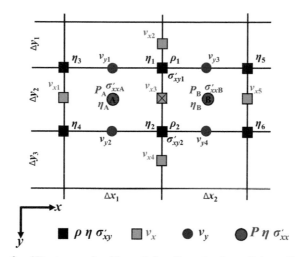

Fig. 7.11. Stencil of a 2D staggered grid used for discretization of the x Stokes equation with a variable viscosity. The crossed square corresponds to the node at which the x Stokes equation is formulated.

where

$$\sigma'_{xy1} = 2\eta_1 \left(\frac{v_{x3} - v_{x2}}{\Delta y_1 + \Delta y_2} + \frac{v_{y3} - v_{y1}}{\Delta x_1 + \Delta x_2} \right),$$

$$\sigma'_{xy2} = 2\eta_2 \left(\frac{v_{x4} - v_{x3}}{\Delta y_2 + \Delta y_3} + \frac{v_{y4} - v_{y2}}{\Delta x_1 + \Delta x_2} \right),$$

$$\sigma'_{xxA} = 2\eta_A \frac{v_{x3} - v_{x1}}{\Delta x_1},$$

$$\sigma'_{xxB} = 2\eta_B \frac{v_{x5} - v_{x3}}{\Delta x_2}.$$

If values of viscosity for the centres of cells (A, B) are not known, they can be computed by averaging the known viscosity values from the basic nodes (cf. black squares in Fig. 7.11):

arithmetic averaging

$$\eta_A = \frac{\eta_1 + \eta_2 + \eta_3 + \eta_4}{4},$$

$$\eta_B = \frac{\eta_1 + \eta_2 + \eta_5 + \eta_6}{4},$$

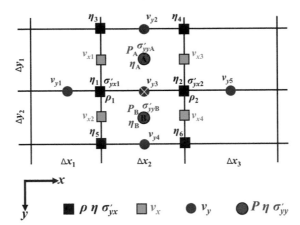

Fig. 7.12. Stencil of a 2D staggered grid used for the discretization of the y Stokes equation with a variable viscosity. The crossed circle corresponds to the node at which the y Stokes equation is formulated.

harmonic averaging

$$\eta_A = \frac{4}{1/\eta_1 + 1/\eta_2 + 1/\eta_3 + 1/\eta_4},$$

$$\eta_B = \frac{4}{1/\eta_1 + 1/\eta_2 + 1/\eta_5 + 1/\eta_6}.$$

y Stokes equation (Fig. 7.12)

$$2\frac{\sigma'_{yyB} - \sigma'_{yyA}}{\Delta y_1 + \Delta y_2} + \frac{\sigma'_{yx2} - \sigma'_{yx1}}{\Delta x_2} - 2\frac{P_B - P_A}{\Delta y_1 + \Delta y_2} = -\frac{\rho_1 + \rho_2}{2}g_y, \qquad (7.6)$$

where

$$\sigma'_{yx1} = 2\eta_1\left(\frac{v_{y3} - v_{y1}}{\Delta x_1 + \Delta x_2} + \frac{v_{x2} - v_{x1}}{\Delta y_1 + \Delta y_2}\right),$$

$$\sigma'_{yx2} = 2\eta_2\left(\frac{v_{y4} - v_{y2}}{\Delta x_2 + \Delta x_3} + \frac{v_{x4} - v_{x3}}{\Delta y_1 + \Delta y_2}\right),$$

$$\sigma'_{yyA} = 2\eta_A\frac{v_{y3} - v_{y2}}{\Delta y_1},$$

$$\sigma'_{yyB} = 2\eta_B\frac{v_{y4} - v_{y3}}{\Delta y_2}.$$

If values of viscosity for the centres of cells (A, B) are not known, they can be computed by for example harmonic averaging of the known viscosity values from the basic nodes (cf. black squares in Fig. 7.12)

$$\eta_A = \frac{4}{1/\eta_1 + 1/\eta_2 + 1/\eta_3 + 1/\eta_4},$$

$$\eta_B = \frac{4}{1/\eta_1 + 1/\eta_2 + 1/\eta_5 + 1/\eta_6}.$$

Note that in all the examples above, the medium is assumed to be incompressible and the deviatoric stress components σ'_{ij} are formulated via the total strain rate tensor components $\dot{\varepsilon}_{ij}$ according to the equations (4.10)–(4.12) and (5.12) as

$$\sigma'_{ij} = 2\eta\dot{\varepsilon}_{ij},$$

where

$$\dot{\varepsilon}_{ij} = \frac{1}{2}\left(\frac{\partial v_i}{\partial x_j} + \frac{\partial v_j}{\partial x_i}\right),$$

i and j are coordinate indexes and x_i and x_j are spatial coordinates.

7.4 Boundary conditions

As we have discussed before, in order to obtain numerical solutions, boundary conditions have to be defined. In mathematics, two types of boundary conditions are most commonly used: *the Dirichlet boundary condition* and *the Neumann boundary condition*. The Dirichlet (or *first-type*) boundary condition specifies the values that a solution Φ needs to take along the boundary of the model domain (e.g. $\Phi = 0$). We already used this type of condition when solving the Poisson equation in Chapter 3. The Neumann (or *second-type*) boundary condition specifies the values that a derivative of the solution Φ needs to satisfy at the boundary (e.g., $\partial\Phi/\partial x = 0$ and $\partial\Phi/\partial y = 0$).

Velocity boundary conditions used for solving Stokes and continuity equations are typically composite and combine both Dirichlet and Neumann conditions. The following boundary conditions are often used in geodynamic modelling:

(1) free slip,
(2) no slip,
(3) free surface,
(4) fast erosion,
(5) infinity-like (external free slip, external no slip, Winkler basement),
(6) prescribed velocity (moving boundary),

(7) periodic,
(8) combined conditions.

(1) A free slip condition requires that the normal velocity component on the boundary is zero and the two other components do not change across the boundary (this condition also implies *zero shear strain rates and shear stresses* along the boundary). For example, for the vertical boundary orthogonal to the x axis, the free slip condition is formulated as follows

$$v_x = 0, \tag{7.7a}$$

$$\frac{\partial v_y}{\partial x} = \frac{\partial v_z}{\partial x} = 0. \tag{7.7b}$$

(2) A no slip condition requires all velocity components on the boundary to be zero

$$v_x = v_y = v_z = 0. \tag{7.8}$$

(3) A free surface condition requires that the boundary is subject to zero parallel shear stress

$$\tau = 0. \tag{7.9}$$

This condition allows the surface to be deformed. Numerical implementation of this condition requires either programming a deformable grid following the deforming surface, or the introduction of a low viscosity layer above the free surface (e.g. Schmeling et al., 2008). The free surface condition may produce a numerical instability related to material advection, which will be discussed in Chapter 8.

(4) A fast erosion condition requires that all the velocity components do not change across the boundary. For example, for the upper model boundary, which is orthogonal to the y axis, the fast erosion condition is formulated as follows

$$\frac{\partial v_x}{\partial y} = \frac{\partial v_y}{\partial y} = \frac{\partial v_z}{\partial y} = 0. \tag{7.10}$$

This condition corresponds to an infinitely fast erosion/deposition at the upper free surface. This surface is always kept horizontal since erosion is so fast that all mountains are instantaneously scraped off and are deposited in valleys. This condition also ensures that the mass in the model is conserved.

(5) An infinity-like condition either mimics the absence of a boundary, or implies that this boundary is located very far away. External free slip (Burg and Gerya, 2005; Gerya et al., 2008b) implies that conditions (7.7a) and (7.7b) are satisfied at a parallel boundary located at the distance ΔL from the actual boundary of the model and the velocity gradient between these two boundaries is constant. For example, the external

free slip condition applied to the lower boundary of the model that is orthogonal to the y axis is:

$$\frac{\partial v_y}{\partial y} \Delta L + v_y = 0, \tag{7.11a}$$

$$\frac{\partial v_x}{\partial y} = 0, \tag{7.11b}$$

$$\frac{\partial v_z}{\partial y} = 0. \tag{7.11c}$$

By analogy, the external no slip condition at the same boundary is formulated as

$$\frac{\partial v_x}{\partial x} \Delta L + v_x = 0, \tag{7.12a}$$

$$\frac{\partial v_y}{\partial y} \Delta L + v_y = 0, \tag{7.12b}$$

$$\frac{\partial v_z}{\partial z} \Delta L + v_z = 0. \tag{7.12c}$$

Note that relations (7.11) and (7.12) ensure global conservation of mass in the computational domain, despite the presence of an 'open' boundary.

Winkler's pliable basement (e.g. Burov et al., 2001; Yamato et al., 2008) assumes isostatic equilibrium at the model bottom and implies that the model overlies an infinite space filled with an inviscid fluid having a small density contrast (e.g. 10 kg/m^3) within the lower part of the model. This is a sort of free surface condition, applied at the lower boundary of the model, which is typically located in the mantle asthenosphere. It assumes that the material underneath the boundary has zero viscosity and moves infinitely fast.

(6) The prescribed velocity condition implies non-zero velocity at a model boundary. When velocity is prescribed orthogonal to the boundary (inward/outward flow), then a compensating outward/inward velocity should be prescribed on other(s) model boundary(ies) in order to ensure mass conservation in the model. In this case, the model boundaries can also be displaced with time in response to the material movement (moving boundary condition, Chapter 21).

(7) Periodic boundary conditions are typically established for paired parallel lateral boundaries of a model and prescribe that all material properties as well as pressure and velocity fields at both sides of each boundary are identical. From a physical point of view, this implies that these two boundaries are open and that flow leaving the model through one boundary immediately re-enters through the opposing side. This condition is often used in mantle convection modelling to simulate part of a spherical/cylindrical shell with a convecting mantle (or mimic it, in Cartesian coordinates).

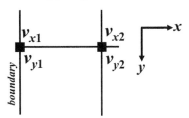

Fig. 7.13. Stencil of a 2D non-staggered grid used for the formulation of no slip and free slip boundary conditions.

(8) Combined conditions represent a mixture between several types of boundary conditions.

All of the described boundary conditions can be time dependent. This could particularly imply that the physical location of the boundary condition may be a function of time (Chapter 21). Boundary conditions can also be applied *inside the model* (Chapter 20).

We will now concentrate on the numerical implementation of the most common, and most simple, free slip and no slip conditions (we will discuss several examples of more complex conditions in Chapters 20 and 21). The numerical implementation of boundary conditions depends on the type of grid.

Non-staggered grid (Fig. 7.13):

$$\text{free slip,} \quad v_{x1} = 0, \quad v_{y1} = v_{y2}, \tag{7.13}$$

$$\text{no slip,} \quad v_{x1} = 0, \quad v_{y1} = 0. \tag{7.14}$$

Staggered grid (Fig. 7.14):

$$\text{free slip,} \quad v_{x1} = 0, \quad v_{y1} = v_{y2}, \tag{7.15}$$

$$\text{no slip,} \quad v_{x1} = 0, \quad v_{y1} = -v_{y2}. \tag{7.16}$$

Condition (7.16) for the vertical velocity v_{y1} implies that zero vertical velocity on the boundary $v_{yb} = 0$ (Fig. 7.14) is averaged from vertical velocities of nearest internal and external nodes as

$$v_{yb} = \frac{v_{y1} + v_{y2}}{2} = 0.$$

7.5 Indexing of unknowns

Another very important issue, in relation to solving the Stokes and continuity equations on a fully staggered grid, is the indexing of the unknowns. This is particularly relevant when

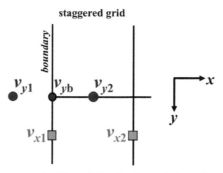

Fig. 7.14. Stencil of a 2D staggered grid used for the formulation of no slip and free slip boundary conditions.

the system of linear equations (global matrix) formulated with finite differences is solved with Gaussian elimination or other direct methods as discussed in Chapter 3. This is a somewhat boring subject but it is extremely important to understand it properly. Remember, *90% of the bugs in your code are made with the indexing* (Bug Rule 6 in the Introduction). Both the possibility of obtaining the solution and the amount of computational work will depend on the method used to index the unknowns (P, v_x and v_y) on the staggered grid. The most optimal staggered grid geometry (Duretz et al., 2011a) includes external velocity nodes, which allows for the second-order accurate implementation of boundary conditions (Figs. 7.14, 7.15). One of the optimal ways of numbering such a grid with an effective numerical resolution of N_x by N_y points is illustrated in Fig. 7.15. The following rules are used for this indexing.

(1) Staggered nodes of the grid are related in a uniform manner (cf. dotted arrows in Fig. 7.15) to the basic nodes of an *extended grid* formed by the intersections of $N_y + 1$ horizontal and $N_x + 1$ vertical gridlines. The extended grid is required to index the external velocity nodes. The same number of unknowns should be related to each basic node (P, v_x and v_y give us three unknowns per basic node, Fig. 7.15). If no staggered node with respective unknowns can be found inside or outside the grid then 'ghost unknowns' are added *for the uniformity of numbering*. Such 'ghost unknowns' are not used in the numerical solution and are set to zeros (cf. nodes with blue indexes in Fig. 7.15).

(2) Basic nodes of the extended grid are indexed (cf. indexes in italics in Fig. 7.15) from 1 to $(N_x + 1) \times (N_y + 1)$, where N_x and N_y is the standard (i.e. non-extended) grid resolution for the model domain (cf. yellow area in Fig. 7.15) in the horizontal and vertical direction, respectively (6 and 5, respectively in Fig. 7.15). The index numbering increases in the direction of the smallest number of gridlines (i.e. in the vertical

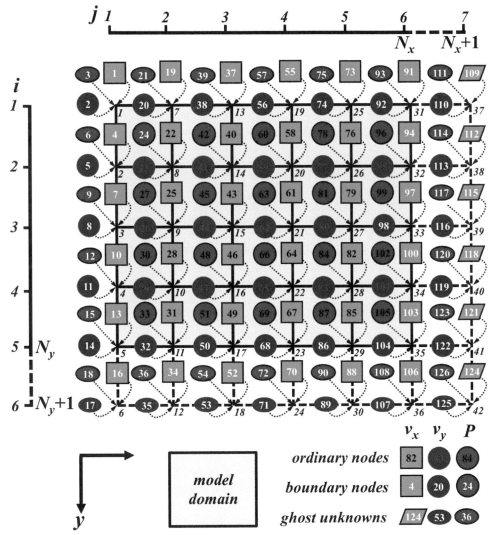

Fig. 7.15. Indexing of unknowns (P, v_x and v_y) for a 6 × 5 2D staggered grid. Dotted arrows show relations of staggered nodes to the basic nodes of the grid. Indexes in italics correspond to numbering of basic nodes. Indexes i and j correspond to numbering of gridlines in the vertical and horizontal directions, respectively.

direction in Fig. 7.15) to ensure a minimal amount of computational work for global matrix inversion

$$in_{node} = (j-1) \times (N_y + 1) + i, \tag{7.17}$$

where in_{node} is the index for the given node, i and j are the indexes of the vertical and horizontal gridlines which are intersecting at the node (Fig. 7.15).

(3) Unknown parameters attached to each basic node are indexed from 1 to $(N_x + 1) \times (N_y + 1) \times 3$, according to the increasing basic node index

$$in_{v_x} = 3 \times in_{node} - 2, \tag{7.18a}$$

$$in_{v_y} = 3 \times in_{node} - 1, \tag{7.18b}$$

$$in_P = 3 \times in_{node}, \tag{7.18c}$$

where in_P, in_{v_x} and in_{v_y} are indexes for P, v_x and v_y related to the given basic node in_{node}.

As one can see in Fig. 7.15, index in_P located in the centre of a grid cell is always bigger than the indexes in_{v_x} and in_{v_y} for the velocity nodes surrounding this cell. This is an important condition since pressure is obtained by solving the continuity equation. An incompressible continuity equation does not initially contain pressure, and the solution is guaranteed by the order of processing during the inversion of the global matrix: the continuity equation (Eq. 7.2) for pressure in a given cell (cf. Fig. 7.9) is processed after Stokes equations (Eqs. 7.5, 7.6) for all surrounding v_x and v_y nodes (cf. Figs. 7.11, 7.12), which contain pressure from this cell. The pressure value should be given only at one of the internal pressure nodes (cf. P node 24 in Fig. 7.15) of the grid such that absolute pressure values can be defined through the pressure gradients present in the momentum equations. We thus need to formulate the boundary condition for a single pressure node inside the model domain.

The idea of obtaining a solution for pressure by formulating an equation that does not contain pressure may sound bizarre. Let us first convince ourselves this is valid by considering the following simple example. We apply Gaussian elimination (Eqs. 3.12– 3.18) to the analogue system of three equations with variables v_x, v_y and P having the global indexes 1, 2 and 3 respectively:

equation A1 (formulated for v_x) $2v_x + 4v_y + 2P = 0$,

equation A2 (formulated for v_y) $3v_x + 9v_y + 6P = 21$,

equation A3 (formulated for P) $v_x + 3v_y = 5$.

Dividing all equations by a number to normalize the coefficient of v_x, we get:

equation B1 $v_x + 2v_y + P = 5$,

equation B2 $v_x + 3v_y + 2P = 7$,

equation B3 $v_x + 3v_y = 5$.

Eliminating v_x by subtracting equation B1 from B2 and B3 yields:

equation B1 $v_x + 2v_y + P = 5,$

equation C2 $v_y + P = 2,$

equation C3 $v_y - P = 0.$

Note that after the elimination operation, P indeed appears in equation C3. Eliminating v_y by subtracting equation C2 from C3 yields:

equation B1 $v_x + 2v_y + P = 5,$

equation C2 $v_y + P = 2,$

equation D3 $-2P = -2.$

Obtaining the solution for P from equation D3

$$P = -2/(-2) = 1.$$

Obtaining the solution for v_y from equation C2

$$v_y = 2 - P = 2 - 1 = 1.$$

Obtaining the solution for v_x from equation B1

$$v_x = 5 - 2v_y - P = 5 - 2 - 1 = 2.$$

So, it works and we obtained all the required solutions, including one for P, by Gaussian elimination (we could also have inferred this from the fact that we have three equations for the three unknowns v_x, v_y and P). Of course, it would be impossible to apply (*without reordering of equations and re-indexing of unknowns*) the same Gaussian elimination approach if the order of global indexes (and respectively equations) was inverted, i.e.

equation 1 (formulated for P) $3v_y + v_x = 5,$

equation 2 (formulated for v_y) $6P + 9v_y + 3v_x = 21,$

equation 3 (formulated for v_x) $2P + 4v_y + 2v_x = 10.$

It should be mentioned, however, that advanced direct solvers (including the '\' command of MATLAB) have internal reordering procedures that allow one to obtain solutions in the latter case as well.

An alternative staggered grid indexing approach defines all velocity nodes used to formulate boundary condition equations for v_x and v_y velocity components (cf. symbols with white indexes in Fig. 7.15) as *ghost nodes* (cf. non-numbered symbols in Fig. 7.16). In this approach, boundary condition equations for velocity are not explicitly added to the

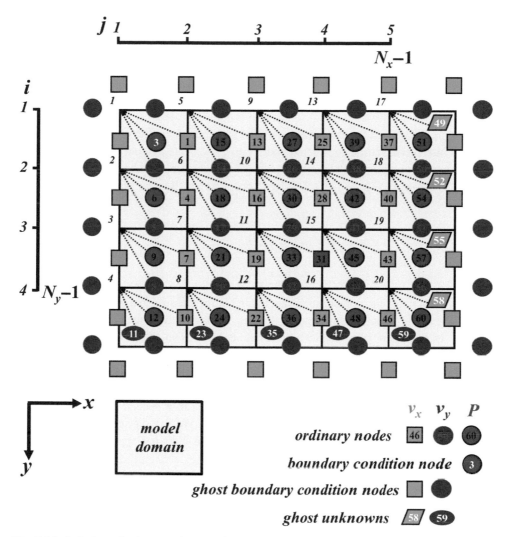

Fig. 7.16. Indexing of unknowns (P, v_x and v_y) for 6 × 5 2D staggered grid with the use of 'ghost velocity nodes'. Dotted arrows show the relationship of the staggered nodes to the basic nodes of the grid (note that only some of the nodes are numbered). Non-numbered symbols correspond to the ghost nodes where velocity boundary condition equations are defined. These boundary condition equations are implicitly used in the numerical solution when formulating momentum and continuity equations for the internal (numbered) nodes of the grid located next to the ghost nodes. Indexes in italics correspond to numbering of basic nodes. Indexes i and j correspond to the numbering of gridlines in the vertical and horizontal directions, respectively.

global matrix and therefore the respective unknowns for the ghost nodes are not indexed. Instead, these boundary condition equations are used in an implicit manner by taking them into account while discretizing the momentum and continuity equations for the internal nodes of the grid located next to the ghost nodes. Values of v_x and v_y velocities in the ghost nodes are recovered from the boundary condition equations after obtaining a global solution for internal nodes. The manner of indexing unknowns is shown in Fig. 7.16 and again is based on a convention that relates staggered nodes to (part of) the basic nodes of the grid, numbered as

$$in_{node} = (j - 1) \times (N_y - 1) + i,$$

where in_{node} is the index for the given node, i and j are indexes for the vertical and horizontal gridlines intersecting at the considered node (Fig. 7.16). Note that only $(N_x - 1) \times (N_y - 1)$ basic nodes are numbered. Unknowns attached for each numbered basic node are indexed respectively from 1 to $(N_x - 1) \times (N_y - 1) \times 3$ according to the increasing basic node index

$$in_{v_x} = 3 \times in_{node} - 2,$$

$$in_{v_y} = 3 \times in_{node} - 1,$$

$$in_P = 3 \times in_{node}.$$

This way of indexing again ensures that the index for pressure in a cell is bigger than the indexes for all the velocities surrounding the cell. The main advantage of this staggered grid structure is that there is a smaller number of unknowns in the global matrix equal to $(N_x - 1) \times (N_y - 1) \times 3$. As before, the boundary condition for pressure needs to be defined in a single cell (*P* node 3 in Fig. 7.16). The implementation of ghost nodes, however, requires additional programming, as the formulation of the momentum and continuity equations for the 'near-boundary' nodes must be done in a special manner in order to take into account various velocity boundary conditions. We will discuss this issue in more detail in Chapters 17 and 18. Please note that the two indexing approaches outlined in Figs. 7.15 and 7.16 *give identical numerical solutions*.

Compared to the stream function approach (Chapter 5), the use of a *pressure-velocity formulation* (also called *primitive variable formulation*) requires three times the number of equations (in 2D) for the same grid resolution. The advantages are, however, that the solution for pressure is obtained and that lower (second) order derivatives are used compared to the fourth-order derivatives required by the stream function based approach (Eq. 5.36).

Programming exercises

Exercise 7.1.
Solve the momentum and continuity equations with finite differences for the case of constant viscosity using a pressure-velocity formulation (Eqs. 5.22, 1.28) on a regular staggered grid (Figs. 7.15, 7.17) of 31×21 points. Program the numerical model for buoyancy-driven flow in a purely vertical gravity field ($g_x = 0$, $g_y = 10$ m/s^2 in Eq. 5.22) for a density structure with two vertical layers (3200 kg/m^3 and 3300 kg/m^3 for the left and

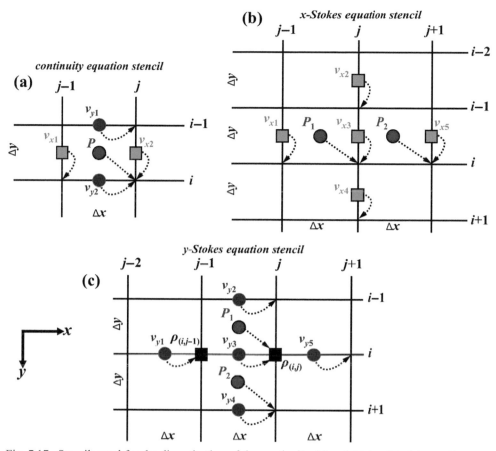

Fig. 7.17. Stencils used for the discretization of the continuity (a) and Stokes (b), (c) equations on a 2D regular staggered grid for models with constant viscosity based on the pressure-velocity formulation. Indexing of gridlines corresponds to basic nodal points. Global indexing of different unknowns (cf. dotted arrows) is made according to Fig. 7.15.

right layers, respectively). The model size is $1000 \times 1500 \text{ km}^2$ (i.e. $1\,000\,000 \times 1\,500\,000 \text{ m}^2$). Use constant viscosity $\eta = 10^{21}$ Pa s for the entire model.

Compose the matrix of coefficients $\{L\}$, the right hand side vector $\{R\}$ and obtain the solution vector $\{S\}$ with the direct solver ($S = L \backslash R$) using relations (7.18) and Figs. 7.15, 7.17 for global indexing of the unknowns. Boundary conditions for the velocity are free slip on all boundaries (Eq. 7.15). Boundary condition for pressure in one cell ($i = 2$ and $j = 2$, Fig. 7.15) is $P = 0$. Do not forget that the condition for all ghost unknowns ($P = 0$, $v_x = 0$, $v_x = 0$) should also be added to the matrix of coefficients $\{L\}$ and right hand side $\{R\}$ (see white numbers in Fig. 7.15 for indexing of these unknowns). Visualize the structure of the sparse matrix $\{L\}$ by using the *spy(L)* command.

The finite difference representation of the momentum and continuity equations in 2D is formulated by analogy to Eqs. (7.2), (7.5) and (7.6) as follows (Fig. 7.17)

$$\eta \frac{v_{x1} - 2v_{x3} + v_{x5}}{\Delta x^2} + \eta \frac{v_{x2} - 2v_{x3} + v_{x4}}{\Delta y^2} - K_{cont} \frac{P'_2 - P'_1}{\Delta x} = 0, \qquad (7.19)$$

$$\eta \frac{v_{y1} - 2v_{y3} + v_{y5}}{\Delta x^2} + \eta \frac{v_{y2} - 2v_{y3} + v_{y4}}{\Delta y^2} - K_{cont} \frac{P'_2 - P'_1}{\Delta y} = -g_y \frac{\rho_{(i,j-1)} + \rho_{(i,j)}}{2}, \qquad (7.20)$$

$$K_{cont} \left(\frac{v_{x2} - v_{x1}}{\Delta x} + \frac{v_{y2} - v_{y1}}{\Delta y} \right) = 0, \qquad (7.21)$$

where the scaled pressure $P' = P/K_{cont}$ and coefficients $K_{cont} = 2\eta/\Delta x + \Delta y$ are used to ensure relatively uniform (i.e. not differing by many orders of magnitude) coefficients in all equations. This scaling helps in obtaining an accurate solution when direct solvers are applied for models with large viscosities. After solving the global matrix for the velocity components and scaled pressure P', the unscaled pressure is computed as $P = P' K_{cont}$.

After obtaining a solution for velocity on staggered nodes, compute the velocity components v_x and v_y for the cell centres (i.e. for 30×20 points) by averaging the velocity from surrounding staggered nodes (use the surrounding velocity nodes for each component, Fig. 7.17a). Plot the results for pressure (*pcolor*) and velocity (*quiver*) on the same diagram with the vertical axis directed downward (*axis ij*).

An example is in **Stokes_continuity_constant_viscosity.m**.

Exercise 7.2.

Modify the previous example for a variable viscosity case. Use different viscosities for the two vertical layers (10^{20} Pa s and 10^{22} Pa s for the left and right layers, respectively). Define the viscosity at the basic nodes (η_s) (Fig. 7.18). Compute viscosity at the centres of cells (η_n) (Fig. 7.18) by using the harmonic average of viscosity values (η_s) at four surrounding basic nodes. Indexing for η_n should be the same as for pressure nodes (Figs. 7.15, 7.18).

The finite difference representation of the continuity equation is the same as before (Eq. 7.21, Fig. 7.17a). Stokes equations are formulated by analogy with Eqs. (7.5) and (7.6) as follows (Fig. 7.18)

$$2\eta_{n(i,j+1)} \frac{v_{x5} - v_{x3}}{\Delta x^2} - 2\eta_{n(i,j)} \frac{v_{x3} - v_{x1}}{\Delta x^2} + \eta_{s(i,j)} \left(\frac{v_{x4} - v_{x3}}{\Delta y^2} + \frac{v_{y4} - v_{y2}}{\Delta x \Delta y} \right)$$

$$- \eta_{s(i-1,j)} \left(\frac{v_{x3} - v_{x2}}{\Delta y^2} + \frac{v_{y3} - v_{y1}}{\Delta x \Delta y} \right) - K_{cont} \frac{P'_2 - P'_1}{\Delta x} = 0, \qquad (7.22)$$

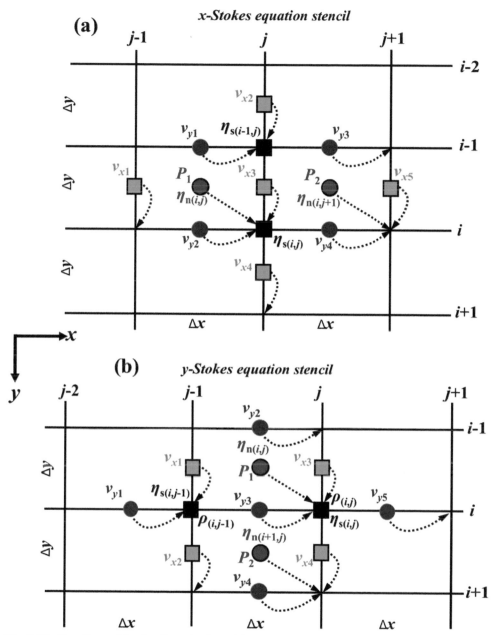

Fig. 7.18. Stencils used for the discretization of the Stokes equations on a 2D regular staggered grid for models with variable viscosity based on the pressure-velocity formulation. Indexing of gridlines corresponds to basic nodal points. Global indexing of different unknowns (cf. dotted arrows) is made according to Fig. 7.15.

$$2\eta_{\mathrm{n}(i+1,j)}\frac{v_{y4}-v_{y3}}{\Delta y^2}-2\eta_{\mathrm{n}(i,j)}\frac{v_{y3}-v_{y2}}{\Delta y^2}+\eta_{\mathrm{s}(i,j)}\left(\frac{v_{y5}-v_{y3}}{\Delta x^2}+\frac{v_{x4}-v_{x3}}{\Delta x \Delta y}\right)$$

$$-\eta_{\mathrm{s}(i,j-1)}\left(\frac{v_{y3}-v_{y1}}{\Delta x^2}+\frac{v_{x2}-v_{x1}}{\Delta x \Delta y}\right)-K_{cont}\frac{P'_2-P'_1}{\Delta y}$$

$$=-g_y\frac{\rho_{(i,j-1)}+\rho_{(i,j)}}{2}, \tag{7.23}$$

where $K_{cont}=2\eta_{min}/(\Delta x+\Delta y)$ (computed with the value of minimal viscosity in the model, $\eta_{\min}$) is again used for scaling the coefficients.

An example is in **Stokes_continuity_variable_viscosity.m**.

8

The advection equation and marker-in-cell method

> **Theory:** Advection equation. Solution methods for continuous and discontinuous variables. Eulerian schemes: upwind differences, flux corrected transport (FCT). Lagrangian schemes: marker-in-cell method. Runge–Kutta advection schemes. Interpolation between markers and nodes. Continuity-based velocity interpolation. 'Sticky air' approach. 'Drunken sailor' instability. Stabilization of free surface.
>
> **Exercises:** Programming of various advection schemes and markers.

8.1 Advection equation

As we already know, the deformation of a continuum changes the spatial distribution of physical properties with time. These changes can be described by the *advection equation*. For a scalar function (A) at an Eulerian point, this equation is written as follows:

$$\frac{\partial A}{\partial t} = -\vec{v} \cdot \mathrm{grad}(A) \tag{8.1a}$$

or in different notation

$$\frac{\partial A}{\partial t} = -v_x \frac{\partial A}{\partial x} - v_y \frac{\partial A}{\partial y} - v_z \frac{\partial A}{\partial z}. \tag{8.1b}$$

For a Lagrangian point, the following advection equation relates changes in its coordinates with material velocities $\vec{v} = (v_x, v_y, v_z)$

$$\frac{Dx_i}{Dt} = v_i, \tag{8.2a}$$

or in different notation

$$\frac{Dx}{Dt} = v_x, \tag{8.2b}$$

$$\frac{Dy}{Dt} = v_y, \tag{8.2c}$$

$$\frac{Dz}{Dt} = v_z, \tag{8.2d}$$

where i is a coordinate index and x_i is a spatial coordinate.

8.2 Eulerian advection methods

The advection equations appear trivial, but this is an apparent simplicity. Solving them numerically *often causes numerical problems*. One such problem is the numerical diffusion of sharp gradients during advection. To illustrate the problem, we solve

$$\frac{\partial \rho}{\partial t} = -v_x \left(\frac{\partial \rho}{\partial x} \right), \tag{8.3}$$

on a regular Eulerian 1D grid with constant spacing between the nodes ($\Delta x = 1$), for the case of constant material velocity ($v_x = 1$) by applying *upwind differences* (i.e., finite differences are taken asymmetrically against the flow direction)

$$\rho_i^{t+\Delta t} = \rho_i^t - v_x \Delta t \frac{\rho_i^t - \rho_{i-1}^t}{\Delta x}, \tag{8.4}$$

and using different values of the time step Δt. Figure 8.1b demonstrates how a sharp density wave (i.e. perturbation) smoothes out (*diffuses*) during the numerical solution of a 1D advection equation obtained using upwind finite differences (Fig. 8.2a).

The intensity of the numerical diffusion depends on the number of numerical steps performed and not on the absolute time of advection (Fig. 8.1b,c). Therefore, smaller time steps give more numerical diffusion for the same total duration of advection (compare Fig. 8.1b and c). On the other hand, to ensure stability of the numerical solution, the time step should be sufficiently small to satisfy the time limitation condition

$$\Delta t \leq \frac{\Delta x}{v_x}, \tag{8.5a}$$

which states that the material should not move for more than one grid step per one time step (this is also called the *Courant condition*). Condition (8.5a) should be satisfied *in every Eulerian point* to prevent oscillations (Fig. 8.1d). Therefore, one should use the minimal ratio between local grid step (which can be variable) and local flow velocity (which can be variable as well) found in a model. In 2D and 3D, this limitation may be even stricter

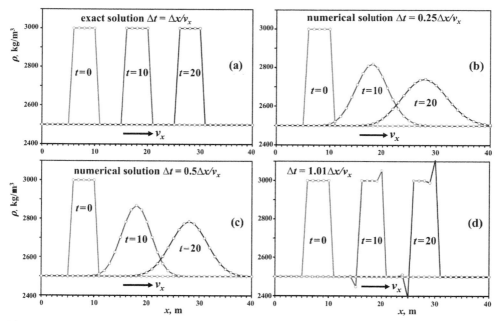

Fig. 8.1. Numerical solutions for the advection of a sharp density wave (i.e. perturbation) obtained using upwind finite differences with different size time steps, Δt. Solutions are obtained with a regular Eulerian 1D grid ($\Delta x = 1$ m) for the case of constant material velocity ($v_x = 1$ m/s).

$$\Delta t \leq \frac{\Delta x}{2v_x} \qquad (8.5b)$$

to prevent artificial oscillations from appearing in the numerical solutions.

Figure 8.3 shows the progress of numerical diffusion during only four time steps. Upwind differences (Eq. 8.4, Fig. 8.2a) are applied along with a constant time step $\Delta t = \frac{1}{2}\Delta x/v_x$. This results in the FD formulation

$$\rho_i^{t+\Delta t} = \rho_i^t - \frac{\rho_i^t - \rho_{i-1}^t}{2} = \frac{\rho_i^t + \rho_{i-1}^t}{2}, \qquad (8.6)$$

which is equivalent to the averaging of density values for the neighbouring grid points and therefore produces strong numerical diffusion (smoothing) of the initial density profile (Fig. 8.3e). The diffusion term, hidden in Eq. (8.4), can be 'exposed' by analysing this equation on the basis of non-diffusive symmetrical *central differences* (Fig. 8.2b)

$$\left(\frac{\partial \rho}{\partial x}\right)_i^{central} = \frac{\rho_{i+1}^t - \rho_{i-1}^t}{2\Delta x}.$$

Equation (8.4) can be reformulated as follows

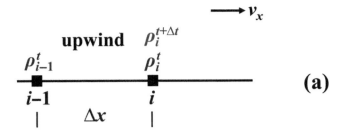

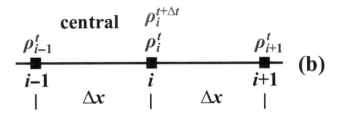

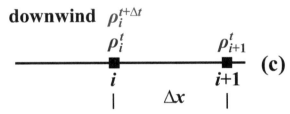

Fig. 8.2. Stencil of a 1D grid used for the discretization of the Eulerian advection equation with upwind (a), central (b) and downwind (c) differences.

$$\rho_i^{t+\Delta t} = \rho_i^t - \frac{v_x \Delta t}{2\Delta x}\left(2\rho_i^t - 2\rho_{i-1}^t + \rho_{i+1}^t - \rho_{i+1}^t\right), \tag{8.7a}$$

$$\frac{\rho_i^{t+\Delta t} - \rho_i^t}{\Delta t} = -v_x \frac{\rho_{i+1}^t - \rho_{i-1}^t}{2\Delta x} + \frac{v_x \Delta x}{2}\left(\frac{\rho_{i-1}^t - 2\rho_i^t + \rho_{i+1}^t}{\Delta x^2}\right), \tag{8.7b}$$

taking that

$$\left(\frac{\partial^2 \rho}{\partial x^2}\right)_i = \frac{\rho_{i-1}^t - 2\rho_i^t + \rho_{i+1}^t}{\Delta x^2}$$

we obtain

$$\left(\frac{\partial \rho}{\partial t}\right)_i = -v_x \left(\frac{\partial \rho}{\partial x}\right)_i^{central} + D\left(\frac{\partial^2 \rho}{\partial x^2}\right)_i, \tag{8.7c}$$

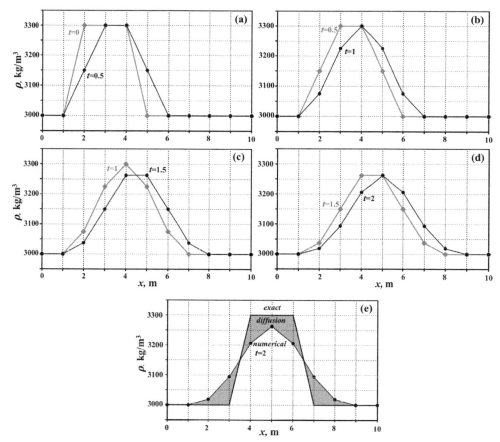

Fig. 8.3. Progress (a)–(d) in the development of numerical diffusion during four time steps ($\Delta t =$ ½$\Delta x/v_x$ = 0.5 s) in the case of 1D advection of a square density wave, with constant velocity (v_x = 1 m/s) on a regular Eulerian grid (Δx = 1 m).

where $D = v_x \Delta x/2$ is the numerical diffusion coefficient. This means that upwind differences inherently contain numerical diffusion compared to central differences. On the other hand, applying central and *downwind differences* for solving the Eulerian advection equation:

$$\text{central FD (Fig. 8.4b)} \quad \rho_i^{t+\Delta t} = \rho_i^t - v_x \Delta t \frac{\rho_{i+1}^t - \rho_{i-1}^t}{2\Delta x}, \tag{8.8}$$

$$\text{downwind FD (Fig. 8.4c)} \quad \rho_i^{t+\Delta t} = \rho_i^t - v_x \Delta t \frac{\rho_{i+1}^t - \rho_i^t}{\Delta x}, \tag{8.9}$$

results in a strong oscillation of the numerical solution compared to the exact one (Fig. 8.4b,c), which is even more dramatic than the numerical diffusion problem which is characteristic for upwind differences (compare Fig. 8.4a,b,c). The oscillations are

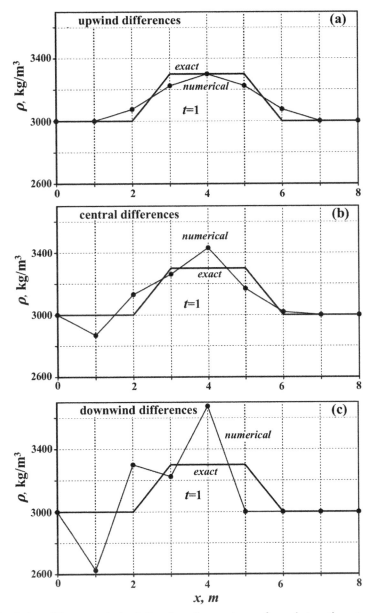

Fig. 8.4. Deviation of the numerical solution from the exact one after only two time steps in the case of upwind (a), central (b), and downwind (c) finite differences for advection of a square wave. The best results are, indeed, obtained by upwind differences, while downwind FD yield strong numerical oscillations. Parameters are as in Fig. 8.3.

caused by the erroneous evaluation of the spatial derivative of density. With both central and downwind differences (in contrast to upwind differences), we take into account the

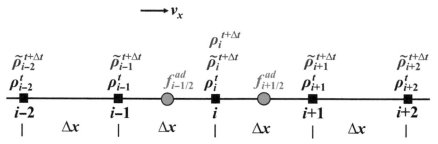

Fig. 8.5. Stencil of a 1D grid used for discretization of the Eulerian advection equation with the FCT algorithm.

density distribution in the *outgoing material flow* which is useless for predicting density distribution in the *incoming material flow.*

One way to minimize numerical diffusion for the Eulerian advection equation is to use higher order numerical schemes such as the *flux-corrected transport* algorithm (*FCT*, Boris and Book, 1973). The FCT is a conservative *shock-capturing* scheme that can, in particular, be used for solving the advection equation. An FCT algorithm consists of two stages, (1) a transport stage and (2) a flux-corrected anti-diffusion stage. The numerical diffusion errors introduced in the first stage are corrected by the anti-diffusion stage. The implementation of the FCT is more complex than for upwind FD and is based on more nodal points (Fig. 8.5). For the case of constant velocity and constant grid spacing, the algorithm of updating advected parameters (e.g. density) on an Eulerian grid is as follows.

(1) Transport stage – using a highly diffusive mass-conservative advection scheme (such as upwind differences, Eqs. 8.4, 8.7) to obtain preliminary values of density ($\tilde{\rho}_i^{t+\Delta t}$) at the next time instant $t+\Delta t$

$$\tilde{\rho}_i^{t+\Delta t} = \rho_i^t - v_x \Delta t \frac{\rho_{i+1}^t - \rho_{i-1}^t}{2\Delta x} + D\Delta t \left(\frac{\rho_{i-1}^t - 2\rho_i^t + \rho_{i+1}^t}{\Delta x^2} \right), \qquad (8.10)$$

where

$$D = \frac{\Delta x^2}{\Delta t} \left[\frac{1}{8} + \frac{1}{2} \left(\frac{v_x \Delta t}{\Delta x} \right)^2 \right]$$

is a numerical diffusion coefficient.

(2) Anti-diffusion stage – correcting the numerical diffusion introduced during the transport stage by defining anti-diffusive fluxes to the left ($f_{i-1/2}^{ad}$), and to the right ($f_{i+1/2}^{ad}$) of the nodal point i,

$$\rho_i^{t+\Delta t} = \widetilde{\rho}_i^{t+\Delta t} - f_{i+1/2}^{ad} + f_{i-1/2}^{ad}, \tag{8.11}$$

where

$$f_{i-1/2}^{ad} = S_{i-1/2} \max\left[0, \min\left(S_{i-1/2}\Delta\widetilde{\rho}_{i-3/2}^{t+\Delta t}, \frac{1}{8}\left|\Delta\widetilde{\rho}_{i-1/2}^{t+\Delta t}\right|, S_{i-1/2}\Delta\widetilde{\rho}_{i+1/2}^{t+\Delta t}\right)\right],$$

$$f_{i+1/2}^{ad} = S_{i+1/2} \max\left[0, \min\left(S_{i+1/2}\Delta\widetilde{\rho}_{i-1/2}^{t+\Delta t}, \frac{1}{8}\left|\Delta\widetilde{\rho}_{i+1/2}^{t+\Delta t}\right|, S_{i+1/2}\Delta\widetilde{\rho}_{i+3/2}^{t+\Delta t}\right)\right],$$

$$\Delta\widetilde{\rho}_{i-3/2}^{t+\Delta t} = \widetilde{\rho}_{i-1}^{t+\Delta t} - \widetilde{\rho}_{i-2}^{t+\Delta t},$$

$$\Delta\widetilde{\rho}_{i-1/2}^{t+\Delta t} = \widetilde{\rho}_{i}^{t+\Delta t} - \widetilde{\rho}_{i-1}^{t+\Delta t},$$

$$\Delta\widetilde{\rho}_{i+1/2}^{t+\Delta t} = \widetilde{\rho}_{i+1}^{t+\Delta t} - \widetilde{\rho}_{i}^{t+\Delta t},$$

$$\Delta\widetilde{\rho}_{i+3/2}^{t+\Delta t} = \widetilde{\rho}_{i+2}^{t+\Delta t} - \widetilde{\rho}_{i+1}^{t+\Delta t},$$

$$S_{i-1/2} = \mathrm{sign}\left(\Delta\widetilde{\rho}_{i-1/2}^{t+\Delta t}\right),$$

$$S_{i+1/2} = \mathrm{sign}\left(\Delta\widetilde{\rho}_{i+1/2}^{t+\Delta t}\right).$$

Here, sign(*A*) is a function that gives −1, 0 and 1 when $A < 0$, $A = 0$ and $A > 0$, respectively.

The FCT algorithm stencil (Fig. 8.5) involves five nodal points in a 1D grid, compared to two nodal points in the case of upwind differences (Fig. 8.2a). FCT stabilizes the numerical diffusion (i.e. the advected wave shape stops changing after some amount of time steps) and gives noticeably better results compared to upwind differences (compare Fig. 8.6a and b). These results, however, are still not perfect and depend on the exact shape of the advected structures (e.g. triangular waves are subjected to a noticeable decrease in amplitude compared to square waves, Fig. 8.6b). It should also be mentioned that various existing higher order Eulerian schemes such as FCT do not eliminate numerical diffusion completely, but rather stabilize it to a certain acceptable level (Fig. 8.6b).

8.3 Marker-in-cell techniques

If numerical diffusion needs to be strongly minimized, Lagrangian and Eulerian-Lagrangian advection algorithms can be used. In geodynamic modelling, very accurate advection of non-diffusive properties such as rock type (composition) with strongly discontinuous (e.g., layering) distribution in space is often required. One of the most popular methods in this case is to combine the use of Lagrangian advecting points (*markers,*

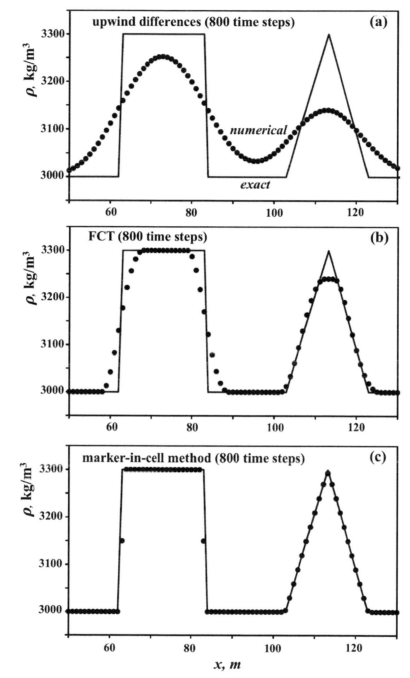

Fig. 8.6. 1D advection of square and triangular density waves with upwind differences (a), FCT (b) and marker-in-cell method (c). The results have been obtained with the programs **Upwind_1D.m**, **FCT_1D.m** and **Markers_1D.m** associated with this chapter.

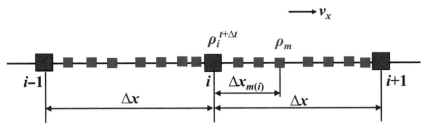

Fig. 8.7. Stencil of a 1D Eulerian-Lagrangian grid used for advection with the marker-in-cell technique. Mobile Lagrangian markers (red squares) move with a prescribed velocity v_x and carry information on material density. At every time step, the density at the Eulerian nodes (blue squares) is interpolated from markers found within two grid cells around the nodes.

tracers or particles) with an immobile, Eulerian grid (e.g. Woidt, 1978; Christensen, 1982; Schmeling, 1987; Weinberg and Schmeling, 1992). In this approach, properties are initially distributed on a large amount of Lagrangian points that are advected according to a given/ computed velocity field. The advected material properties (e.g. density) are then interpolated from the displaced Lagrangian points to the Eulerian grid (Fig. 8.7) by using a weighted-distance averaging such as the following linear interpolation formula

$$\rho_i^{t+\Delta t} = \frac{\sum\limits_{m} \rho_m w_{m(i)}}{\sum\limits_{m} w_{m(i)}}, \qquad (8.12)$$

$$w_{m(i)} = 1 - \frac{\Delta x_{m(i)}}{\Delta x},$$

where $w_{m(i)}$ is the weight of the mth marker for the ith node, $\Delta x_{m(i)}$ denotes the distance from the mth marker to the ith node. In 1D, the density at an Eulerian node is interpolated only with the markers found in the two surrounding cells (i.e. within one grid step distance from the node). For a more local interpolation, a shorter limiting distance from the node (i.e. within half grid step distance) with fewer markers can be used.

This method is called the *marker-in-cell* (*MIC*) technique. Obviously, the results of pure advection (e.g. of density, Fig. 8.6c) with MIC are not subjected to any numerical diffusion (with the exception of non-accumulating interpolation errors between Lagrangian markers and Eulerian nodes) since markers always retain their original density values and only change positions with time.

8.4 Marker advection schemes

In order to move a Lagrangian marker A, different advection schemes can be used. The most simple is the *first-order accurate* advection scheme

$$x_A^{t+\Delta t} = x_A^t + v_{xA}\Delta t, \tag{8.13a}$$

$$y_A^{t+\Delta t} = y_A^t + v_{yA}\Delta t, \tag{8.13b}$$

where x_A^t and y_A^t are the coordinates of marker A at the current time (t); $x_A^{t+\Delta t}$ and $y_A^{t+\Delta t}$ are the coordinates of the same marker at the next moment in time ($t + \Delta t$); v_{xA} and v_{yA} are components of the material velocity vector at the point A, at time t. The velocity of the Lagrangian point A can significantly change during the displacement if there is a strong spatial variation of the velocity field. In this case, a first-order advection scheme will not be very accurate. This problem can be rectified, by either using smaller time steps (Δt), or by using *higher order advection schemes*.

One of the most popular methods in geodynamic modelling is the *Runge–Kutta advection scheme*

$$x_A^{t+\Delta t} = x_A^t + v_x^{eff}\Delta t, \tag{8.14a}$$

$$y_A^{t+\Delta t} = y_A^t + v_y^{eff}\Delta t, \tag{8.14b}$$

where v_x^{eff} and v_y^{eff} are components of the effective material velocity vector for the point A over the period between the current (t) and next ($t + \Delta t$) moments of time. Components of the effective material velocity are computed by using the material velocity at several different points in space (varying from 2 to 4, depending on the order of the scheme).

The *second-order Runge–Kutta scheme* uses two points (A and B)

$$v_x^{eff} = v_{xB}, \tag{8.15a}$$

$$v_y^{eff} = v_{yB}, \tag{8.15b}$$

where the coordinates of point B are computed as

$$x_B = x_A^t + v_{xA}\frac{\Delta t}{2}, \quad y_B = y_A^t + v_{yA}\frac{\Delta t}{2}. \tag{8.15c}$$

The *third-order Runge–Kutta scheme* uses three points (A, B and C)

$$v_x^{eff} = \frac{1}{6}(v_{xA} + 4v_{xB} + v_{xC}), \tag{8.16a}$$

$$v_y^{eff} = \frac{1}{6}(v_{yA} + 4v_{yB} + v_{yC}), \tag{8.16b}$$

where the coordinates of points B and C are computed as

$$x_B = x_A^t + v_{xA}\frac{\Delta t}{2}, \quad y_B = y_A^t + v_{yA}\frac{\Delta t}{2}, \tag{8.16c}$$

$$x_C = x_A^t + (2v_{xB} - v_{xA})\Delta t, \quad y_C = y_A^t + (2v_{yB} - v_{yA})\Delta t. \tag{8.16d}$$

And finally, the *classical fourth-order Runge–Kutta scheme* uses four points (*A*, *B*, *C* and *D*)

$$v_x^{eff} = \frac{1}{6}(v_{xA} + 2v_{xB} + 2v_{xC} + v_{xD}), \tag{8.17a}$$

$$v_y^{eff} = \frac{1}{6}(v_{yA} + 2v_{yB} + 2v_{yC} + v_{yD}), \tag{8.17b}$$

where the coordinates of points *B*, *C* and *D* are computed as

$$x_B = x_A^t + v_{xA}\frac{\Delta t}{2}, \quad y_B = y_A^t + v_{yA}\frac{\Delta t}{2}, \tag{8.17c}$$

$$x_C = x_A^t + v_{xB}\frac{\Delta t}{2}, \quad y_C = y_A^t + v_{yB}\frac{\Delta t}{2}, \tag{8.17d}$$

$$x_D = x_A^t + v_{xC}\Delta t, \quad y_D = y_A^t + v_{yC}\Delta t. \tag{8.17e}$$

The last advection scheme is very accurate in space (fourth-order accuracy) but less accurate in time (first-order accuracy) if the velocity field for the current moment of time (i.e. *A* configuration velocity field) is used for *B*, *C* and *D* points. An alternative algorithm is to use the Runge–Kutta scheme in both space and time: all markers are first displaced to their *B* points and then material properties are re-interpolated to Eulerian nodes and a new *B* configuration velocity field is computed by solving the momentum and continuity equations. This *B* velocity field is then used for moving markers to their *C* points for computing a *C* configuration velocity field etc. Obviously this approach is computationally more expensive.

Runge–Kutta advection schemes (8.13)–(8.17) are two dimensional. For 3D advection problems, extra coordinates (z_A, z_B, z_C, z_D) and velocity components (v_{zA}, v_{zB}, v_{zC}, v_{zD}, v_z^{eff}) have to be added to these schemes (write 3D Runge–Kutta advection schemes as an exercise).

8.5 Interpolation of physical parameters between markers and nodes

Various interpolation schemes can be used to interpolate physical properties (e.g., density, viscosity, heat capacity) from the Lagrangian markers to the Eulerian nodes. The following standard first-order accurate bilinear scheme is often used to calculate an interpolated value of a parameter $B_{(i,j)}$ for the *ij*th node using values (B_m) assigned to all markers found in the four surrounding cells (Fig. 8.8)

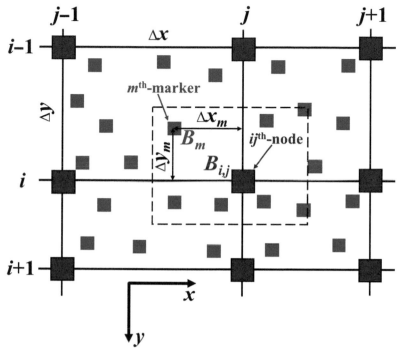

Fig. 8.8. Stencil of a 2D grid used for the interpolation of physical properties from the markers to the nodes (Gerya and Yuen, 2003a). The dashed boundary indicates the area from which markers are used for interpolating properties to node (i,j) in the case of a local interpolation scheme.

$$B_{i,j} - \frac{\sum_m B_m w_{m(i,j)}}{\sum_m w_{m(i,j)}}, \qquad (8.18)$$

$$w_{m(i,j)} = \left(1 - \frac{\Delta x_m}{\Delta x}\right) \times \left(1 - \frac{\Delta y_m}{\Delta y}\right),$$

where $w_{m(i,j)}$ represents a distance-dependent weight of the mth marker for the ijth node; Δx_m and Δy_m are the distances from the mth marker to the ijth node. It is worth mentioning that the use of higher order interpolation schemes (e.g. Fornberg, 1995) produces undesirable numerical fluctuations in scalar, vector and tensor properties interpolated at the proximity of sharp transitions. This scenario frequently occurs in geodynamic models, hence the first-order interpolation is preferred. A more local interpolation from markers to nodes can again be obtained by using fewer markers located within a limited (e.g. half grid step, see dashed rectangle in Fig. 8.8) range of the vertical and horizontal distances around an Eulerian node. Such a local interpolation of properties from markers to nodes has in

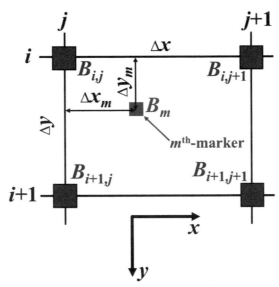

Fig. 8.9. Stencil of a 2D grid used for the interpolation of physical properties from nodes to markers (Gerya and Yuen, 2003a).

many cases (but not always) an effect similar to increasing the Eulerian grid resolution (Duretz et al., 2011a).

Interpolation of physical parameters (e.g., velocity, pressure) from the Eulerian nodes to Lagrangian markers and other geometrical points is also commonly required. One of the simplest 2D interpolation methods is to use values of the physical parameter B, defined at the four Eulerian nodes surrounding a given marker (or any other geometric point). An effective value of the parameter B for the mth marker can then be calculated using the first-order bilinear interpolation scheme as follows

$$B_m = B_{i,j}\left(1 - \frac{\Delta x_m}{\Delta x}\right)\left(1 - \frac{\Delta y_m}{\Delta y}\right) + B_{i,j+1}\frac{\Delta x_m}{\Delta x}\left(1 - \frac{\Delta y_m}{\Delta y}\right)$$
$$+ B_{i+1,j}\left(1 - \frac{\Delta x_m}{\Delta x}\right)\frac{\Delta y_m}{\Delta y} + B_{i+1,j+1}\frac{\Delta x_m \Delta y_m}{\Delta x \Delta y},$$

$$(8.19)$$

where B_m denotes the value of the parameter B for the mth marker.

In the case of a regularly spaced grid, the indexes of the four nodal points that surround a given marker (Fig. 8.9) can be calculated directly from marker coordinates by normalizing them to a grid step size (Exercise 8.2). In the case of an irregularly spaced grid, these indexes have to be found on the basis of some efficient search procedure; for example a bisection procedure can be used, which allows the indexes of the two nearest vertical and

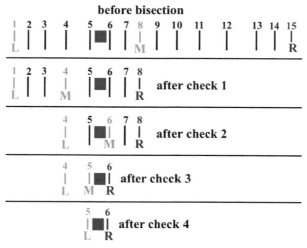

Fig. 8.10. Finding two nearest grid lines around the marker (red square) by a bisection algorithm in the case of an irregularly spaced grid. The search starts by having the first (line 1) and last (line 15) lines of the grid as leftmost (L = 1) and rightmost (R = 15) limits for the bisection. At each check, the middle line M = (L+R)/2 is defined and its horizontal position is compared with that of the marker. Depending on the result of the comparison, line M becomes either L (when M is to the left of the marker) or R (when M is to the right of the marker). The check continues until the difference in the index between L and R becomes one. The required number of checks (*n*) is given by the relation $2^n = N$, where *N* is the number of grid lines.

horizontal grid lines that bound the cell which contains the marker to be defined (Fig. 8.10, Exercise 8.3). An alternative, faster approach (which however requires additional memory) is to store the unique cell index for each marker and update it at every time step by checking only the nearest grid lines.

It should be mentioned that performing interpolation between nodes and markers does introduce numerical diffusion. This problem becomes particularly significant when it is required to interpolate a time-dependent physical parameter (e.g., temperature or stress) back and forth between the markers and nodes. Such diffusion can however be minimized by *interpolating increments rather than total values of the time-dependent parameter* from the nodes to the markers (Figs. 8.9, 8.11):

$$
\begin{aligned}
B_m^{t+\Delta t} = B_m^t &+ \left(B_{i,j}^{t+\Delta t} - B_{i,j}^t\right)\left(1 - \frac{\Delta x_m}{\Delta x}\right)\left(1 - \frac{\Delta y_m}{\Delta y}\right) + \left(B_{i,j+1}^{t+\Delta t} - B_{i,j+1}^t\right)\frac{\Delta x_m}{\Delta x}\left(1 - \frac{\Delta y_m}{\Delta y}\right) \\
&+ \left(B_{i+1,j}^{t+\Delta t} - B_{i+1,j}^t\right)\left(1 - \frac{\Delta x_m}{\Delta x}\right)\frac{\Delta y_m}{\Delta y} + \left(B_{i+1,j+1}^{t+\Delta t} - B_{i+1,j+1}^t\right)\frac{\Delta x_m \Delta y_m}{\Delta x \Delta y},
\end{aligned}
\tag{8.20}
$$

where indices *t* and *t*+Δ*t* correspond to the current and next time instant, respectively.

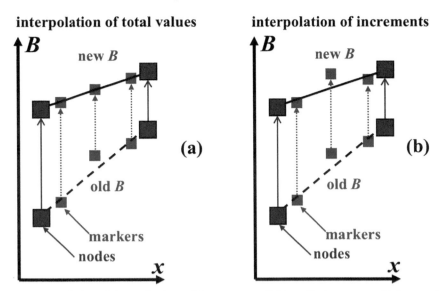

Fig. 8.11. Interpolation of the total values of the parameter B (a) and its increments (b) from nodes to markers. Note that the interpolation of increments does not smooth out subgrid variations of B onto the markers.

8.6 Continuity-based velocity interpolation

One potential problem with marker advection is that simple bilinear interpolation of velocity components to markers does not always respect accurately the *continuity condition* (e.g., $\mathrm{div}(\vec{v}) = 0$ condition for incompressible fluid). In the case of large deformation, this may result in an artificial local convergence or divergence of markers producing significant gaps in their distribution (Fig. 8.12). This effect is particularly strong in the case of simple shear parallel to cell diagonals (Fig. 8.12b). We can minimize this problem by using various *continuity-based higher order velocity interpolation schemes* (e.g. Jenny et al., 2001; Meyer and Jenny, 2004; Wang et al., 2015; Pusok et al., 2017). These schemes ensure that not only velocities but also normal strain rates (i.e., velocity divergence components) are properly interpolated to markers, thereby satisfying the continuity condition. One efficient approach for the staggered grid consists in the use of average v_x and v_y velocity components computed at cell centres (i.e., at pressure nodes). The velocity boundary conditions have to be applied to these average velocities by using external cells outside of the grid (cf. external pressure nodes, Fig. 7.15). Bilinear interpolation of the averaged velocity components from the cell centres (including external cells) to moving markers respects the continuity condition notably better than the direct interpolation of these components from staggered velocity nodes (cf., Figs. 8.12b,c). This is because the velocity averaging increases the order of the scheme from the first to the second (i.e. interpolation stencil increases from one to two cells) in the x direction for v_x velocity and in the y direction

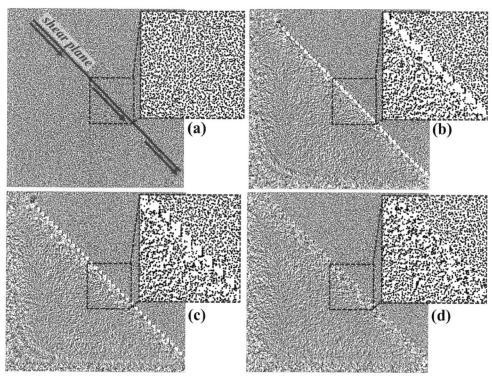

Fig. 8.12. Influence of velocity interpolation schemes on marker distribution. (a) Initial configuration. A grid resolution of 41 × 41 nodes with 40 000 randomly distributed markers is used. Red solid line shows the position of the diagonal two-cell wide shear plane prescribed with the use of internal velocity boundary conditions. Red arrows show the velocity direction prescribed below the plane. Velocity atop the plane is set to zero. (b), (c), (d) Marker distribution after 1000 time steps with marker displacement limit of 0.1 grid step per time step. (b) Velocity components are interpolated to markers from staggered nodes. (c) Averaged velocity components are interpolated to markers from cell centres (pressure nodes). (d) Velocity components are interpolated to markers from both staggered nodes and cell centres weighted as 2/3 and 1/3, respectively. The models are computed with the code **Markers_divergence.m** associated with this chapter.

for v_y velocity. Various combinations between the staggered and averaged velocity components can also be used. A particularly efficient empirical scheme combines 1/3 of velocity interpolated from the cell centres with 2/3 of velocity interpolated from staggered nodes (Fig. 8.12d).

8.7 'Sticky air' approach

The marker-in-cell method can also be used for a simplified implementation of the free surface condition based on the 'sticky air' method (e.g. Matsumoto and Tomoda,

1983; Gerya and Yuen, 2003b; Schmeling et al., 2008; Crameri et al. 2012a). With this method, a low density, low viscosity fluid layer is introduced above the crust at the top of the model. The fluid/crust interface operates as an internal free surface given that viscosity of the fluid layer is taken sufficiently low and its thickness is sufficiently large (Crameri et al., 2012a). Density of the fluid layer corresponds to either air $(1 \ kg/m^3)$ or water $(1000 \ kg/m^3)$. On the other hand, fluid viscosity is typically taken many orders of magnitudes larger than that of air or water – this is why it is called 'sticky air'. Viscosity and thickness of the 'sticky air' should simply ensure that it always deforms at very low stresses, thereby not creating any noticeable resistance to the deformation of the 'air'/crust interface, which thus behaves as an internal free surface. It is found that the 'sticky air' approach works well if the following condition is satisfied (Crameri et al., 2012a)

$$\frac{\eta_{st}}{\eta_{ch}} \left(\frac{L}{h_{st}} \right)^3 \ll 1, \tag{8.21}$$

where η_{st} and h_{st} are the viscosity and thickness of the 'sticky air' layer, and η_{ch} and L are the characteristic viscosity and length scale of the model, respectively. Deformation of the internal free surface is enabled by advection of 'sticky air' and crust markers. Topography of the 'air'/crust interface can also be traced by using various Eulerian or Lagrangian approaches, which will be discussed in Chapter 21.

The main problem with the internal free surface (as well as with any other free surface conditions) is that it may become numerically unstable during deformation and start to oscillate back and forth around an equilibrium free surface configuration (Fig. 8.13a). In geodynamic modelling, this peculiar free surface behaviour is often called *'drunken sailor'* instability (Kaus et al., 2010). The main reason for the 'drunken sailor' instability is inaccurate (*explicit*) time integration of the *moving free surface position*, which does not take into account changes in material velocity *during the time step*. Reducing this instability to a reasonably low level often requires significant reduction (up to 30 times compared to the free slip condition without the 'sticky' air layer, Kaus et al., 2010) of the marker displacement magnitude and respectively of the time step, which thus significantly slows down calculations with a free surface.

In the case of the 'sticky air' approach, the problem of 'drunken sailor' instability can be efficiently resolved by taking into account *implicitly* advection-related density changes at Eulerian nodes (Kaus et al., 2010; Duretz et al., 2011a). In order to do so, we can approximate the density value of the gravity term in the Stokes equation for the end (rather than for the beginning) of the time step based on the Eulerian advection equation (8.1)

$$\rho_{\Delta t} = \rho + \frac{\partial \rho}{\partial t} \Delta t = \rho - v_x \frac{\partial \rho}{\partial x} \Delta t - v_y \frac{\partial \rho}{\partial y} \Delta t, \tag{8.22}$$

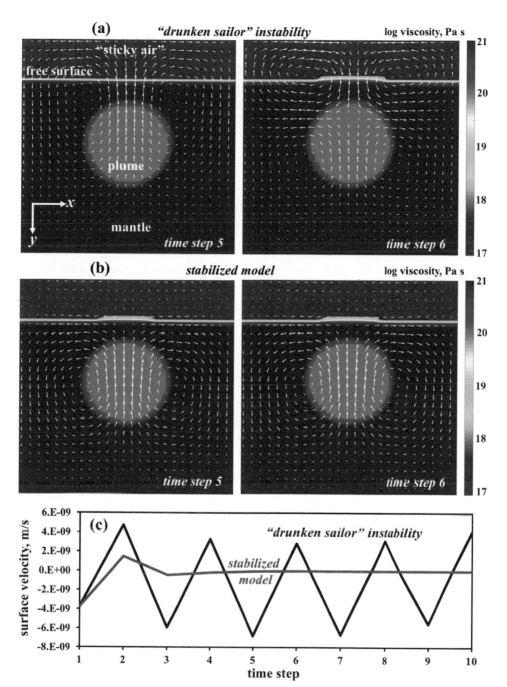

Fig. 8.13. Behaviour of free surface in the case of the 'sticky air' method. (a) Unstable model showing surface velocity oscillations ('drunken sailor instability'). (b) Stabilized model. (c) Comparison of surface velocity evolution atop the plume for the unstable (black line) and stabilized (red line) models. Grid resolution is 51 × 51 nodes with 40 000 randomly distributed markers. Model size is 500 × 500 km². Plume radius is 100 km. Density of materials: 'sticky air' 1 kg/m³, mantle 3300 kg/m³, plume 3200 kg/m³. Marker displacement per time step is limited by a 0.5 grid step for both models. White arrows in (a), (b) show velocity distribution. The models are computed with the code **Sticky_air.m** associated with this chapter.

where ρ and $\rho_{\Delta t}$ are density in Eulerian nodes for the beginning and the end of the time step Δt, respectively, v_x, v_y are (yet unknown) material velocity components. In the case of purely vertical gravity, the y Stokes equation is then modified as

$$\frac{\partial \sigma'_{yy}}{\partial y} + \frac{\partial \sigma'_{yx}}{\partial x} - \frac{\partial P}{\partial y} + \rho_{\Delta t} g_y = 0, \tag{8.23a}$$

or

$$\frac{\partial \sigma'_{yy}}{\partial y} + \frac{\partial \sigma'_{yx}}{\partial x} - \frac{\partial P}{\partial y} - g_y \Delta t \left(v_x \frac{\partial \rho}{\partial x} + v_y \frac{\partial \rho}{\partial y} \right) = -\rho g_y, \tag{8.23b}$$

which assumes that the influence of changes of deviatoric stress gradients and pressure gradients during the time step for the force balance at Eulerian nodes is negligible compared to the influence of density changes ($\rho_{\Delta t} - \rho$). Components of the density gradient $\partial \rho / \partial x$ and $\partial \rho / \partial y$ have to be evaluated at the same v_y velocity nodes, where the y Stokes equation is discretized. Test calculations show (Fig. 8.13) that the proposed free surface stabilization efficiently eliminates the problem of 'drunken sailor' instability and allows the use of large time steps in combination with the 'sticky air' approach.

Programming exercises

Exercise 8.1.
Program and compare the simple Eulerian advection schemes (upwind, central and downwind FD, Eqs. (8.4), (8.8), (8.9)) in 1D for the case illustrated in Fig. 8.6. The model resolution is 151 nodal points. Other parameters are the same as in Fig. 8.6. An example is in **Upwind_1D.m**.

Exercise 8.2.
Program and compare the FCT method and the marker-in-cell schemes in 1D for the same model. Use 5 markers per cell (200 markers for the entire model) and 'recycle' markers which leave the model from one side by adding them to the other side (periodic boundary condition). To recycle markers, update their coordinate as follows. For markers that leave the model through the right boundary and appear at the left boundary, set

$$x_m^{recycled} = x_m - L. \tag{8.24a}$$

For markers that leave the model through the left boundary and enter it at the right boundary, set

$$x_m^{recycled} = x_m + L, \tag{8.24b}$$

where x_m, $x_m^{recycled}$ are marker coordinates before and after recycling, respectively. Use the following formula to define the index j (smallest index is 1) of the nearest node to the left of the marker from its coordinate

$$j = \mathrm{int}\left(\frac{x_m}{\Delta x}\right) + 1, \qquad (8.25)$$

where x_m is the marker coordinate, Δx is the nodal (Eulerian) grid space, int() is a function which returns the integer part of a value. Possible MATLAB implementation of Eq. (8.25) is

$$j = \mathrm{fix}(xm/dx) + 1.$$

Examples are in **FCT_1D.m** and **Markers_1D.m**.

Exercise 8.3.
Modify the previous marker-based code by introducing non-uniform distances between nodal points and markers and by using a bisection algorithm (Fig. 8.10) to define indices of the two nearest nodes. Prescribe a slightly variable velocity on the nodal points and interpolate it to markers when displacing them. An example is in **Markers_1Dirregular.m**.

Exercise 8.4.
Modify the 2D model with variable viscosity (Exercise 7.2) by including the advection of density and viscosity fields with markers. Use a 500×500 km^2 model with 51×51 grid and free slip boundaries. Initial model geometry corresponds to the mantle ($\eta = 10^{21}$ Pa s, $\rho = 3300$ kg/m^3) with a central circular plume of 100 km radius ($\eta = 10^{20}$ Pa s, $\rho = 3200$ kg/m^3). Create a grid of 200×200 markers (i.e., 4×4 markers per cell) with small (up to 1/2 of the marker grid distance) random displacements (*rand*) relative to regular positions. Randomization is introduced to prevent the opening of big gaps between markers during the simulation. Such gaps often occur if regular marker grids are used (e.g. due to pure shear related stretching of the distances between regularly distributed markers). Save the horizontal and vertical coordinates for every marker and assign them with the density and viscosity depending on the position in either the mantle or the plume. An alternative approach, which requires less memory, is to assign every marker with a *material type index* depending on the initial position (i.e. 1 and 2 for the mantle and plume, respectively). Then the marker density and viscosity (and potentially any other material-dependent property) can be estimated based on the material type index. Interpolate the marker density and viscosity (η_s in Fig. 7.19) to the basic nodes of the grid using Eq. (8.18). Write a loop over the markers and add the density and viscosity of each marker to the four surrounding nodes (Fig. 8.9). For computing i and j indices for the upper left node next to the marker (Fig. 8.9), apply Eq. (8.25) separately for each coordinate. Compute the viscosity for the centres of cells (η_n in Fig. 7.17) by averaging the viscosity from the four surrounding basic nodes with harmonic average (an alternative way is to interpolate this viscosity directly from markers based on Eq. 8.18). After obtaining a velocity field, define a time step in such

a manner that the marker displacements do not exceed half the grid step. Interpolate the v_x and v_y velocity components from the staggered nodes (Eq. 8.19) to the markers and displace them using a first-order accurate scheme (Eq. 8.13). Note that for staggered nodes, Eq. (8.25) is modified since these nodes are shifted by half of the grid distance relative to the basic ones (Fig. 7.15)

$$i = \text{int}\left(\frac{y_m + \Delta y/2}{\Delta y}\right) + 1 \quad \text{for } v_x \text{ nodes,} \tag{8.26a}$$

$$j = \text{int}\left(\frac{x_m + \Delta x/2}{\Delta x}\right) + 1 \quad \text{for } v_y \text{ nodes.} \tag{8.26b}$$

After displacing all the markers, go to the next time step and interpolate the density and viscosity to the nodes using the new marker positions. An example is in **Stokes_Continuity_Markers.m**.

Exercise 8.5.
Update the 2D code developed above to use the classical fourth-order Runge–Kutta scheme (Eqs. 8.14, 8.17) for marker displacement. An example is in **Stokes_Continuity_Markers_Runge_Kutta.m**.

Exercise 8.6.
Update the 2D code developed above by programming a continuity-based higher order velocity interpolation scheme discussed in this chapter. Compute the average v_x and v_y velocity components at cell centres (i.e., at pressure nodes) inside the grid. Compute the velocity at the external pressure points (Fig. 7.15) by applying the (free slip) velocity boundary conditions to the computed average velocities. Move markers with velocities interpolated from both staggered nodes (with weight of 2/3) and pressure nodes (with weight of 1/3). Use both internal and external nodes for velocity interpolation. An example is in **Stokes_continuity_based_advection.m**.

Exercise 8.7.
Update the 2D code developed above by defining with markers a 100 km thick top 'sticky air' layer ($\eta = 10^{17}$ Pa s, $\rho = 1$ kg/m³). Interpolate density from markers to v_y nodes (ρ_v in Fig. 8.14). Implement free surface stabilization with Eq. (8.23) by using a modified y Stokes stencil (Fig. 8.14)

$$2\eta_{\text{n}(i+1,j)}\frac{v_{y4} - v_{y3}}{\Delta y^2} - 2\eta_{\text{n}(i,j)}\frac{v_{y3} - v_{y2}}{\Delta y^2} + \eta_{\text{s}(i,j)}\left(\frac{v_{y5} - v_{y3}}{\Delta x^2} + \frac{v_{x4} - v_{x3}}{\Delta x \Delta y}\right)$$

$$- \eta_{\text{s}(i,j-1)}\left(\frac{v_{y3} - v_{y1}}{\Delta x^2} + \frac{v_{x2} - v_{x1}}{\Delta x \Delta y}\right) - K_{cont}\frac{P'_2 - P'_1}{\Delta y}$$

$$- g_y \Delta t \left[(v_{x1} + v_{x2} + v_{x3} + v_{x4})\frac{\rho_{\text{v}(i,j+1)} - \rho_{\text{v}(i,j-1)}}{8\Delta x}\right.$$

$$\left. + v_{y3}\frac{\rho_{\text{v}(i+1,j)} - \rho_{\text{v}(i-1,j)}}{2\Delta x}\right] = -g_y \rho_{\text{v}(i,j)}, \tag{8.27}$$

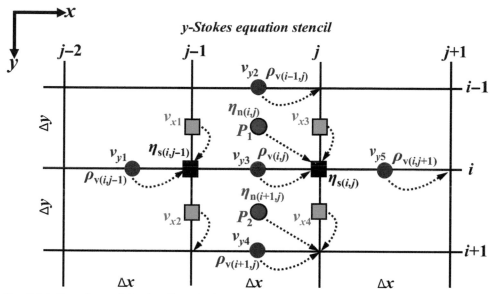

Fig. 8.14. Stencils used for the discretization of the stabilized *y*-Stokes equation (8.23) on a 2D regular staggered grid for models with variable viscosity based on pressure-velocity formulation. Indexing of gridlines corresponds to basic nodal points. Global indexing of different unknowns (cf. dotted arrows) is made according to Fig. 7.15.

where $K_{cont} = 2\eta_{\min}/(\Delta x + \Delta y)$. Set up the initial time step $\Delta t = 0$. Use Δt from the previous time step in the *y* Stokes equation (8.27). Compare solutions with and without the free surface stabilization (Fig. 8.13). An example is in **Sticky_air.m**.

9

The heat conservation equation

> **Theory:** Fourier's law of heat conduction. Heat conservation equation and its derivation. Radioactive, viscous, adiabatic and latent heating. Heat conservation equation for the case of constant thermal conductivity and its relation to the Poisson equation. Heat diffusion time scale.
>
> **Exercises:** Computing shear and adiabatic heating distribution for buoyancy-driven flow.

9.1 Fourier's law of heat conduction

Heat transport plays a crucial role in geodynamic processes and is often intrinsically coupled to deformation, as for example in mantle convection, granitic cupola growth, subduction etc. Let us first study the equations relevant to heat transport processes. The most basic one is Fourier's law of heat conduction, which relates the heat flux q, (W/m^2) to the temperature gradient $\partial T/\partial x (K/m)$

$$q = -k\frac{\partial T}{\partial x}, \tag{9.1}$$

where k $(W/(m\,K))$ is the thermal conductivity. Thermal conductivity may depend on P, T, composition and structure of the material. The heat flux q is the amount of heat that passes through a unit surface area, per unit time. As we know, heat is always transferred from a hot body to a colder one. This is reflected by the minus sign in the right part of Eq. (9.1), which implies that heat fluxes in the direction of decreasing temperature, i.e. the heat flux is positive when the temperature gradient $\partial T/\partial x$ is negative. In three dimensions, the heat flux is a vector that can be decomposed into three components

$$\vec{q} = (q_x,\ q_y,\ q_z).$$

In this case, Fourier's law relates heat fluxes in different directions to the respective temperature gradient components

$$\vec{q} = -k\nabla T \text{ or } q_i = -k\frac{\partial T}{\partial x_i},$$ (9.2)

where i is a coordinate index and x_i is a spatial coordinate, or

$$q_x = -k\frac{\partial T}{\partial x},$$

$$q_y = -k\frac{\partial T}{\partial y},$$

$$q_z = -k\frac{\partial T}{\partial z}.$$

9.2 Heat conservation equation

In order to predict changes in temperature due to heat transport, the *heat conservation equation*, also called *temperature equation*, has to be solved. This equation describes the balance of heat in a continuum and relates temperature changes to heat *generation*, *advection* and *conduction*. The Lagrangian temperature equation has the following form

$$\rho C_P \frac{DT}{Dt} = -\frac{\partial q_i}{\partial x_i} + H,$$ (9.3a)

where the repeated index i means a *summation* of derivatives (i.e., divergence) of heat flux components by respective coordinates (x, y, z) so that

$$\rho C_P \frac{DT}{Dt} = -\frac{\partial q_x}{\partial x} - \frac{\partial q_y}{\partial y} - \frac{\partial q_z}{\partial z} + H,$$ (9.3b)

or by using Eq. (9.2)

$$\rho C_P \frac{DT}{Dt} = \frac{\partial}{\partial x}\left(k\frac{\partial T}{\partial x}\right) + \frac{\partial}{\partial y}\left(k\frac{\partial T}{\partial y}\right) + \frac{\partial}{\partial z}\left(k\frac{\partial T}{\partial z}\right) + H,$$ (9.3c)

where ρ is density (kg/m^3); C_P is heat capacity at constant pressure (*isobaric heat capacity*, J/(kg K)); H is volumetric heat production/consumption (W/m^3). In fact, the product (ρC_P could be considered as *volumetric isobaric heat capacity* (J/(kg m^3)), which is also a useful physical quantity that we will use in the following chapters. DT/Dt is the substantive time derivative of temperature corresponding to the standard Lagrangian-Eulerian relation, which was already discussed in Chapters 1 and 5

$$\frac{DT}{Dt} = \frac{\partial T}{\partial t} + \vec{v} \cdot \text{grad}(T).$$

For example, in 3D

$$\frac{DT}{Dt} = \frac{\partial T}{\partial t} + v_x \frac{\partial T}{\partial x} + v_y \frac{\partial T}{\partial y} + v_z \frac{\partial T}{\partial z}.$$

Accordingly, the temperature equation in an Eulerian form can be written as follows

$$\rho C_P \left(\frac{\partial T}{\partial t} + \vec{v} \cdot \text{grad}(T) \right) = -\frac{\partial q_i}{\partial x_i} + H, \tag{9.4}$$

or in complete 3D form as,

$$\rho C_P \left(\frac{\partial T}{\partial t} + v_x \frac{\partial T}{\partial x} + v_y \frac{\partial T}{\partial y} + v_z \frac{\partial T}{\partial z} \right) = -\frac{\partial q_x}{\partial x} - \frac{\partial q_y}{\partial y} - \frac{\partial q_z}{\partial z} + H,$$

or by using Eq. (9.2) as

$$\rho C_P \left(\frac{\partial T}{\partial t} + v_x \frac{\partial T}{\partial x} + v_y \frac{\partial T}{\partial y} + v_z \frac{\partial T}{\partial z} \right) = \frac{\partial}{\partial x} \left(k \frac{\partial T}{\partial x} \right) + \frac{\partial}{\partial y} \left(k \frac{\partial T}{\partial y} \right) + \frac{\partial}{\partial z} \left(k \frac{\partial T}{\partial z} \right) + H.$$

The Lagrangian heat conservation equation can be derived by analysing heat fluxes through a small moving rectangular Lagrangian (material) volume of mass m and with dimensions Δx, Δy and Δz (Fig. 9.1). Let us assume that the initial temperature of this volume is T_0. Based on the orientation of coordinate axes, heat comes into the volume through the boundaries A, C and E and leaves the volume through the opposite boundaries B, D and F. In addition, some amount of heat ΔQ_{int} is generated inside the volume. Heat fluxes and heat generation change the amount of heat in the volume and after a small period of time Δt, the volume temperature changes to T_1. The amount of heat ΔQ required to change the temperature can be computed from the following thermodynamic relation

$$\Delta Q = m C_P \Delta T = m C_p (T_1 - T_0). \tag{9.5a}$$

In accordance with the energy conservation principle, this amount of heat should match the bulk effect of various heat sources and sinks in the volume such that

$$\Delta Q = \Delta Q_{int} + \Delta Q_A - \Delta Q_B + \Delta Q_C - \Delta Q_D + \Delta Q_E - \Delta Q_F, \tag{9.5b}$$

where ΔQ_A, ΔQ_B, ΔQ_C, ΔQ_D, ΔQ_E and ΔQ_F represent the amounts of heat that flux through the respective boundaries during the period of time Δt. These can be computed according to the definition of heat fluxes as

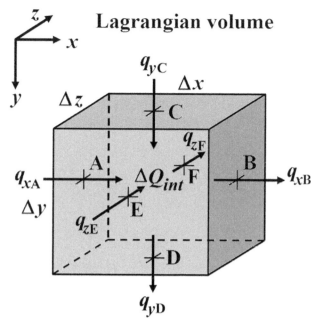

Fig. 9.1. Lagrangian elementary volume considered for derivation of the respective form of the heat conservation equation. Arrows show heat flux components responsible for *heat fluxes through* respective boundaries (A, B, C, D, E and F).

$$\Delta Q_A = q_{xA}\Delta y\Delta z\Delta t,$$
$$\Delta Q_B = q_{xB}\Delta y\Delta z\Delta t,$$
$$\Delta Q_C = q_{yC}\Delta x\Delta z\Delta t,$$
$$\Delta Q_D = q_{yD}\Delta x\Delta z\Delta t,$$
$$\Delta Q_E = q_{zE}\Delta x\Delta y\Delta t,$$
$$\Delta Q_F = q_{zF}\Delta x\Delta y\Delta t,$$

(9.6)

where q_{xA}, q_{xB}, q_{xC}, q_{xD}, q_{xE} and q_{zF} are the heat flux vector components responsible for *heat transport through* respective boundaries (Fig. 9.1).

Equating the right hand sides of Eqs. (9.4) and (9.5) and dividing through by Δt and the volume $V = \Delta x\Delta y\Delta z$, we obtain the following equation for the energy conservation in a Lagrangian volume (verify as an exercise)

$$\frac{m}{V}C_P\frac{\Delta T}{\Delta t} = -\frac{(q_{xB} - q_{xA})}{\Delta x} - \frac{(q_{yD} - q_{yC})}{\Delta y} - \frac{(q_{zF} - q_{zE})}{\Delta z} + \frac{\Delta Q_{int}}{V\Delta t},$$

(9.7a)

or in a different notation

$$\frac{m}{V} C_P \frac{\Delta T}{\Delta t} = -\frac{\Delta q_{xBA}}{\Delta x} - \frac{\Delta q_{yDC}}{\Delta y} - \frac{\Delta q_{zFE}}{\Delta z} + \frac{\Delta Q_{int}}{V \Delta t},$$ (9.7b)

where Δq_{xBA}, Δq_{yDC} and Δq_{zFE} are the changes (differences) in respective heat fluxes between respective boundaries. Taking into account the relationships $\rho = m/V$, and $H = (1/V)(DQ_{int}/Dt)$, and further assuming that Δt, Δx, Δy and Δz all tend towards zero, the differences in Eq. (9.7b) can be replaced by derivatives and we obtain the Lagrangian heat conservation equation

$$\rho C_P \frac{DT}{Dt} = -\frac{\partial q_x}{\partial x} - \frac{\partial q_y}{\partial y} - \frac{\partial q_z}{\partial z} + H.$$

9.3 Heat generation and consumption

There are several types of heat generation/consumption processes that should be taken into account in the temperature equation

$$\rho C_P \frac{DT}{Dt} = -\frac{\partial q_i}{\partial x_i} + H_r + H_s + H_a + H_L,$$ (9.8)

where i is a coordinate index, x_i is a spatial coordinate and H_r, H_s, H_a and H_L are the radioactive, shear, adiabatic and latent heat productions (W/m^3), respectively.

The radioactive heat production (*radioactive heating*) (H_r) is due to the decay of radioactive elements that are present in rocks. The amount of radioactive heat production depends strongly on the type of rock and typical, easy-to-remember values are: 2×10^{-6} W/m^3 for granite (continental crust), 2×10^{-7} W/m^3 for basalt (oceanic crust) and 2×10^{-8} W/m^3 for peridotite (mantle) (Turcotte and Schubert, 2002).

The shear heat production (*shear heating*) (H_s) is related to the dissipation of mechanical energy during irreversible non-elastic (e.g., viscous and/or plastic) deformation and can be calculated via the deviatoric stresses and strain rates as follows

$$H_s = \sigma'_{ij} \dot{\varepsilon}'_{ij},$$ (9.9a)

where i and j are coordinate indexes (x, y, z) and the repeated ij indexes denote summation. In the case of 3D viscous deformation of an incompressible fluid Eq. (9.9a) becomes

$$H_s = \sigma'_{xx} \dot{\varepsilon}_{xx} + \sigma'_{yy} \dot{\varepsilon}_{yy} + \sigma'_{zz} \dot{\varepsilon}_{zz} + 2 \left(\sigma'_{xy} \dot{\varepsilon}_{xy} + \sigma'_{xz} \dot{\varepsilon}_{xz} + \sigma'_{yz} \dot{\varepsilon}_{yz} \right).$$ (9.9b)

Shear heating is always positive (or zero) since for irreversible deformation the signs of the respective σ'_{ij} and $\dot{\varepsilon}'_{ij}$ are identical and their products are thus always positive. In the case of viscous deformation, $\dot{\varepsilon}'_{ij} = \sigma'_{ij}/2\eta$ (Eq. 5.11), and shear heating can be computed as

$$H_s = \frac{1}{2\eta}\sigma'_{ij}{}^2. \tag{9.9c}$$

The adiabatic heat production/consumption (adiabatic heating/cooling) (H_a) is related to changes in pressure and can be calculated via pressure changes as

$$H_a = T\alpha\frac{DP}{Dt}, \tag{9.10}$$

where DP/Dt is the substantive time derivative of pressure and α is thermal expansion (which is typically positive, Chapter 2). In contrast to shear and radioactive heating, adiabatic effects can be either positive or negative. It is known from thermodynamics that the temperature of a substance under conditions of no thermal exchange typically increases with increasing pressure and decreases with decreasing pressure, which thus directly reflects the sign of DP/Dt. The effects of adiabatic heating can be very significant in cases of strong changes in pressure. Therefore, this type of heat production/consumption is important for mantle convection in the Earth and other sufficiently large planets characterized by very significant mantle pressure variations (on the order of several tens to hundreds of GPa).

The latent heat production/consumption (latent heating/cooling) (H_L) is due to phase transformations in rocks subjected to changes in pressure and temperature. A very common example is the latent heat effect in the case of melting, which is negative (heat sink, $H_L<0$); the effect is positive (heat production, $H_L>0$) for crystallization.

9.4 Simplified temperature equations

In a complete form, the temperature equation looks quite complicated, but at least it does not 'hide' three equations in one, in contrast to the momentum equation. In the case of constant thermal conductivity ($k = $ const), the temperature equation simplifies to

$$\rho C_P\frac{DT}{Dt} = k\frac{\partial^2 T}{\partial x^2} + k\frac{\partial^2 T}{\partial y^2} + k\frac{\partial^2 T}{\partial z^2} + H_r + H_s + H_a + H_L \tag{9.11a}$$

or

$$\rho C_P\frac{DT}{Dt} = k\nabla^2 T + H_r + H_s + H_a + H_L, \tag{9.11b}$$

where $\nabla^2 T$ is the Laplacian of temperature.

When the internal heat production is negligible and there is no advection of material (purely conductive heat transport), the temperature equation takes a form which is similar to the Poisson equation

$$\frac{\partial T}{\partial t} = \kappa \nabla^2 T, \tag{9.12}$$

where $\kappa = k/\rho C_P$ is thermal diffusivity (m^2/s).

If the temperature does not change with time, heat conservation is described by a steady-state temperature equation. The steady-state Eulerian temperature equation $\partial T/\partial t = 0$ corresponds to the case when the temperature remains constant in an immobile, Eulerian observation point, while the temperature at Lagrangian material points can change. In this case, the temperature equation becomes

$$\rho C_P \left(\vec{v} \cdot \text{grad}(T) \right) = -\frac{\partial q_i}{\partial x_i} + H_r + H_s + H_a + H_L, \tag{9.13}$$

where i is a coordinate index and x_i is a spatial coordinate. This form of the equation is frequently used for computing equilibrium temperature profiles across a deforming medium, for example in the case of steady magma flow in a channel. The steady-state Lagrangian temperature equation $DT/Dt = 0$ corresponds to the case when the temperature does not change at Lagrangian points but can vary at Eulerian observation points according to the purely advective heat transport,

$$\frac{\partial T}{\partial t} + \vec{v} \cdot \text{grad}(T) = 0. \tag{9.14}$$

In this case, the temperature equation is as follows

$$-\frac{\partial q_i}{\partial x_i} + H_r + H_s + H_a + H_L = 0. \tag{9.15}$$

The steady-state Eulerian-Lagrangian temperature equation ($\partial T/\partial t = 0$ and $DT/Dt = 0$) holds for the case when no displacement of the medium occurs, pressure and temperature are constant and therefore $H_s = 0$, $H_a = 0$ and $H_L = 0$. This equation has the simple form

$$-\frac{\partial q_i}{\partial x_i} + H_r = 0, \tag{9.16}$$

and is often used for the calculation of steady-state geotherms that characterize changes of temperature with depth in a layered sequence of rocks with variable radioactive heat production.

Simplified steady-state temperature equations are often used for obtaining analytical solutions, which are used for testing the accuracy of numerical codes (Chapter 20). Indeed

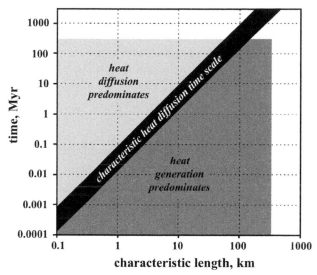

Fig. 9.2. Time scales for different thermal regimes calculated according to the equation $t_{diff} = L^2/\kappa$ with $\kappa = 10^{-6}$ m^2 s^{-1}. Shaded areas show length and time scales characteristic of collisional orogens (Burg and Gerya, 2005).

some analytical solutions also exist for more complicated non-steady-state (*transient*) problems (e.g. Tikhonov and Samarsky, 1972; Shukla, 2005), which will be further discussed in Chapter 20.

9.5 Heat diffusion time scales

One important aspect that can be analysed analytically concerns the *time scales* of heat diffusion processes. Heat generated within any region is spread by conduction (i.e. diffused) on a characteristic time scale (t_{diff}) that depends on the width L of the region according to

$$t_{diff} = \frac{L^2}{\kappa}. \tag{9.17}$$

Since thermal diffusivity is relatively constant for the majority of rocks ($\kappa \approx 10^{-6}$ m^2/s), Eq. (9.17) provides a useful tool to quickly evaluate a characteristic duration of various heating/cooling processes in nature. This equation suggests that the duration of heat dissipation via conduction grows as the square of the width (L) of the region. For instance, although the shear heat produced within a 100 metre wide shear zone dissipates in only ~1000 yr, the heat generated within a 1 km wide shear zone requires about 100 000 yr for a similar degree of conductive cooling (Fig. 9.2).

Analytical exercises

Exercise 9.1.
Integrate Eq. (9.16) in order to calculate the steady-state temperature profile across the continental crust with radioactive heat production $H_r = 1 \times 10^{-6}$ W/m^3, if the temperature at the surface is 300 K and the temperature at the bottom of the crust is 700 K. Take the thermal conductivity of the crust to be $k = 2$ W/(m K).

Exercise 9.2.
Compute from Eq. (9.13) the steady-state Eulerian temperature profile across the magmatic channel described in Exercise 5.1 (Fig. 5.2). Assume the temperature at the channel walls, as well as the temperature gradient along the channel, to be constant. Use $T_0 = 1300$ K and $\partial T / \partial y = 1$ K/m for the analysed horizontal section across the channel. Take the thermal conductivity to be $k = 2$ W/(m K) and the isobaric heat capacity as $C_P = 1000$ J/(kg K).

Programming exercise

Exercise 9.3.
Use the plume model (Exercise 8.7) to compute the strain rate components and deviatoric stress components as follows (Fig. 9.3a,b)

$$\dot{\varepsilon}_{xy(i,j)} = \frac{1}{2} \left(\frac{v_{x(i+1,j)} - v_{x(i,j)}}{\Delta y} + \frac{v_{y(i,j+1)} - v_{y(i,j)}}{\Delta x} \right), \tag{9.18}$$

$$\sigma'_{xy(i,j)} = 2\eta_{s(i,j)} \dot{\varepsilon}_{xy(i,j)}, \tag{9.19}$$

$$\dot{\varepsilon}_{xx(i,j)} = \frac{v_{x(i,j)} - v_{x(i,j-1)}}{\Delta x}, \tag{9.20}$$

$$\sigma'_{xx(i,j)} = 2\eta_{n(i,j)} \dot{\varepsilon}_{xx(i,j)}. \tag{9.21}$$

Compute the $\sigma'_{xy}\dot{\varepsilon}_{xy}$ term for the pressure nodes (see the red circle in Fig. 9.3c) by averaging its values computed at the four surrounding basic nodes (see solid squares in Fig. 9.3c). It is important to compute the average of $\sigma'_{xy}\dot{\varepsilon}_{xy}$ products (rather than a product of averaged σ'_{xy} and $\dot{\varepsilon}_{xy}$) to ensure a correct non-negative value of shear heating.

Compute and visualize (with *pcolor* MATLAB function) the shear heating distribution for all internal pressure nodes by using the following equation based on Eq. (9.9)

$$H_s = 2\sigma'_{xx}\dot{\varepsilon}_{xx} + 2\sigma'_{xy}\dot{\varepsilon}_{xy}. \tag{9.22}$$

Equation (9.22) is valid for an incompressible medium in 2D ($\dot{\varepsilon}'_{xx} = \dot{\varepsilon}_{xx} = -\dot{\varepsilon}'_{yy} = -\dot{\varepsilon}_{yy}$, $\sigma'_{xx} = -\sigma'_{yy}$).

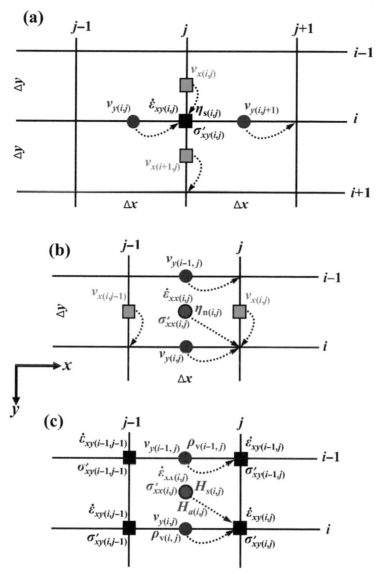

Fig. 9.3. Stencils used for computing strain rates, deviatoric stresses, shear and adiabatic heating. (a) Shear strain rate and deviatoric stress. (b) Normal strain rate and deviatoric stress. (c) Shear and adiabatic heating. Indexing of different parameters (dotted arrows) is made according to Fig. 7.15.

Compute and visualize (with *pcolor* MATLAB function) the adiabatic heating for the same buoyancy-driven flow using the following approximate formula derived from Eq. (9.10) under the assumption that $DP/Dt \approx (\partial P/\partial y)\,v_y \approx \rho g_y v_y$,

$$H_a = T\alpha v_y \rho g_y. \tag{9.23}$$

Use $T = 1300$ K and $\alpha = 3 \times 10^{-5}$ 1/K and compute the adiabatic heating for the internal pressure nodes (see the red circle in Fig. 9.3c). Compute the vertical velocity component $v_y = (v_{y(i,j)} + v_{y(i-1,j)})/2$ and density $\rho = (\rho_{v(i,j)} + \rho_{v(i-1,j)})/2$ for these nodes from the two nearest staggered v_y nodes (see blue circles in Fig. 9.3c). Compare the magnitudes of the shear and adiabatic heating. An example is in **Shear_adiabatic_heating.m**.

10

Numerical solution of the heat conservation equation

> **Theory:** Discretization of the heat conservation equation with finite differences. Explicit and implicit solution schemes of the heat conservation equation. Conservative and non-conservative discretization schemes. Advection of temperature with Eulerian methods, numerical diffusion. Advection of temperature with markers. Subgrid diffusion. Thermal boundary conditions.
> **Exercises:** Solving the heat conservation equation in the case of constant and variable thermal conductivity with explicit and implicit solution schemes. Programming various thermal boundary conditions. Advecting temperature with upwind differences and markers.

10.1 Explicit and implicit formulation of the temperature equation

We now start with the numerical formulation and solution of the temperature equation. Discretization of this equation with finite differences can be done in an *explicit* and an *implicit* manner. In order to understand the differences, let us consider an example of heat diffusion in a non-deforming medium with constant thermal conductivity (k) and constant volumetric heat capacity (ρC_P)

$$\frac{\partial T}{\partial t} = \frac{k}{\rho C_P} \Delta T. \tag{10.1}$$

In 2D, this discretization is as follows.
Explicit FD (Fig. 10.1):

$$\frac{T_3 - T_3^o}{\Delta t} = \frac{k}{\rho C_P} \left(\frac{T_1^o - 2T_3^o + T_5^o}{\Delta x^2} + \frac{T_2^o - 2T_3^o + T_4^o}{\Delta y^2} \right). \tag{10.2}$$

This form is called explicit because the new temperature (T) for the next time instant $t + \Delta t$, where Δt is the time step, can be explicitly calculated from the known temperatures (T^o) at the current time instant t

139

explicit 5-point cross

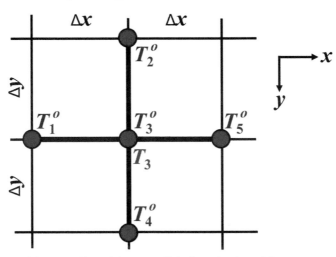

Fig. 10.1. Stencil of the 2D grid used for an explicit discretization of the temperature equation with constant thermal conductivity.

$$T_3 = T_3^o + \frac{k\Delta t}{\rho C_P}\left(\frac{T_1^o - 2T_3^o + T_5^o}{\Delta x^2} + \frac{T_2^o - 2T_3^o + T_4^o}{\Delta y^2}\right). \qquad (10.3)$$

The explicit formulation does not require composing and solving a global system of equations and is therefore very convenient to program. However, this formulation has a strong limitation on the time step that can be used in the calculations. The time step must satisfy the condition

$$\Delta t < \frac{\min(\Delta x^2, \Delta y^2, \Delta z^2)}{2\kappa}, \qquad (10.4)$$

where $\kappa = k/\rho C_P$ is thermal diffusivity and Δx, Δy, Δz is the grid spacing in respective directions. This limitation means that the number of time steps increases as the square of the decrease in the grid spacing, which can be quite inconvenient for high-resolution thermal models. If larger time steps are indeed employed, numerical oscillations occur that increase with the number of time steps (Fig. 10.2, see also program example **Explicit_implicit_1D.m**).

The implicit finite difference discretization is given by (Fig. 10.3)

$$\frac{T_3 - T_3^o}{\Delta t} = \frac{k}{\rho C_P}\left(\frac{T_1 - 2T_3 + T_5}{\Delta x^2} + \frac{T_2 - 2T_3 + T_4}{\Delta y^2}\right). \qquad (10.5)$$

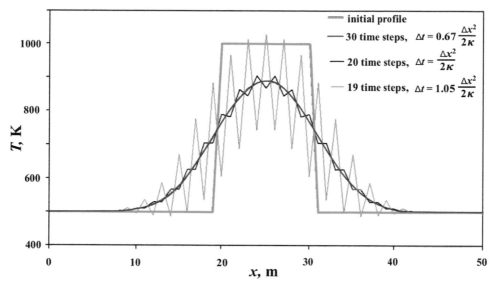

Fig. 10.2. Oscillations of explicit numerical solution of Eq. (10.1) due to the use of a too large time step. An example is in **Explicit_implicit_1D.m**.

implicit 5-point cross

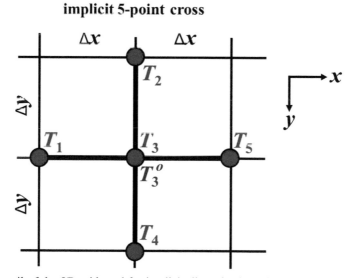

Fig. 10.3. Stencil of the 2D grid used for implicit discretization of the temperature equation with constant thermal conductivity.

This form is called implicit because the new temperature (T) for the next time instant $t+\Delta t$ cannot be explicitly calculated from the temperatures (T^o) known from the current time instant t. In order to obtain new temperatures, the global system of equations written for all points of the model has to be solved

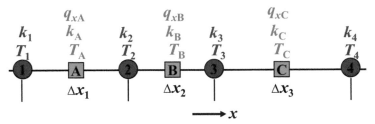

Fig. 10.4. 1D staggered grid used for the discretization of the temperature equation with variable thermal conductivity. 1, 2, 3, 4 are basic nodes (circles) of the grid where the temperature equations are formulated. A, B, C are additional nodes (squares) of the grid where the heat fluxes are defined.

$$\frac{T_3}{\Delta t} - \frac{k}{\rho C_P}\left(\frac{T_1 - 2T_3 + T_5}{\Delta x^2} + \frac{T_2 - 2T_3 + T_4}{\Delta y^2}\right) = \frac{T_3^o}{\Delta t}. \tag{10.6}$$

It is important to mention that the implicit formulation places no limitation on the size of the time step (in the absence of internal heat sources and advective terms). Indeed, very large implicit time steps do not necessarily guarantee an accurate solution (since the time derivative in Eq. (10.5) is only first order accurate in time).

10.2 Conservative finite differences

Conservative finite difference discretization should be used in cases when the heat conservation equation contains a variable thermal conductivity. Such finite differences ensure *conservation of heat fluxes*, thereby allowing an accurate numerical solution. In a general sense, this is analogous to the formulation of conservative finite differences for the Stokes equation with variable viscosity, which was described in Chapter 7. Below, examples of non-conservative and conservative finite differences are compared for the 1D heat conservation equation (Fig. 10.4)

$$\rho C_P \frac{DT}{Dt} = -\frac{\partial q_x}{\partial x},$$

$$\text{where } q_x = -k\frac{\partial T}{\partial x}.$$

An *erroneous non-conservative* FD formulation of the heat flux terms, in either the explicit or the implicit formulation, for the two basic nodes 2 and 3 can for example be obtained (we can 'arrive' at this equation by assuming erroneously that all we need to do is to use Eq. (10.5) and put different thermal conductivities for different nodes):

$$\text{node 2} \quad \left(\frac{\partial q}{\partial x}\right)_2 = -k_2\frac{(T_3 - T_2)/\Delta x_2 - (T_2 - T_1)/\Delta x_1}{(\Delta x_1 + \Delta x_2)/2} \tag{10.7a}$$

$$\text{node 3} \quad \left(\frac{\partial q}{\partial x}\right)_3 = -k_3 \frac{(T_4 - T_3)/\Delta x_3 - (T_3 - T_2)/\Delta x_2}{(\Delta x_2 + \Delta x_3)/2} \tag{10.7b}$$

which implicitly means that the formulations of the horizontal heat flux q_{xB} for temperature equations at nodes 2 and 3 are different due to the different thermal conductivities k_2 and k_3:

$$\text{node 2} \quad q_{xB} = -k_2 \frac{T_3 - T_2}{\Delta x_2}$$

$$\text{node 3} \quad q_{xB} = -k_3 \frac{T_3 - T_2}{\Delta x_2}.$$

This implies that heat flux is *not conserved* and artificially 'jumps' between basic nodes in response to the difference in the thermal conductivity at these nodes.

On the other hand, a ***proper conservative*** FD formulation of the heat flux term in either the explicit or implicit temperature equations for the two basic nodes 2 and 3 is given by

$$\text{node 2} \quad \left(\frac{\partial q}{\partial x}\right)_2 = 2\frac{q_{xB} - q_{xA}}{(\Delta x_1 + \Delta x_2)} \quad \text{or} \quad \left(\frac{\partial q}{\partial x}\right)_2 = -\frac{k_B(T_3 - T_2)/\Delta x_2 - k_A(T_2 - T_1)/\Delta x_1}{(\Delta x_1 + \Delta x_2)/2}, \tag{10.8a}$$

$$\text{node 3} \quad \left(\frac{\partial q}{\partial x}\right)_3 = 2\frac{q_{xC} - q_{xB}}{(\Delta x_2 + \Delta x_3)} \quad \text{or} \quad \left(\frac{\partial q}{\partial x}\right)_3 = -\frac{k_C(T_4 - T_3)/\Delta x_3 - k_B(T_3 - T_2)/\Delta x_2}{(\Delta x_2 + \Delta x_3)/2}, \tag{10.8b}$$

which imply that the expressions for heat flux q_{xB} at nodes 2 and 3 are identical:

$$q_{xB} = -k_B \frac{T_3 - T_2}{\Delta x_2}.$$

Thus, a conservative FD formulation of the temperature equation (either explicit or implicit) is based on the following three formal rules that are analogous to the rules discussed in Chapter 7 for the Stokes equation with variable viscosity.

(1) The temperature equation is initially discretized in terms of heat fluxes at *basic nodes* of the grid (cf. nodes 2, 3, Fig. 10.4),

$$\text{node 2} \quad \left(\rho Cp \frac{DT}{Dt}\right)_2 = 2\frac{q_{xB} - q_{xA}}{(\Delta x_1 + \Delta x_2)},$$

$$\text{node 3} \quad \left(\rho Cp \frac{DT}{Dt}\right)_3 = 2\frac{q_{xC} - q_{xB}}{(\Delta x_2 + \Delta x_3)}.$$

(2) These heat fluxes are formulated for *additional (heat flux) nodes* of the grid (cf. nodes A, B, C, Fig. 10.4)

$$\text{node A}\quad q_{x\text{A}} = -k_\text{A}\frac{T_2 - T_1}{\Delta x_1},$$

$$\text{node B}\quad q_{x\text{B}} = -k_\text{B}\frac{T_3 - T_2}{\Delta x_2},$$

$$\text{node C}\quad q_{x\text{C}} = -k_\text{C}\frac{T_4 - T_3}{\Delta x_3}.$$

Note that we have to use thermal conductivity values k_A, k_B and k_C for the additional nodes (A, B, C) at the locations where the heat fluxes are defined. If these values are not known, they can be computed by for example arithmetic averaging of known thermal conductivity values from the basic nodes (1, 2, 3, 4)

$$k_\text{A} = \frac{k_1 + k_2}{2},$$

$$k_\text{B} = \frac{k_2 + k_3}{2},$$

$$k_\text{C} = \frac{k_4 + k_3}{2}.$$

Another possibility is to use harmonic averaging

$$k_\text{A} = \frac{2k_1 k_2}{k_1 + k_2},$$

$$k_\text{B} = \frac{2k_2 k_3}{k_2 + k_3},$$

$$k_\text{C} = \frac{2k_3 k_4}{k_3 + k_4}.$$

The harmonic average formula can be derived from the condition that the heat flux to the left of the additional nodes must equal the flux to the right of these nodes. The derivation for node B is done *under the assumption* that the thermal conductivities between nodes 2 and B, and between B and 3 remain constant, and are equal to k_2 and k_3, respectively. Then the following equation can be formulated,

$$q_{x\text{B}} = -2k_2\frac{T_\text{B} - T_2}{\Delta x_2},$$

$$q_{xB} = -2k_3 \frac{T_3 - T_B}{\Delta x_2},$$

$$q_{xB} = -k_B \frac{T_3 - T_2}{\Delta x_2}.$$

Solving these equations with respect to T_B and k_B gives (verify as an exercise)

$$T_B = \frac{T_2 k_2 + T_3 k_3}{k_2 + k_3} \quad \text{and} \quad k_B = \frac{2k_2 k_3}{k_2 + k_3}.$$

Derivations for nodes A and C can be done similarly (derive as an exercise).

(3) Identical formulations of heat fluxes are used for the temperature equation at different basic nodes.

It is important to mention that conservative finite differences are formulated *in terms of the thermal conductivity (k) and not in terms of the thermal diffusivity* $\kappa = k/\rho C_P$, otherwise one can obtain artificial variations in heat fluxes due to spatial variations of the volumetric heat capacity (ρC_P). Both density and heat capacity values should always be taken for the basic node at which this equation is formulated.

By applying these rules in 2D, the following conservative implicit FD formulation can be derived for the Lagrangian temperature equation (Fig. 10.5)

$$\rho C_{P_3} \left(\frac{DT}{Dt} \right)_3 = -\left(\frac{\partial q_x}{\partial x} \right)_3 - \left(\frac{\partial q_y}{\partial y} \right)_3 + H_3, \tag{10.9a}$$

$$\rho C_{P_3} \frac{T_3 - T_3^o}{\Delta t} = -2 \frac{q_{xB} - q_{xA}}{\Delta x_1 + \Delta x_2} - 2 \frac{q_{yD} - q_{yC}}{\Delta y_1 + \Delta y_2} + H_3, \tag{10.9b}$$

$$\rho C_{P_3} \frac{T_3}{\Delta t} + 2 \frac{q_{xB} - q_{xA}}{\Delta x_1 + \Delta x_2} + 2 \frac{q_{yD} - q_{yC}}{\Delta y_1 + \Delta y_2} = H_3 + \rho C_{P_3} \frac{T_3^o}{\Delta t}, \tag{10.9c}$$

where

$$q_{xA} = -k_A \frac{T_3 - T_1}{\Delta x_1},$$

$$q_{xB} = -k_B \frac{T_5 - T_3}{\Delta x_2},$$

$$q_{yC} = -k_C \frac{T_3 - T_2}{\Delta y_1},$$

$$q_{yD} = -k_D \frac{T_4 - T_3}{\Delta y_2}.$$

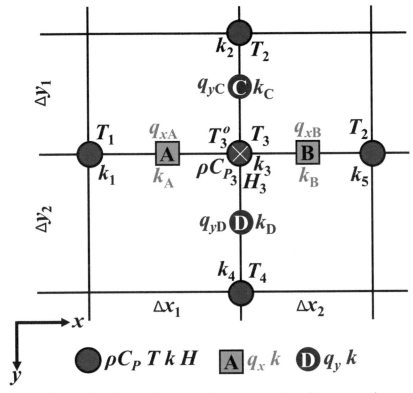

Fig. 10.5. Stencil of a 2D grid used for the implicit discretization of the Lagrangian temperature equation with variable thermal conductivity. The crossed circle corresponds to the node for which the temperature equation is formulated.

If values of thermal conductivity for heat flux nodes (A, B, C, D) are not known, they can be computed by averaging known thermal conductivity values from the basic nodes (cf. green squares in Fig. 10.5). For example,

$$\text{arithmetic average } k_A = \frac{k_1 + k_3}{2}, \ k_B = \frac{k_3 + k_5}{2}, \ k_C = \frac{k_2 + k_3}{2}, \ k_D = \frac{k_3 + k_4}{2},$$

$$\text{harmonic average } k_A = \frac{2k_1 k_3}{k_1 + k_3}, \ k_B = \frac{2k_3 k_5}{k_3 + k_5}, \ k_C = \frac{2k_2 k_3}{k_2 + k_3}, \ k_D = \frac{2k_3 k_4}{k_3 + k_4}.$$

Obviously, conservative 2D formulations can also be explicit (derive as an exercise).

10.3 Advection of temperature with Eulerian methods

If the temperature equation is solved in an Eulerian form for a deforming/moving medium, then the advective term $\vec{v} \cdot \text{grad}(T)$ is present in the temperature equation

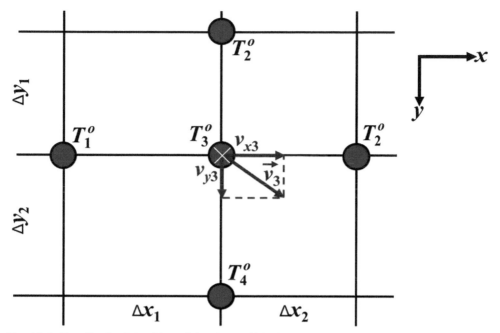

Fig. 10.6. Stencil of a 2D grid used for the explicit discretization of the advective term for the Eulerian temperature equation.

$$\rho C_P \left(\frac{\partial T}{\partial t} + \vec{v} \cdot \mathrm{grad}(T) \right) = -\frac{\partial q_i}{\partial x_i} + H, \qquad (10.10)$$

where i is a coordinate index and x_i is a spatial coordinate. When discretizing this term, explicit finite differences are typically used, based on temperature and velocity at the current time instant. A common approach is to use explicit asymmetric 'upwind' differences (Chapter 8), i.e. to perform the differencing *against the direction* of material flow vector (Fig. 10.6)

$$\vec{v}_3 \cdot \mathrm{grad}(T^o)_3 = v_{x3} \left(\frac{\partial T^o}{\partial x} \right)_3 + v_{y3} \left(\frac{\partial T^o}{\partial y} \right)_3, \qquad (10.11a)$$

$$
\begin{aligned}
\left(\frac{\partial T^o}{\partial x} \right)_3 &= \frac{T_3^o - T_1^o}{\Delta x_1} \quad \text{when } v_{x3} > 0 \text{ and} \\
\left(\frac{\partial T^o}{\partial x} \right)_3 &= \frac{T_5^o - T_3^o}{\Delta x_2} \quad \text{when } v_{x3} < 0,
\end{aligned}
\qquad (10.11b)
$$

$$\left(\frac{\partial T^o}{\partial y}\right)_3 = \frac{T_3^o - T_2^o}{\Delta y_1} \quad \text{when} \quad v_{y3} > 0 \quad \text{and}$$

$$\left(\frac{\partial T^o}{\partial y}\right)_3 = \frac{T_4^o - T_3^o}{\Delta y_2} \quad \text{when} \quad v_{y3} < 0. \tag{10.11c}$$

Such explicit advection terms are simply added to the right hand side of the temperature equation (10.9). Including Eulerian advective terms in the temperature equation imposes an additional restriction on the time step given by the *Courant condition* (Eq. 8.5).

It should be mentioned that advective terms could also be formulated implicitly. Irrespective of the formulation, these terms always introduce *artificial numerical diffusion* of temperature on the Eulerian grid (Chapter 8). This problem is not relevant for slow flow because *real physical diffusion* is typically faster than numerical diffusion and the latter can be neglected. Numerical diffusion, however, becomes relevant for models with rapid advection (e.g. in subduction models). This problem can be minimized by: (i) using more complicated, higher order Eulerian advection FD schemes (e.g., FCT, Chapter 8) and (ii) advecting temperature with Lagrangian points (*method of characteristics, method of markers*) which we will discuss below.

10.4 Advection of temperature with markers

In order to avoid numerical diffusion of temperature one can use the Lagrangian form of heat conservation equation (Chapter 9) and advect temperature with the marker-in-cell technique described in Chapter 8. The temperature equation is then formulated in a Lagrangian form and is discretized with Eq. (10.9).

Temperature is originally prescribed on Lagrangian markers and interpolated to Eulerian nodes at the given time instant t by using the following *heat conservative form* of bilinear interpolation equation (8.18)

$$T_{i,j}^o = \frac{\displaystyle\sum_m T_m^o \rho C_{Pm} w_{m(i,j)}}{\displaystyle\sum_m \rho C_{Pm} w_{m(i,j)}}, \tag{10.12}$$

$$w_{m(i,j)} = \left(1 - \frac{\Delta x_m}{\Delta x}\right) \times \left(1 - \frac{\Delta y_m}{\Delta y}\right),$$

where $T_{i,j}^o$ is temperature of the ijth node, T_m^o and ρC_{Pm} are respectively temperature and volumetric isobaric heat capacity of the mth marker, $w_{m(i,j)}$ represents a statistical weight of the mth marker at the ijth node; Δx_m and Δy_m are the distances from the mth marker to the ijth node. After interpolation, boundary conditions should be applied to the interpolated values of $T_{i,j}^o$ at the model boundaries. Please note that it is better to directly define ρC_{Pm}

values on markers and interpolate these values to Eulerian nodes as $\rho C_{P(i,j)}$ with Eq. (8.18) rather than interpolate separately $\rho_{m(i,j)}$ and $C_{Pm(i,j)}$.

After solving the Lagrangian temperature equation on the Eulerian nodes (Fig. 10.5) the changes in the effective temperature field for the Eulerian nodes are calculated as

$$\Delta T_{i,j} = T_{i,j} - T_{i,j}^{o}, \tag{10.13}$$

where $T_{i,j}$ is the new (i.e. for the time instant $t+\Delta t$) temperature of the *ij*th node.

The corresponding temperature increments for the markers ΔT_m are then interpolated from the nodes using relation (8.19) (Fig. 8.9) in order to calculate new marker temperatures T_m as

$$T_m = T_m^{o} + \Delta T_m. \tag{10.14}$$

The interpolation of the calculated *temperature changes* from the Eulerian nodal points to the Lagrangian markers reduces numerical diffusion in an efficient manner (Chapter 8). This method does not produce any smoothing of the temperature distribution between adjacent markers (Fig. 8.11), thus resolving the thermal structure of a numerical model in much finer detail.

However, a main problem with treating advection-diffusion processes using the simple incremental update scheme of Eq. (10.14) is that small-scale variations of the thermal structures may appear on a *subgrid scale* (i.e. differences in temperature between closely located markers). These variations cannot be damped out by grid-scale corrections of Eq. (10.14). For example, in the case of strong chaotic mixing of markers (e.g. due to thermal convection), Eq. (10.14) may produce numerical oscillations of the temperature field assigned to the adjacent markers (Fig. 10.7a). These oscillations do not damp out with time based on a characteristic heat diffusion time scale, as would be the case if physical diffusion were active. The introduction of a consistent *subgrid diffusion operation* is the way to correct this problem. In order to define this operation, we decompose temperature changes computed from Eq. (10.13) into a *subgrid part* $\Delta T_{i,j}^{subgrid}$ and a *remaining part* $\Delta T_{i,j}^{remaining}$ such that

$$\Delta T_{i,j} = \Delta T_{i,j}^{subgrid} + \Delta T_{i,j}^{remaining}. \tag{10.15}$$

In order to compute the subgrid part, we apply subgrid diffusion on the markers over a characteristic local heat diffusion time scale t_{diff} (Fig. 9.2) and then interpolate the respective temperature changes back to nodes. Subgrid temperature changes on markers are computed as follows

$$\Delta T_m^{subgrid} = \left(T_{m(nodal)}^{o} - T_m^{o} \right) \left[1 - exp\left(-d\frac{\Delta t}{t_{diff}} \right) \right], \tag{10.16}$$

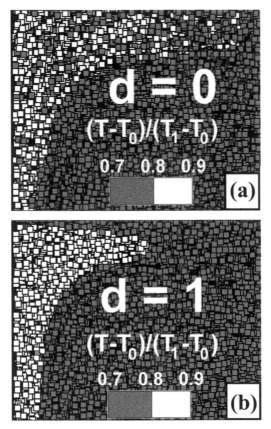

Fig. 10.7. Temperature structure of the markers for a zoomed-in area of the numerical model of convection (Gerya and Yuen, 2003a). Different colours of markers correspond to different values of temperature assigned to the markers. T_1 and T_0 are the maximal and minimal temperatures for the experiment. (a) and (b) show the results calculated without ($d = 0$) and with ($d = 1$) subgrid diffusion (see Eq. 10.16), respectively.

$$t_{diff} = \frac{\rho C_{Pm}}{k_m \left(2/\Delta x^2 + 2/\Delta y^2\right)}$$

where t_{diff} is defined for the corresponding cell of the grid where the marker is located (Fig. 8.9); k_m and ρC_{Pm} are respectively the thermal conductivity and volumetric isobaric heat capacity of a given marker, d is a dimensionless numerical diffusion coefficient (one can use empirical values in the range $0 \leq d \leq 1$). $T^o_{m(nodal)}$ is interpolated for the marker from $T^o_{i,j}$ values for nodes using the relation (8.19) (Fig. 8.9). Equation (10.16) is derived by analysing analytically temperature relaxation in a 2D grid stencil (Fig. 10.5).

After obtaining $\Delta T^{subgrid}_m$ for all markers, $\Delta T^{subgrid}_{i,j}$ are computed by interpolation from markers to nodes using Eq. (10.12) (Fig. 8.8)

$$\Delta T_{i,j}^{subgrid} = \frac{\sum\limits_{m} \Delta T_{m}^{subgrid} \rho C_{Pm} w_{m(i,j)}}{\sum\limits_{m} \rho C_{Pm} w_{m(i,j)}}. \tag{10.17}$$

Then $\Delta T_{i,j}^{remaining}$ is computed for the nodes from Eq. (10.15)

$$\Delta T_{i,j}^{remaining} = \Delta T_{i,j} - \Delta T_{i,j}^{subgrid}. \tag{10.18}$$

Finally, the new corrected marker temperatures $T_{m}^{corrected}$ are computed according to the modified relation (10.14) that now takes into account Eqs. (10.15)–(10.18) and thus removes the non-physical subgrid oscillations

$$T_{m}^{corrected} = T_{m}^{o} + \Delta T_{m}^{subgrid} + \Delta T_{m}^{remaining}, \tag{10.19}$$

where $\Delta T_{m}^{subgrid}$ is given by Eq. (10.16) and $\Delta T_{m}^{remaining}$ is interpolated from nodal values of $\Delta T_{i,j}^{remaining}$ to markers according to standard bilinear interpolation (Eq. 8.19, Fig. 8.9).

Equation (10.16) requires the decay of differences between marker temperature values T_{m}^{o} and interpolated nodal temperature values $T_{m(nodal)}^{o}$ on the characteristic time scale (Δt_{diff}) of local heat diffusion on Eulerian nodes. It is important to emphasize that the subgrid diffusion does not change the total temperature increments $\Delta T_{i,j}$ computed on nodal points from the heat conservation equation. Instead it splits them into two parts $\Delta T_{i,j}^{subgrid}$ and $\Delta T_{i,j}^{remaining}$. By introducing a subgrid diffusion operation, unrealistic subgrid oscillations are self-consistently removed (see Fig. 10.7b) over the characteristic local heat diffusion time scale. Realistic subgrid variations will, however, be preserved by this scheme if they are related, for example, to the rapid mixing by advection dominating flows. In this case, a refinement of the Eulerian grid will be required to accurately resolve these variations.

It is also important to mention that subgrid diffusion is a method for correcting *small* non-physical subgrid oscillations that appear on the markers due to mechanical mixing processes and should only be applied in case of strong marker advection and mixing. It is not the way to remove any arbitrary discrepancy between the marker and nodal temperature fields. Such discrepancies can appear, for example, in the initial temperature distribution due to prescribing sharp temperature fronts on a fine marker grid, which cannot be properly resolved by a notably coarser Eulerian grid. Large initial temperature discrepancies between markers and nodes should be corrected by interpolation of total nodal temperatures (rather than their changes) to markers with the use of Eq. (8.19) during the first time step.

10.5 Thermal boundary conditions

In order to solve the temperature equation numerically, thermal boundary conditions have to be specified. The following boundary conditions are frequently used in geodynamic modelling:

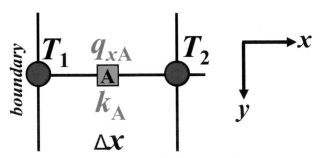

Fig. 10.8. Stencil of a 2D grid used for the discretization of the thermal boundary conditions.

(1) constant temperature
(2) insulating boundary (zero heat flux, symmetry condition)
(3) constant heat flux
(4) infinity-like conditions (external constant temperature)
(5) periodic boundary
(6) combined boundary conditions.

From a mathematical point of view, condition (1) is the *Dirichlet boundary condition* whereas conditions (2) and (3) are the *Neumann boundary conditions*. Conditions (4), (5) and (6) are composite and combine both Dirichlet and Neumann conditions. Numerical examples of different boundary conditions are shown below (Fig. 10.8).

(1) *A constant temperature condition* implies that the temperature at a boundary is assigned a given value (which may change both along the boundary and in time)

$$T = \text{const}(x, y, z, t)$$

(10.20a)

or in discretized form (Fig. 10.8)

$$T_1 = \text{const}(x, y, z, t).$$

(10.20b)

This condition is typically applied at the lower and upper boundaries of geodynamic models.

(2) *An insulating boundary condition (no heat flux, lateral symmetry condition)* means that heat does not flux through a boundary, which means (from Eq. 9.1) that no temperature gradient exists across this boundary, i.e.

$$q_x = -k \frac{\partial T}{\partial x} = 0,$$

(10.21a)

or

$$\frac{\partial T}{\partial x} = 0,$$

(10.21b)

or in discretized form (Fig. 10.8)

$$T_1 - T_2 = 0. \tag{10.21c}$$

This symmetry condition is used at the lateral boundaries for almost every 2D and 3D Cartesian geodynamic model.

(3) *A constant heat flux condition* does not limit the temperature values at a boundary, but prescribes a heat flux across the boundary (this heat flux can be time and coordinate dependent)

$$q_x = -k\frac{\partial T}{\partial x} = \text{const}(x, y, z, t), \tag{10.22a}$$

or in discretized form (Fig. 10.8)

$$k_A \frac{T_1 - T_2}{\Delta x} = \text{const}(x, y, z, t). \tag{10.22b}$$

(4) *Infinity-like conditions* either mimic the absence of a thermal boundary or imply that this boundary is located very far away. For example, the external constant temperature condition (Burg and Gerya, 2005; Gerya et al., 2008b) implies that conditions (10.20a) and (10.20b) are satisfied at a parallel boundary located at the distance ΔL from the actual boundary of the model, and that the temperature gradient between these two boundaries is constant

$$\frac{\partial T}{\partial x} = \frac{T - T_{external}}{\Delta L}, \tag{10.23a}$$

or in discretized form (Fig. 10.8)

$$\frac{T_2 - T_1}{\Delta x} = \frac{T_1 - T_{external}}{\Delta L} \tag{10.23b}$$

or

$$T_1 - T_2 \frac{\Delta L}{\Delta L + \Delta x} = \frac{\Delta x}{\Delta L + \Delta x} T_{external} \tag{10.23c}$$

where $T_{external} = \text{const}\ (x, y, z, t)$ is the prescribed temperature at the parallel external boundary.

(5) Periodic boundary conditions are typically established for paired parallel lateral boundaries of a model and imply that temperature fields *at both sides of each boundary* are identical. The physical meaning and usage of this condition is the same as for the respective mechanical boundary condition (Chapter 7).

(6) Combined conditions represent a mixture between several types of boundary conditions.

Thermal boundary conditions can also be applied *inside the model*.

Please note that for the Neumann boundary conditions (2) and (3) it is actually better to place the model boundary at the heat flux points (cf. point A, Fig. 10.8) such that some temperature points will be actually located outside of the model (similarly to e.g. external velocity points, Fig. 7.15). This would increase accuracy of the boundary conditions from the first to the second order.

Programming exercises

Exercise 10.1.
Write a program to solve the temperature equation in 2D, in both explicit and implicit form (Figs. 10.1 and 10.3, respectively). Use a regular grid of 51 × 31 points. The model size is 1000×1500 km² (i.e. 1 000 000 × 1 500 000 m²). Use constant thermal conductivity $k = 3$ W/(m K), density $\rho = 3200$ kg/m³ and heat capacity $C_P = 1000$ J/(kg K) for the entire model. Test Eq. (10.4) for the time step limitation in the case when explicit FD is used. The initial setup corresponds to a background temperature of 1000 K with a rectangular thermal wave (1300 K) in the middle ('wave' means sharp perturbation of the temperature field). Global indexing of the unknowns in the implicit case is the same as for the 2D Poisson equation (Exercise 3.2)

$$k = N_y \times (j - 1) + i, \tag{10.24}$$

where k is the index in the global matrix computed from horizontal (j) and vertical (i) geometrical indices and N_y is the number of nodes in the vertical direction.

Try using constant temperature (1000 K, Eq. 10.20), and insulating boundary conditions (Eq. 10.21) at all boundaries. An example is in **Explicit_Implicit2D.m**.

Exercise 10.2.
Modify the previous example to take into account variable thermal conductivity k and volumetric heat capacity ρC_P using a conservative finite difference formulation (Eq. 10.9, Fig. 10.5) for a uniform grid (i.e. $\Delta x_1 = \Delta x_2 = \Delta x$ and $\Delta y_1 = \Delta y_2 = \Delta y$ in Eq. 10.9) and no heat production ($H_3 = 0$)

$$
\rho C_{P_3} \frac{T_3}{\Delta t} - \frac{(k_3 + k_5)(T_5 - T_3) - (k_1 + k_3)(T_3 - T_1)}{2\Delta x^2}
$$
$$
- \frac{(k_3 + k_4)(T_4 - T_3) - (k_2 + k_3)(T_3 - T_2)}{2\Delta y^2} = \rho C_{P_3} \frac{T_3^o}{\Delta t}. \tag{10.25}
$$

Use different physical properties for the area of the temperature wave ($k = 10$ W/(m K), $\rho = 3300$ kg/m³, $C_P = 1100$ J/(kg K)) and the surrounding medium ($k = 3$ W/(m K), $\rho = 3200$ kg/m³, $C_P = 1000$ J/(kg K)). Define k and ρC_P values at the temperature points. An example is in **Variable_conductivity.m**.

Exercise 10.3.

Modify the two previous examples by adding the advective terms into the temperature equation (Eq. 10.11) and by using a uniform velocity field $v_x = v_y = 10^{-9}$ m/s in the entire model. Test Eq. (8.5) for the time step limitation in 1D and refine it for 2D. Do not forget to also change the thermal conductivity k and volumetric heat capacity ρC_P values at internal nodes with time by solving the Eulerian advection equation with upwind differences (Chapter 8). Experiment with different velocities to see advection- and conduction-dominated regimes. Observe that the numerical diffusion is clearly visible when the chosen velocity is large (e.g. $v_x = v_y = 10$ m/s). In this case, the time scale for the experiment will be far below the characteristic thermal diffusion time scale (compute it by using a prescribed size of the wave and a total time in your experiment, Eq. 9.17, Fig. 9.2) and thus the moving temperature wave should remain largely unchanged. Examples are in **Conduction_advection2D.m** and **Variable_conductivity_advection2D.m**.

Exercise 10.4.

Modify Exercise 10.3 by adding temperature advection with a marker-in-cell approach combined with subgrid diffusion (Eqs. 10.14–10.19). Use an implicit solution of the Lagrangian temperature equation (10.25). Use marker routines from Exercises 8.4 and 8.5. Interpolate thermal conductivity k, volumetric heat capacity ρC_P and temperature T_m^o from markers to nodes at every time step. Use Eq. (8.8) for k and ρC_P and Eq. (10.12) for T_m^o. Move the markers with the prescribed constant velocity and recycle them when they leave the model. Use Eqs. (8.24a,b) for changing the horizontal and vertical coordinates of the markers with time. Do not forget to match nodal and marker temperature fields during the first time step (i.e., interpolate new temperature values $T_{i,j}$ from nodes to markers). Also, do not forget to apply the boundary conditions for the nodal temperatures $T_{i,j}^o$ interpolated from markers at every time step. Experiment with different velocities to see advection- and conduction-dominated regimes. Compare these results with Exercise 10.3 for the case when the chosen velocities are large, both without ($d = 0$ in Eq. 10.16) and with ($d = 1$ in Eq. 10.16) subgrid diffusion. An example is in **Variable_conductivity_markers2D.m**.

11

2D thermomechanical code structure

> **Theory:** Principal steps of a coupled thermomechanical solution with finite differences and marker-in-cell techniques. Organization of a thermomechanical code for the case of viscous, multi-component flows. Adding self-gravity. Handling free planetary surface with a weak layer approach.
> **Exercises:** Building 2D thermomechanical codes.

11.1 What do we expect from geodynamic codes?

Before describing possible structures for thermomechanical codes, let us discuss what we actually expect from a state of the art twenty-first century, numerical geodynamic modelling tool. Today, as numerical modelling of geodynamic and planetary processes is in the 'new millennium' (although it is only around 50 years old, see Introduction), geoscientists are targeting realistic modelling of lithospheric, mantle and planetary dynamics and plate tectonics (e.g. Gerya and Yuen, 2003a, 2007; Moresi et al, 2007; Zhong et al., 2007; Tackley, 2008; Crameri et al., 2012b; Gerya, 2011, 2014b and references therein). The rheology of crustal and mantle rocks depends strongly on the temperature, strain rate, volatile content, grain size and the fluid pressure. Physical and dynamical circumstances imposed by the sharply varying viscosity represent a major challenge for solving the momentum equation in geodynamics, unlike those found in the oceanographic or atmospheric sciences. Another complication is due to the variable thermal conductivity in the heat conservation equation. The thermal conductivity of various crustal and mantle rocks is notably different and is also a strong function of temperature, pressure and mineralogy which causes numerical difficulties compared to the constant thermal conductivity situation. Finally, all physical properties of rocks, including viscosity and conductivity, vary strongly with the rock chemical composition and/or mineralogy. In various geodynamic situations, these result in sharp fronts involving *multi-component flows* (i.e. flows composed of many rock types with contrasting compositions and physical properties). Therefore we should consider at least three important technical requirements for numerical geodynamic modelling:

(1) the ability to conserve stresses under conditions that involve sharply discontinuous viscosity distributions;
(2) the ability to conserve heat fluxes under conditions that involve sharply varying conductivity and temperature gradients at thermal or chemical layers with temperature-dependent conductivity;
(3) the ability to conserve physical properties, such as temperature field, chemical species, viscosity and density in flows with a strong advection character.

11.2 Thermomechanical code structure

Since we want to address all the above requirements in our state of the art '*all-in-one*' thermomechanical code, let us discuss in detail how this can be achieved by using a marker-in-cell algorithm, combined with conservative staggered finite differences for primitive variable (pressure-velocity) formulation. The code structure should reflect the physical relations of momentum, continuity, temperature and advection equations. For example, the momentum and continuity equations have to be solved simultaneously (as we always did) to obtain values for velocity that are present in both equations. The temperature equation requires values of adiabatic and shear heating that are computed from the velocity, pressure, stress and strain rate fields. Therefore, the temperature equation can only be solved after solving the momentum and continuity equations. The advection equation accounts for temporal changes in spatial distribution of various physical parameters (including temperature) and requires a velocity field. This equation should thus be solved after solving the momentum, continuity and temperature equations.

The flow chart in Fig. 11.1 gives an example of a structure for a numerical, thermo-mechanical viscous 2D code that uses staggered finite differences and the marker-in-cell technique (*SFD-MIC*) to solve the momentum, continuity and temperature equations. The principal steps of the algorithm are as follows.

(1) Calculate the scalar physical properties (η_m, ρ_m, α_m, ρC_{Pm}, k_m, etc.) for each marker and interpolate these properties, as well as advected temperature from the markers to Eulerian nodes (Chapters 8, 10). Apply boundary conditions for the nodal temperatures interpolated from markers.
(2) Discretize the 2D momentum and continuity equations with a pressure-velocity formulation on a staggered grid by composing and solving the global matrix problem with a direct method, which is used because of its stability and high accuracy (Chapter 7).
(3) Define a suitable displacement time step Δt_m for markers (typically limiting maximal displacement to 0.01–1.0 of the minimal grid step) based on the velocity field computed in Step 2 (Chapter 8).
(4) Calculate the shear and adiabatic heating terms $H_{s(i,j)}$ and $H_{a(i,j)}$ at the Eulerian nodes (Chapter 9).

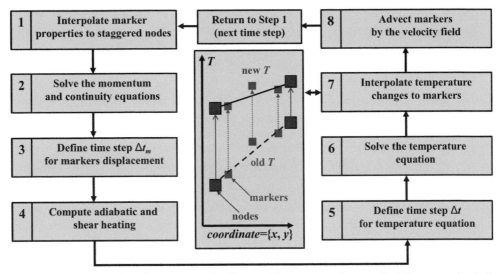

Fig. 11.1. Flow chart that gives an example of a possible structure of a numerical thermomechanical viscous 2D code, which employs staggered finite differences and marker-in-cell technique (SFD-MIC) for solving the momentum, continuity and temperature equations (Gerya and Yuen, 2003a).

(5) Define a suitable time step Δt for the temperature equation. We take the smallest time step of three time step limiters: given absolute time step limit; given optimal marker displacement time step limit (see Step 3); given absolute nodal temperature change limit (typically 1–20 K) (Chapter 10).

(6) Discretize the temperature equation in a Lagrangian formulation, with implicit time stepping, and solve the global matrix problem with a direct method (Chapter 10).

(7) Interpolate the calculated nodal temperature changes (see Step 7 in Fig. 11.1) from the Eulerian nodes to the markers and calculate new marker temperatures $T_m^{corrected}$ taking into account physical diffusion on a subgrid (marker) level (Chapter 10).

(8) Use a fourth-order in space, first-order in time classical Runge–Kutta scheme combined with *continuity-based velocity interpolation* (Chapter 8) to advect all markers throughout the mesh according to the globally calculated velocity field (see Step 2). Return to Step 1 to perform the next time step.

An important aspect of solving thermomechanical viscous problems is that we have to provide an initial condition for temperature before starting the computation. Figure 11.2 shows the geometry of an irregularly spaced, fully staggered numerical grid corresponding to the algorithm outlined above. The irregularly spaced grid is extremely useful in handling geodynamic situations with localization phenomena and multiple scale features which will be further discussed in Chapters 18, 20 and 21. Note that by doing the programming exercises for previous chapters *we have already implemented* (surprise, surprise ... ☺) one-by-one all the steps of the above computational algorithm and we will now discuss in full detail how to connect these separate steps in order to create a state of the art code out of our 'embryonic' 2D codes.

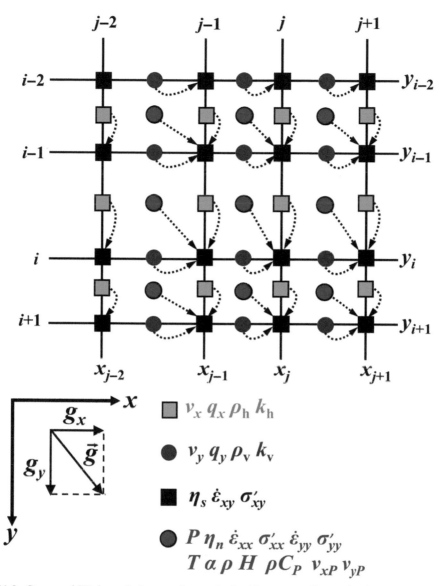

Fig. 11.2. Staggered 2D, irregularly spaced numerical grid corresponding to the algorithm presented in Fig. 11.1.

Steps 1, 7, 8: Interpolation of physical parameters between markers and nodes

According to our marker-in-cell approach, the temperature field and the rock type are represented by values assigned for the markers initially distributed on a fine regular marker mesh with a small ($\leq$ half of the marker grid distance) random displacement. Other scalar properties such as density, viscosity, thermal conductivity etc. are computed for each marker at every time step in accordance to the rock type associated with each marker. This approach allows us to minimize the amount of storage associated with markers to only 3 floating point values (two coordinates and temperature) and one integer (rock type). The number of different rock types in one experiment is typically limited to a few tens at most. There are several equations for computing material properties (viscosity, density, etc.) associated with each rock type, which allow us to compute these properties for a marker of a given type, based on the local temperature, pressure and strain rate at this marker. The effective values of all these properties at various Eulerian nodal points are computed from the markers at each time step by using a bilinear interpolation (Eq. 8.18). Note that viscosity for shear (η_s) and normal (η_n) stresses is interpolated separately to the basic nodes and cell centres, respectively (Fig. 11.2). Similarly, density and thermal conductivity are interpolated to both v_x and v_y nodes (Fig. 11.2). Temperature, density, thermal expansion, volumetric heat capacity, radioactive and latent heating (Chapter 9) are also interpolated to pressure nodes.

Local interpolation schemes could be used for viscosity interpolation that take into account markers found within half a grid spacing distance from the nodes, in both the horizontal and vertical directions (see dashed boundary in Fig. 8.8). The statistical weights of these markers thus vary from 0.25 to 1 (one can also renormalize these weights to vary from 0 to 1). The local interpolation of viscosity typically (but not always) allows for a more accurate solution of the momentum equation in the case of a strongly variable viscosity (Gerya and Yuen, 2007; Schmeling et al., 2008; Duretz et al., 2011a). In cases when we require a more accurate solution of the temperature equation with strongly variable thermal conductivity, localized interpolation schemes from markers could also be used for the thermal conductivity k (see blue circles and green squares in Fig. 11.2). Thermal boundary conditions should be applied to the obtained nodal values after interpolating the temperature from markers. This precludes accumulating an interpolation error, which will otherwise grow along the model boundaries as the number of time steps increases.

We use a bilinear interpolation procedure (Fig. 8.9, Eq. 8.19) for interpolating scalar properties (including calculated temperature changes), vectors and tensors from the corresponding Eulerian nodal points (see different types of Eulerian nodes in Fig. 11.2) to the markers. Equation (8.19) is used uniformly, for interpolating stresses, strain rates, pressure, temperature and other properties from respective nodal points to markers. A higher order continuity-based interpolation scheme (Chapter 8) is used for velocity, which involves both staggered v_x, v_y nodes and averaged v_{xP}, v_{yP} velocity components calculated at cell centres (Fig. 11.2). Since our staggered grid represents, in fact, the superposition of four simple rectangular grids corresponding to different scalar fields, vectors and tensors (see four different symbols for grid points in Fig 11.2), these Eulerian grids are used individually for interpolating the respective field variables. Defining the indices of the four nodal points that surround a given marker is done on the basis of the bisection procedure (Fig. 8.10).

Step 2: Solving the momentum and continuity equations

In 2D, we have two Stokes equations of slow, viscous incompressible flow in a uniform gravity field

$$x \text{ Stokes equation} \quad \frac{\partial \sigma'_{xx}}{\partial x} + \frac{\partial \sigma'_{xy}}{\partial y} - \frac{\partial P}{\partial x} = -\rho g_x, \tag{11.1}$$

$$y \text{ Stokes equation} \quad \frac{\partial \sigma'_{yy}}{\partial y} + \frac{\partial \sigma'_{yx}}{\partial x} - \frac{\partial P}{\partial y} = -\rho g_y, \tag{11.2}$$

$$\sigma'_{xx} = 2\eta\dot{\varepsilon}_{xx},$$

$$\sigma'_{yy} = 2\eta\dot{\varepsilon}_{yy},$$

$$\sigma'_{xy} = \sigma'_{yx} = 2\eta\dot{\varepsilon}_{xy},$$

$$\dot{\varepsilon}_{xx} = \frac{\partial v_x}{\partial x},$$

$$\dot{\varepsilon}_{yy} = \frac{\partial v_y}{\partial y},$$

$$\dot{\varepsilon}_{xy} = \frac{1}{2}\left(\frac{\partial v_x}{\partial y} + \frac{\partial v_y}{\partial x}\right).$$

Since in 2D by definition $\sigma'_{yy} = -\sigma'_{xx}$ and for incompressible fluid $\dot{\varepsilon}_{yy} = -\dot{\varepsilon}_{xx}$, we can also avoid any nodal storage for σ'_{yy} and $\dot{\varepsilon}_{yy}$.

The conservation of mass is given by the incompressible, 2D continuity equation

$$\frac{\partial v_x}{\partial x} + \frac{\partial v_y}{\partial y} = 0. \tag{11.3}$$

We use the standard procedure described in Chapter 7 (Figs. 7.11, 7.12, Eqs. 7.5, 7.6) for the formulation of FD schemes to represent the momentum equations (11.1) and (11.2) in a stress-conservative form.

The x Stokes equation (11.1) is discretized at the horizontal velocity node $v_{x(i,j)}$ (cf., dotted arrows in Fig. 11.2 for indexing of different parameters)

$$2\frac{\sigma'_{xx(i,j+1)} - \sigma'_{xx(i,j)}}{x_{j+1} - x_{j-1}} + \frac{\sigma'_{xy(i,j)} - \sigma'_{xy(i-1,j)}}{y_i - y_{i-1}} - 2\frac{P_{(i,j+1)} - P_{(i,j)}}{x_{j+1} - x_{j-1}} = -\rho_{h(i,j)}g_x, \tag{11.4}$$

where

$$\sigma'_{xy(i-1,j)} = 2\eta_{s(i-1,j)}\left(\frac{v_{x(i,j)} - v_{x(i-1,j)}}{y_i - y_{i-2}} + \frac{v_{y(i-1,j+1)} - v_{y(i-1,j)}}{x_{j+1} - x_{j-1}}\right),$$

$$\sigma'_{xy(i,j)} = 2\eta_{s(i,j)} \left(\frac{v_{x(i+1,j)} - v_{x(i,j)}}{y_{i+1} - y_{i-1}} + \frac{v_{y(i,j+1)} - v_{y(i,j)}}{x_{j+1} - x_{j-1}} \right),$$

$$\sigma'_{xx(i,j)} = 2\eta_{n(i,j)} \frac{v_{x(i,j)} - v_{x(i,j-1)}}{x_j - x_{j-1}},$$

$$\sigma'_{xx(i,j+1)} = 2\eta_{n(i,j+1)} \frac{v_{x(i,j+1)} - v_{x(i,j)}}{x_{j+1} - x_j},$$

where i and j indexes denote, respectively, the vertical and horizontal indexes of the nodal points corresponding to the different physical parameters (Fig. 11.2) within the staggered grid.

The y Stokes equation (11.2) is discretized at the vertical velocity node $v_{y(i,j)}$

$$2\frac{\sigma'_{yy(i+1,j)} - \sigma'_{yy(i,j)}}{y_{i+1} - y_{i-1}} + \frac{\sigma'_{xy(i,j)} - \sigma'_{xy(i,j-1)}}{x_j - x_{j-1}} - 2\frac{P_{(i+1,j)} - P_{(i,j)}}{y_{i+1} - y_{i-1}} = -\rho_{v(i,j)} g_y, \tag{11.5}$$

where $\sigma'_{xy(i,j)}$ is given above for Eq. (11.4),

$$\sigma'_{xy(i,j-1)} = 2\eta_{s(i,j-1)} \left(\frac{v_{x(i+1,j-1)} - v_{x(i,j-1)}}{y_{i+1} - y_{i-1}} + \frac{v_{y(i,j)} - v_{y(i,j-1)}}{x_j - x_{j-2}} \right),$$

$$\sigma'_{yy(i,j)} = 2\eta_{n(i,j)} \frac{v_{y(i,j)} - v_{y(i-1,j)}}{y_i - y_{i-1}},$$

$$\sigma'_{yy(i+1,j)} = 2\eta_{n(i+1,j)} \frac{v_{y(i+1,j)} - v_{y(i,j)}}{y_{i+1} - y_i}.$$

The continuity equation (11.3) is discretized at the pressure node $P_{i,j}$

$$\frac{v_{x(i,j)} - v_{x(i,j-1)}}{x_j - x_{j-1}} + \frac{v_{y(i,j)} - v_{y(i-1,j)}}{y_i - y_{i-1}} = 0. \tag{11.6}$$

After composing all equations, we solve the resulting global matrix problem by an accurate, direct method in order to obtain the solution for velocity and pressure, which was explained in detail in Chapter 7.

Steps 4-7: Solving the temperature equation

In order to avoid numerical diffusion of temperature, we use the Lagrangian form of heat conservation equation and advect the temperature with markers with the technique described in Chapter 10. The temperature equation is formulated in 2D for the case of variable thermal conductivity and takes into account heat generation H from variable sources including radioactive (H_r), adiabatic (H_a), shear (H_s) and latent (H_L) heat production:

$$\rho C_P \frac{DT}{Dt} = -\frac{\partial q_x}{\partial x} - \frac{\partial q_y}{\partial y} + H_r + H_a + H_s + H_L, \qquad (11.7)$$

where

$$q_x = -k\frac{\partial T}{\partial x},$$

$$q_y = -k\frac{\partial T}{\partial y},$$

$$H_r = \text{const},$$

$$H_a = T\alpha\frac{DP}{Dt} = T\alpha\left(\frac{\partial P}{\partial x}v_x + \frac{\partial P}{\partial y}v_y\right),$$

$$H_s = \sigma'_{xx}\dot{\varepsilon}_{xx} + \sigma'_{yy}\dot{\varepsilon}_{yy} + \sigma'_{xy}\dot{\varepsilon}_{xy} + \sigma'_{yx}\dot{\varepsilon}_{yx} = 2\sigma'_{xx}\dot{\varepsilon}_{xx} + 2\sigma'_{xy}\dot{\varepsilon}_{xy}.$$

This equation takes into account the adiabatic heating term, which is in some contradiction with using the incompressible fluid approximation in the momentum and continuity equations (Eqs. 11.1–11.3). However, this is a common simplification (called the *extended Boussinesq approximation*) in numerical geodynamic modelling. It is frequently adopted because of the very small thermal expansion and compressibility of crustal and mantle rocks in the absence of phase transformations (Chapter 2). The calculation of H_a can be simplified by neglecting deviations (which are relatively small in most cases) of the dynamic pressure gradients $\partial P/\partial x$ and $\partial P/\partial y$ from ρg_x and ρg_y values

$$H_a \approx T\alpha\rho\left(g_x v_x + g_y v_y\right).$$

In the version of thermomechanical staggered grid shown in Fig. 11.2, temperature points are located in cell centres and include external nodes located outside of the model domain (cf. pressure points in Fig. 7.15). This grid configuration allows for more accurate (second-order) formulation of thermal boundary conditions for heat fluxes that are located in respective staggered velocity points (Fig. 11.2). Boundary conditions for temperature at the external temperature nodes should be formulated similarly to the external velocity points (Chapter 7, derive as an exercise). We use a standard formal procedure (Fig. 10.5, Eq. 10.9) for the formulation of a heat flux conservative FD scheme in order to discretize the temperature equation (11.7)

$$\rho C_{P(i,j)} \frac{T_{i,j} - T^o_{i,j}}{\Delta t} = -\frac{q_{x(i,j)} - q_{x(i,j-1)}}{x_j - x_{j-1}} - \frac{q_{y(i,j)} - q_{y(i-1,j)}}{y_i - y_{i-1}} \tag{11.8}$$
$$+ H_{r(i,j)} + H_{a(i,j)} + H_{s(i,j)} + H_{L(i,j)}$$

or, by grouping the known parameters on the right hand side,

$$\rho C_{P(i,j)} \frac{T_{i,j}}{\Delta t} + \frac{q_{x(i,j)} - q_{x(i,j-1)}}{x_j - x_{j-1}} + \frac{q_{y(i,j)} - q_{y(i-1,j)}}{y_i - y_{i-1}}$$

$$= \rho C_{P(i,j)} \frac{T^o_{i,j}}{\Delta t} + H_{r(i,j)} + H_{a(i,j)} + H_{s(i,j)} + H_{L(i,j)}$$

where

$$q_{x(i,j-1)} = -2k_{h(i,j-1)} \frac{T_{i,j} - T_{i,j-1}}{x_j - x_{j-2}},$$

$$q_{x(i,j)} = -2k_{h(i,j)} \frac{T_{i,j+1} - T_{i,j}}{x_{j+1} - x_{j-1}},$$

$$q_{y(i-1,j)} = -2k_{v(i-1,j)} \frac{T_{i,j} - T_{i-1,j}}{y_i - y_{i-2}},$$

$$q_{y(i,j)} = -2k_{v(i,j)} \frac{T_{i+1,j} - T_{i,j}}{y_{i+1} - y_{i-1}},$$

$$H_{a(i,j)} = T^o_{i,j} \alpha_{i,j} \rho_{i,j} \left(\frac{v_{x(i,j)} + v_{x(i,j-1)}}{2} g_x + \frac{v_{y(i,j)} + v_{y(i,j-1)}}{2} g_y \right),$$

$$H_{s(i,j)} = 2\sigma'_{xx(i,j)} \dot{\varepsilon}_{xx(i,j)} + \tfrac{1}{2}\sigma'_{xy(i-1,j-1)} \dot{\varepsilon}_{xy(i-1,j-1)} + \tfrac{1}{2}\sigma'_{xy(i,j-1)} \dot{\varepsilon}_{xy(i,j-1)}$$

$$+ \tfrac{1}{2}\sigma'_{xy(i-1,j)} \dot{\varepsilon}_{xy(i-1,j)} + \tfrac{1}{2}\sigma'_{xy(i,j)} \dot{\varepsilon}_{xy(i,j)}$$

where T^o and T are temperature values for the current (t) and next ($t + \Delta t$) time instant, respectively; $H_{r(i,j)}$, $H_{s(i,j)}$, $H_{a(i,j)}$, $\alpha_{i,j}$, $\rho_{i,j}$, $\rho C_{P(i,j)}$, $\sigma_{xy(i,j)}$, $\dot{\varepsilon}_{xy(i,j)}$, $v_{x(i,j)}$ etc. are values of the corresponding parameters for various nodes of the staggered grid (Fig. 11.2).

To solve the temperature equation, we solve the global matrix problem with a direct method. The matrix also contains the linear equations associated with the thermal boundary conditions. The overall numbering of $T_{i,j}$ for the *global temperature matrix* takes into account that $(N_x + 1) \times (N_y + 1)$ temperature nodes (including external ones) are present in the staggered grid

$$in_T = (j - 1) \times (N_y + 1) + i. \tag{11.9}$$

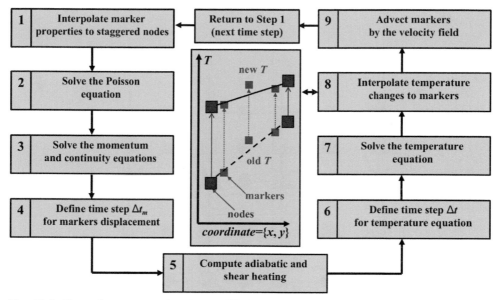

Fig. 11.3. Flow chart representing the modified algorithm of the numerical thermomechanical viscous 2D code allowing the modelling of self-gravitating planetary bodies with a fully staggered Cartesian grid (Gerya and Yuen, 2007). Note the new Step 2 compared to Fig. 11.1.

For advection of temperature, we use the same marker-in-cell technique (Chapter 10) as is used for advection of other material properties. The interpolation of the calculated temperature changes, from the Eulerian nodal points to the moving markers, reduces numerical diffusion in an efficient manner (Chapter 10). Non-physical small scale (subgrid) temperature oscillations appearing on markers (Fig. 10.7a) in the case of strong chaotic mixing are removed (Fig. 10.7b) by using a consistent subgrid diffusion operation (Eqs. 10.15–10.19).

11.3 Adding self-gravity and free planetary surface

For the case when we want to model the dynamics of self-gravitating planetary bodies, the numerical algorithms presented above can be easily modified to take a variable gravity field and a free planetary surface into account. This is done by solving the Poisson equation for gravity potential (Chapter 2) after computing the nodal density field from markers, and before solving the momentum and continuity equations (Fig. 11.3). A staggered grid corresponding to this new algorithm is shown in Fig. 11.4.

In this grid, the gravitational potential Φ (Chapter 2) is defined at the cell centres (i.e. at the pressure nodes, including external ones) and computed according to the 2D Poisson equation

$$\frac{\partial^2 \Phi}{\partial x^2} + \frac{\partial^2 \Phi}{\partial y^2} = 4K\pi G\rho, \qquad (11.10)$$

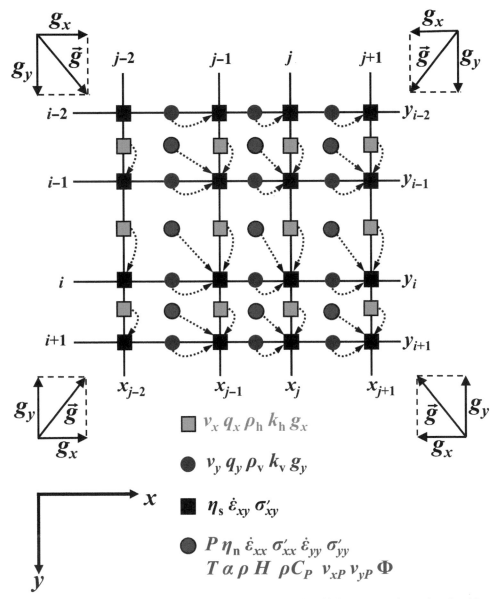

Fig. 11.4. Staggered 2D irregularly spaced numerical grid, which corresponds to the algorithm presented in Fig. 11.3.

where $G = 6.672 \times 10^{-11}$ (Nm^2)/kg^2 is the gravitational constant and K depends on the 3D geometry of the self-gravitating body modelled in 2D ($K = 1$ and $K = 2/3$ stand for cylindrical and spherical geometries, respectively). The factor $K = 2/3$ scales the 2D gravity field inside a cylinder of constant density ρ

$$\Phi(r)_{cylindrical} = \pi G \rho r^2, \quad g(r)_{cylindrical} = -\frac{\partial \Phi(r)_{cylindrical}}{\partial r} = -2\pi G \rho r,$$

to a 3D gravity field inside a sphere of the same density

$$\Phi(r)_{spherical} = \frac{2}{3}\pi G \rho r^2, \quad g(r)_{spherical} = -\frac{\partial \Phi(r)_{spherical}}{\partial r} = -\frac{4}{3}\pi G \rho r,$$

where r is the distance from the centre of the cylinder/sphere. It should be mentioned that this simplified scaling does not allow the exact reproduction of a spherical gravity field in 2D. In particular, the gravitational acceleration is noticeably overestimated *outside the self-gravitating body* since it is proportional to $1/r$ for a cylindrical gravity field and to $1/r^2$ for a spherical one. Our scaling approach allows us to capture changes in an *internal gravity field* that acts on a self-gravitating body with a changing internal density distribution (Lin et al., 2009). In many cases this is sufficient for the purposes of modelling internal planetary processes.

Discretizing Eq. (11.10) in 2D is rather simple and uses a 5-node stencil typical for approximating the Poisson equation with finite differences

$$2\frac{\Phi_{(i,j+1)} - \Phi_{(i,j)}}{(x_{j+1} - x_{j-1})(x_j - x_{j-1})} - 2\frac{\Phi_{(i,j)} - \Phi_{(i,j-1)}}{(x_j - x_{j-2})(x_j - x_{j-1})}$$

$$- 2\frac{\Phi_{(i+1,j)} - \Phi_{(i,j)}}{(y_{i+1} - y_{i-1})(y_i - y_{i-1})} - 2\frac{\Phi_{(i,j)} - \Phi_{(i-1,j)}}{(y_i - y_{i-2})(y_i - y_{i-1})}$$

$$= 4K\pi G \rho_{(i,j)}, \tag{11.11}$$

where the i and j indexes denote, respectively, the horizontal and vertical positions of the nodal points corresponding to the different physical parameters within the staggered grid (cf. dotted arrows in Fig. 11.4). We solve the global matrix problem by a direct method. In this matrix, the Poisson equation (11.11) is combined with linear equations for the boundary conditions defined at all cell centres located around a *circular equipotential boundary*. The overall numbering of unknowns for the *global gravity potential matrix* is the same as for temperature (Eq. 11.9).

The gravitational acceleration vector components are then defined at respective Eulerian nodes (see blue circles and green squares in Fig. 11.4) by numerical differentiation

$$g_{x(i,j)} = -2\frac{\Phi_{(i,j+1)} - \Phi_{(i,j)}}{x_{j+1} - x_{j-1}}, \tag{11.12}$$

$$g_{y(i,j)} = -2\frac{\Phi_{(i+1,j)} - \Phi_{(i,j)}}{y_{i+1} - y_{i-1}}. \tag{11.13}$$

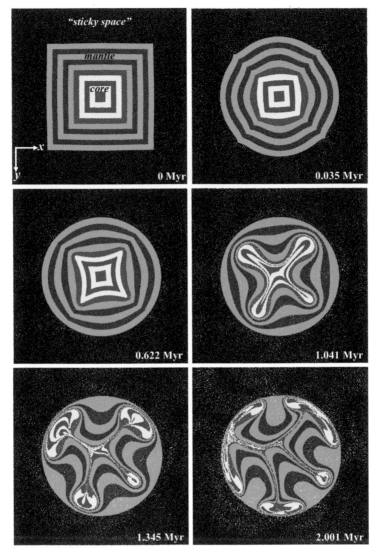

Fig. 11.5. Numerical modelling of a self-gravitating planetary body using the spherical-Cartesian method. The model is computed with the code **i2planet.m** associated with this chapter.

The new, locally defined gravity acceleration components $g_{x(i,j)}$ and $g_{y(i,j)}$ are then included in the right hand side of the x Stokes and y Stokes equations (11.4) and (11.5) instead of the globally defined g_x and g_y values, respectively.

Numerical modelling of deformation of a self-gravitating planetary body requires computation of the gravity field which changes with time in response to variations in density distribution inside the planet. Changes in shape of the planet and the related planetary surface deformation should also be considered. In order to tackle these requirements, one can use a 'spherical-Cartesian' approach (Honda et al., 1993;

Gerya and Yuen, 2007; Lin et al., 2009) that allows the computation of self-gravitating bodies of arbitrary form on Cartesian grids including the presence of a free planetary surface.

(1) The body is surrounded by the weak medium 'sticky space' (e.g. Fig. 11.5) of very low density (≤ 1 kg/m^3) and low viscosity allowing for a high (10^1–10^6) viscosity contrast at the planetary surface.
(2) The gravity field is computed by solving the Poisson equation for the gravitational potential (Eq. 11.11) inside a circular equipotential boundary (Chapter 2) based on the density distribution portrayed by the markers at each time step.
(3) While solving the momentum equation, the components of the gravitational acceleration vector are computed locally by numerical differentiation of the gravitational potential (Eqs. 11.12, 11.13) at the corresponding nodal points.

As seen in Fig. 11.5, the spontaneously formed planetary surface is numerically stable under conditions of very strong internal deformation inside the planet. In addition, a spontaneously forming spherical/cylindrical shape of the body is characteristic for a stable density distribution (i.e. when density increases toward the core of the body, see final stages of Fig. 11.5). No evidence for non-spherical Cartesian grid dependence of this stable shape was discerned (Lin et al., 2009).

Programming exercises

Exercise 11.1.

Program a thermomechanical code (Figs. 11.1, 11.2) based on a regularly spaced staggered grid for the case of variable viscosity and thermal conductivity. Combine the thermal and mechanical solutions programmed for Exercise 10.4 and Exercise 8.7, respectively. Use the same model setup with the mantle plume and the 'sticky air' layer and prescribe different temperature, thermal conductivity, volumetric heat capacity, thermal expansion and radiogenic heat production for the mantle ($T = 1500$ K, $k = 3$ W/(m K), $\rho C_P = 3.3 \times 10^6$ J/(m^3 K), $\alpha = 3 \times 10^{-5}$ 1/K, $H_r = 2 \times 10^{-8}$ W/m^3), plume ($T = 1800$ K, $k = 2$ W/(m K), $\rho C_P = 3.2 \times 10^6$ J/(m^3 K), $\alpha = 2 \times 10^{-5}$ 1/K, $H_r = 3 \times 10^{-8}$ W/m^3) and 'sticky air' ($T = 273$ K, $k = 300$ W/(m K), $\rho C_P = 3.3 \times 10^6$ J/(m^3 K), $\alpha = 0$, $H_r = 0$). Use insulating boundary conditions at the left and right boundaries and constant temperature conditions at the top ($T = 273$ K) and bottom ($T = 1500$ K). Include radioactive, shear and adiabatic heating terms (Eq. 11.8) in the temperature equation (shear and adiabatic heating were already programmed for Exercise 9.3). Take the initial time step 10^{10} s. Program an increase of the time step by 20% per time step. Restrict the time step by limiting the maximal marker displacement by 0.5 of the grid step per time step. Restrict the time step by limiting the maximal temperature changes to 20 K per time step. If these changes are bigger, then reduce the time step proportionally and repeat the solution of the temperature equation for the second time (solving the temperature equation is computationally inexpensive compared to the momentum and continuity equations). An example is in **i2vis.m**.

Exercise 11.2.
Modify Exercise 11.1 by adding self-gravitation (Figs. 11.3, 11.4, Eqs. 11.10–11.13).
Define boundary condition $\Phi = 0$ for all nodes located at the distance greater than half
model size from the centre of the model, which is the assumed distance to the equipotential
boundary $\Phi = 0$. Use a 10 000 × 10 000 km^2 model with 101 × 101 grid (4 × 4 markers per
cell) and insulating no slip boundaries. Initial model geometry corresponds to the
6000 × 6000 km^2 'squared planet' ($\eta = 10^{21}$ Pa s, $\rho = 3300$ kg/m^3, $T = 1500$ K, $k = 3$ W/
(m K), $\rho C_P = 3.3 \times 10^6$ J/(m^3 K), $\alpha = 3 \times 10^{-5}$ 1/K, $H_r = 2 \times 10^{-8}$ W/m^3) with a 3000 ×
3000 km^2 low density core ($\eta = 10^{17}$ Pa s, $\rho = 3000$ kg/m^3, $T = 1800$ K, $k = 2$ W/(m K), $\rho C_P = 3.0 \times 10^6$ J/(m^3 K), $\alpha = 2 \times 10^{-5}$ 1/K, $H_r = 2 \times 10^{-7}$ W/m^3). The planet is surrounded by the
'sticky space' ($\eta = 10^{17}$ Pa s, $\rho = 0$ kg/m^3, $T = 273$ K, $k = 3000$ W/(m K), $\rho C_P = 3.3 \times 10^6$ J/(m^3 K), $\alpha = 0$, $H_r = 0$). Use the same conditions for the time step as in
Exercise 11.1. Look at the changes in the planetary shape with time toward a stable self-
gravitating geometry. Observe changes in the internal structure of the planet related to the
instability of the low density core (Fig. 11.5). An example is in **i2planet.m**.

12

Elasticity and plasticity

Theory: Elastic rheology. Rotation of elastic stresses. Maxwell visco-elastic rheology. Plastic rheology. Plastic yielding criterion. Plastic flow potential. Plastic flow rule. Visco-elasto-plastic rheology.
Exercises: Stress buildup/relaxation with a visco-elastic Maxwell rheology, elastic stress rotation programming.

12.1 Why should we care about elasticity and plasticity?

As mentioned in the Introduction, rocks behave elastically on a relatively short time scale ($<10^4$ years) and, therefore, modelling of relatively fast processes within the Earth's crust and mantle (e.g. magma intrusion) should take into account the elastic properties of rocks. On the other hand, rocks at cold temperatures can also be subjected to localized *brittle* (at low pressure) and *plastic* (at higher pressure) deformation, which leads to shear zones and fracture zones in natural rock complexes. Therefore, if we want to account for this broad range of geodynamic conditions in our models, we should generally consider the *visco-elasto-plastic rheology* of rocks and be able to model such a complex rheology with our thermomechanical numerical codes. This chapter discusses elastic and plastic rheological behaviours and compares them to the viscous rheology. Please note that in geology viscous and plastic deformation are often used as synonyms, which is thus different from the continuum mechanics convention used here.

12.2 Elastic rheology

The elastic rheology assumes proportionality of stress and strain (Fig. 12.1). This is expressed by Hooke's law

$$\tau = E\gamma, \tag{12.1}$$

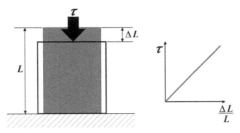

Fig. 12.1. Relationship between applied stress τ and deformation ΔL, of an elastic body with initial length L.

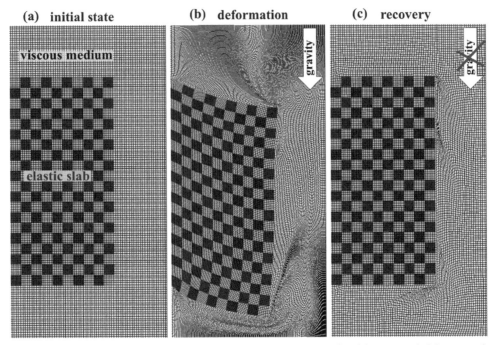

Fig. 12.2. Reversible deformation of initially unstressed (a) elastic slab surrounded by a weak viscous medium (Gerya and Yuen, 2007). Deformation of the slab in (b) is caused by a vertical gravity field. When gravity is 'removed', the deformed slab recovers its original shape (c) while the surrounding medium remains deformed since viscous deformation is irreversible.

where τ is applied stress, $\gamma = \Delta L/L$ is elastic strain (i.e. displacement ΔL normalized to the initial length L of the deforming body) and E is the proportionality (elasticity) coefficient. In contrast to viscous deformation, elastic deformation is reversible: if the load applied on an elastic body is removed, the body recovers its original state (Fig. 12.2). This shape recovery effect is reproduced with a spring.

For an isotropic body in 3D, the elastic relationship is written in tensorial form as

$$\sigma_{ij} = \lambda \delta_{ij} \varepsilon_{kk} + 2\mu \varepsilon_{ij}, \tag{12.2}$$

$$\varepsilon_{ij} = \frac{1}{2}\left(\frac{\partial u_i}{\partial x_j} + \frac{\partial u_j}{\partial x_i}\right), \tag{12.3}$$

$$\varepsilon_{kk} = \varepsilon_{xx} + \varepsilon_{yy} + \varepsilon_{zz}, \tag{12.4}$$

where x_i and x_j are the spatial coordinates (x, y, z), σ_{ij} are components of stress, u_i and u_j are components of the *material displacement* vector $\vec{u} = (u_x, u_y, u_z)$ so that $v_i = Du_i/Dt$, where v_i are components of the velocity vector $\vec{v} = (v_x, v_y, v_z)$, ε_{ij} are components of the strain tensor so that $\dot{\varepsilon}_{ij} = D\varepsilon_{ij}/Dt$, where $\dot{\varepsilon}_{ij}$ are components of the strain rate tensor (Chapter 4), ε_{kk} is volumetric strain (*cubical dilatation*) and λ and μ are two elastic parameters termed *Lamé's constants* (which depend on pressure, temperature and composition). The material displacement characterizes the absolute amount of movement (i.e. is similar to ΔL in Fig. 12.1), while the strain tensor ε_{ij} reflects the relative intensity of the deformation (i.e. is similar to $\Delta L/L$ in Fig. 12.1).

Formulating pressure through mean normal stress (Chapter 4), we can obtain the following relations

$$P = -\frac{\sigma_{kk}}{3} = -\frac{\sigma_{xx} + \sigma_{yy} + \sigma_{zz}}{3}, \tag{12.5}$$

$$P = -\frac{3\lambda + 2\mu}{3}\left(\varepsilon_{xx} + \varepsilon_{yy} + \varepsilon_{zz}\right) = -B\varepsilon_{kk}, \tag{12.6}$$

$$B = \lambda + \frac{2}{3}\mu, \tag{12.7}$$

where B is the *bulk modulus* (see Chapter 2) also called *incompressibility*; it establishes the relation between mean stress (pressure) and volumetric strain. Accordingly, the deviatoric stresses σ'_{ij} can be formulated as

$$\sigma'_{ij} = \sigma_{ij} - \frac{\sigma_{kk}}{3}\delta_{ij} = \sigma_{ij} + P\delta_{ij}, \tag{12.8}$$

$$\sigma'_{ij} = \sigma_{ij} - B\varepsilon_{kk}\delta_{ij} = 2\mu\left(\varepsilon_{ij} - \frac{1}{3}\delta_{ij}\varepsilon_{kk}\right), \tag{12.9}$$

$$\sigma'_{ij} = 2\mu\varepsilon'_{ij}, \tag{12.10}$$

$$\varepsilon'_{ij} = \varepsilon_{ij} - \frac{1}{3}\delta_{ij}\varepsilon_{kk}, \tag{12.11}$$

where ε'_{ij} are components of the deviatoric strain tensor and μ is the *shear modulus* or *rigidity*, which is one of the Lamé constants. The elastic shear modulus establishes the relationship between the deviatoric stress and deviatoric strain for the elastic rheology (μ is

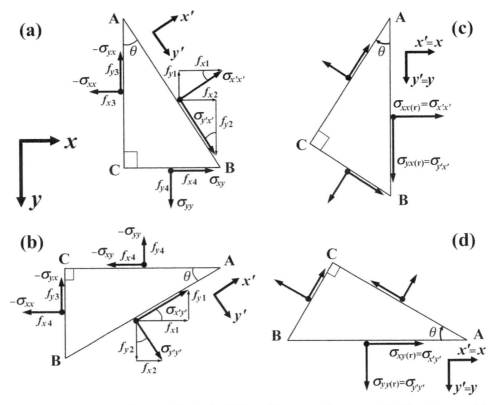

Fig. 12.3. Force balance for a small triangle ABC used for computing σ_{xx}, σ_{yx} (a), (c) and σ_{yy}, σ_{xy} (b), (d) stress components after a clockwise solid body rotation of the triangle by an angle θ around point A.

thus somewhat similar to the shear viscosity η that defines the analogous relationship between the deviatoric stress and the deviatoric strain rate for a viscous rheology, see Eq. 5.11).

12.3 Rotation of elastic stresses

An important peculiarity of stress behaviour in a deforming elastic medium consists in changes in local stress orientation due to the rotation of material points. This rotation is caused by a *rigid body rotation component*, which is present in the velocity field and changes the orientation of principal stress axes for moving Lagrangian points. Stress rotation changes stress tensor components but not the stress invariants P and σ_{II}.

The method of computing various elastic stress components after rotation of a Lagrangian point by an angle θ can be derived on the basis of analysing the force balance for a small triangle ABC, as shown in Fig. 12.3. Let us take a small triangle with two sides parallel to the x and y axes oriented as shown in Fig. 12.3a. The force equilibrium condition requires that forces acting on the outside of the triangle balance each other and that the resulting force is zero

$$f_x = f_{x1} + f_{x2} + f_{x3} + f_{x4} = 0, \tag{12.12}$$

$$f_y = f_{y1} + f_{y2} + f_{y3} + f_{y4} = 0, \tag{12.13}$$

where f are forces acting on different sides of the triangle from the outside. These forces can be computed from shear and normal stresses as discussed in Chapter 4 (note that arrows shown on side AC correspond to the counterforce part of σ_{xx} and σ_{yx} which therefore have a minus sign)

$$|\text{AC}| = |\text{AB}| \cos(\theta),$$

$$|\text{BC}| = |\text{AB}| \sin(\theta),$$

$$f_{x1} - \sigma_{x'x'} |\text{AB}| \cos(\theta),$$

$$f_{x2} = \sigma_{y'x'} |\text{AB}| \sin(\theta),$$

$$f_{x3} = -\sigma_{xx} |\text{AC}| = -\sigma_{xx} |\text{AB}| \cos(\theta),$$

$$f_{x4} - \sigma_{xy} |\text{BC}| = \sigma_{xy} |\text{AB}| \sin(\theta),$$

$$f_{y1} = -\sigma_{x'x'} |\text{AB}| \sin(\theta),$$

$$f_{y2} = \sigma_{y'x'} |\text{AB}| \cos(\theta),$$

$$f_{y3} = -\sigma_{yx} |\text{AC}| = -\sigma_{yx} |\text{AB}| \cos(\theta),$$

$$f_{y4} = \sigma_{yy} |\text{BC}| = \sigma_{yy} |\text{AB}| \sin(\theta),$$

where $|\text{AB}|$, $|\text{AC}|$ and $|\text{BC}|$ are the lengths of the respective triangle sides. Then, Eqs. (12.12) and (12.13) can be converted to yield

$$\frac{f_x}{|\text{AB}|} = \sigma_{x'x'} \cos(\theta) + \sigma_{y'x'} \sin(\theta) - \sigma_{xx} \cos(\theta) + \sigma_{xy} \sin(\theta) = 0, \tag{12.14}$$

$$\frac{f_y}{|\text{AB}|} = -\sigma_{x'x'} \sin(\theta) + \sigma_{y'x'} \cos(\theta) - \sigma_{yx} \cos(\theta) + \sigma_{yy} \sin(\theta) = 0. \tag{12.15}$$

By multiplying Eq. (12.14) by $\cos(\theta)$ and Eq. (12.15) by $\sin(\theta)$, we obtain

$$\sigma_{x'x'} \cos^2(\theta) + \sigma_{y'x'} \sin(\theta)\cos(\theta) - \sigma_{xx} \cos^2(\theta) + \sigma_{xy} \sin(\theta)\cos(\theta) = 0, \tag{12.16}$$

$$-\sigma_{x'x'} \sin^2(\theta) + \sigma_{y'x'} \cos(\theta)\sin(\theta) - \sigma_{yx} \cos(\theta)\sin(\theta) + \sigma_{yy} \sin^2(\theta) = 0. \tag{12.17}$$

By subtracting Eq. (12.17) from Eq. (12.16) we get

$$\sigma_{x'x'}\left(\cos^2(\theta) + \sin^2(\theta)\right) - \sigma_{xx}\cos^2(\theta)$$
$$+ \sigma_{xy}\cos(\theta)\sin(\theta) + \sigma_{yx}\cos(\theta)\sin(\theta) - \sigma_{yy}\sin^2(\theta) = 0,$$

which can be further simplified to

$$\sigma_{x'x'} = \sigma_{xx}\cos^2(\theta) + \sigma_{yy}\sin^2(\theta) - \sigma_{xy}\sin(2\theta) \qquad (12.18)$$

by using $\sigma_{yx} = \sigma_{xy}$ and the trigonometric relations $\sin^2(\theta) + \cos^2(\theta) = 1$ and $2\sin(\theta)\cos(\theta) = \sin(2\theta)$.

Similarly, by multiplying Eq. (12.14) by $\sin(\theta)$ and Eq. (12.15) by $\cos(\theta)$, and adding them to each other, we obtain the following expression for $\sigma_{y'x'}$

$$\sigma_{y'x'} = \frac{1}{2}\left(\sigma_{xx} - \sigma_{yy}\right)\sin(2\theta) + \sigma_{xy}\cos(2\theta) \qquad (12.19)$$

using the trigonometric relation $\cos^2(\theta) - \sin^2(\theta) = \cos(2\theta)$ (verify as an exercise).

Obviously, after the triangle has been rotated clockwise by an angle θ around point A (Fig. 12.3c), the AB side becomes parallel to the y axis and $\sigma_{y'x'}$, $\sigma_{y'x'}$, will correspond to the respective stress components $\sigma_{yx(r)}$, $\sigma_{yx(r)}$, for the rotated system.

Similarly, by analysing Fig. 12.3b,d, the following expressions for the corrected stress components $\sigma_{y'y'}$ and $\sigma_{x'y'}$ can be obtained (verify as an exercise):

$$\sigma_{y'y'} = \sigma_{xx}\sin^2(\theta) + \sigma_{yy}\cos^2(\theta) + \sigma_{xy}\sin(2\theta), \qquad (12.20)$$

$$\sigma_{x'y'} = \frac{1}{2}\left(\sigma_{xx} - \sigma_{yy}\right)\sin(2\theta) + \sigma_{xy}\cos(2\theta). \qquad (12.21)$$

Equations (12.19) and (12.21) are obviously equivalent since $\sigma_{yx} = \sigma_{xy}$. Note that rotation does not change the first stress invariant (mean normal stress, pressure) and thus

$$\sigma_{x'x'} + \sigma_{y'y'} = \sigma_{xx} + \sigma_{yy} = -2P.$$

Equations for rotating deviatoric normal stresses are similar to Eqs. (12.18) and (12.20) since subtracting mean stress does not change the form of these expressions, hence

$$\sigma'_{x'x'} = \sigma'_{xx}\cos^2(\theta) + \sigma'_{yy}\sin^2(\theta) - \sigma'_{xy}\sin(2\theta), \qquad (12.22)$$

$$\sigma'_{y'y'} = \sigma'_{xx}\sin^2(\theta) + \sigma'_{yy}\cos^2(\theta) + \sigma'_{xy}\sin(2\theta). \qquad (12.23)$$

The condition $\sigma'_{x'x'} + \sigma'_{y'y'} = 0$ is also satisfied due to $\sigma'_{xx} + \sigma'_{yy} = 0$.

If the angle θ is very small and tends to 0, then $\cos(\theta)$ tends to 1, $\sin^2(\theta)$ tends to 0, and $\sin(\theta)$ tends to θ. In this case, Eqs. (12.18)–(12.23) can be simplified to the *Jaumann co-rotation formulas*, which are often used in numerical modelling of elastic problems to account for effects of rigid body rotation of stresses:

$$\sigma'_{x'x'} = \sigma'_{xx} - \sigma'_{xy}2\theta, \tag{12.24}$$

$$\sigma'_{y'y'} = \sigma'_{yy} + \sigma'_{xy}2\theta, \tag{12.25}$$

$$\sigma'_{x'y'} = \sigma'_{xy} + \left(\sigma'_{xx} - \sigma'_{yy}\right)\theta. \tag{12.26}$$

In a complex velocity field, the intensity of rotation is defined by the local rotation rate ω, which can be computed from the local velocity field as

$$\omega = \frac{\partial\theta}{\partial t} = \frac{1}{2}\left(\frac{\partial v_y}{\partial x} - \frac{\partial v_x}{\partial y}\right). \tag{12.27}$$

Note that the expressions for stress rotation and formulation of ω depend on

- the convention for normal stresses – whether they are taken to be positive (here) or negative (e.g. Turcotte and Schubert, 2002) under extension,
- orientation of the x and y axes,
- whether the clockwise rotation direction is taken to be positive (here) or negative (e.g. Turcotte and Shubert, 2002).

In particular, Eq. (12.27) may become invalid if the definitions of stresses, axes and rotation are different from those used here. In this case, Eqs. (12.18)–(12.27) should not be used 'automatically', but should rather be *re-derived* on the basis of a similar analysis.

In 3D, the rotation rate is represented by a *rotation rate tensor* ω with components defined as

$$\omega_{ij} = \frac{1}{2}\left(\frac{\partial v_i}{\partial x_j} - \frac{\partial v_j}{\partial x_i}\right), \tag{12.28}$$

where i and j are coordinate indexes; x_i and x_j are spatial coordinates and the view at the ij-plane should be taken in the direction of the third axis. In contrast to *symmetric* stress and strain rate tensors $\left(\text{i.e. } \sigma_{ij} = \sigma_{ji}, \dot{\varepsilon}_{ij} = \dot{\varepsilon}_{ji}\right)$, the rotation rate tensor is *anti-symmetric* (i.e., $\omega_{ij} = -\omega_{ji}$) and the diagonal components of this tensor are always equal to zero (i.e., $\omega_{xx} = \omega_{yy} = \omega_{zz} = 0$). One possible way of computing both total and deviatoric stress rotation in 3D is to use the general form of the *Jaumann stress rate:*

total stress

$$\dot{\sigma}_{ij(Jaumann)} = \omega_{ik}\sigma_{kj} - \sigma_{ik}\omega_{kj}, \tag{12.29}$$

deviatoric stress

$$\dot{\sigma}'_{ij(Jaumann)} = \omega_{ik}\sigma'_{kj} - \sigma'_{ik}\omega_{kj}, \tag{12.30}$$

where $\dot{\sigma}'_{ij(Jaumann)}$ is the rate of change for the rotating σ_{ij} stress component in *the moving non-rotating reference frame* and the repeated index k indicates a summation. Using Eq. (12.30) in 3D for example, the σ'_{xx} deviatoric stress component yields

$$\dot{\sigma}'_{xx(Jaumann)} = \omega_{xx}\sigma'_{xx} + \omega_{xy}\sigma'_{yx} + \omega_{xz}\sigma'_{zx} - \sigma'_{xx}\omega_{xx} - \sigma'_{xy}\omega_{yx} - \sigma'_{xz}\omega_{zx}, \tag{12.31a}$$

$$\dot{\sigma}'_{xx(Jaumann)} = \omega_{xy}\sigma'_{yx} + \omega_{xz}\sigma'_{zx} - \sigma'_{xy}\omega_{yx} - \sigma'_{xz}\omega_{zx}, \tag{12.31b}$$

$$\dot{\sigma}'_{xx(Jaumann)} = 2\sigma'_{xy}\omega_{xy} + 2\sigma'_{xz}\omega_{xz}, \tag{12.31c}$$

where

$$\omega_{xx} = \frac{1}{2}\left(\frac{\partial v_x}{\partial x} - \frac{\partial v_x}{\partial x}\right) = 0, \quad \omega_{xy} = \frac{1}{2}\left(\frac{\partial v_x}{\partial y} - \frac{\partial v_y}{\partial x}\right) = -\omega_{yx}$$

and

$$\omega_{xz} = \frac{1}{2}\left(\frac{\partial v_x}{\partial z} - \frac{\partial v_z}{\partial x}\right) = -\omega_{zx},$$

according to Eq. (12.28). Similar derivations can also be done for other deviatoric stress components (verify these as an exercise):

$$\dot{\sigma}'_{yy(Jaumann)} = 2\sigma'_{yx}\omega_{yx} + 2\sigma'_{yz}\omega_{yz}, \tag{12.32}$$

$$\dot{\sigma}'_{zz(Jaumann)} = 2\sigma'_{zx}\omega_{zx} + 2\sigma'_{zy}\omega_{zy}, \tag{12.33}$$

$$\dot{\sigma}'_{xy(Jaumann)} = \dot{\sigma}'_{yx(Jaumann)} = \left(\sigma'_{xx} - \sigma'_{yy}\right)\omega_{yx} + \omega_{xz}\sigma'_{zy} - \sigma'_{xz}\omega_{zy}, \tag{12.34}$$

$$\dot{\sigma}'_{xz(Jaumann)} = \dot{\sigma}'_{zx(Jaumann)} = \left(\sigma'_{xx} - \sigma'_{zz}\right)\omega_{zx} + \omega_{xy}\sigma'_{yz} - \sigma'_{xy}\omega_{yz}, \tag{12.35}$$

$$\dot{\sigma}'_{yz(Jaumann)} = \dot{\sigma}'_{zy(Jaumann)} = \left(\sigma'_{yy} - \sigma'_{zz}\right)\omega_{zy} + \omega_{yx}\sigma'_{xz} - \sigma'_{yx}\omega_{xz}. \tag{12.36}$$

In 2D, Eqs. (12.31)–(12.33) are equivalent to previously derived Eqs. (12.24)–(12.27) since $\omega = \omega_{yx} = -\omega_{xy}$.

It is worth mentioning that in continuum mechanics, in addition to the Jaumann stress rate, typically used in geodynamic modelling, there are a large variety of other *objective stress rate* formulations such as the Truesdell rate, the Green–Naghdi rate, the Oldroyd rate, the convective rate etc. (e.g. Shabana, 2008). However, the other objective derivatives (beside Jaumann) do not preserve the deviatoric property of a tensor. Hence, using them in our case is not straightforward as our formulations assume a splitting of stress into a deviatoric and a homogeneous (pressure) part.

It should also be mentioned that numerical calculations of stress rotation based on the Jaumann stress rate are less accurate than the analytical formulas (12.18)–(12.21) for 2D finite angle rotation. Accuracy of the calculation can be improved by applying a smaller time step and/or using higher order Runge–Kutta integration schemes *in stress space* (e.g. Farrington et al., 2014; Popov et al., 2014a; Kaus et al., 2016; Popov, personal communication). Another efficient way to improve the accuracy of 3D stress rotation calculations is to extend the 2D analytical finite angle stress rotation approach to 3D (e.g. Rubinstein and Atluri, 1983; Popov et al., 2014a; Kaus et al., 2016). One optimal numerical 3D finite angle stress rotation algorithm has been recently proposed and tested by Popov et al. (2014a, personal communication), which is based on the work of Rubinstein and Atluri (1983). According to this approach, 3D rotation is represented by a vorticity pseudo-vector ($\vec{\omega}$), which has three components:

$$\omega_x = \frac{1}{2}\left(\frac{\partial v_z}{\partial y} - \frac{\partial v_y}{\partial z}\right), \quad \omega_y = \frac{1}{2}\left(\frac{\partial v_x}{\partial z} - \frac{\partial v_z}{\partial x}\right), \quad \omega_z = \frac{1}{2}\left(\frac{\partial v_y}{\partial x} - \frac{\partial v_x}{\partial y}\right). \tag{12.37}$$

The 3D algorithm can then be summarized as follows (Popov et al., 2014a, personal communication).

(1) Compute the vorticity vector magnitude:

$$\omega_{mag} = \sqrt{\omega_x^2 + \omega_y^2 + \omega_z^2}. \tag{12.38}$$

(2) Compute the unit rotation vector $\vec{n}$, which also has three components:

$$n_x = \frac{\omega_x}{\omega_{mag}}, \quad n_y = \frac{\omega_y}{\omega_{mag}}, \quad n_z = \frac{\omega_z}{\omega_{mag}}. \tag{12.39}$$

(3) Integrate the incremental rotation angle:

$$\theta = \omega_{mag}\Delta t. \tag{12.40}$$

(4) Evaluate the rotation matrix using the Euler–Rodrigues formula:

$$R_{mat} = \cos(\theta)\begin{pmatrix} 1 & 0 & 0 \\ 0 & 1 & 0 \\ 0 & 0 & 1 \end{pmatrix} + \sin(\theta)\begin{pmatrix} 0 & -n_z & n_y \\ n_z & 0 & -n_x \\ -n_y & n_x & 0 \end{pmatrix} + \left(1 - \cos(\theta)\right)\begin{pmatrix} n_x n_x & n_x n_y & n_x n_z \\ n_y n_x & n_y n_y & n_y n_z \\ n_z n_x & n_z n_y & n_z n_z \end{pmatrix}. \tag{12.41}$$

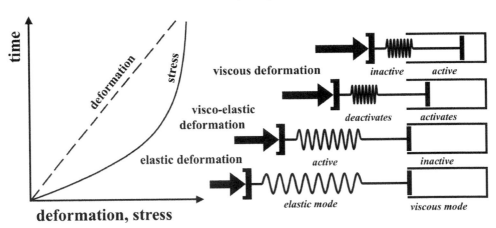

deformation, stress

Fig. 12.4. Schematic representation of the Maxwell visco-elastic rheology. The solid line shows a typical pattern of visco-elastic stress buildup under the condition of a linearly growing deformation (dashed line). Deformation initially starts in an elastic mode (see shortening of the spring) but with the growing stress, viscous deformation activates and becomes dominant (see movement of the dashpot) and stress stabilizes. The length of the black arrows reflects the magnitude of the applied stress at different moments in time.

(5) Compute the rotated stress matrix by multiplying R_{mat}, σ' and R_{mat}^{T} matrices:

$$\sigma'_{rotated} = R_{mat} \times \sigma' \times R_{mat}^{T}, \tag{12.42}$$

where R_{mat}^{T} is the *transpose* of the matrix R_{mat} (R_{mat}^{T} is obtained by reflecting the elements of the matrix R_{mat} along its main diagonal).

This simple 3D stress rotation algorithm is very efficient and has superior accuracy compared to other methods (Popov et al., 2014a; Kaus et al., 2016).

12.4 Maxwell visco-elastic rheology

A visco-elastic rheology is obtained by combining viscous (Eq. 5.11) and elastic (Eq. 12.10) rheological relations under certain physical assumptions (e.g. Turcotte and Schubert, 2002, Chapter 7, Section 7–10). In numerical geodynamic modelling, Maxwell visco-elastic rheology is the most commonly used type; it is based on the assumption that both viscous and elastic deformations are happening under the same applied deviatoric stress σ'_{ij} such that the bulk deviatoric strain rate $\dot{\varepsilon}'_{ij}$ can be represented as a sum of viscous $\dot{\varepsilon}'_{ij(viscous)}$ and elastic $\dot{\varepsilon}'_{ij(elastic)}$ strain rates (see Fig. 12.4 for the relationship between the viscous and elastic deformations of a Maxwell body)

$$\dot{\varepsilon}'_{ij} = \dot{\varepsilon}'_{ij(viscous)} + \dot{\varepsilon}'_{ij(elastic)}, \tag{12.43}$$

which can then be obtained from the rheological relations (5.11) and (12.10) under the assumption that the shear modulus in Eq. (12.10) is constant:

$$\dot{\varepsilon}'_{ij(viscous)} = \frac{1}{2\eta}\sigma'_{ij},\tag{12.44}$$

$$\dot{\varepsilon}'_{ij(elastic)} = \frac{D\varepsilon'_{ij(elastic)}}{Dt} = \frac{D}{Dt}\left(\frac{\sigma'_{ij}}{2\mu}\right) = \frac{1}{2\mu}\frac{D\sigma'_{ij}}{Dt},\tag{12.45}$$

$$\dot{\varepsilon}'_{ij} = \frac{1}{2\eta}\sigma'_{ij} + \frac{1}{2\mu}\frac{D\sigma'_{ij}}{Dt},\tag{12.46}$$

where $D\sigma'_{ij}/Dt$ is the *objective co-rotational time derivative* of the deviatoric stress component σ'_{ij} in a *moving non-rotating Lagrangian reference frame* which accounts for the effects of stress rotation discussed above.

It should be noted that Eq. (12.45) in fact re-defines the shear modulus μ as

$$\frac{D\sigma'_{ij}}{Dt} = 2\mu\dot{\varepsilon}'_{ij},\tag{12.47}$$

which (at variable μ) differs from its original definition by Eq. (12.10). Similarly, the bulk modulus can be re-defined compared to Eq. (12.6) as

$$\frac{DP}{Dt} = -B\dot{\varepsilon}_{kk},\tag{12.48}$$

which also directly relates (through the continuity equation, Chapter 1) the bulk modulus B and the compressibility β (Chapter 2)

$$B = \frac{1}{-\dot{\varepsilon}_{kk}}\frac{DP}{Dt} = \frac{1}{-\text{div}(\vec{v})}\frac{DP}{Dt} = \frac{1}{\frac{Dln(\rho)}{Dt}}\frac{DP}{Dt} = \frac{1}{\frac{\partial ln(\rho)}{\partial P}\frac{DP}{Dt}}\frac{DP}{Dt} = \frac{1}{\beta}.\tag{12.49}$$

The modified definitions of shear and bulk moduli by Eqs. (12.47)–(12.49) are consistent with the *Biot non-linear elasticity theory of incremental deformation* (Biot, 1965). These definitions are more convenient for modelling deformation of complex visco-elastic and visco-elasto-plastic materials, such as rocks, under conditions of broad variations of pressure and temperature causing variations of elastic moduli.

12.5 Plastic rheology

The plastic rheology assumes that an absolute shear stress limit σ_{yield} exists for a body and after reaching this limit *plastic yielding* occurs (Fig. 12.5). Like viscous deformation, plastic yielding is irreversible, but the pattern of deformation is notably different

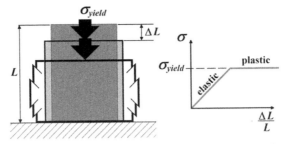

Fig. 12.5. Relationship between the applied stress σ and deformation ΔL, of an elastic-plastic body with initial length L. Elastic deformation changes to plastic yielding after reaching a stress limit σ_{yield}.

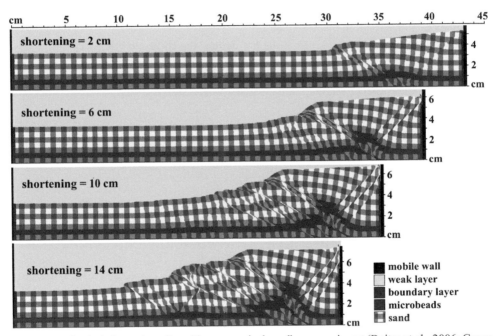

Fig. 12.6. Plastic deformation of sand in a numerical sandbox experiment (Buiter et al., 2006; Gerya and Yuen, 2007). Irreversible localized plastic deformation forms multiple, highly deformed shear zones separating relatively undeformed blocks.

(Fig. 12.6): plastic creep is localized and forms multiple highly deformed shear zones separating relatively undeformed blocks.

The plastic strength σ_{yield} of a rock generally depends on the mean stress of the solid (P^s) and on the pore fluid pressure (P^f, we will discuss this in more detail in Chapter 16). Based on the simplified Griffith theory (Griffith, 1924; Cai, 2010; Keller et al., 2013) it could be assumed that:

for dry rocks

$$\sigma_{yield} = \sigma_c + \gamma_{int}P \quad \text{when} \quad P > \frac{\sigma_c - \sigma_t}{1 - \gamma_{int}} \quad \text{(confined fractures)}, \qquad (12.50a)$$

$$\sigma_{yield} = \sigma_t + P \quad \text{when} \quad P < \frac{\sigma_c - \sigma_t}{1 - \gamma_{int}} \quad \text{(tensile fractures)}, \qquad (12.50b)$$

and for fluid-bearing rocks

$$\sigma_{yield} = \sigma_c + \gamma_{int}\left(P^t - P^f\right) \quad \text{when} \quad P^t - P^f > \frac{\sigma_c - \sigma_t}{1 - \gamma_{int}} \quad \text{(confined fractures)} \qquad (12.51a)$$

$$\sigma_{yield} = \sigma_t + \left(P^t - P^f\right) \quad \text{when} \quad P^t - P^f < \frac{\sigma_c - \sigma_t}{1 - \gamma_{int}} \quad \text{(tensile fractures)} \qquad (12.51b)$$

$$P^t = (1 - \phi)P^s + \phi P^f, \qquad (12.52)$$

where $\sigma_{yield} \geq 0$ (*non-negative strength* requirement), $\gamma_{int} = \sin(\varphi)$ is the *internal friction coefficient* (φ is the *angle of internal friction*), $P^t = P$ is *total pressure*, ϕ is fluid volume fraction (porosity), σ_c and σ_t are respectively *compressive strength* and *tensile strength* ($2 \leq (\sigma_c/\sigma_t) \leq 8$, Cai, 2010; Keller et al., 2013). The compressive strength σ_c is related to the material *cohesion* $C = \sigma_c/\cos(\varphi)$. For the majority of dry fractured crystalline rocks, γ_{int} is independent of composition and varies from 0.85 at $P < 200$ MPa to 0.60 at higher pressure (Byerlee's law, Byerlee, 1978; Brace and Kohlstedt, 1980). The plastic strength of dry rocks thus strongly increases with pressure to a limit of several GPa. The strength is limited by the *Peierls mechanism* of plastic deformation (Evans and Goetze, 1979; Kameyama et al., 1999; Karato, 2008).

The Peierls mechanism is a temperature-dependent mode of plastic deformation (also called exponential creep), which takes over from the dislocation creep mechanism at elevated stresses (typically above 0.1 GPa). Rheological relationships (flow law) for Peierls creep are commonly represented as (Katayama and Karato, 2008)

$$\dot{\varepsilon}_{II} = A_{Peierls}\sigma_{II}{}^2 \exp\left\{ -\frac{E_a + PV_a}{RT} \left[1 - \left(\frac{\sigma_{II}}{\sigma_{Peierls}}\right)^k \right]^q \right\}, \qquad (12.53)$$

where $\sigma_{Peierls}$, $A_{Peierls}$, E_a, V_a, k and q are experimentally determined parameters (Chapter 6): $\sigma_{Peierls}$ is the Peierls stress that limits the strength of the material and is similar to σ_{yield} in Eq. (12.50), $A_{Peierls}$ is a material constant for Peierls creep (Pa^{-2} s^{-1}), E_a is the activation energy (J/mol) and V_a is the activation volume (J/Pa), the exponents k and q depend on the shape and geometry of obstacles that limit the dislocation motion. Microscopic models show that k and q should have the following range $0 < k \leq 1$, $1 \leq q \leq 2$ (Kocks et al., 1975). In contrast to other types of plasticity, Peierls creep is already activated at stresses that are notably lower than the actual strength of material given by

$\sigma_{Peierls}$. This deformation mechanism is very important, in particular for deformation of subducting slabs characterized by lowered temperature and elevated stresses compared to the surrounding mantle (e.g. Duretz et al., 2011b; Karato et al., 2001), or for lithospheric-scale shear localization (Kaus and Podladchikov, 2006).

Further information about various types of plasticity used in geosciences such as Mohr–Coulomb, Von Mises, Drucker–Prager and Treska models can be found in the books of Turcotte and Schubert (2002) and Ranalli (1995).

12.6 Visco-elasto-plastic rheology

In nature, the general behaviour of rocks is visco-elasto-plastic, which can be formulated by decomposing the bulk deviatoric strain rate $\dot{\varepsilon}'_{ij}$ into the three respective components:

$$\dot{\varepsilon}'_{ij} = \dot{\varepsilon}'_{ij(viscous)} + \dot{\varepsilon}'_{ij(elastic)} + \dot{\varepsilon}'_{ij(plastic)}, \tag{12.54}$$

where

$$\dot{\varepsilon}'_{ij(viscous)} = \frac{1}{2\eta} \sigma'_{ij}, \tag{12.55}$$

$$\dot{\varepsilon}'_{ij(elastic)} = \frac{1}{2\mu} \frac{D\sigma'_{ij}}{Dt}, \tag{12.56}$$

$$\dot{\varepsilon}'_{ij(plastic)} = 0 \text{ for } \sigma_{II} < \sigma_{yield},$$
$$\dot{\varepsilon}'_{ij(plastic)} = \chi \frac{\partial G_{plastic}}{\partial \sigma'_{ij}} = \chi \frac{\sigma'_{ij}}{2\sigma_{II}} \text{ for } \sigma_{II} = \sigma_{yield}, \tag{12.57}$$

$$G_{plastic} = \sigma_{II}, \tag{12.58}$$

$$\sigma_{II} = \sqrt{\frac{1}{2} \sigma'_{ij}{}^2}, \tag{12.59}$$

where $D\sigma'_{ij}/Dt$ is the objective co-rotational time derivative of the deviatoric stress component σ'_{ij}, Eq. (12.57) is the *plastic flow rule*, σ_{yield} is the *plastic yield strength* for a given rock, σ_{II} is the second invariant of the deviatoric stress tensor (Eq. 4.5), $G_{plastic}$ is the *plastic flow potential*, which reflects the amount of mechanical energy per unit volume that supports plastic deformation, $Pa = N/m^2 = J/m^3$, and χ is the plastic multiplier, which satisfies *the Drucker–Prager* plastic yielding condition

$$\sigma_{II} = \sigma_{yield}. \tag{12.60}$$

The plastic multiplier is a *variable scaling coefficient*, which connects, in a uniform way, components of the plastic strain rate $\dot{\varepsilon}'_{ij(plastic)}$ with the deviatoric stress components when the yielding condition (12.60) is reached. This coefficient is unknown a priori and should be determined locally at each moment of time by solving Eqs. (12.54)–(12.60), based on local values of stresses σ'_{ij}, strain rates $\dot{\varepsilon}'_{ij}$, viscosity η and shear modulus μ. Based on Eqs. (12.57)–(12.60), we can conclude that the plastic multiplier χ is equal to the double of the second invariant of the deviatoric plastic strain rate tensor $\dot{\varepsilon}_{II(plastic)}$ (derive as an exercise):

$$\chi = 2\dot{\varepsilon}_{II(plastic)}, \tag{12.61}$$

$$\dot{\varepsilon}_{II(plastic)} = \sqrt{\frac{1}{2}\dot{\varepsilon}'^{2}_{ij(plastic)}}. \tag{12.62}$$

Plastic flow rule formulation (Eq. 12.57) includes deviatoric stress and strain rate components only and, consequently, the plastic potential formulation (Eq. 12.58) is the same for both *dilatant* (i.e. increasing volume during plastic deformation) and *non-dilatant* materials. For plastic deformation of dilatant materials, this formulation is, therefore, combined with the equation describing volumetric changes:

$$\Gamma_{plastic} = \text{div}(\vec{v}) = 2\sin(\psi)\dot{\varepsilon}_{II(plastic)}, \tag{12.63}$$

where ψ is the dilatation angle, which generally depends on total plastic strain.

Analytical exercise

Exercise 12.1.
Derive the equation for visco-elastic stress buildup/relaxation with time, using the Maxwell model (Eq. 12.46) under conditions of constant strain rate $\dot{\varepsilon}'_{ij}$, viscosity η, shear modulus μ, and no stress rotation involved such that $D\sigma'_{ij}/Dt = D\sigma'_{ij}/Dt$. Take the initial state of stress to be given by σ'_{0ij} and integrate Eq. (12.46) (now reformulated in terms of stress σ'_{ij}) to obtain the analytical solution. Reformulate the resulting equation in terms of Maxwell relaxation time

$$t_{Maxwell} = \frac{\eta}{\mu}, \tag{12.64}$$

which defines the characteristic time scale for visco-elastic stress relaxation.

Programming exercises

Exercise 12.2.
Use the analytical formula from the previous example to compute and compare stress-time curves for the following parameters: (1) $\sigma'_{0ij} = 0$ Pa, $\dot{\varepsilon}'_{ij} = 10^{-14}$ 1/s, $\eta = 10^{21}$ Pa s, $\mu = 10^{10}$ Pa; (2) $\sigma'_{0ij} = 10^{8}$ Pa, $\dot{\varepsilon}'_{ij} = 10^{-14}$ 1/s, $\eta = 10^{21}$ Pa s, $\mu = 10^{10}$ Pa; (3) $\sigma'_{0ij} = 0$ Pa, $\dot{\varepsilon}'_{ij} = 10^{-15}$ 1/s,

$\eta = 10^{21}$ Pa s, $\mu = 10^{10}$ Pa; (4) $\sigma'_{0ij} = 0$ Pa, $\dot{\varepsilon}'_{ij} = 10^{-14}$ 1/s, $\eta = 10^{22}$ Pa s, $\mu = 10^{10}$ Pa; (5) $\sigma'_{0ij} = 0$ Pa, $\dot{\varepsilon}'_{ij} = 10^{-14}$ 1/s, $\eta = 10^{21}$ Pa s, $\mu = 10^{11}$ Pa; (6) $\sigma'_{0ij} = 0$ Pa, $\dot{\varepsilon}'_{ij} = 10^{-14}$ 1/s, $\eta = 10^{22}$ Pa s, $\mu = 10^{11}$ Pa. Try to understand how the different parameters control the stress buildup/ relaxation. An example is in **Viscoelastic_stress.m**.

Exercise 12.3.

Compute the visco-elasto-plastic stress buildup and observe the changes in the viscous, elastic and plastic strain rates with time when using the parameters from case (4) of the previous example. Assume the condition that the visco-elastic stress σ'_{ij} must not exceed the yield stress limit of 1.5×10^8 Pa. Use Eq. (12.55) to compute the viscous strain rate. Use Eqs. (12.54), (12.56) and (12.57) to compute elastic and plastic strain rates. Consider that after reaching the yielding limit, the stress in the visco-elasto-plastic material should not change anymore and therefore $D\sigma'_{ij}/Dt = D\sigma'_{ij}/Dt = 0$. An example is in **Viscoelastoplastic_strain_rate.m**.

Exercise 12.4.

Modify Exercise 6.3 by adding Peierls creep for the high stress region ($>10^8$ Pa). Compute the effective viscosity for this region by analogy to Eqs. (6.16)–(6.18) as follows

$$\frac{1}{\eta_{eff}} = \frac{1}{\eta_{diff}} + \frac{1}{\eta_{disl}} + \frac{1}{\eta_{Peierls}}, \tag{12.65}$$

where $\eta_{Peierls}$ is Peierls creep viscosity defined on the basis of Eqs. (12.53) and (6.4) as

$$\eta_{Peierls} = \frac{1}{2A_{Peierls}\sigma_{II}} \exp\left\{ \frac{E_a + PV_a}{RT} \left[1 - \left(\frac{\sigma_{II}}{\sigma_{Peierls}} \right)^k \right]^q \right\}. \tag{12.66}$$

Use the following Peierls creep parameters (Evans and Goetze, 1979; Katayama and Karato, 2008): dry olivine, $k = 1$, $q = 2$, $A_{Peierls} = 10^{-4.2}$ Pa^{-2} s^{-1}, $E_a = 540\ 000$ J/mol, $\sigma_{Peierls} = 9.1 \times 10^9$ Pa; wet olivine, $k = 1$, $q = 2$, $A_{Peierls} = 10^{-4.2}$ Pa^{-2} s^{-1}, $\sigma_{Peierls} = 2.9 \times 10^9$ Pa; $E_a = 4\ 300\ 000$ J/mol. Note that the activation energy E_a for Peierls creep is the same as the activation energy for respective dislocation creep (Table 6.1). Note that the stress σ_{II} used for computing effective viscosity with Eq. (12.66) should always be limited by $\sigma_{Peierls}$, which corresponds to the upper strength limit. An example is in **Peierls_creep.m**.

Exercise 12.5.

Compare different formulas for the case of 2D stress rotation with constant angular velocity $\omega = 1$. Use different numbers of time steps per one revolution time interval ($t = 2\pi/\omega$) and compare differences between the initial ($\sigma'_{xx} = 10^6$ Pa, $\sigma'_{yy} = -10^6$ Pa, $\sigma'_{xy} = 0$) and final stress states for the following approaches: (A) analytical (Eqs. 12.21–12.23), (B) Jaumann (Eqs. 12.24–12.26), (C) Jaumann with effective stress rate (Eqs. 12.31, 12.32, 12.34 with $\omega_{yx} = -\omega_{xy} = \omega$, $\omega_{xz} = \omega_{zx} = \omega_{yz} = \omega_{zy} = 0$, $\sigma'_{xz} = \sigma'_{yz} = \sigma'_{zz} = 0$) computed with a fourth-order (in stress space) Runge–Kutta method (Chapter 8) and (D) 3D finite angle rotation

(Eqs. 12.37–12.42 with $\omega_x = \omega_y = 0$, $\omega_z = \omega$, $\sigma'_{xz} = \sigma'_{yz} = \sigma'_{zz} = 0$). In the last case use matrix multiplication (*) and transpose (') to compute the rotated stress matrix

$$\text{SIGMArotated} = \text{Rmat} * \text{SIGMA} * \text{Rmat'}.$$

Use the analytical solution as a reference to evaluate error for the other stress rotation methods. An example is in **Stress_rotation.m**.

Exercise 12.6.

Combine the previous exercise with stress advection and rotation on markers in a squared 1×1 m^2 model area. Define a rotational $v_x = \omega(0.5 - y)$, $v_y = \omega(x - 0.5)$, $\omega = 1$ (Exercise 4.1, Fig. 4.4) velocity field around the model centre on a regular staggered 51×51 grid with 4×4 randomly distributed markers per cell. Compute $\omega_{yx(i,j)} = \frac{1}{2}\left(\partial v_y / \partial x - \partial v_x / \partial y\right)$ for basic nodal points using finite differences (i.e., similarly to $\dot{\varepsilon}_{xy(i,j)}$ in Eq. (9.18), Fig. 9.3a). Define deviatoric stress components on the markers. Move markers located within 0.5 m distance from the model centre by the prescribed nodal velocity field (Exercise 8.6) and rotate their stresses using ω_{yx} values interpolated from the grid at each time step. Trace the evolution of the changes in coordinates and stresses for a selected marker. Test different stress rotation approaches. An example is in **Stress_rotation_markers.m**.

13

2D implementation of visco-elasto-plasticity

> **Theory:** Numerical implementation of visco-elasto-plastic rheology. Organization of a thermomechanical code in the case of a 2D, visco-elasto-plastic medium. Numerical treatment of plasticity. Visco-elasto-plastic iterations. Visco-plastic rheology.
>
> **Exercises:** Programming 2D thermomechanical codes with a visco-elasto-plastic rheology.

13.1 Viscous-like reformulation of visco-elasto-plasticity

One way of reformulating the visco-elasto-plastic rheological model (12.54) for easy implementation into a viscous code (which we have already programmed) is based on using finite differences in time. The deviatoric stress σ'_{ij} is expressed as a function of the total deviatoric strain rate $\dot{\varepsilon}'_{ij}$ from the visco-elasto-plastic constitutive relationship (12.54), by using first-order finite differences in time to represent the objective co-rotational time derivatives $D\sigma'_{ij}/Dt$ of visco-elastic stresses

$$\frac{\underset{\frown}{D}\sigma'_{ij}}{Dt} = \frac{\sigma'_{ij} - \sigma'^{o}_{ij}}{\Delta t}, \tag{13.1}$$

$$\sigma'_{ij} = 2\eta_{vp}\dot{\varepsilon}'_{ij}Z + \sigma'^{o}_{ij}(1 - Z), \tag{13.2}$$

$$Z = \frac{\Delta t\mu}{\Delta t\mu + \eta_{vp}}, \tag{13.3}$$

$$\begin{aligned} \eta_{vp} &= \eta \quad \text{for} \quad \sigma_{II} < \sigma_{yield}, \text{ and} \\ \eta_{vp} &= \eta\frac{\sigma_{II}}{\eta\chi + \sigma_{II}} = \eta\frac{\sigma_{II}}{2\eta\dot{\varepsilon}_{II(plastic)} + \sigma_{II}}, \quad \text{for} \quad \sigma_{II} = \sigma_{yield}, \end{aligned} \tag{13.4}$$

where Δt is the computational time step, σ'^{o}_{ij} indicates the deviatoric stress from the previous time step corrected for advection and rotation (Chapter 12), Z is the visco-elasticity

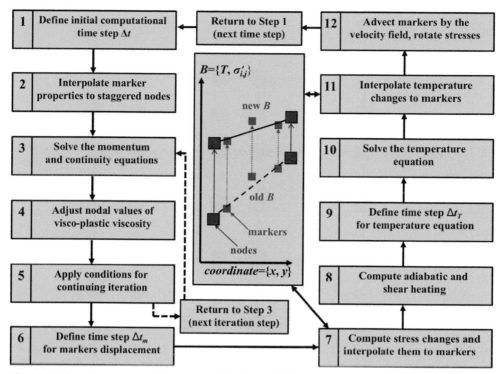

Fig. 13.1. Flow chart representing an example of a possible structure for a numerical thermomechanical visco-elasto-plastic 2D code, which uses staggered finite differences and marker-in-cell technique (SFD-MIC) for solving Stokes, continuity and temperature equations.

factor and $\eta_{vp} \leq \eta$ is a viscosity-like Lagrangian parameter (*effective visco-plastic viscosity*) that characterizes the intensity of the plastic deformation ($\eta_{vp} = \eta$ when no plastic yielding occurs). Equation (13.2) can also be applied to a more simple visco-elastic Maxwell rheology by using the condition that $\eta_{vp} = \eta$.

13.2 Structure of visco-elasto-plastic thermomechanical code

After obtaining the *constitutive relationship* (Eq. 13.2) between the stress and strain rate, we can now formulate the momentum, continuity, and temperature equations for the case of visco-elasto-plastic deformation in 2D and implement these equations in a thermomechanical visco-elasto-plastic modelling algorithm (Fig. 13.1). The algorithm is largely based on our viscous thermomechanical code structure discussed in Chapter 11 (Fig. 11.1).

The flow chart in Fig. 13.1 gives an example of a *possible* structure of a numerical thermomechanical visco-elasto-plastic 2D code, which uses *staggered finite differences and marker-in-cell technique* (SFD-MIC) to solve the momentum, continuity and temperature equations. The principal steps of the algorithm are as follows.

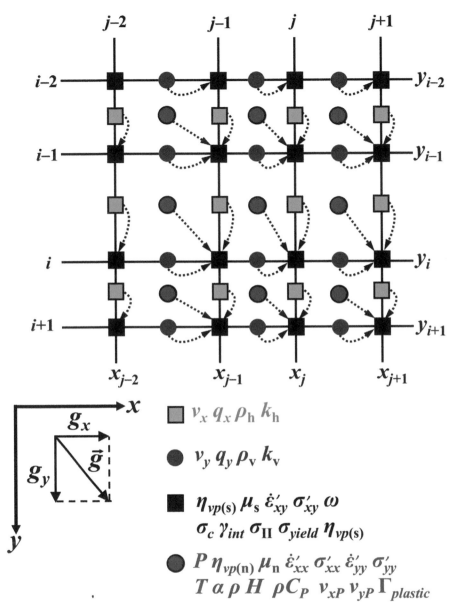

Fig. 13.2. Staggered 2D irregularly spaced numerical grid corresponding to the algorithm presented in Fig. 13.1.

(1) Define a suitable computational time step Δt for the momentum and continuity equations.

(2) Compute and/or interpolate the material properties (η_{vp}, μ, ρ, ρC_P, k, etc.) as well as scalars and tensors (T, σ'_{ij}, etc.) from the markers to the Eulerian nodes (Fig. 13.2).

(3) Discretize the 2D Stokes and continuity equations and compute the velocity and pressure by solving the global matrix problem with a direct method.

(4) and (5) Adjust iteratively (this is *global iteration* that also includes step 3) visco-plastic viscosity η_{vp} on the Eulerian nodes to satisfy the plastic yielding condition (13.4) to a desirable accuracy. These are optional steps performed when an accurate representation of plastic yielding is required.

(6) Define a suitable displacement time step Δt_m for markers (typically limiting the maximal displacement to 0.01–1.0 of the *minimal grid step*), which can be generally smaller than or equal to the computational time step Δt (see Step 1).

(7) Calculate the stress changes (Eq. 13.2) on the Eulerian nodes *for the displacement step* Δt_m (see Step 4), interpolate these changes to the markers and calculate new tensor values associated with the markers (see central panel in Fig. 13.1).

(8) Calculate the shear and adiabatic heating terms $H_{s(i,j)}$ and $H_{a(i,j)}$ at the Eulerian nodes from the computed velocity, pressure, strain rate and stress fields (see Step 3).

(9) Define a suitable time step Δt_T for the temperature equation. One can use a minimum time step value satisfying the following conditions: a given absolute time step limit on the order of a minimal characteristic thermal diffusion time scale for the processes being modelled; a given optimal marker displacement time step limit (see Step 3); a given absolute nodal temperature change limit (typically 1–20 K) (Chapter 10). The temperature equation can be solved with the preliminary displacement time step Δt_m to define possible temperature changes.

(10) Discretize the temperature equation implicitly in time and solve the global matrix problem by a direct method. The temperature equation can be solved in several steps when $\Delta t_T < \Delta t_m$.

(11) Interpolate the calculated nodal temperature changes (see central panel in Fig. 13.1) from the Eulerian nodes, to the markers, and calculate new marker temperatures.

(12) Advect all markers throughout the mesh according to the globally calculated velocity field (see Step 3). Components of the stress tensor defined on the markers are recomputed analytically to account for any local stress rotation (Chapter 12).

An important aspect of solving thermomechanical visco-elasto-plastic problems is that we have to provide initial conditions for stress and temperature before starting the computation.

Figure 13.2 shows the geometry of an irregularly spaced, fully staggered numerical grid corresponding to the algorithm. The visco-elasto-plastic code can be developed on the basis of viscous thermomechanical code (Chapter 11). Therefore, we will concentrate in the following sections on the new modifications that were made to the code described in Chapter 11.

Step 1: Defining a suitable computational time step

Note that the computational time step Δt, for the momentum and continuity equations, and the displacement time step Δt_m, for markers, are generally independent of each other and can only be related by the condition $\Delta t_m \leq \Delta t$. For an elastic medium, the velocity field numerically computed from the Stokes equation strongly depends on the value of Δt. This time step influences the *numerical viscosity* $\eta_{numerical}$, which can be derived from Eqs. (13.2), (13.3)

$$\eta_{numerical} = \eta_{vp}Z = \frac{\eta_{vp}\mu\Delta t}{\eta_{vp} + \mu\Delta t}.$$

If Δt is much less then the *Maxwell relaxation time*

$$t_{Maxwell} = \frac{\eta_{vp}}{\mu} \qquad (13.5)$$

of a visco-elastic/visco-elasto-plastic medium, then this medium behaves purely elastically, $\eta_{vp} \gg \mu\Delta t$ and the numerical viscosity depends linearly on the time step

$$\eta_{numerical} = \mu\Delta t. \qquad (13.6)$$

In this situation, the smaller we take the time step, the smaller the computational viscosity becomes, which results in larger velocities. Clearly this causes numerical problems, since if Δt tends to zero, the velocities will tend towards infinity. Physically meaningful solutions can be obtained in two ways. The first approach is to take an extremely small time step (on the order of seconds or less) and introduce inertial forces into the system by using Navier–Stokes rather than the Stokes equations (we will discuss this approach in Chapters 14 and 15). Numerical models in this case will only be able to address the seismic modes of deformation within a very limited period of time (hours, days), which is obviously not satisfactory if we want to model geodynamic processes that last for millions and even billions of years (e.g. mantle convection). The second approach (e.g. Kaus and Becker, 2007) is (when possible) to choose Δt in such a manner that it will be significantly shorter than the Maxwell time of the rheologically strongest materials that are present in the numerical experiment (e.g. a high viscosity lithosphere), but significantly larger than the Maxwell time of any rheologically weak materials (e.g. low viscosity asthenosphere). In this case, the weak materials satisfy the condition $\eta_{vp} \ll \mu\Delta t$ and they will behave purely viscously in a broad range of Δt, such that their numerical viscosity is independent of the time step

$$\eta_{numerical} = \eta_{vp}, \qquad (13.7)$$

which will tend to stabilize the velocity field computed inside the model. For realistic variations in viscosity $10^{16} < \eta < 10^{26}$ (Pa s) and shear modulus $10^{10} < \mu < 10^{11}$ (Pa) of solid rocks, this approach allows one to have geologically relevant computational time steps on the order of 10^1–10^5 years. In the case of global visco-elasto-plastic iteration on Eulerian nodes (which we will discuss in Section 13.3), the computational time step should also be suitable (small enough) to satisfy accurately the plastic yielding condition.

It should be pointed out that the actual numerical viscosity contrast in experiments with visco-elastic materials

$$\min\left(\eta_{vp(min)}, \Delta t\mu_{min}\right) \leq \eta_{numerical} \leq \min\left(\eta_{vp(max)}, \Delta t\mu_{max}\right), \tag{13.8}$$

is typically reduced compared to purely viscous experiments

$$\eta_{vp(min)} \leq \eta_{numerical} \leq \eta_{vp(max)}. \tag{13.9}$$

This contrast can be further reduced by decreasing the computational time step Δt. This actually makes the visco-elastic numerical experiments computationally 'easier' than the purely viscous ones. On the other hand, the visco-elastic numerical solutions become equivalent to the viscous one if we choose very large computational time steps, which are bigger than the maximal Maxwell relaxation time (Eq. 13.5) estimated locally (e.g. on markers) within the model.

Step 2: Interpolation of scalar fields, vectors and tensor fields

According to our algorithmic approach, the temperature field as well as components of the σ'_{ij} and (when needed) other tensors are represented by values assigned to markers. The effective values of all these parameters at the Eulerian nodal points are interpolated from the markers at each time step. As in the viscous code, one could use either standard 4-cell or local 1-cell interpolation schemes (Fig. 8.8) for some parameters: viscosity, shear modulus and deviatoric stress components. In order to avoid non-physical 'rigid boundary' effects on interfaces between *high-η-low-μ* and *low-η-high-μ* materials, one should use a harmonic average rather than an arithmetic mean for the shear modulus (see Eq. 8.18, Fig. 8.8 for notation)

$$\mu_{(i,j)} = \frac{\sum_m w_{m(i,j)}}{\sum_m \frac{1}{\mu_m} w_{m(i,j)}}. \tag{13.10}$$

As in the viscous code, both the visco-plastic viscosity η_{vp} and shear modulus μ are defined at basic nodal points and cell centres (cf. solid squares and red circles in Fig. 13.2) when used for computing the normal and shear components of the deviatoric stress tensor. The viscosity, shear modulus and the respective stress and (when needed) strain rate components for these nodal points are interpolated from the surrounding markers (Fig. 8.8).

Step 3: Solving the momentum and continuity equations

In the case of 2D visco-elasto-plastic compressible material, the solution of the Stokes equations (11.1) and (11.2) does not change. The only alteration concerns the expressions for the deviatoric stresses and strain rates:

$$\sigma'_{xx} = 2\eta_{vp}\dot{\varepsilon}'_{xx}\frac{\mu\Delta t}{\mu\Delta t + \eta_{vp}} + \sigma'^{o}_{xx}\frac{\eta_{vp}}{\mu\Delta t + \eta_{vp}} = -\sigma'_{yy}, \tag{13.11}$$

$$\sigma'_{xy} = 2\eta_{vp}\dot{\varepsilon}'_{xy}\frac{\mu\Delta t}{\mu\Delta t + \eta_{vp}} + \sigma'^{o}_{xy}\frac{\eta_{vp}}{\mu\Delta t + \eta_{vp}} = \sigma'_{yx}, \tag{13.12}$$

$$\dot{\varepsilon}'_{xx} = \frac{1}{2}\left(\frac{\partial v_x}{\partial x} - \frac{\partial v_y}{\partial y}\right) = -\dot{\varepsilon}'_{yy}, \tag{13.13}$$

$$\dot{\varepsilon}'_{xy} = \frac{1}{2}\left(\frac{\partial v_x}{\partial y} + \frac{\partial v_y}{\partial x}\right) = \dot{\varepsilon}'_{yx}. \tag{13.14}$$

Equation (13.13) assumes that $\dot{\varepsilon}_{zz} = (\dot{\varepsilon}_{xx} + \dot{\varepsilon}_{yy})/2$ and thus $\dot{\varepsilon}_{kk} = 3(\dot{\varepsilon}_{xx} + \dot{\varepsilon}_{yy})/2$, $\dot{\varepsilon}'_{zz} = 0$ (Eqs. 4.11, 4.12).

The conservation of mass is approximated by a compressible time-dependent 2D continuity equation (we will discuss this further in Chapter 14),

$$\frac{\partial v_x}{\partial x} + \frac{\partial v_y}{\partial y} = \Gamma_{plastic}, \tag{13.15}$$

which takes into account volumetric changes during plastic deformation of dilatant materials (Eqs. 12.62, 12.63).

The conservative FD representation of the momentum equations (Eqs. 11.4, 11.5) with free surface stabilization (Chapter 8) modified for a visco-elasto-plastic rheology with Eqs. (13.11)–(13.14) is as follows (see Fig. 13.2 for the indexing of the grid points) for the x Stokes equation (11.1), (11.4):

$$\frac{4\Delta t}{x_{j+1} - x_{j-1}}\left(\frac{\eta_{vp(n)(i,j+1)}\dot{\varepsilon}'_{xx(i,j+1)}\mu_{n(i,j+1)}}{\mu_{n(i,j+1)}\Delta t + \eta_{vp(n)(i,j+1)}} - \frac{\eta_{vp(n)(i,j)}\dot{\varepsilon}'_{xx(i,j)}\mu_{n(i,j)}}{\mu_{n(i,j)}\Delta t + \eta_{vp(n)(i,j)}}\right)$$

$$+ \frac{2\Delta t}{y_i - y_{i-1}}\left(\frac{\eta_{vp(s)(i,j)}\dot{\varepsilon}'_{xy(i,j)}\mu_{s(i,j)}}{\mu_{s(i,j)}\Delta t + \eta_{vp(s)(i,j)}} - \frac{\eta_{vp(s)(i-1,j)}\dot{\varepsilon}'_{xy(i-1,j)}\mu_{s(i-1,j)}}{\mu_{s(i-1,j)}\Delta t + \eta_{vp(s)(i-1,j)}}\right) - 2\frac{P_{i,j+1} - P_{i,j}}{x_{j+1} - x_{j-1}}$$

$$- g_x\Delta t\left(v_{x(i,j)}\frac{P_{h(i,j+1)} - P_{h(i,j-1)}}{x_{j+1} - x_{j-1}} + \frac{\left(v_{y(i-1,j)} + v_{y(i-1,j+1)} + v_{y(i,j)} + v_{y(i,j+1)}\right)\left(P_{h(i+1,j)} - P_{h(i-1,j)}\right)}{2(y_{i+1} + y_i - y_{i-1} - y_{i-2})}\right)$$

$$= -P_{h(i,j)}g_x - \frac{2}{x_{j+1} - x_{j-1}}\left(\frac{\sigma'^{o}_{xx(i,j+1)}\eta_{vp(n)(i,j+1)}}{\mu_{n(i,j+1)}\Delta t + \eta_{vp(n)(i,j+1)}} - \frac{\sigma'^{o}_{xx(i,j)}\eta_{vp(n)(i,j)}}{\mu_{n(i,j)}\Delta t + \eta_{vp(n)(i,j)}}\right)$$

$$- \frac{1}{y_i - y_{i-1}}\left(\frac{\sigma'^{o}_{xy(i,j)}\eta_{vp(s)(i,j)}}{\mu_{s(i,j)}\Delta t + \eta_{vp(s)(i,j)}} - \frac{\sigma'^{o}_{xy(i-1,j)}\eta_{vp(s)(i-1,j)}}{\mu_{s(i-1,j)}\Delta t + \eta_{vp(s)(i-1,j)}}\right), \tag{13.16}$$

$$\dot{\varepsilon}'_{xx(i,j)} = \frac{1}{2}\left(\frac{v_{x(i,j)} - v_{x(i,j-1)}}{x_j - x_{j-1}} - \frac{v_{y(i,j)} - v_{y(i-1,j)}}{y_i - y_{i-1}}\right),$$

$$\dot{\varepsilon}'_{xx(i,j+1)} = \frac{1}{2}\left(\frac{v_{x(i,j+1)} - v_{x(i,j)}}{x_{j+1} - x_j} - \frac{v_{y(i,j+1)} - v_{y(i-1,j+1)}}{y_i - y_{i-1}}\right),$$

$$\dot{\varepsilon}'_{xy(i-1,j)} = \frac{v_{x(i,j)} - v_{x(i-1,j)}}{y_i - y_{i-2}} + \frac{v_{y(i-1,j+1)} - v_{y(i-1,j)}}{x_{j+1} - x_{j-1}},$$

$$\dot{\varepsilon}'_{xy(i,j)} = \frac{v_{x(i+1,j)} - v_{x(i,j)}}{y_{i+1} - y_{i-1}} + \frac{v_{y(i,j+1)} - v_{y(i,j)}}{x_{j+1} - x_{j-1}},$$

where σ'^o_{xy} and σ'^o_{xx} are the deviatoric stress tensor components from the previous time step (corrected for advection and rotation), which are interpolated from markers; Δt is the computational time step. Discretization of the y Stokes equation (11.5) is analogous to Eq. (13.16) (derive as an exercise).

The *compressible* continuity Eq. (13.15) is discretized as follows:

$$\frac{v_{x(i,j)} - v_{x(i,j-1)}}{x_j - x_{j-1}} + \frac{v_{y(i,j)} - v_{y(i-1,j)}}{y_i - y_{i-1}} = \Gamma_{plastic(i,j)}, \tag{13.17}$$

where $\Gamma_{plastic(i,j)}$ is the volumetric effect of dilatant plastic deformation (e.g. Eq. 12.63) interpolated from markers.

As in the viscous code, the global matrix is solved by a highly accurate, direct method and the numbering of unknowns in this global matrix remains the same.

Step 7: Interpolation of stress changes from nodes to markers

After defining the material displacement time step Δt_m, new stresses and stress increments for the Eulerian nodes are calculated at the respective nodal points according to Eq. (13.2) as

$$\sigma'_{xx(i,j)} = 2\eta_{vp(n)(i,j)}\dot{\varepsilon}'_{xx(i,j)}\frac{\mu_{n(i,j)}\Delta t_m}{\mu_{n(i,j)}\Delta t_m + \eta_{vp(n)(i,j)}} + \sigma'^o_{xx(i,j)}\frac{\eta_{vp(n)(i,j)}}{\mu_{n(i,j)}\Delta t_m + \eta_{vp(n)(i,j)}}, \tag{13.18}$$

$$\sigma'_{xy(i,j)} = 2\eta_{vp(s)(i,j)}\dot{\varepsilon}'_{xy(i,j)}\frac{\mu_{s(i,j)}\Delta t_m}{\mu_{s(i,j)}\Delta t_m + \eta_{vp(s)(i,j)}} + \sigma'^o_{xy(i,j)}\frac{\eta_{vp(s)(i,j)}}{\mu_{s(i,j)}\Delta t_m + \eta_{vp(s)(i,j)}}, \tag{13.19}$$

$$\Delta\sigma'_{xx(i,j)} = \sigma'_{xx(i,j)} - \sigma'^o_{xx(i,j)}, \tag{13.20}$$

$$\Delta\sigma'_{xy(i,j)} = \sigma'_{xy(i,j)} - \sigma'^o_{xy(i,j)}. \tag{13.21}$$

The corresponding deviatoric stress increments for the markers are then added from the nodes using a standard first-order interpolation scheme (Fig. 8.9, Eq. 8.19) and the newly updated values of stress components $\sigma'_{xx(m)}$ and $\sigma'_{xy(m)}$ are thus obtained for markers. Other deviatoric stress components are not stored on markers and are obtained by using the standard relations $\sigma'_{yy} = -\sigma'_{xx}$ and $\sigma'_{yx} = \sigma'_{xy}$.

The interpolation of the calculated stress increments from the Eulerian nodal points to the Lagrangian markers is similar to the temperature update strategy described in Chapter 10 and effectively reduces the problem of numerical diffusion and non-physical subgrid oscillations of stresses. It uses a subgrid stress relaxation operation occurring over a characteristic Maxwell time. In order to define this operation, stress increments computed from Eqs. (13.20), (13.21) are decomposed into a subgrid part $\Delta\sigma'^{subgrid}_{xx(i,j)}$, $\Delta\sigma'^{subgrid}_{xy(i,j)}$ and a remaining part $\Delta\sigma'^{remaining}_{xx(i,j)}$, $\Delta\sigma'^{remaining}_{xy(i,j)}$ so that

$$\Delta\sigma'_{xx(i,j)} = \Delta\sigma'^{subgrid}_{xx(i,j)} + \Delta\sigma'^{remaining}_{xx(i,j)}, \tag{13.22}$$

$$\Delta\sigma'_{xy(i,j)} = \Delta\sigma'^{subgrid}_{xy(i,j)} + \Delta\sigma'^{remaining}_{xy(i,j)}. \tag{13.23}$$

In order to compute the subgrid part, we apply a subgrid stress relaxation on the markers by using a characteristic, local Maxwell visco-elastic relaxation time scale $t_{Maxwell(m)}$ and then interpolating the respective stress changes back to the nodes. Subgrid stress changes on the markers are computed as follows:

$$\Delta\sigma'^{subgrid}_{xx(m)} = \left(\sigma'^{o}_{xx(m)nodal} - \sigma'^{o}_{xx(m)}\right)\left[1 - \exp\left(-d_{ve}\frac{\Delta t_m}{t_{Maxwell(m)}}\right)\right], \tag{13.24}$$

$$\Delta\sigma'^{subgrid}_{xy(m)} = \left(\sigma'^{o}_{xy(m)nodal} - \sigma'^{o}_{xy(m)}\right)\left[1 - \exp\left(-d_{ve}\frac{\Delta t_m}{t_{Maxwell(m)}}\right)\right], \tag{13.25}$$

where $t_{Maxwell(m)} = \eta_{vp(m)}/\mu_m$ is defined for each marker; d_{ve} is a dimensionless, numerical visco-elastic relaxation coefficient (one can use empirical values in the range of $0 \le d_{ve} \le 1$); $\sigma'^{o}_{xx(m)nodal}$ and $\sigma'^{o}_{xy(m)nodal}$ are interpolated for any given marker from respectively $\sigma'^{o}_{xx(i,j)}$ and $\sigma'^{o}_{xy(i,j)}$ from the nodal values using Eq. (8.19) (Fig. 8.9).

After obtaining $\Delta\sigma'^{subgrid}_{xx(m)}$ and $\Delta\sigma'^{subgrid}_{xy(m)}$ for all markers, $\Delta\sigma'^{subgrid}_{xx(i,j)}$ and $\Delta\sigma'^{subgrid}_{xy(i,j)}$ are computed by interpolation from the markers to the nodes using Eq. (8.18) (Fig. 8.8).

Then $\Delta\sigma'^{remaining}_{xx(i+1/2,j+1/2)}$ and $\Delta\sigma'^{remaining}_{xy(i,j)}$ are computed for the nodes from Eqs. (13.22), (13.23):

$$\Delta\sigma'^{remaining}_{xx(i,j)} = \Delta\sigma'_{xx(i,j)} - \Delta\sigma'^{subgrid}_{xx(i,j)}, \tag{13.26}$$

$$\Delta\sigma'^{remaining}_{xy(i,j)} = \Delta\sigma'_{xy(i,j)} - \Delta\sigma'^{subgrid}_{xy(i,j)}. \tag{13.27}$$

Finally, new corrected marker stresses $\sigma'^{corrected}_{xx(m)}$ and $\sigma'^{corrected}_{xy(m)}$ are computed according to the following relation:

$$\sigma_{xx(m)}^{'corrected} = \sigma_{xx(m)}^{'o} + \Delta\sigma_{xx(m)}^{'subgrid} + \Delta\sigma_{xx(m)}^{'remaining}, \tag{13.28}$$

$$\sigma_{xy(m)}^{'corrected} = \sigma_{xy(m)}^{'o} + \Delta\sigma_{xy(m)}^{'subgrid} + \Delta\sigma_{xy(m)}^{'remaining}, \tag{13.29}$$

where $\Delta\sigma_{xx(m)}^{'subgrid}$ and $\Delta\sigma_{xy(m)}^{'subgrid}$ are given by Eqs. (13.24) and (13.25), respectively, and $\Delta\sigma_{xx(m)}^{'remaining}$ and $\Delta\sigma_{xy(m)}^{'remaining}$ are interpolated from nodal values of $\Delta\sigma_{xx(i,j)}^{'remaining}$ and $\Delta\sigma_{xy(i,j)}^{'remaining}$ to the markers according to standard bilinear interpolation (Eq. 8.19, Fig. 8.9).

Equations (13.24) and (13.25) require the decay of differences between marker stress values $\sigma_{xx(m)}^{'o}$, $\sigma_{xy(m)}^{'o}$ and the interpolated nodal stress values $\sigma_{xx(m)nodal}^{'o}$ and $\sigma_{xy(m)nodal}^{'o}$ on the characteristic local, Maxwell visco-elastic relaxation time scale $t_{Maxwell(m)}$. It is important to emphasize that the subgrid relaxation does not change the values of stress increments $\Delta\sigma_{xx(i,j)}^{'}$ and $\Delta\sigma_{xy(i,j)}^{'}$ computed on nodal points from Eqs. (13.20) and (13.21), respectively, but instead splits them into two parts $\Delta\sigma_{xx(i,j)}^{'subgrid}$, $\Delta\sigma_{xy(i,j)}^{'subgrid}$ and $\Delta\sigma_{xx(i,j)}^{'remaining}$ and $\Delta\sigma_{xy(i,j)}^{'remaining}$. Introducing the subgrid relaxation operation removes unrealistic subgrid stress oscillations over the characteristic local Maxwell visco-elastic relaxation time. Similarly to temperature, realistic subgrid variations will be preserved by this scheme. The subgrid stress diffusion scheme should, however, only be applied in the case of strong marker advection and mixing.

Steps 8–11: Solving the temperature equation

Numerical techniques for discretizing and solving the temperature equation are identical to those used in the viscous code of Chapter 11 (Eq. 11.8). An important difference, however, occurs in computing the shear heating term. Since elastic deformation is reversible and does not contribute to mechanical energy dissipation, this deformation has to be excluded from shear heating calculation and therefore

$$H_s = 2\sigma_{xx}^{'}\left(\dot{\varepsilon}_{xx}^{'} - \dot{\varepsilon}_{xx(elastic)}^{'}\right) + 2\sigma_{xy}^{'}\left(\dot{\varepsilon}_{xy}^{'} - \dot{\varepsilon}_{xy(elastic)}^{'}\right). \tag{13.30}$$

With Eqs. (12.54), (13.1)–(13.3) this can be further transformed to (derive as an exercise)

$$H_s = 2\sigma_{xx}^{'}\left(\dot{\varepsilon}_{xx}^{'} - \frac{1}{2\mu}\frac{D\sigma_{xx}^{'}}{Dt}\right) + 2\sigma_{xy}^{'}\left(\dot{\varepsilon}_{xy}^{'} - \frac{1}{2\mu}\frac{D\sigma_{xy}^{'}}{Dt}\right), \tag{13.31}$$

$$H_s = 2\sigma_{xx}^{'}\left(\dot{\varepsilon}_{xx}^{'} - \frac{\sigma_{xx}^{'} - \sigma_{xx}^{'o}}{2\mu\Delta t}\right) + 2\sigma_{xy}^{'}\left(\dot{\varepsilon}_{xy}^{'} - \frac{\sigma_{xy}^{'} - \sigma_{xy}^{'o}}{2\mu\Delta t}\right), \tag{13.32}$$

$$H_s = \frac{\sigma_{xx}'^2}{\eta_{vp}} + \frac{\sigma_{xy}'^2}{\eta_{vp}}. \tag{13.33}$$

Finally, in the FD representation, the shear heating at temperature nodal points (cell centres, see red circles in Fig. 13.2) can be computed as follows

$$H_{s(i,j)} = \frac{\sigma_{xy(i,j)}'^2}{\eta_{vp(n)(i,j)}} + \frac{1}{4}\left(\frac{\sigma_{xy(i-1,j-1)}'^2}{\eta_{vp(s)(i-1,j-1)}} + \frac{\sigma_{xy(i,j-1)}'^2}{\eta_{vp(s)(i,j-1)}} + \frac{\sigma_{xy(i-1,j)}'^2}{\eta_{vp(s)(i-1,j)}} + \frac{\sigma_{xy(i,j)}'^2}{\eta_{vp(s)(i,j)}} \right), \tag{13.34}$$

where $\sigma_{xx(i,j)}'$ and $\sigma_{xy(i,j)}'$ are defined by Eqs. (13.18) and (13.19).

Step 12: Rotation of stresses

Another difference from the viscous algorithm consists in computing local stress changes due to the local rotation of markers. To do this, the rotation rate

$$\omega = \frac{1}{2}\left(\frac{\partial v_y}{\partial x} - \frac{\partial v_x}{\partial y} \right)$$

is defined at the same basic grid points as the shear strain rate

$$\dot{\varepsilon}_{xy}' = \frac{1}{2}\left(\frac{\partial v_x}{\partial y} + \frac{\partial v_y}{\partial x} \right)$$

(see solid squares in Fig. 13.2) and is similarly computed from the velocity field via finite differences

$$\omega_{(i,j)} = \frac{v_{y(i,j+1)} - v_{y(i,j)}}{x_{j+1} - x_{j-1}} - \frac{v_{x(i+1,j)} - v_{x(i,j)}}{y_{i+1} - y_{i-1}}. \tag{13.35}$$

Deviatoric stresses on markers are recomputed before advecting them according to analytical formulas (12.21) and (12.22) by using a first-order accurate scheme in space and time to compute rotation angles for the markers θ_m,

$$\theta_m = \omega_m \Delta t_m, \tag{13.36}$$

where ω_m is interpolated from the basic nodes using a standard interpolation formula (Eq. 8.19, Fig. 8.9).

13.3 Visco-elasto-plastic iterations

It is important to mention that computation of visco-elasto-plastic solutions may require additional iterations in order to adjust stresses, strain rates and visco-plastic viscosity

(η_{vp}) fields on markers and nodes. These iterations can be *local* (i.e. done individually for every marker) and *global* (i.e. involving repeated global solving of the momentum and continuity equations and re-computing properties for markers and nodes). It is most optimal to do these iterations before the displacement of markers by the velocity field and then advect iteratively computed marker values of effective visco-plastic viscosity $\eta_{vp(m)}$.

Local visco-elastic iterations on markers. These are commonly needed when the effective viscosity of the medium is non-Newtonian and depends on stresses (e.g. due to the dislocation creep). Note that, in the case of visco-elastic deformation, the effective non-Newtonian ductile viscosity η_m (Chapter 6) should be formulated in *terms of the deviatoric stresses* (Eq. 6.5a) and *not in terms of the strain rates* (Eq. 6.5b), since these strain rates are *not purely viscous* and include elastic and plastic deformation. In order to avoid numerical oscillations, Eq. (6.5a) should be formulated on markers in terms of the future visco-elastic stresses predicted locally from Eq. (13.2) and should be based on the current marker's stresses $\sigma_{xx(m)}^{'corrected}$, $\sigma_{xy(m)}^{'corrected}$, strain rates $\dot{\varepsilon}_{xx(m)}^{'}$, $\dot{\varepsilon}_{xy(m)}^{'}$ and shear modulus μ_m. Since the combination of Eqs. (6.5a) and (13.2) is non-linear, local iterations for each marker are needed to obtain self-consistent values of $\sigma_{xx(m)}^{'}$, $\sigma_{xy(m)}^{'}$, $\sigma_{II(m)}$ and η_m. We will use such local visco-elastic iterations on markers in Chapter 21 for exploring various geodynamic models with non-Newtonian rock rheology.

Computing visco-plastic viscosity for markers. The described visco-elastic iterations should be done *before* we check the plastic yielding condition (13.4) on the marker, which requires we use the marker's pressure P_m and consistent values of visco-elastic stresses $\sigma_{xx(m)}^{'}$, $\sigma_{xy(m)}^{'}$ and corresponding $\sigma_{II(m)}$. If $\sigma_{II(m)}$ is larger than $\sigma_{yield(m)}$ for a given marker, a new value of the viscosity-like parameter, $\eta_{vp(m)}$, should be computed to satisfy the condition $\sigma_{II(m)} = \sigma_{yield(m)}$ for that marker. A simple way to do this consists in re-formulation of stress evolution equations (13.11), (13.12) for the marker by assuming constant shear modulus μ_m and strain rates $\dot{\varepsilon}_{xx(m)}^{'}$, $\dot{\varepsilon}_{xy(m)}^{'}$ for the marker:

$$\sigma_{xx(m)}^{'} = \frac{\eta_{vp(m)}}{\mu_m \Delta t_m + \eta_{vp(m)}} \left(2\mu_m \Delta t_m \dot{\varepsilon}_{xx(m)}^{'} + \sigma_{xx(m)}^{'o} \right), \tag{13.37}$$

$$\sigma_{xy(m)}^{'} = \frac{\eta_{vp(m)}}{\mu_m \Delta t_m + \eta_{vp(m)}} \left(\mu_m \Delta t_m \dot{\varepsilon}_{xy(m)}^{'} + \sigma_{xy(m)}^{'o} \right), \tag{13.38}$$

$$\sigma_{II(m)} = \sqrt{\sigma_{xx(m)}^{'2} + \sigma_{xy(m)}^{'2}} = \frac{\eta_{vp(m)}}{\mu_m \Delta t_m + \eta_{vp(m)}} \sigma_{II(m)elastic}, \tag{13.39}$$

$$\sigma_{II(m)elastic} = \sqrt{\left(2\mu_m \Delta t_m \dot{\varepsilon}_{xx(m)}^{'} + \sigma_{xx(m)}^{'o} \right)^2 + \left(2\mu_m \Delta t_m \dot{\varepsilon}_{xy(m)}^{'} + \sigma_{xy(m)}^{'o} \right)^2}, \tag{13.40}$$

where $\sigma_{II(m)elastic}$ is the second invariant for a purely elastic stress buildup. By applying the Drucker–Prager plastic yielding condition (12.50)

$$\sigma_{yield(m)} = max\left(\sigma_{c(m)} + \gamma_{int(m)}P_m, 0\right) = \sigma_{II(m)}, \qquad (13.41)$$

where $\sigma_{yield(m)} \geq 0$ (*non-negative strength* requirement), and assuming constant marker pressure P_m, friction coefficient $\gamma_{int(m)}$ and compressive strength $\sigma_{c(m)}$ we can obtain the following expression for the effective visco-plastic viscosity of the marker (please derive as an exercise)

$$\eta_{vp(m)} = \mu_m \Delta t_m \frac{\sigma_{yield(m)}}{\sigma_{II(m)elastic} - \sigma_{yield(m)}}. \qquad (13.42)$$

Computed values of $\eta_{vp(m)}$ could be advected with markers and used to interpolate nodal visco-plastic viscosity values $\eta_{vp(n)(i,j)}$ and $\eta_{vp(s)(i,j)}$ at the next time step.

Global visco-elasto-plastic iterations on markers. In contrast to computing a non-Newtonian viscosity, Eqs. (13.37)–(13.42) can be explicitly applied to each marker without performing any local iteration. However, this procedure strongly affects the marker visco-plastic viscosity $\eta_{vp(m)}$ and thereby nodal visco-plastic viscosity values $\eta_{vp(n)(i,j)}$ and $\eta_{vp(s)(i,j)}$. This may change the global pressure-velocity solution, which in turn affects values of P_m, $\dot{\varepsilon}'_{xx(m)}$ and $\dot{\varepsilon}'_{xy(m)}$ used in Eqs. (13.37)–(13.42). Therefore, global iterations that involve solving the momentum and continuity equations (Fig. 13.1, Steps 3–5), are often required. One way of performing this iteration is to repeat cycles of global solutions on the nodal points and make local re-adjustments on the markers (without displacing them) by re-evaluating their visco-plastic viscosity with Eqs. (13.37)–(13.42) when the condition

$$\sigma_{II(m)elastic} > \sigma_{yield(m)} \qquad (13.43)$$

is satisfied locally for the marker. The new value of the marker viscosity $\eta_{vp(m)}$ computed with Eq. (13.42) is used for the next iteration if the condition

$$\eta_{vp(m)} < \eta_m \qquad (13.44)$$

is satisfied.

This global iteration method, called *Picard iteration,* allows some adjustment of the yielding condition (13.41) for the markers. On the other hand, *it is not possible to satisfy the plastic yielding condition for all markers* so that for each marker

$$\sigma_{II(m)} - \sigma_{yield(m)} = \sqrt{\left(\sigma'^{corrected}_{xx(m)}\right)^2 + \left(\sigma'^{corrected}_{xy(m)}\right)^2} - \left(C_m + \gamma_{int(m)}P_m\right) = 0, \qquad (13.45)$$

where $\sigma'^{corrected}_{xx(m)}$ and $\sigma'^{corrected}_{xy(m)}$ are current corrected marker stresses computed according to Eqs. (13.28), (13.29). This problem arises because the number of *independent*

Drucker–Prager plastic yielding conditions that could be applied in this non-linear system is likely to be limited by the number of independent nodal pressure values (i.e., by the number of pressure points inside the grid), which is much smaller than the number of markers. Therefore, an error minimization criterion should be used, for example minimum average mean square error of condition (13.45) for markers

$$\Delta\sigma_{yield(markers)} = \frac{1}{N_{yield(markers)}} \sqrt{\sum_m \left(\sigma_{II(m)} - \sigma_{yield(m)} \right)^2} = \text{min}, \qquad (13.46)$$

where $N_{yield(markers)}$ is the number of markers for which conditions (13.43) and (13.44) are both satisfied. This criterion implies that the global Picard iteration could stop after reaching some stable level of error defined by Eq. (13.46). This level is *not known a priori* and can change from one time step to the other. It should also be mentioned that the Picard iteration on markers is generally not very efficient in lowering $\Delta\sigma_{yield(markers)}$ compared to global visco-elasto-plastic iterations on Eulerian nodes, which are described below. On the other hand, visco-elasto-plastic solutions produced with the use of different computational approaches may often look relatively similar and only differ in some details (Fig. 13.3).

Global visco-elasto-plastic iterations on Eulerian nodes. In contrast to markers, it is possible to accurately satisfy the plastic yielding condition at the basic nodal points of the staggered grid by global Picard iteration. In this case, the Drucker–Prager plastic yielding condition (12.57) should be discretized at these points as (Fig. 13.2)

$$\sigma_{yield(i,j)} = \sigma_{II(i,j)}, \qquad (13.47)$$

$$\sigma_{yield(i,j)} = \text{max}\left[\sigma_{c(i,j)} + \gamma_{int(i,j)} \frac{1}{4} \left(P_{(i,j)} + P_{(i+1,j)} + P_{(i,j+1)} + P_{(i+1,j+1)} \right), 0 \right], \qquad (13.48)$$

$$\sigma_{II(i,j)} = \sqrt{ \left(\sigma'_{xy(i,j)} \right)^2 + \frac{1}{16} \left(\sigma'_{xx(i,j)} + \sigma'_{xx(i+1,j)} + \sigma'_{xx(i,j+1)} + \sigma'_{xx(i+1,j+1)} \right)^2 }, \qquad (13.49)$$

where $\sigma_{yield(i,j)} \geq 0$ (*non-negative strength* requirement), $\sigma_{c(i,j)}$ and $\gamma_{int((i,j)}$ are values of respectively the compressive strength and internal friction coefficient at the basic nodal points, which should be interpolated from surrounding markers using the standard relation (8.18) (Fig. 8.8). The Picard iteration is performed by repeating solution of the Stokes and continuity equations and evaluating $\sigma_{II(i,j)elastic}$ at each basic node as

$$\sigma_{II(i,j)elastic} = \frac{\mu_{s(i,j)}\Delta t + \eta_{s(i,j)}}{\eta_{s(i,j)}} \sigma_{II(i,j)}, \qquad (13.50)$$

where Δt is the computational time step and $\eta_{(s)(i,j)}$ is the value of the ductile viscosity for the node. If the condition

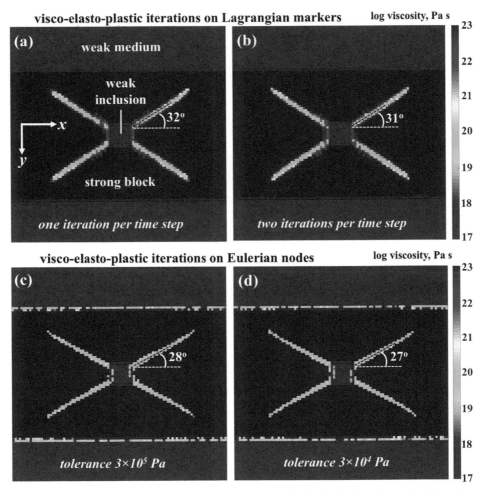

Fig. 13.3. Development of *plastic shear bands* around a weak inclusion inside a shortening visco-elasto-plastic block in the absence of gravity. Model setup corresponds to Exercise 13.2 (Fig. 13.4). (a),(b),(c) and (d) show solutions obtained by different methods for the same moment of time (507 years). Please note that in the case of visco-elasto-plastic Picard iteration on Eulerian nodes, shear band angles are very close to the characteristic *Coulomb angle* $45° − \varphi/2 = 26.5°$, where $\varphi = 37°$ ($\gamma_{int} = 0.6$) is the internal friction angle used in the numerical experiment. In the case of iterations on Lagrangian markers, shear band angles are in between the Coulomb angle and the *Arthur angle* $45° − (\varphi + \psi)/4 = 35.75°$, where $\psi = 0°$ is the dilatation angle used in the numerical experiment. The models are computed with the codes (a) **Viscoelastoplastic2D.m**, (b) **Viscoelastoplastic2D_iterations.m**, (c),(d) **i2elvis.m** associated with this chapter.

$$\sigma_{II(i,j)elastic} > \sigma_{yield(i,j)} \qquad (13.51)$$

is satisfied, a new value of the visco-plastic viscosity is computed at the node as

$$\eta_{vp(s)(i,j)} = \mu_{s(i,j)} \Delta t \frac{\sigma_{yield(i,j)}}{\sigma_{II(i,j)elastic} - \sigma_{yield(i,j)}} \tag{13.52}$$

which is used for the next iteration if the condition

$$\eta_{vp(s)(i,j)} < \eta_{(s)(i,j)} \tag{13.53}$$

is satisfied. For the convergence of iteration, visco-plastic viscosity at cell centres $\eta_{vp(n)(i,j)}$ (Fig. 13.2) should be computed by using the harmonic average of viscosity from four surrounding basic nodes (Fig. 13.2)

$$\eta_{vp(n)(i,j)} = \frac{4}{1/\eta_{vp(s)(i,j)} + 1/\eta_{vp(s)(i-1,j)} + 1/\eta_{vp(s)(i,j-1)} + \eta_{vp(s)(i-1,j-1)}}. \tag{13.54}$$

The global plastic yielding condition error is calculated at basic nodal points from Eq. (13.47)

$$\Delta\sigma_{yield(nodal)} = \frac{1}{N_{yield(nodal)}} \sqrt{\sum_i \sum_j \left(\sigma_{II(i,j)} - \sigma_{yield(ij)} \right)^2}, \tag{13.55}$$

where $N_{yield(nodal)}$ is the number of basic points at which either the previously computed or new value of $\eta_{vp(s)(i,j)}$ satisfies condition (13.53). The Picard iteration is interrupted when a certain a priori defined desirable low level of $\Delta\sigma_{yield(nodal)}$ is achieved. When this level is not achieved after a certain maximal number of Picard iterations, the computational time step Δt is reduced and the iteration is restarted from the beginning (i.e., from the previous physically meaningful state of stresses and visco-plastic viscosities on the grid). If needed, a computer accuracy solution could be achieved with this *adaptive time stepping method* but in this case the computational time step Δt will typically be strongly reduced and the number of iterations will be large. More efficient global iteration procedures such as *the Newton–Raphson method* (e.g. Belytschko et al., 2000; Souza de Neto et al., 2009) could also be used (e.g. Popov and Sobolev, 2008) instead of the Picard iteration.

Finally, it is important to mention that high strain rate shear zones formed by plastic yielding localization (Figs. 12.6, 13.3) are *Lagrangian features* related to specific material points (and not to immobile Eulerian nodes), which should be thus advected with the material flow. In the case of global iterations on Eulerian nodes, this advection is taken into account by advection of stresses and material properties (including η_s and $\eta_{vp(s)}$) with markers and interpolation of them to the nodes. Before the advection, the adjusted visco-plastic viscosity *at the plastically yielding Eulerian nodes* (i.e., for which $\eta_{vp(s)(i,j)} < \eta_{s(i,j)}$) should be interpolated (Fig. 8.9) to markers by using a harmonic average, which is more sensitive to the viscosity reduction

$$\left[\begin{array}{l} \eta_{vp(m)} = 1 / \dfrac{1}{\eta_{vp(s)(i,j)}} \left(1 - \dfrac{\Delta x_m}{\Delta x}\right) \left(1 - \dfrac{\Delta y_m}{\Delta y}\right) + \dfrac{1}{\eta_{vp(s)(i,j+1)}} \dfrac{\Delta x_m}{\Delta x} \left(1 - \dfrac{\Delta y_m}{\Delta y}\right) \\[3mm] + \dfrac{1}{\eta_{vp(s)(i+1,j)}} \left(1 - \dfrac{\Delta x_m}{\Delta x}\right) \dfrac{\Delta y_m}{\Delta y} + \dfrac{1}{\eta_{vp(s)(i+1,j+1)}} \dfrac{\Delta x_m \Delta y_m}{\Delta x \Delta y} \, . \end{array} \right]$$ (13.56)

In the case of visco-elasto-plastic deformation, displacement of markers should typically be rather small (0.1–0.001 grid step per time step) to accurately resolve the development of shear zones in the visco-elasto-plastic continuum.

13.4 Visco-plastic rheology

Visco-plastic rheology is a simplified treatment of visco-elasto-plastic material assuming that elastic effects are negligible and can be ignored on the long time scales of geodynamic processes. Governing rheological relations for this material can be obtained from Eqs. (12.54)–(12.60) under the assumption that $\dot{\varepsilon}'_{ij(elastic)} = 0$. The total deviatoric strain rate $\dot{\varepsilon}'_{ij}$ is then decomposed into the viscous and plastic components:

$$\dot{\varepsilon}'_{ij} = \dot{\varepsilon}'_{ij(viscous)} + \dot{\varepsilon}'_{ij(plastic)},$$ (13.57)

where

$$\dot{\varepsilon}'_{ij(viscous)} = \frac{1}{2\eta} \sigma'_{ij},$$ (13.58)

$$\begin{array}{ll} \dot{\varepsilon}'_{ij(plastic)} = 0 & \text{for} \quad \sigma_{\text{II}} < \sigma_{yield}, \\[3mm] \dot{\varepsilon}'_{ij(plastic)} = \chi \dfrac{\sigma'_{ij}}{2\sigma_{\text{II}}} & \text{for} \quad \sigma_{\text{II}} = \sigma_{yield}. \end{array}$$ (13.59)

The visco-plastic constitutive relations become (derive as an exercise)

$$\sigma'_{ij} = 2\eta_{vp} \dot{\varepsilon}'_{ij},$$ (13.60)

$$\eta_{vp} = \frac{\sigma_{\text{II}}}{2\dot{\varepsilon}_{\text{II}}},$$ (13.61)

$$\begin{array}{ll} \eta_{vp} = \eta & \text{for} \quad \sigma_{\text{II}} < \sigma_{yield}, \\[3mm] \eta_{vp} = \eta \dfrac{\sigma_{\text{II}}}{\eta \chi + \sigma_{\text{II}}} = \dfrac{\sigma_{yield}}{2\dot{\varepsilon}_{\text{II}}} & \text{for} \quad \sigma_{\text{II}} = \sigma_{yield}, \end{array}$$ (13.62)

which are in fact very similar to a non-Newtonian viscous model (Chapter 6) and could be relatively easily implemented into a viscous thermomechanical numerical modelling algorithm (Chapter 11).

It has however been demonstrated recently (Spiegelman et al., 2016) that the simplicity of the visco-plastic model is apparent and some problems arise with the solvability of 2D incompressible Stokes and continuity equations for pressure-dependent Drucker–Prager visco-plasticity. Analysis suggests that in this case the numerical solution can become ill posed and does not converge in terms of accurately reproducing the yielding condition (13.47) on the grid. In contrast to the more complex visco-elasto-plastic problems, the global error level $\Delta\sigma_{yield(nodal)}$ (Eq. 13.55) for pressure-dependent visco-plastic solutions remains relatively high and does not decrease with increasing number of global iterations. This is probably related to the fact that in the absence of elasticity, no adaptive time stepping could be used to improve the accuracy of the pressure-dependent, time step independent visco-plastic yielding condition.

On the other hand, visco-plastic rheologies are commonly used in numerical geodynamic modelling of lithospheric processes, which show prominent spontaneous strain localization phenomena resulting in the formation of various patterns of shear zones (e.g. Buiter et al., 2006, 2016). These shear zones are typically one to two grid steps wide and the pattern is typically non-unique and changes with changing model resolution (e.g. Buiter et al., 2006, 2016) that is also true in the case of visco-elasto-plastic rheology. On the other hand, shear zone angles and some other characteristics (e.g., taper angle, gross-scale strain pattern etc.) seem to be relatively robust and agree with both theoretical predictions and analogue models (e.g. Buiter et al., 2006, 2016). Therefore it is important to describe an efficient marker in cell algorithm, which could be used for solving visco-plastic thermomechanical problems.

This algorithm is in fact almost identical to the one used for solving viscous thermo-mechanical problems (Chapter 11) and the only addition concerns computing the visco-plastic viscosity for markers $\eta_{vp(m)}$ before advecting them by the velocity field (i.e. in between Steps 7 and 8 in Fig. 11.1). This is done by calculating first the non-Newtonian ductile viscosity η_m (Chapter 6) formulated in *terms of the second strain rate invariant* (Eq. 6.5b) and then evaluating plastic yielding condition (13.62) for the next time step as

$$2\eta_m \dot{\varepsilon}_{\text{II}(m)} \le \sigma_{yield(m)}. \qquad (13.63)$$

If this condition is not satisfied, visco-plastic viscosity of the marker is computed from Eq. (13.62).

An important aspect of treating visco-plasticity with staggered finite differences and marker-in-cell techniques concerns numerical diffusion of strain rates. The problem arises from the fact that strain rates are computed on the nodes using the interpolated, visco-plastic viscosity parameter η_{vp}, which in turn depends on strain rates. Interpolation of the mutually dependent parameters $\dot{\varepsilon}'_{xx}$, $\dot{\varepsilon}'_{xy}$ and η_{vp} back and forth between markers and nodes introduces systematic numerical diffusion. It has the same origin as in the case discussed in Chapter 8 (Fig. 8.11a), when interpolating total values (rather than increments) of temperature or stress back and forth between markers and nodes. The diffusion defocuses plastic deformation zones, which widen, so deviating notably from strongly localized deformation patterns as reported in rocks. Numerical diffusion can be restricted by using a stress-based approach in which strain rates used in Eq. (13.62) for each marker are not interpolated from the surrounding nodes directly but are computed from interpolated nodal stresses $(\sigma'_{xx}, \sigma'_{xy})$.

In order to restrict the numerical diffusion of shear zones, the following stress-based approach for computing of the effective strain rate invariant $\dot{\varepsilon}_{II(m)}$ for the markers is used

$$\dot{\varepsilon}_{II(m)} = \left(\frac{\sqrt{\sigma'^2_{xx(m)} + \sigma'^2_{xy(m)}}}{2\eta_{vp(m)}} \right)^S \left(\sqrt{\dot{\varepsilon}'^2_{xx(m)} + \dot{\varepsilon}'^2_{xy(m)}} \right)^{1-S}, \qquad (13.64)$$

where $S = 0.9$–0.95 is an empirical power exponent, $\eta_{vp(m)}$ is the value computed for the same marker during the previous time step and $\sigma'_{xx(m)}, \sigma'_{xy(m)}, \dot{\varepsilon}'_{xx(m)}, \dot{\varepsilon}'_{xy(m)}$ values are interpolated to the marker from respective nodal points by using the standard bilinear interpolation scheme (Eq. 18.19). This approach can be easily implemented on markers and allows focusing of shear zones to within one to two grid cells (see program example **Viscoplastic2D.m** associated with this chapter).

In contrast to visco-elasto-plastic models where stress builds up gradually due to elastic deformation, visco-plastic models produce an instantaneous stress growth. Therefore, in the beginning of the calculations, several tens to hundreds of iterations with very small time step and negligible marker displacement are needed to adjust the visco-plastic viscosity field to applied initial and boundary conditions.

Programming exercises

Exercise 13.1.
Add elasticity to the viscous thermomechanical code developed in Exercise 11.1. Use an incompressible continuity equation and computational time step $\Delta t = 10^{10}$ s. Use shear modulus $\mu = 5 \times 10^{10}$ Pa, $\mu = 3 \times 10^{10}$ Pa, and $\mu = 2 \times 10^{10}$ Pa for the mantle, plume and 'sticky air', respectively. Program update of stresses on markers and subgrid diffusion of stresses. Program solution of the temperature equation in several steps for the case when $\Delta t_T < \Delta t_m$. An example is in **Viscoelastic2D.m**.

Exercise 13.2.
Add plasticity to the code. Modify the model setup for shortening in the absence of gravity $(g_x = g_y = 0)$ of a visco-elasto-plastic block embedded in a weak medium and containing a weak squared inclusion (Fig. 13.4). The model is 100×100 km^2 in size with a resolution of 101×101 nodes and 4×4 randomly distributed markers per cell. Initial temperature in the entire model is 273 K and the insulation condition is applied at all boundaries. Use the following material parameters: block (100×60 km^2), $\mu = 10^{11}$ Pa, $\eta = 10^{23}$ Pa s, $\sigma_c = 10^8$ Pa, $\gamma_{int} = 0.6$; 10×10 km^2, squared inclusion and weak medium, $\mu = 10^{11}$ Pa, $\eta = 10^{17}$ Pa s, $\sigma_c = 10^7$ Pa, $\gamma_{int} = 0$. Other material properties of the block, inclusion and weak medium are the same as in the previous exercise for the mantle, plume and sticky air, respectively.

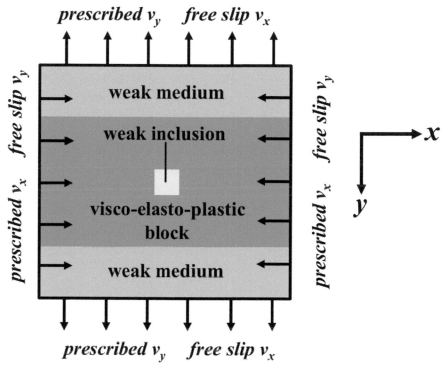

Fig. 13.4. Numerical setup for shortening of a visco-elasto-plastic block in the absence of gravity.

Program a constant horizontal shortening and vertical extension rate of 5×10^{-9} m/s applied at the vertical and horizontal boundaries respectively (Fig. 13.4, this condition is mass conservative and corresponds to a bulk shortening strain rate of 10^{-13} m/s). Apply 10^8 Pa as the boundary condition for pressure. Use a computational time step of 5×10^7 seconds and a marker displacement step of 0.01 of the grid step. Program plastic yielding on markers (before displacing them) by adjusting their visco-plastic viscosity $\eta_{vp(m)}$ using Eqs. (13.37)–(13.42). Limit this viscosity by upper (10^{23} Pa s) and lower (10^{17} Pa s) *viscosity cutoff values*. Interpolate this newly computed visco-plastic viscosity to nodal points at the next time step. Compute and visualize changes of $\Delta\sigma_{yield(markers)}$ for different time steps. Try different values of the internal friction coefficient for the block (from 0 to 0.7) and see how the orientation of the shear bands changes. An example is in **Viscoelastoplastic2D.m.**

Exercise 13.3.
Modify the 2D code programmed for the previous exercise by adding global visco-elasto-plastic Picard iterations of visco-plastic viscosity on Eulerian nodes (Eqs. 13.47–13.54). Interpolate viscosity $\eta_{(s)(i,j)}$, visco-plastic viscosity $\eta_{vp(s)(i,j)}$, compressive strength $\sigma_{c(i,j)}$ and friction coefficient $\gamma_{int(i,j)}$ from markers to basic nodes at each time step and then adjust the interpolated nodal visco-plastic viscosity $\eta_{vp(s)(i,j)}$ using Picard iteration

(use $\eta_{vp(s)(i,j)} = \eta_{(s)(i,j)}$ for the first time step). Interpolate the adjusted visco-plastic viscosity $\eta_{vp(s)(i,j)}$ back to markers before advection using Eq. (13.56). Set a tolerance level 3×10^5 Pa for $\Delta\sigma_{yield(nodal)}$ (Eq. 13.55). Reduce the computational time step Δt by a factor of 2 if this level is not achieved in 200 iterations and restart the iteration from the original nodal values of $\eta_{vp(s)(i,j)}$ stored at the beginning of the time step. Repeat this operation until the tolerance level is achieved. In the following time steps, attempt to increase Δt by a factor 1.1 (at each time step) until the original value of $\Delta t = 5 \times 10^7$ seconds is achieved. Visualize changes of $\Delta\sigma_{yield(nodal)}$ during iterations. Try different values for the tolerance level and the maximal number of iterations to see how the numerical solution changes. Compare this solution to the previous exercise. An example is in **i2elvis.m**.

14

2D thermomechanical modelling of inertial processes

> **Theory:** Numerical implementation of inertia and elastic compressibility. Organization of a thermomechanical code in the case of 2D, visco-elasto-plastic deformation with inertia. Thermomechanical iterations.
> **Exercises:** Programming a 2D thermomechanical code with inertia.

14.1 Why is it important to consider inertia?

As we discussed in Chapter 5, in highly viscous flows, the inertial forces are negligible with respect to viscous resistance and gravitational forces. Consequently, the majority of long term geodynamic processes are too slow to be affected by inertia. This situation changes, however, when the viscosity of the medium is strongly reduced (e.g. in some magmatic processes involving low viscosity magma) or when elastic deformation plays a dominant role (e.g. during earthquakes). In addition, there are some intrinsically fast processes of geodynamic significance, such as meteorite impacts, volcanic eruptions, landslides, avalanches, seismic wave propagation etc., in which inertia is playing a dominant role. Therefore it would be good to model (at least some of) these processes with the relatively simple numerical techniques described in the previous chapters. In this chapter, we aim to discuss one possible way of extending the scope of our geodynamical visco-elasto-plastic codes to modelling fast processes with inertia.

14.2 Reformulation of governing equations

In order to account for inertia forces we should obviously include inertial terms into the momentum conservation equation. We can thus use the Lagrangian Navier–Stokes equations (Chapter 5)

$$x\ \text{Navier–Stokes equation}\ \frac{\partial \sigma'_{xx}}{\partial x} + \frac{\partial \sigma'_{xy}}{\partial y} - \frac{\partial P}{\partial x} + \rho g_x = \rho \frac{D v_x}{Dt}, \quad (14.1)$$

y Navier–Stokes equation $\dfrac{\partial \sigma'_{yy}}{\partial y} + \dfrac{\partial \sigma'_{yx}}{\partial x} - \dfrac{\partial P}{\partial y} + \rho g_y = \rho \dfrac{Dv_y}{Dt},$ (14.2)

where Dv_x/Dt and Dv_y/Dt are components of the material acceleration vector.

In addition, the Lagrangian mass conservation equation (Chapter 1) should account for changes in density caused by rapid pressure and temperature variations

$$\frac{\partial v_x}{\partial x} + \frac{\partial v_y}{\partial y} + \beta \frac{DP}{Dt} = \Gamma_{plastic} + \alpha \frac{DT}{Dt},$$ (14.3)

$$\alpha = -\frac{\partial \ln(\rho)}{\partial T},$$ (14.4)

$$\beta = \frac{\partial \ln(\rho)}{\partial P},$$ (14.5)

where $\Gamma_{plastic}$ is the volumetric effect of dilatant plastic deformation (Eq. 12.63), α and β are respectively thermal expansion and compressibility coefficients, DT/Dt and DP/Dt are Lagrangian time derivatives of temperature and pressure, respectively.

In contrast, the Lagrangian energy conservation equation remains unchanged and should simply contain shear (H_s) and adiabatic (H_a) heating terms, which couple it to both the momentum and the continuity equations (14.1)–(14.5)

$$\rho C_P \frac{DT}{Dt} = \frac{\partial}{\partial x}\left(k\frac{\partial T}{\partial x}\right) + \frac{\partial}{\partial y}\left(k\frac{\partial T}{\partial y}\right) + H_r + H_a + H_s,$$ (14.6)

$$H_s = 2\sigma'_{xx}\dot{\varepsilon}'_{xx(visco\_plastic)} + 2\sigma'_{xy}\dot{\varepsilon}'_{xy(visco\_plastic)},$$ (14.7)

$$H_a = \alpha T \frac{DP}{Dt}.$$ (14.8)

Visco-elasto-plastic rheology discussed in the previous chapters (Eqs. 12.54–12.59, 13.1–13.4) requires some elaboration. In order to account for opening of fractures, the tensile yielding condition (12.50b) should be taken into account at low pressures

$$\sigma_{yield} = \sigma_c + \gamma_{int}P \quad \text{when} \quad P > \frac{\sigma_c - \sigma_t}{1 - \gamma_{int}} \text{ (confined fractures)},$$ (14.9a)

$$\sigma_{yield} = \sigma_t + P \quad \text{when} \quad P < \frac{\sigma_c - \sigma_t}{1 - \gamma_{int}} \text{ (tensile fractures)},$$ (14.9b)

$$2 \leq \frac{\sigma_c}{\sigma_t} \leq 8,$$ (14.9c)

where $\sigma_{yield} \geq 0$ (non-negative strength requirement), σ_c and σ_t are respectively compressive and tensile strength. We should also take into account that intense brittle/plastic deformation processes can lead to damage or even fragmentation of an intact material. It is therefore useful to introduce strain weakening of the material so that the strength (σ_c, σ_t) and/or internal friction coefficient (γ_{int}) decrease with increasing plastic strain $\varepsilon_{plastic}$

$$\sigma_c = \sigma_{c0} \text{ when } \varepsilon_{plastic} < \varepsilon_0, \tag{14.10a}$$

$$\sigma_c = \sigma_{c0} + \left(\varepsilon_{plastic} - \varepsilon_0\right) \frac{\sigma_{c1} - \sigma_{c0}}{\varepsilon_1 - \varepsilon_0} \text{ when } \varepsilon_0 \le \varepsilon_{plastic} \le \varepsilon_1, \tag{14.10b}$$

$$\sigma_c = \sigma_{c1} \text{ when } \varepsilon_{plastic} > \varepsilon_1, \tag{14.10c}$$

$$\sigma_t = \sigma_{t0} \text{ when } \varepsilon_{plastic} < \varepsilon_0, \tag{14.11a}$$

$$\sigma_t = \sigma_{t0} + \left(\varepsilon_{plastic} - \varepsilon_0\right) \frac{\sigma_{t1} - \sigma_{t0}}{\varepsilon_1 - \varepsilon_0} \text{ when } \varepsilon_0 \le \varepsilon_{plastic} \le \varepsilon_1, \tag{14.11b}$$

$$\sigma_t = \sigma_{t1} \text{ when } \varepsilon_{plastic} > \varepsilon_1 \tag{14.11c}$$

$$\gamma_{int} = \gamma_{int(0)} \text{ when } \varepsilon_{plastic} < \varepsilon_0, \tag{14.12a}$$

$$\gamma_{int} = \gamma_{int(0)} + \left(\varepsilon_{plastic} - \varepsilon_0\right) \frac{\gamma_{int(1)} - \gamma_{int(0)}}{\varepsilon_1 - \varepsilon_0} \text{ when } \varepsilon_0 \le \varepsilon_{plastic} \le \varepsilon_1, \tag{14.12b}$$

$$\gamma_{int} = \gamma_{int(1)} \text{ when } \varepsilon_{plastic} > \varepsilon_1, \tag{14.12c}$$

$$\varepsilon_{plastic} = \int \dot{\varepsilon}_{II(plastic)} dt, \tag{14.13a}$$

$$\dot{\varepsilon}_{II(plastic)} = \frac{D\varepsilon_{plastic}}{Dt} = \frac{\sigma_{II}}{2} \left(\frac{1}{\eta_{vp}} - \frac{1}{\eta}\right), \tag{14.13b}$$

where σ_{c0} and σ_{c1} are respectively the initial and final compressive strength, σ_{t0} and σ_{t1} are respectively the initial and final tensile strength, $\gamma_{int(0)}$ and $\gamma_{int(1)}$ are respectively the initial and final friction coefficient, ε_0 and ε_1 are respectively the lower and upper limits for the strain weakening interval, $\dot{\varepsilon}_{II(plastic)}$ is the second invariant of the plastic strain rate (Eq. 12.62), η and η_{vp} are respectively the ductile viscosity and the effective visco-plastic viscosity (Eq. 13.4).

14.3 Structure of visco-elasto-plastic thermomechanical code with inertia

In order to account for changes in the governing equations, we need to modify slightly the thermomechanical algorithm presented in the previous chapter (Fig. 13.1). First, we need to take into account that momentum and mass conservation equations (14.1)–(14.3) are strongly coupled to and should be solved together with the temperature equation (14.6). This could be done by including the solution of the temperature equation in the global visco-elasto-plastic Picard iteration on Eulerian nodes discussed in the previous chapter

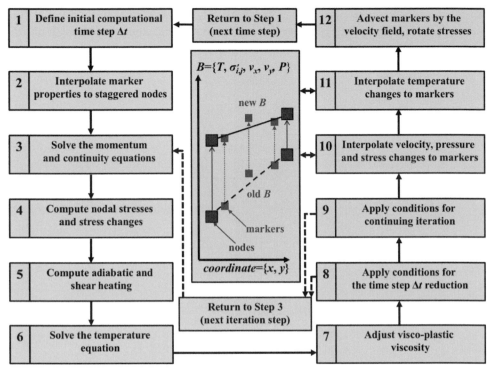

Fig. 14.1. Flow chart representing an example of a possible structure for a numerical thermomechanical compressible visco-elasto-plastic 2D code with inertia, which uses staggered finite differences and marker-in-cell technique (SFD-MIC) for solving Navier–Stokes, continuity and temperature equations.

(Exercise 13.3). This visco-elasto-plastic iteration will thus become *thermomechanical iteration*. Second, we should use an adaptive time step Δt, which should be the same for all conservation equations as well as for the displacement of markers. The time step should be thus adapted during the thermomechanical iteration based on several criteria limiting maximal magnitudes for stress, temperature, pressure, velocity and marker coordinate change. Finally, Lagrangian markers should additionally advect material pressure, velocity and plastic strain that are needed for solving of the governing equations (14.1)–(14.13).

The flow chart in Fig. 14.1 gives an example of a possible structure of a numerical compressible thermomechanical visco-elasto-plastic 2D code with inertia based on staggered finite differences and marker-in-cell technique (SFD-MIC). The principal steps of the algorithm are as follows.

(1) Define an initial value of the time step Δt for the momentum, continuity and temperature equations. The Δt value from the previous time step can be used as an initial guess, which could be increased, decreased or remain unchanged depending on the expected model dynamics.

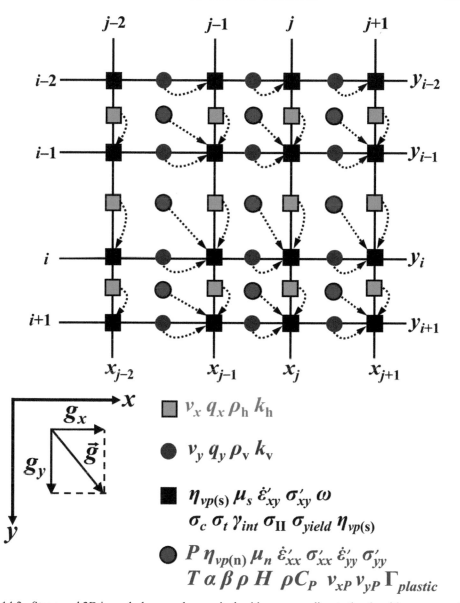

Fig. 14.2. Staggered 2D irregularly spaced numerical grid corresponding to the algorithm presented in Fig. 14.1.

(2) Compute and/or interpolate the material properties (σ_c, σ_t, γ_{int}, η, η_{vp}, μ, ρ, ρC_P, k, α, β), as well as scalars, vectors and tensors (T, P, v_x, v_y, σ'_{ij}) from the markers to the Eulerian nodes (Fig. 14.2).

(3) Discretize the 2D Navier–Stokes and continuity equations and compute the velocity and pressure by solving the global matrix problem with a direct method.

(4) Calculate the new stress and stress changes on the Eulerian nodes.

(5) Calculate the shear and adiabatic heating terms $H_{s(i,j)}$ and $H_{a(i,j)}$ at the Eulerian nodes from the computed velocity, pressure, strain rate and stress fields (see Step 3).

(6) Solve the temperature equation implicitly in time by a direct method.

(7) Adjust visco-plastic viscosity η_{vp} by applying plastic yielding conditions (13.47)–(13.53), (14.9) on basic nodes (cf. black squares in Fig. 14.2). Calculate the global error for plastic yielding $\Delta\sigma_{yield(nodal)}$ (Eq. 13.55).

(8) Apply conditions for the time step (Δt) reduction based on prescribed maximal magnitudes for stress, temperature, pressure, velocity and marker coordinate change per time step as well as based on the prescribed maximal number of thermomechanical iterations. If the time step is reduced, thermomechanical iteration is restarted at Step 3 from the previous physically meaningful state of temperature and visco-plastic viscosity.

(9) Apply conditions for continuing the thermomechanical iteration based on the tolerance level for $\Delta\sigma_{yield(nodal)}$ and for the magnitude of nodal temperature, pressure, velocity etc. changes between iteration steps. If these interruption criteria are not achieved, the calculations are repeated from Step 3.

(10) Interpolate calculated nodal velocity and stress changes to the markers and calculate new values of these parameters for the markers (see central panel in Fig. 14.1).

(11) Interpolate the calculated nodal temperature changes (see central panel in Fig. 14.1) from the Eulerian nodes to the markers, and calculate new marker temperatures.

(12) Advect all markers throughout the mesh according to the calculated velocity field (see Step 3). Components of the deviatoric stress tensor defined on the markers are recomputed analytically to account for any local stress rotation (Chapter 12).

An important aspect of solving thermomechanical visco-elasto-plastic problems with inertia is that we have to provide initial conditions not only for stress and temperature but also for pressure and velocity before starting the computation. All these properties are advected with markers (including v_x, v_y velocity components), which allow their time derivatives to be computed in the moving non-rotating Lagrangian reference frame. Remarkably, the geometry of an irregularly spaced, fully staggered numerical grid used in the previous chapter (Fig. 13.2) is applicable to this new algorithm with added thermal expansion and compressibility for the pressure points (cf. red circles in Fig. 14.2). In the following sections, new modifications needed to introduce inertia, elastic compressibility and thermal expansion into the visco-elasto-plastic code (Exercise 13.3) will be described.

Step 1: Defining an optimal computational time step

In problems involving inertia, the computational time step Δt is typically unknown a priori and should be adapted depending on the model state. Therefore, in the beginning of a numerical experiment, Δt should be set to some 'reasonable' value by trying probing solutions for any new model setup. It is also useful to define upper (Δt_{max}) and lower (Δt_{min}) limits for the time step changes during the calculations. During the calculations, current Δt values could be defined on the basis of past Δt values from the previous time steps, for example

$$\Delta t = \min[\Delta t_{\max}, \max\left(\Delta t_{\min}, a\Delta t_{-1}(\Delta t_{-1}/\Delta t_{-2})^b\right)], \quad (14.14)$$

where Δt_{-1} and Δt_{-2} are Δt values for the last and pre-last time steps, respectively, $a \geq 1$ and $0 \leq b \leq 1$ are empirical parameters for Δt adaptation. Please note that the Δt value defined by Eq. (14.14) is only an initial guess, which will be adapted (reduced) during the thermo-mechanical iteration (see Step 8).

Step 2: Interpolation of marker properties to nodes

An important modification of the algorithm described in Chapter 13 concerns interpolation of different velocity components from markers to respective staggered velocity nodal points. In order to do that, the following *momentum conservative form* of bilinear interpolation equation (8.18) should be used

$$v_{x(i,j)} = \frac{\sum_m v_{x(m)} \rho_m w_{m(i,j)}}{\sum_m \rho_m w_{m(i,j)}}, \quad (14.15a)$$

$$v_{y(i,j)} = \frac{\sum_m v_{y(m)} \rho_m w_{m(i,j)}}{\sum_m \rho_m w_{m(i,j)}}, \quad (14.15b)$$

$$w_{m(i,j)} = \left(1 - \frac{\Delta x_m}{\Delta x}\right) \times \left(1 - \frac{\Delta y_m}{\Delta y}\right),$$

where $v_{x(i,j)}$ and $v_{y(i,j)}$ are the respective nodal velocity components, $v_{x(m)}$ and $v_{y(m)}$ are the respective marker velocity components, ρ_m is the density of the marker. Please note that the weight of the mth marker $w_{m(i,j)}$ should be computed separately for the v_x and v_y staggered nodes. Similarly to the interpolated temperature (Chapter 10), boundary conditions should be applied to the interpolated values of $v_{x(i,j)}$ and $v_{y(i,j)}$ at the model boundaries. It is also important for the accuracy of the inertia calculations (Zhigadla, 2015) to use a non-local 4-cell scheme (Fig. 8.8) for the material density interpolation to v_x and v_y velocity nodes, which is then used for the inertial and gravity terms in the Navier–Stokes equations (14.1), (14.2).

Step 3: Solving the momentum and continuity equations

In the case of 2D inertial deformation of visco-elasto-plastic compressible material, the visco-elasto-plastic constitutive relationships (13.11)–(13.14) remain unchanged and the only alteration of the previously used Stokes equations concerns addition of respective discretized inertial terms

$$\frac{Dv_x}{Dt} = \frac{v_x - v_x^o}{\Delta t},$$

(14.16a)

$$\frac{Dv_y}{Dt} = \frac{v_y - v_y^o}{\Delta t},$$

(14.16b)

where Δt is the computational time step, v_x^o, v_y^o indicate the velocity components from the previous time step corrected for advection. Consequently, the conservative FD representation of the x Navier–Stokes equation (14.1) is only slightly different from Eq. (13.16) (see Fig. 14.2 for the indexing of the grid points):

$$\frac{4\Delta t}{x_{j+1} - x_{j-1}} \left(\frac{\eta_{vp(n)(i,j+1)}\dot{\varepsilon}'_{xx(i,j+1)}\mu_{n(i,j+1)}}{\mu_{n(i,j+1)}\Delta t + \eta_{vp(n)(i,j+1)}} - \frac{\eta_{vp(n)(i,j)}\dot{\varepsilon}'_{xx(i,j)}\mu_{n(i,j)}}{\mu_{n(i,j)}\Delta t + \eta_{vp(n)(i,j)}} \right)$$

$$+ \frac{2\Delta t}{y_i - y_{i-1}} \left(\frac{\eta_{vp(s)(i,j)}\dot{\varepsilon}'_{xy(i,j)}\mu_{s(i,j)}}{\mu_{(s)(i,j)}\Delta t + \eta_{vp(s)(i,j)}} - \frac{\eta_{vp(s)(i-1,j)}\dot{\varepsilon}'_{xy(i-1,j)}\mu_{s(i-1,j)}}{\mu_{(s)(i-1,j)}\Delta t + \eta_{vp(s)(i-1,j)}} \right) - 2\frac{P_{i,j+1} - P_{i,j}}{x_{j+1} - x_{j-1}} - \rho_{x(i,j)}\frac{v_{x(i,j)}}{\Delta t}$$

$$- g_x\Delta t \left(v_{x(i,j)}\frac{\rho_{h(i,j+1)} - \rho_{h(i,j)}}{x_{j+1} - x_{j-1}} + \frac{\left(v_{y(i-1,j)} + v_{y(i-1,j+1)} + v_{y(i,j)} + v_{y(i,j+1)} \right)\left(\rho_{h(i+1,j)} - \rho_{h(i-1,j)} \right)}{2(y_{i+1} + y_i - y_{i-1} - y_{i-2})} \right)$$

$$= -\rho_{h(i,j)} \left(g_x + \frac{v_{x(i,j)}^o}{\Delta t} \right) - \frac{2}{x_{j+1} - x_{j-1}} \left(\frac{\sigma'^o_{xx(i,j+1)}\eta_{vp(n)(i,j+1)}}{\mu_{(n)(i,j+1)}\Delta t + \eta_{vp(n)(i,j+1)}} - \frac{\sigma'^o_{xx(i,j)}\eta_{vp(n)(i,j)}}{\mu_{(n)(i,j)}\Delta t + \eta_{vp(n)(i,j)}} \right)$$

$$- \frac{1}{y_i - y_{i-1}} \left(\frac{\sigma'^o_{xy(i,j)}\eta_{vp(s)(i,j)}}{\mu_{s(i,j)}\Delta t + \eta_{vp(s)(i,j)}} - \frac{\sigma'^o_{xy(i-1,j)}\eta_{vp(s)(i-1,j)}}{\mu_{s(i-1,j)}\Delta t + \eta_{vp(s)(i-1,j)}} \right),$$

(14.17)

$$\dot{\varepsilon}'_{xx(i,j)} = \frac{1}{2} \left(\frac{v_{x(i,j)} - v_{x(i,j-1)}}{x_j - x_{j-1}} - \frac{v_{y(i,j)} - v_{y(i-1,j)}}{y_i - y_{i-1}} \right),$$

$$\dot{\varepsilon}'_{xx(i,j+1)} = \frac{1}{2} \left(\frac{v_{x(i,j+1)} - v_{x(i,j)}}{x_{j+1} - x_j} - \frac{v_{y(i,j+1)} - v_{y(i-1,j+1)}}{y_i - y_{i-1}} \right),$$

$$\dot{\varepsilon}'_{xy(i-1,j)} = \frac{v_{x(i,j)} - v_{x(i-1,j)}}{y_i - y_{i-2}} + \frac{v_{y(i-1,j+1)} - v_{y(i-1,j)}}{x_{j+1} - x_{j-1}},$$

$$\dot{\varepsilon}'_{xy(i,j)} = \frac{v_{x(i+1,j)} - v_{x(i,j)}}{y_{i+1} - y_{i-1}} + \frac{v_{y(i,j+1)} - v_{y(i,j)}}{x_{j+1} - x_{j-1}},$$

where v_x^o is the v_x velocity component from the previous time step, which is interpolated from advected markers. Discretization of the y Navier–Stokes equation (14.2) is analogous to Eq. (14.17) (derive as an exercise).

The time-dependent compressible continuity Eq. (14.3) is discretized as follows

$$\frac{v_{x(i,j)} - v_{x(i,j-1)}}{x_j - x_{j-1}} + \frac{v_{y(i,j)} - v_{y(i-1,j)}}{y_i - y_{i-1}} + \beta_{i,j}\frac{P_{i,j}}{\Delta t} = \Gamma_{plastic(i,j)} + \beta_{i,j}\frac{P_{i,j}^o}{\Delta t} + \alpha_{i,j}\frac{T_{i,j} - T_{i,j}^o}{\Delta t},$$

(14.18)

where $P_{i,j}^o$, $T_{i,j}^o$, $\Gamma_{plastic(i,j)}$, $\alpha_{i,j}$ and $\beta_{i,j}$ are respectively the old pressure (i.e., from the previous time step), old temperature, volumetric effect of dilatant plastic deformation, thermal expansion and compressibility coefficients interpolated from advected markers. Because of the presence of old pressure in the continuity equation (14.18), the boundary condition for pressure is not needed. Instead, pressure in a boundary condition cell could be adjusted before solving the continuity equation by subtracting from $P_{i,j}^o$ nodal and P_m^o marker pressure fields a constant ΔP_{BC}

$$\Delta P_{BC} = P_{BC(computed)} - P_{BC(required)}, \qquad (14.19)$$

where $P_{BC(computed)}$ and $P_{BC(required)}$ are respectively the computed (i.e., interpolated from markers) and required pressure in the boundary condition cell. This subtraction will allow adjusting the numerical solution for pressure.

As before, the global matrix for the momentum and continuity equations is solved by a direct method and the numbering of unknowns in this global matrix remain the same. Please note, that the right hand side of Eq. (14.18) contains an unknown value of the new temperature $T_{i,j}$, which should be adjusted during the thermomechanical iteration (Fig. 14.1, $T_{i,j}^o$ could be used instead of $T_{i,j}$ for the first iteration). In the case of very short time steps an *explicit formulation* (cf. discussion in Chapter 10) could be used for all conservation equations, which does not require composing and solving of global matrices.

Steps 5 and 6: Solving of the temperature equation

Numerical techniques for discretizing and solving the temperature equation are identical to those used in the visco-elasto-plastic code of Chapter 13 (Eqs. 11.8, 13.34). The only alteration concerns computing of the adiabatic heating term (Eq. 14.8), which directly uses the time derivative of pressure, thereby providing coupling between the temperature equation (14.6) and the momentum and continuity equations (14.1–14.3)

$$H_{a(i,j)} = T_{i,j}^o \alpha_{i,j} \frac{P_{i,j} - P_{i,j}^o}{\Delta t}. \qquad (14.20)$$

Step 8: Applying conditions for the time step reduction

The single value of the time step $\Delta t = \Delta t_T = \Delta t_m$ used in all conservation equations and for the marker advection should thus satisfy all limitations previously applied to different time steps (Chapter 13) so that

$$\Delta t = \min\{\Delta t_{\max}, \max[\Delta t_{\min}, \min(\Delta t, \Delta t_m, \Delta t_T)]\}. \qquad (14.21)$$

In addition to limiting the magnitude of the marker displacement and temperature changes, pressure, stress and velocity changes should also be controlled. In particular, limiting the magnitude of stress and pressure changes is very important to accurately resolve the transformation of the kinetic energy of motion into the potential energy of stresses and vice versa (Zhigadla, 2015). Another important aspect is that the time step adaptation should also be associated with adaptation of the lower cutoff viscosity limit (η_{min}) to ensure proper behaviour of low viscosity materials (such as air or water) included in the calculation. These materials should behave predominantly viscously (rather than elastically) so that condition $\eta \ll \mu\Delta t$ for them is satisfied

$$\eta_{min} = \min(\eta_{weak}, c\Delta t\mu_{weak}), \tag{14.22}$$

where η_{weak} and μ_{weak} are the viscosity and shear modulus of the weak material and $c \leq 10^{-3}$ is an empirical coefficient, which should also ensure that the computational viscosity contrast in the calculation is not too large (Eq. 13.8).

Step 10: Interpolation of velocity, pressure and stress changes from nodes to markers

After completing the thermomechanical iteration, velocity and pressure increments for the Eulerian nodes are calculated at the respective nodal points

$$\Delta v_{x(i,j)} = v_{x(i,j)} - v^o_{x(i,j)}, \tag{14.23}$$

$$\Delta v_{y(i,j)} = v_{y(i,j)} - v^o_{y(i,j)}, \tag{14.24}$$

$$\Delta P_{i,j} = P_{i,j} - P^o_{i,j}, \tag{14.25}$$

where $v^o_{x(i,j)}$, $v^o_{y(i,j)}$, $P^o_{i,j}$ and $v_{x(i,j)}$, $v_{y(i,j)}$, $P_{i,j}$ are values of respective parameters for the current (t) and next ($t+\Delta t$) time instant, respectively. These increments are then interpolated to the markers similarly to the deviatoric stress components and temperature (Chapters 10, 13).

Subgrid diffusion for pressure could be based on the same principle of local visco-elastic relaxation as for the deviatoric stress components (Eq. 13.22). Pressure increments computed from Eq. (14.25) are decomposed into a subgrid part $\Delta P^{subgrid}_{i,j}$ and a remaining part $\Delta P^{remaining}_{i,j}$

$$\Delta P_{i,j} = \Delta P^{subgrid}_{i,j} + \Delta P^{remaining}_{i,j}. \tag{14.26}$$

The subgrid part is computed for each marker

$$\Delta P_m^{subgrid} = \left(P_{m(nodal)}^o - P_m^o \right) \left[1 - \exp\left(-d_{ve} \frac{\Delta t}{t_{Maxwell(m)}} \right) \right], \qquad (14.27)$$

where $t_{Maxwell(m)} = \eta_{vp(m)}/\mu_m$ is defined for each marker; $0 \le d_{ve} \le 1$ is a dimensionless, numerical visco-elastic relaxation coefficient; $P_{m(nodal)}^o$ is interpolated for any given marker from $P_{i,j}^o$ at the surrounding nodes using relation (8.19) (Fig. 8.9). After obtaining $\Delta P_m^{subgrid}$ for all markers, $\Delta P_{i,j}^{subgrid}$ are computed by interpolation from the markers to the nodes using Eq. 8.18 (Fig. 8.8). Then $\Delta P_{i,j}^{remaining}$ are computed for the nodes from Eq. (14.26)

$$\Delta P_{i,j}^{remaining} = \Delta P_{i,j} - \Delta P_{i,j}^{subgrid}. \qquad (14.28)$$

Finally, new corrected marker pressures $P_{(m)}^{corrected}$ are computed as

$$P_m^{corrected} = P_m^o + \Delta P_m^{subgrid} + \Delta P_m^{remaining}, \qquad (14.29)$$

where $\Delta P_m^{subgrid}$ is given by Eq. (14.26), and $\Delta P_m^{remaining}$ is interpolated from nodal values of $\Delta P_{i,j}^{remaining}$ to the markers with the standard bilinear interpolation (Eq. 8.19, Fig. 8.9).

Subgrid diffusion for velocity can be calculated from the characteristic local momentum diffusion time scale t_M in a viscous medium

$$t_M = \frac{L^2}{v}, \qquad (14.30)$$

$$v = \frac{\eta}{\rho}, \qquad (14.31)$$

where v is *kinematic viscosity* (also called *momentum diffusivity*, m^2/s) and L is the width of the region. Please note that these relations are quite similar to the characteristic local thermal diffusion time scale discussed in Chapters 9 and 10.

Velocity increments computed from Eqs. (14.23), (14.24) are decomposed into a subgrid part $\Delta v_{x(i,j)}^{subgrid}$, $\Delta v_{y(i,j)}^{subgrid}$ and a remaining part $\Delta v_{x(i,j)}^{remaining}$, $\Delta v_{y(i,j)}^{remaining}$

$$\Delta v_{x(i,j)} = \Delta v_{x(i,j)}^{subgrid} + \Delta v_{x(i,j)}^{remaining}, \qquad (14.32)$$

$$\Delta v_{y(i,j)} = \Delta v_{y(i,j)}^{subgrid} + \Delta v_{y(i,j)}^{remaining}. \qquad (14.33)$$

The subgrid part is computed for each marker

$$\Delta v_{x(m)}^{subgrid} = \left(v_{x(m)nodal}^o - v_{x(m)}^o \right) \left[1 - \exp\left(-d_M \frac{\Delta t}{t_{M(m)}} \right) \right], \qquad (14.34)$$

$$\Delta v_{y(m)}^{subgrid} = \left(v_{y(m)nodal}^{o} - v_{y(m)}^{o} \right) \left[1 - \exp\left(-d_M \frac{\Delta t}{t_{M(m)}} \right) \right], \tag{14.35}$$

$$t_{M(m)} = \frac{\rho_m}{\eta_{vp(m)nodal}(2/\Delta x^2 + 2/\Delta y^2)},$$

where $t_{M(m)}$ is defined for the corresponding cell of the grid where the marker is located (Fig. 8.9); $\eta_{vp(m)\text{nodal}}$ and ρ_m are respectively visco-plastic viscosity (interpolated from basic nodes, Eq. 13.56) and density of a given marker; $0 \le d_M \le 1$ is a dimensionless, numerical momentum diffusion coefficient; $v_{y(m)\text{nodal}}^{o}$, $v_{y(m)\text{nodal}}^{o}$ are interpolated for any given marker from $v_{x(i,j)}^{o}$, $v_{y(i,j)}^{o}$ at the surrounding staggered velocity nodes using relation (8.19) (Fig. 8.9). After obtaining $\Delta v_{x(m)}^{subgrid}$, $\Delta v_{y(m)}^{subgrid}$ for all markers, $\Delta v_{x(i,j)}^{subgrid}$, $\Delta v_{y(i,j)}^{subgrid}$ are computed by interpolation from the markers to the respective velocity nodes using a momentum con-servative interpolation approach (Eq. 14.15, Fig. 8.8). Then $\Delta v_{x(i,j)}^{remaining}$, $\Delta v_{y(i,j)}^{remaining}$ are computed for the nodes from Eqs. (14.32), (14.33)

$$\Delta v_{x(i,j)}^{remaining} = \Delta v_{x(i,j)} - \Delta v_{x(i,j)}^{subgrid}, \tag{14.36}$$

$$\Delta v_{y(i,j)}^{remaining} = \Delta v_{y(i,j)} - \Delta v_{y(i,j)}^{subgrid}. \tag{14.37}$$

Finally, new corrected marker velocities $v_{x(m)}^{corrected}$, $v_{y(m)}^{corrected}$ are computed as

$$v_{x(m)}^{corrected} = v_{x(m)}^{o} + \Delta v_{x(m)}^{subgrid} + \Delta v_{x(m)}^{remaining}, \tag{14.38}$$

$$v_{y(m)}^{corrected} = v_{y(m)}^{o} + \Delta v_{y(m)}^{subgrid} + \Delta v_{y(m)}^{remaining}, \tag{14.39}$$

where $\Delta v_{x(m)}^{subgrid}$, $\Delta v_{y(m)}^{subgrid}$ are given by Eqs. (14.34), (14.35) and $\Delta v_{x(m)}^{remaining}$, $\Delta v_{y(m)}^{remaining}$ are interpolated from nodal values of $\Delta v_{x(i,j)}^{remaining}$, $\Delta v_{y(i,j)}^{remaining}$ to the markers with the standard bilinear interpolation (Eq. 8.19, Fig. 8.9).

Numerical treatment of deviatoric stresses remains the same and follows Eqs. (13.18)–(13.29), in which the marker displacement time step Δt_m is equal to the adjusted computa-tional time step Δt. As before, the subgrid pressure, temperature, velocity and stress diffusion schemes should only be applied in the case of strong marker advection and mixing.

Step 12: Advection of markers

Advection of markers remains the same as before, although the applied continuity-based Runge–Kutta advection scheme (Chapter 8) is only first-order accurate in time and does not account for velocity changes during the time step. This could be improved by further reducing the time step Δt and/or using higher order advection schemes in time, which however, will be computationally expensive.

14.4 Using the code with and without inertia

The described numerical algorithm is quite universal and allows modelling both fast and slow thermomechanical processes. In particular, all types of 'slow models' described in the previous chapter can also be computed with this new approach. This could be achieved by using a relatively large computational time step Δt that will automatically deactivate inertial terms in the Navier–Stokes equations (14.1, 14.2) and make them operate as the usual visco-elasto-plastic Stokes equations. In this case, the computational time step Δt and marker displacement time step Δt_m could be unequal ($\Delta t_m < \Delta t$). The deactivation of the inertial terms will also automatically damp both *pressure waves* (*P-waves*) and *shear waves* (*S-waves*) in the models.

In the case of modelling of impact processes associated with large variation in pressure and temperature, the density, isobaric heat capacity, thermal expansion and compressibility coefficients cannot be assumed constant and should be computed from equations of state for different materials (Chapter 2). This could be done by using analytical and/or numerical differentiation procedures based on standard thermodynamic relationships (Exercise 2.1).

Finally, the described algorithm allows modelling of the generation and propagation of seismic waves (Zhigadla, 2015; Herrendörfer et al., 2018). The accuracy of these calculations is, however, relatively low due to the simple first-order discretization schemes used for the time derivatives of velocity and pressure in the Navier–Stokes and continuity equations, respectively. This simple *backward Euler scheme* produces an artificial reduction (numerical damping) of the amplitude of seismic waves with time. In addition, relatively low order spatial approximation is used in our finite difference schemes, which causes artificial dispersion of seismic waves. Compared to more specialized higher order (in both space and time) methods used for seismic wave modelling, our simple numerical scheme requires much higher spatial resolution and shorter time steps to resolve the wave propagation accurately.

Programming exercises

Exercise 14.1.

Add inertia and compressibility to the visco-elasto-plastic thermomechanical code developed in Exercise 13.3. Use compressibility $\beta = 10^{-11}$ 1/Pa for all materials. Run the same experiment with the same time stepping criteria and tolerance for accuracy of the plastic yielding condition. Set tolerance levels 3×10^5 Pa and 1×10^{-3} K for maximal changes in nodal pressure and temperature, respectively, between iteration steps. Limit maximal stress changes per time step by 3×10^8 Pa. Use $c = 10^{-3}$ in Eq. (14.21) to change the lower

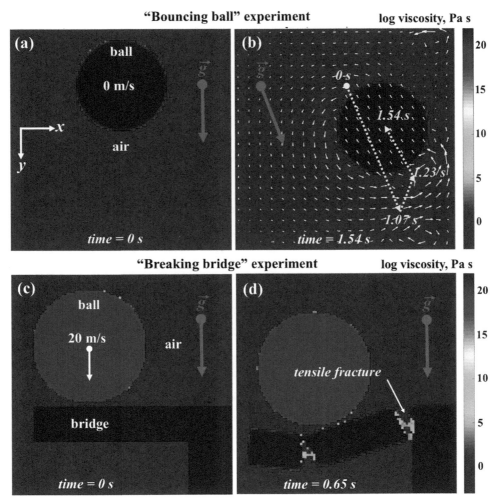

Fig. 14.3. Numerical model setups and results for 'bouncing ball' (a),(b) and 'breaking bridge' (c),(d) experiments. The gravity field in (b) is inclined (orange arrow, $g_x = 4$ m/s^2, $g_y = 10$ m/s^2), dotted arrows show the trajectory of the ball centre. The models are computed with the codes (a),(b) **Inertia2D.m** and (c),(d) **i2elvisNS.m** associated with this chapter.

viscosity cutoff limit with changing Δt. Compare results of previous incompressible experiments with the new compressible code at different values of $\beta = 10^{-10}$–10^{-12} 1/Pa. An example is in **Compressible2D.m**.

Exercise 14.2.
Modify the model setup from the previous exercise to make a 'bouncing ball' experiment (Fig. 14.3a). The model size is 10×10 m^2 with resolution 101 × 101 points and 4 × 4 markers per cell. Program a free fall of a ball (circular object, $\eta = 10^{22}$ Pa s, $\rho = 3000$ kg/m^3, $k = 2$ W/(m K)) surrounded by 'sticky air' ($\eta = 10^{-3}$ Pa s, $\rho = 1$ kg/m^3, $k = 3000$ W/(m K)) in

the vertical gravity field ($g_x = 0$, $g_y = 10$ m/s^2) and its collision with the lower boundary. Apply 10^5 Pa as the boundary condition for pressure. Use $c = 10^{-12}$ in Eq. (14.22). Use an initial computational time step of $\Delta t = 0.1$ s. Use $\mu = 10^8$ Pa, $\beta = 10^{-8}$ 1/Pa, $\alpha = 3 \times 10^{-5}$ 1/K, $\gamma_{int} = 0.6$, $\sigma_c = 10^8$ Pa, for all materials. Try different limits (10^5–10^6 Pa) for maximal pressure change per time step and see how this will affect the computation. Try to change the orientation of gravity ($g_x = 4$ m/s^2, $g_y = 10$ m/s^2) and see how this will affect the trajectory of the ball (Fig. 14.3a,b). An example is in **Inertia2D.m**.

Exercise 14.3.
Modify the model setup from the previous exercise to make a 'breaking bridge' experiment (Fig. 14.3c). Add the tensile yielding condition (14.9b) and plastic strain weakening (Eqs. 14.9–14.13) to the code. Increase the model size to 100×100 m^2 (grid resolution remains 101×101 points and 4×4 markers per cell). Modify the model setup for the collision of a ball ($\eta = 10^{20}$ Pa s, $\rho = 3000$ kg/m^3, $k = 2$ W/(m K), $\sigma_{c0} = \sigma_{c1} = 2 \times 10^8$ Pa, $\sigma_{c0} = \sigma_{c1} = 10^8$ Pa, $\gamma_{int(0)} = \gamma_{int(1)} = 0.6$, $\varepsilon_0 = 0$, $\varepsilon_1 = 1$) with a semi-bridge ($\eta = 10^{22}$ Pa s, $\rho = 3000$ kg/m^3, $k = 3$ W/(m K), $\sigma_{c0} = 6 \times 10^7$ Pa, $\sigma_{c1} = 6 \times 10^6$ Pa, $\sigma_{t0} = 3 \times 10^7$ Pa, $\sigma_{t1} = 3 \times 10^5$ Pa, $\gamma_{int(0)} = \gamma_{int(1)} = 0.6$, $\varepsilon_0 = 0$, $\varepsilon_1 = 0.1$). Use the same properties for the air as before (plastic parameters are $\sigma_{c0} = \sigma_{c1} = 3 \times 10^5$ Pa, $\sigma_{t0} = \sigma_{t1} = 3 \times 10^5$ Pa, $\gamma_{int(0)} = \gamma_{int(1)} = 0.6$, $\varepsilon_0 = 0$, $\varepsilon_1 = 1$). Set the initial vertical velocity of the ball to 20 m/s (on markers) and re-interpolate marker velocity from nodes at the first time step (i.e., similarly to temperature) to avoid large initial marker-node velocity discrepancy. Use an initial computational time step of $\Delta t = 0.001$ s. Vary strain weakening limits ε_0 and ε_1 and check their influence on the collision (Fig. 14.3d). An example is in **i2elvisNS.m**.

15

Seismo-thermomechanical modelling

> **Theory:** What is seismo-thermomechanical modelling? Rate-dependent friction. Rate- and state-dependent friction. Regularized rate- and state-dependent friction formulation. Invariant plasticity-like reformulation of rate- and state-dependent friction. Adaptive time stepping. Organization of a seismo-thermomechanical code. Visco-elasto-plastic iterations.
> **Exercises:** Programming a 2D seismo-thermomechanical code.

15.1 What is seismo-thermomechanical modelling?

Seismo-thermomechanical (STM) modelling is a very recent branch of geodynamic modelling, which investigates the development of earthquakes and their time series (*seismic cycles*) during various geodynamic processes using a continuum-based visco-elasto-plastic approach with inertia. This approach is quite similar to the one described in the previous chapter with one important difference that an *invariant re-formulation* of *rate-dependent* and *rate- and state-dependent friction* is used instead of Drucker–Prager plasticity.

15.2 Numerical modelling of earthquakes

Earthquakes happen repeatedly in various active tectonic settings and pose a great societal and economic hazard as they cause severe damage in increasingly populated areas. This has been demonstrated by recent large earthquakes, such as the 2011 Tohoku earthquake (subduction zone), the 2016 Nepal earthquake (continental collision zone), the 1999 Izmit earthquake (strike-slip fault) and the 2016–2017 Central Italian earthquakes (normal fault). Therefore it is crucial to understand the dynamics and physical controls of destructive earthquakes and their complex cyclic occurrence in different tectonic settings. Our thus far limited understanding of seismic cycles at different tectonic plate boundaries and fault systems is a result of their complex geometry and rheology, spatial inaccessibility, and the *limited observational timespan* over which geophysical measurements and seismological data are available. As a result, the currently available global record of seismological observations has at least two main limitations (Herrendörfer et al., 2018). First, each fault setting is unique in its set of tectonic characteristics. Hence, the observed seismicity is the result of the combined effect of various parameters, whose isolated role remains hidden. Second, each fault zone is currently at a different stage within its seismic cycle. Consequently, we need to somehow merge the inherently different snapshots from different fault zones together.

To overcome the observational limitations, numerical modelling is a key tool, since it can cover long time periods with many earthquake cycles and it can investigate the role of a single parameter at a time (e.g. Lapusta et al., 2000; Wang, 2007; van Dinther et al., 2013a, b). Furthermore, models can provide quantitative insights into the physical processes that are active along the unreachable, deeply buried parts of a fault system. Modelling can also be used for testing hypotheses based on observations or theoretical considerations and for generating testable predictions for future research and observations (e.g. Herrendörfer et al., 2015; Dal Zilio et al., 2018).

Numerical modelling of earthquakes has contributed in recent decades to our understanding of important processes leading to and resulting from an earthquake (Lapusta and Barbot, 2012). These processes, which are often summarized to the term *earthquake cycle* (or *seismic cycle*), include: inter-seismic loading, the nucleation of an earthquake, dynamic rupture propagation (co-seismic stage), and post-seismic deformation. Ideally, seismic cycle models should relate self-consistently *very slow, long-term geodynamic and tectonic processes* (tens of thousands to millions of years) to *very fast, short-term earthquake processes* (milliseconds to hours). In order to achieve that, at least three key ingredients should be combined (Wang, 2007): (1) a rate-dependent friction, (2) slow tectonic loading, and (3) visco-elastic stress relaxation. A comprehensive model should thereby account for a slowly loaded, dynamically evolving fault system embedded in a three-dimensional, heterogeneous visco-elasto-plastic medium. It needs to incorporate a fault constitutive model that is in agreement with laboratory experiments and that allows for different (slow and rapid) fault slip modes. Furthermore, it should resolve all relevant physical processes, whose time scales range from millions of years (geodynamic time scale) to milliseconds (co-seismic time scale) (e.g. Wang, 2007; van Dinther et al., 2013a; Sobolev and Muldashev, 2017; Herrendörfer et al., 2018).

A modelling approach that incorporates all these challenging ingredients is currently under development by different research groups (e.g. Lapusta et al., 2000; Wang, 2007; van Dinther et al., 2013a,b; Sobolev and Muldashev, 2017; Herrendörfer et al., 2018). On the one hand, classical simulations of earthquake cycles (e.g. Rice, 1993; Ben-Zion and Rice, 1997; Lapusta et al., 2000; Liu and Rice, 2007; Lapusta and Liu, 2009) typically use a predefined discrete (i.e. infinitely thin) fault embedded in a homogeneous elastic medium. This fault does not form spontaneously and its orientation and geometry do not change with time, which is in contrast to evolving faults (shear zones) formed in long-term visco-elasto-plastic geodynamic models (e.g., Figs. 12.6, 13.3). On the other hand, classical geodynamic models with self-consistently forming and propagating faults do not account for inertial forces and use very large time steps, which do not allow resolving of the earthquake cycle.

This is why it is very tempting to try to combine the advantages of earthquake cycle simulations and geodynamic simulations into a single continuum-based seismo-thermomechanical (STM) approach (van Dinther et al., 2013a,b; Sobolev and Muldashev, 2017; Herrendörfer et al., 2018).

In this chapter, we will discuss two STM approaches for modelling earthquake cycles with staggered finite difference and marker-in-cell technique developed by van Dinther et al. (2013a) for rate-dependent friction and by Herrendörfer et al. (2018) for rate- and state-dependent friction. These STM approaches are in continued development and will probably have important implications for future earthquake cycle modelling research.

15.3 Rate-dependent friction

Rate-dependent friction (e.g. Ampuero and Ben-Zion, 2008) implies that the internal friction coefficient along a fault strongly depends on the fault slip rate and can either increase (*rate strengthening*) or decrease (*rate weakening*) with increasing this rate. In order to model such behaviour along spontaneously forming faults in a continuum framework, a rate-dependent internal friction coefficient could be introduced into the Drucker–Prager plastic yielding condition (Eqs. 12.50a, 12.60) by assuming that the fault thickness in numerical experiments is always equal to the characteristic size of our grid cell Δx (van Dinther et al., 2013a) (Fig. 13.3)

$$\sigma_{II} = \sigma_{yield} = \sigma_c + \gamma_{eff} P, \tag{15.1}$$

$$\gamma_{eff} = \gamma_d + \frac{\gamma_s - \gamma_d}{1 + V_p/V_c}, \tag{15.2}$$

$$V_p = 2\dot{\varepsilon}_{II(plastic)} D, \tag{15.3}$$

$$\dot{\varepsilon}_{II(plastic)} = \frac{\sigma_{II}}{2}\left(\frac{1}{\eta_{vp}} - \frac{1}{\eta}\right), \tag{15.4}$$

where σ_c is compressive strength, γ_{eff}, γ_s, γ_d are respectively effective, static and dynamic internal friction coefficients, $D = \Delta x$ is the assumed fault thickness, V_p is the plastic slip rate along the fault, V_c is the characteristic slip rate at which half of the friction change occurs, $\dot{\varepsilon}_{II(plastic)}$ is the second invariant of the local plastic strain rate, η and η_{vp} are respectively the ductile viscosity and effective visco-plastic viscosity (Eq. 13.4). Depending on the relationship between γ_s and γ_d, the effective friction γ_{eff} could be either *rate weakening* (if $\gamma_s > \gamma_d$) or *rate strengthening* (if $\gamma_s < \gamma_d$).

Numerical implementation of rate-dependent friction is relatively simple and can be performed with either the marker-based or grid-based plasticity treatment approaches with adaptive time stepping discussed in Chapters 13 and 14. Due to the inter-dependence of σ_{yield}, $\dot{\varepsilon}_{II(plastic)}$ and η_{vp}, iterations are needed to compute new values of the visco-plastic viscosity with Eqs. (13.42), (13.52) (see example in **Rate_dependent_friction.m** associated with this chapter). These iterations are especially relevant if materials with rate-strengthening friction ($\gamma_s < \gamma_d$) are used for calculations. It is also important to note that rate-dependent friction behaviour is characterized by an *abrupt transition* from the inter-seismic to the co-seismic phase, which is associated with a step-like, many order of magnitude reduction in the computational time step associated with similar increase in material velocities. Nevertheless, this relatively simple continuum approach is very useful in modelling and analysing earthquake cycles in various geodynamic settings in a simplified manner, and thereby making quantitative and qualitative predictions on large-scale characteristics of seismicity (e.g. van Dinther et al., 2013b, 2014; Herrendoerfer et al., 2015; Dal Zilio et al., 2018).

15.4 Rate- and state-dependent friction

It is typically assumed that slip along faults is governed by the *rate- and state-dependent friction (RSF)* (Dieterich, 1978, 1979; Ruina, 1983). This friction formulation is based on *slide-hold-slide* and *velocity stepping* rock experiments, initially conducted by Dieterich (1972, 1978) and confirmed in subsequent studies (see review by Marone, 1998). In this formulation, the friction γ_{eff} relates shear stress τ_s to normal stress σ_n at the fault and depends on slip rate V and state Θ as

$$\tau_s = \gamma_{eff}\sigma_n = \left[\gamma_0 + a\ln\left(\frac{V}{V_0}\right) + b\ln\left(\frac{\Theta V_0}{L}\right)\right]\sigma_n, \tag{15.5}$$

where L is the characteristic slip distance and γ_0 is the reference friction coefficient defined at an arbitrary steady-state slip rate V_0. The rate-dependent term $a\ln(V/V_0)$ is called the instantaneous viscosity-like *direct effect* (Rice and Ruina, 1983), because it represents the immediate response of γ_{eff}, and hence τ_s, to a change in V, which is proportional to a. The state-dependent term $b\ln(\Theta V_0/L)$ is referred to as the *evolution effect* as it is described by the evolving state variable Θ, which has dimension of time (s).

Different evolution laws have been proposed to parameterize the change in Θ as a function of time (Bhattacharya et al., 2015): the aging law (Ruina, 1983), which is based on the observation of time-dependent healing at stationary contact (Dieterich, 1972), the slip law and the Nagata law (Nagata et al., 2012). Earthquake cycle simulations commonly apply the aging (or slowness) law (e.g. Lapusta et al., 2000; Liu and Rice, 2007) defined as

$$\frac{\partial \Theta}{\partial t} = 1 - \frac{V\Theta}{L}. \tag{15.6}$$

At constant slip velocity V, state Θ evolves towards a steady state $\partial\Theta/\partial t = 0$, $\Theta_{ss} = L/V$ (e.g. Rice and Ruina, 1983). The corresponding steady-state friction coefficient is defined as (derive as an exercise)

$$\gamma_{eff} = \gamma_0 + (a - b)\ln\left(\frac{V}{V_0}\right). \tag{15.7}$$

Based on Eq. (15.7) it is easy to understand that friction could be either *rate weakening* (if $a < b$) or *rate strengthening* (if $a > b$).

Rate- and state-dependent friction in its canonical form (Eq. 15.5) is somewhat inconvenient to use in the continuum framework since it implies that slip velocity V should always be greater than zero, which is not possible to satisfy in the case of emerging and propagating faults. This problem, however, does not exist for the more recent regularized RSF formulation, which will be discussed below.

15.5 Regularized rate- and state-dependent friction

Despite its long history, rate- and state-dependent friction is still empirical and lacks a general physical explanation. Nevertheless, some physical interpretations of the terms in Eq. (15.5) have been made in the past. Θ has been related to the age of the asperities along a sliding surface (Dieterich, 1981; Dieterich and Kilgore, 1994), to porosity (Sleep, 1995) and to grain size (Moore et al., 2019). L was shown to correlate with the roughness of the frictional surface and the particle size of the gouge (Dieterich, 1979, 1981). The direct effect has been interpreted as an Arrhenius type thermally activated rate process of forward and backward dislocation jumps at asperity contacts (e.g. Chester, 1994). This interpretation has led to the regularized version of rate- and state-dependent friction (Lapusta et al., 2000; Rice et al., 2001), which is also more convenient to use in the continuum framework

$$\tau_s = \text{arcsinh}\left[\frac{V}{2V_0}\exp\left(\frac{b}{a}\ln\left(\frac{\Theta V_0}{L}\right) + \frac{\gamma_0}{a}\right)\right]a\sigma_n. \tag{15.8}$$

The regularized version overcomes the deficiency that Eq. (15.5) is ill posed at $V = 0$ and can lead to negative friction for $V \ll V_0$. The difference between the standard and regularized versions of RSF becomes negligible for V approaching V_0. Furthermore, recent studies have made progress in deriving a micro-physical model that would ultimately allow the scaling of laboratory results to nature (e.g. Chen and Spiers, 2016; Moore et al., 2019).

One limitation of RSF is that it is based on experiments conducted at slip rates much lower than seismic rates. The logarithmic weakening of RSF with slip velocity is weaker than the weakening observed in high slip rate experiments. Such additional weakening mechanisms have been explored in earthquake cycle simulations (see review by Lapusta and Barbot, 2012). Nevertheless, results from earthquake cycle simulations with RSF equation (15.8) agree with observations in terms of slip rate, stress drop, amount of slip, rupture speed, and accelerating post-seismic slip after an earthquake (see review by Lapusta and Barbot, 2012). Not only earthquake slip, but also different parts of the slip spectrum including slow slip transients observed in nature (Peng and Gomberg, 2010), may be explained using the rate and state fault model, as confirmed by laboratory experiments (Leeman et al., 2016) and numerical simulations (Liu and Rice, 2007).

It is important to note that past studies have identified length and time scales associated with RSF, which are crucial to numerically resolve the earthquake cycle in space and time. The first length scale is the minimum size of slipping patch required for the transition from stable sliding to unstable slip along a rate-weakening fault, which is called the *nucleation length* (Rice and Ruina, 1983; Ruina, 1983; Dieterich, 1992; Rice, 1993; Rice et al., 2001; Lapusta, 2003; Rubin and Ampuero, 2005; Ampuero and Rubin, 2008; Kaneko et al., 2008). The second length scale, which is smaller than the first, is the *cohesive zone size* at the dynamically progagating rupture front, along which the stress linearly drops with slip from its maximum to dynamic value (Cocco and Bizzarri, 2002; Cocco et al., 2004; Day et al., 2005; Lapusta and Liu, 2009). Resolving these length scales with a fine enough spatial resolution is important for accurate modelling of earthquakes (e.g. Rice. 1993; Ben-Zion and Rice, 1993, 1997; Lapusta et al., 2000; Lapusta and Liu, 2009). In a typical earthquake cycle, the slip velocities change from less than 10^{-9} m/s in the inter-seismic period, and 10^{-8}–10^{-3} m/s during the accelerating nucleation and decelerating post-seismic phases, to 0.01–100 m/s during the dynamic rupture propagation. Lapusta et al. (2000) showed that this range of slip rate is resolvable due to the existence of the direct effect of RSF. To capture and resolve this more than nine orders of magnitude range in slip velocities in time, these authors derived an adaptive time step, which is inversely proportional to the slip rate and is a function of constitutive parameters.

15.6 Invariant reformulation of regularized rate- and state-dependent friction

There are three key differences between the concepts that underlie the rate- and state-dependent friction equations (15.5), (15.8) used in classical seismic cycle simulations and the yielding condition (15.1) used in our continuum approach (Herrendörfer et al., 2018). The first difference concerns the definition of deformation for slip on the one hand and for plastic strain on the other hand. In classical seismic cycle simulations, discontinuous brittle deformation occurs in the form of slip along a predefined fault plane (i.e. within an infinitely thin deformation zone). In contrast, in continuum mechanics, plastic deformation is treated as strain, which is a form of volumetric deformation. Plastic deformation can occur everywhere and spontaneously localize into a shear zone or fault zone of finite thickness, whose location and orientation are allowed to change through time (Figs. 12.6, 13.3). This difference in the concepts of slip and strain requires that the slip rate V is related to the plastic strain rate (e.g. to $\dot{\varepsilon}_{II(plastic)}$, Eq. 15.3). Chester (1994), Sleep (1997) and Noda and Shimamoto (2012) have proposed scaling inelastic strain rates by the thickness of the fault zone D. Here we follow the same approach by relating the slip rate to the plastic strain rate using Eq. (15.3).

Secondly, the fault strength is evaluated differently. Both shear stress τ_s and normal stress σ_n, which determine the shear strength for rate- and state-dependent friction, are related to the orientation of the predefined fault. In contrast, in the continuum approach it is common to use scalars that are invariant of the coordinate system and are able to adapt to spontaneous fault evolution. The yielding function, such as Eq. (15.1), is therefore defined as a function of pressure (P) and second invariant of the deviatoric stress tensor (σ_{II}) instead of σ_n and τ_s, respectively. In addition, in continuum mechanics, material has a certain residual strength at zero pressure, which is related to its compressive strength σ_c (Eq. 15.1) and hence to cohesion $C = \sigma_c/cos(\varphi)$. In rate- and state-dependent friction, in contrast, the fault is treated as a broken material with zero cohesion. Alternatively, Marone et al. (1992) discussed the option that cohesion is part of the state variable Θ.

The third key difference lies in the question of when plastic deformation or slip becomes active. In Eq. (15.1), the yield strength is usually treated as a threshold value of stress. Plastic deformation begins only if the second invariant of the deviatoric stress tensor is equal to the yield strength, and it stops as soon as the stress falls below the strength. Thus, slip velocity changes discontinuously around the yield strength. In contrast, a key assumption underlying the rate- and state-dependent friction equation (15.5) is that some plastic slip always occurs if the shear stress is larger than zero. The difference between these concepts was already noted by Nakatani (2001). He interprets the term $\sigma_n[\gamma_0 + b\ln(\Theta V_0/L)]$ in Eq. (15.5) as interface strength. This interface strength is similar to a threshold value in the sense that slip rates become noticeable only when stress approaches it. Following Nakatani (2001), rate- and state-dependent friction can be regarded as a smooth version of the classical yield strength. The smoothness of the rate- and state-dependent yield strength is proportional to a. Consequently, if a tends to zero, the rate- and state-dependent friction framework approaches the classical notion of yield strength.

By taking all these differences between RSF and the usual definition of the yield strength into account, we can arrive at the invariant regularized formulation of rate- and state-dependent friction (Herrendörfer et al., 2018)

$$\sigma_{II} = \sigma_{yield} = \sigma_c + \text{arcsinh}\left[\frac{V_p}{2V_0}\exp\left(\frac{b}{a}\ln\left(\frac{\Theta V_0}{L}\right) + \frac{\gamma_0}{a}\right)\right]aP \qquad (15.9)$$

with the aging evolution law

$$\frac{\partial \Theta}{\partial t} = 1 - \frac{V_p\Theta}{L}. \qquad (15.10)$$

We can reformulate these two equations in a different form, which is more convenient for numerical treatment (derive as an exercise)

$$\sigma_{II} = \sigma_{yield} = \sigma_c + \text{arcsinh}\left[\frac{V_p}{2V_0}\exp\left(\frac{b\Omega + \gamma_0}{a}\right)\right]aP, \qquad (15.11)$$

$$\frac{\partial \Omega}{\partial t} = \frac{V_0 \exp(-\Omega) - V_p}{L}, \tag{15.12}$$

$$\Omega = \ln\left(\frac{\Theta V_0}{L}\right), \tag{15.13}$$

where Ω is the modified dimensionless state parameter.

15.7 Adaptive time stepping

Adaptive time stepping is a crucial part of STM modelling since it allows accurate resolution of both the initiation and the dynamics of earthquakes. Following Herrendörfer et al. (2018), we require that the time step length is the minimum of the time steps needed to resolve the state weakening (Δt_w) and healing (Δt_h) and to limit marker displacement per grid cell (Δt_m) as

$$\Delta t = \zeta \, min(\Delta t_w, \Delta t_h, \Delta t_m), \tag{15.14}$$

where $\zeta < 1$ is the time step factor and the condition (15.14) is applied over the entire model domain.

To accurately resolve the weakening stage (i.e., decrease of the state parameter Θ or Ω), we can use the adaptive time step Δt_w developed by Lapusta et al. (2000) and Lapusta and Liu (2009), which is inversely proportional to the slip rate

$$\Delta t_w = X_{state} \frac{L}{V_p}, \tag{15.15}$$

$$X_{state} = min\left(1 - \frac{(b-a)P}{KL}, 0.2\right) \text{ when } \xi < 0, \tag{15.16a}$$

$$X_{state} = min\left(\frac{aP}{KL - (b-a)P}, 0.2\right) \text{ when } \xi > 0, \tag{15.16b}$$

$$\xi = \frac{1}{4}\left(\frac{KL}{aP} - \frac{b-a}{a}\right)^2 - \frac{KL}{aP}, \tag{15.17}$$

$$K = \frac{2\mu}{\pi \Delta x (1 - v)}, \tag{15.18}$$

$$v = \frac{3B - 2\mu}{6B + 2\mu}, \tag{15.19}$$

where the factor $X_{state} \leq 0.2$ implies that the state parameter should not decrease for more than 20% per time step, v is Poisson's ratio and B is bulk modulus ($B = 1/\beta$, where β is compressibility, Chapter 12).

For resolving the healing phase (i.e., increase of the state parameter Θ or Ω), the healing time step limit Δt_h could be formulated as (Herrendörfer et al., 2018)

$$\Delta t_h = X_{state} \min \left(\frac{L}{V_p}, \frac{L}{V_0} \exp(\Omega) \right), \tag{15.20a}$$

or

$$\Delta t_h = X_{state} \min \left(\frac{L}{V_p}, \Theta \right), \tag{15.20b}$$

where the factor $X_{state} \leq 0.2$ implies that the state parameter should not increase for more than 20% per time step. We can also additionally control increasing or decreasing of the state parameter (Θ or Ω) by limiting its maximal change per time step (i.e., similarly to limiting maximal marker displacement)

$$\Delta t \left| \frac{1}{\Theta} \frac{\partial \Theta}{\partial t} \right| \leq X_{state}, \tag{15.21}$$

or

$$\Delta t \left| \frac{\partial \Omega}{\partial t} \right| \leq X_{state}, \tag{15.22}$$

where $X_{state} \leq 0.2$ is the imposed limit for changes of the state parameter given by Eqs. (15.10), (15.12).

Limitations for the marker displacement time step Δt_m are imposed in the same manner as before. In the case of STM modelling, the marker displacement limit is typically rather small (0.01–1% *of the minimal grid step*), which allows accurate resolution of the dynamics of individual seismic events.

15.8 Visco-elasto-plastic iterations with rate- and state-dependent friction

Global visco-elasto-plastic Picard iteration on Eulerian basic nodes described in Chapter 13 needs some modification to optimize it for rate- and state-dependent friction. The following approach of computing new values of state parameter Ω and visco-plastic viscosity η_{vp} is suggested (Herrendörfer et al., 2018, verify as an exercise, cf. Fig. 15.1 for the location of different points).

(1) Compute the second stress invariant $\sigma_{II(i,j)}$ from Eq. (13.49).
(2) Compute the plastic slip rate from Eq. (15.11)

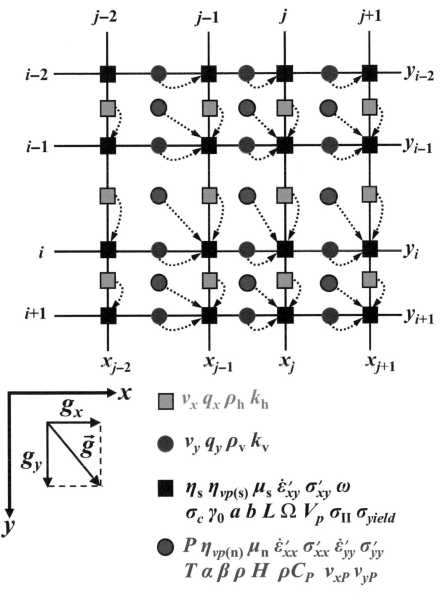

Fig. 15.1. Staggered 2D irregularly spaced numerical grid corresponding to the seismo-thermomechanical code with rate- and state-dependent friction. The algorithm for this code is presented in Fig. 14.1 with some modifications discussed in the text of this chapter.

$$V_{p(i,j)} = 2V_0 \sinh\left(\frac{4\max\left(\sigma_{\mathrm{II}(i,j)} - \sigma_{c(i,j)}, 0\right)}{a_{(i,j)}\left(P_{(i,j)} + P_{(i+1,j)} + P_{(i,j+1)} + P_{(i+1,j+1)}\right)}\right) \exp\left(-\frac{b_{(i,j)}\Omega_{(i,j)} + \gamma_{0(i,j)}}{a_{(i,j)}}\right).$$

$$(15.23)$$

(3) Compute a new value for the state parameter $\Omega_{(i,j)}$ from Eq. (15.12) by using an analytical solution derived by Lapusta and Liu (2009) under the assumption of constant $V_{p(i,j)}$ during the time step:

$$\Omega_{(i,j)} = \ln\left[\exp\left(\Omega_{0(i,j)}\right)\left(1 - \frac{V_{p(i,j)}\Delta t}{L_{(i,j)}}\right) + \frac{V_0\Delta t}{L_{(i,j)}}\right] \text{ if } \frac{V_p\Delta t}{L_{(i,j)}} \leq 10^{-6}, \tag{15.24a}$$

$$\Omega_{(i,j)} = \ln\left[\frac{V_0}{V_{p(i,j)}} + \left(\exp\left(\Omega_{0(i,j)}\right) - \frac{V_0}{V_{p(i,j)}}\right)\exp\left(-\frac{V_{p(i,j)}\Delta t}{L_{(i,j)}}\right)\right] \text{ if } \frac{V_p\Delta t}{L_{(i,j)}} > 10^{-6}. \tag{15.24b}$$

(4) Compute the yield strength from Eq. (15.11) with the updated $V_{p(i,j)}$ and $\Omega_{(i,j)}$ values

$$\sigma_{yield(i,j)} = \sigma_{c(i,j)} + \text{arcsinh}\left[\frac{V_{p(i,j)}}{2V_0}\exp\left(\frac{b_{(i,j)}\Omega_{(i,j)} + \gamma_{0(i,j)}}{a_{(i,j)}}\right)\right]a_{(i,j)}\frac{P_{(i,j)} + P_{(i+1,j)} + P_{(i,j+1)} + P_{(i+1,j+1)}}{4}. \tag{15.25}$$

(5) Compute the second plastic strain rate invariant from Eq. (15.3)

$$\dot{\varepsilon}_{\text{II}(plastic)(i,j)} = \frac{V_{p(i,j)}}{2D}. \tag{15.26}$$

(6) Compute a new visco-plastic viscosity from Eq. (13.4)

$$\eta_{vp(s)(i,j)} = \eta_{s(i,j)}\frac{\sigma_{yield(i,j)}}{2\eta_{s(i,j)}\dot{\varepsilon}_{\text{II}(plastic)(i,j)} + \sigma_{yield(i,j)}}. \tag{15.27}$$

(7) Compute the difference $\sigma_{\text{II}(i,j)} - \sigma_{yield(i,j)}$ and update the global error $\Delta\sigma_{yield(nodal)}$ with Eq. (13.55).

(8) Repeat iteration until $\Delta\sigma_{yield(nodal)}$ becomes smaller than a predefined tolerance level.

This iteration strategy shows stable convergence with Picard iteration, which is typically better than the convergence for Drucker–Prager plasticity and rate-dependent friction.

15.9 Structure of seismo-thermomechanical code

It should be noted that the algorithm developed in the previous chapter for modelling of visco-elasto-plastic inertial deformation (Fig. 14.1) essentially remains the same and only experiences three rather minor modifications. The first one consists in adding more criteria for the adaptive time stepping, which allows evolution of the state variable (Eqs. 15.14–15.22) to be resolved properly. The second modification concerns the method of performing visco-elasto-plastic Picard iteration, which is described above (Eqs. 15.23–15.27).

The third modification concerns update of nodal stresses, pressure, velocity and state parameter for advection and rotation (in the case of stresses). Since STM models typically require very small marker displacements (0.01–1% of the grid step), contributions from advection are very small compared to changes caused by other time-dependent processes. As a result, advection-related changes of the nodal values become much smaller than the error associated with interpolation of these quantities from markers to nodes (i.e. error due to Eq. 8.18). In order to minimize this interpolation error, small effects could be added to nodal points as (very small) increments instead of re-interpolation of bulk parameter quantities from markers to nodes after advection. The following approach could be used (Herrendörfer et al., 2018):

$$B_{a(i,j)} = B_{(i,j)} + \Delta B_{m(i,j)}, \tag{15.28}$$

$$\Delta B_{m(i,j)} = B_{m1(i,j)} - B_{m0(i,j)}, \tag{15.29}$$

where $B_{a(i,j)}$ is the nodal value of parameter B corrected for advection (and rotation, in the case of stresses), $B_{(i,j)}$ is the initial value of this parameter obtained from solving differential equations on Eulerian nodes, $B_{m0(i,j)}$ and $B_{m1(i,j)}$ are nodal B values interpolated from respective updated quantities of markers by using Eq. (8.18) before and after advection (and stress rotation), respectively. Nodal visco-plastic viscosity values do not need such correction since they are iteratively updated at each time step. This modified advection scheme is *only applicable for very small marker displacement steps* (otherwise it can produce numerical oscillation) and can only be recommended for deviatoric stresses, pressure, velocity and state parameter, since their interpolation error may notably influence the STM solution.

Alternatively, one could use a *semi-Lagrangian scheme* (*SLS*, e.g., Spiegelman and Katz, 2006; Ismail-Zadeh and Tackley, 2010) and trace positions of Lagrangian points overlapping with respective Eulerian nodes *back in time to their departure points*. At the departure points, the advected values of deviatoric stresses, pressure, velocity and state parameter can then be obtained by using bilinear interpolation (Fig. 8.9, Eq. 8.19) from the respective surrounding Eulerian nodes. The Lagrangian point back-tracing can be done with the same continuity-based advection schemes as used for markers (Chapter 8) by applying a negative value of the displacement time step Δt_m. This SLS scheme is somewhat similar to the upwind differences (Chapter 8) and is therefore slightly numerically diffusive (example is in the code **i2elvisSTM_backtracing.m** associated with this chapter).

It should also be noted that at very small time steps needed for the accurate STM solution, both material advection and stress rotation effects are negligible and could be ignored. In this case, the use of active markers carrying evolving material properties is not needed and an Eulerian node-based algorithm could be used for restricted time intervals (example is in the code **i2elvisSM.m** associated with this chapter). On the other hand, passive markers could be used for tracing inter-seismic, co-seismic and post-seismic displacements. The accuracy of the displacement calculations on Lagrangian markers is many orders of magnitude higher than the Eulerian grid resolution and is mainly affected by the interpolation error from nodes to markers. Higher order continuity-based Runge–Kutta interpolation schemes are recommended (Chapter 8).

15.10 Transition from thermomechanical to seismo-thermomechanical calculations

During STM modelling it is often necessary to first run simulations in thermomechanical mode with large time step and simple Drucker–Prager plasticity and then turn this into seismo-thermomechanical calculations with rate-dependent or rate- and state-dependent friction and adaptive time step. This is done to rapidly reach a certain 'mature' geodynamic model stage with well-established fault patterns and then investigate the seismicity associated with this stage (e.g. van Dinther et al., 2013b, 2014; Dal Zilio et al., 2018). Transition to the STM modelling regime should be performed gradually and needs adjustment of friction parameters. In the case of rate-dependent friction, we can use gradual reduction of the time step and introduction of γ_{eff} (Eqs. 15.1, 15.2) instead of γ_{int} (Eqs. 12.50a) into Picard iteration. In the case of rate- and state-dependent friction, we can instead compute an 'equilibrium value' of state parameter $\Omega_{(i,j)}$ for yielding basic nodes using Eqs. (15.23), (15.26) and the local plastic strain rate $\dot{\varepsilon}_{\mathrm{II}(plastic)(i,j)}$ for Drucker–Prager plasticity (Eq. 15.4) (derive as an exercise)

$$\Omega_{(i,j)} = \frac{a_{(i,j)}}{b_{(i,j)}} \ln \left[\frac{2V_0 \, \sinh \left(\dfrac{4 \, max\left(\sigma_{\mathrm{II}(i,j)} - \sigma_{c(i,j)}, 0 \right)}{a_{(i,j)} \left(P_{(i,j)} + P_{(i+1,j)} + P_{(i,j+1)} + P_{(i+1,j+1)} \right)} \right)}{D\sigma_{\mathrm{II}(i,j)} \left(\dfrac{1}{\eta_{vp(s)(i,j)}} - \dfrac{1}{\eta_{s(i,j)}} \right)} \right] - \frac{\gamma_{0(i,j)}}{b_{(i,j)}}, \qquad (15.30)$$

where $\eta_{vp(s)(i,j)}$ is the visco-plastic viscosity computed for the Drucker–Prager plasticity model. We can then use this computed value of state parameter to start iterations with the rate- and state-dependent friction.

15.11 Dependence of seismo-thermomechanical calculations on the grid step

Due to the use of scaling $D = \Delta x$ between the characteristic fault width D and the grid step Δx, results of STM calculations are typically independent of the grid step size and converge with increasing grid resolution (van Dinther et al., 2013a; Herrendörfer et al., 2018). An exception concerns modelling of a gradual fault localization process in case of rate- and state-dependent friction. In this case, the fault width D evolves (decreases) with time and should be thus parameterized.

One possible approach could be to scale D with the state variable Θ or slip rate V_p as

$$D = \min[D_{\max}, \max(\Delta x, K_\Theta \Theta V_0)], \tag{15.31}$$

or

$$D = \max\left[\Delta x, D_{\max}\left(1 - K_V \frac{V_p}{V_0}\right)\right], \tag{15.32}$$

where $D_{\max}$ and Δx are respectively maximal and minimal thicknesses of the fault in the model, K_Θ and K_V are dimensionless positive coefficients. $D_{\max}$ and K_Θ or $D_{\max}$ and K_V are two additional parameters of a continuum-based RSF model of a localizing fault. These parameters are thus added to the classical set of RSF variables (γ_0, a, b, V_0 and Θ) describing the behaviour of a fully localized discrete fault. Testing of Eqs. (15.31), (15.32) for a strike-slip fault localizing from a Gaussian perturbation of Θ suggests that the time of the first seismic event is nearly independent of the model resolution (see example in **i2elvisSM.m** associated with this chapter).

An alternative approach could be based on deriving an invariant RSF formulation from grain-scale creep processes (Moore et al., 2019) by combining grain size- and stress-dependent viscous flow laws (Chapter 6, Eqs. 6.3–6.5) with grain size evolution models (e.g., Hall and Parmentier, 2003 and references therein).

Programming exercise

Exercise 15.1.

Add rate- and state-dependent friction to the visco-elasto-plastic thermomechanical code with inertia developed in Exercise 14.1. Program the 150×150 km^2 model setup shown in Fig. 15.2 with a 3 km thick strike-slip fault ($\Omega = -1$) located in between two elastic blocks ($\Omega = 40$, $a = 0.011$, $b = 0.017$). Model resolution is 101×101 points and 4×4 markers per cell (this model resolution is rather low but sufficient for qualitative testing purposes). The model has no gravity ($g_x = g_y = 0$). Within the fault, prescribe an 86 km wide central rate-weakening zone ($a = 0.011$, $b = 0.017$) surrounded by two 32 km wide rate-strengthening zones ($a = 0.011$, $b = 0.001$). Use $\sigma_c = 0$, $L = 0.01$, $\gamma_0 = 0.2$, $\alpha = 0$ 1/K, $\beta = 2 \times 10^{-11}$ 1/Pa, $\mu = 3 \times 10^{10}$ Pa, $\rho = 2700$ kg/m^3, $\eta = 10^{23}$ Pa s, $k = 3$ W/(m K), $T = 273$ K for all materials. Define a reference velocity $V_0 = 4 \times 10^{-9}$ m/s. Thermal boundary conditions are insulation at all boundaries. Velocity boundary conditions are: $v_y = 0$ at all boundaries, $v_x = 2 \times 10^{-9}$ m/s at the top, $v_x = -2 \times 10^{-9}$ m/s at the bottom and $\partial v_x/\partial x = 0$ at the left and right boundaries. Use $D = \Delta y$ (fault thickness) and $P = 5$ MPa as the initial condition for

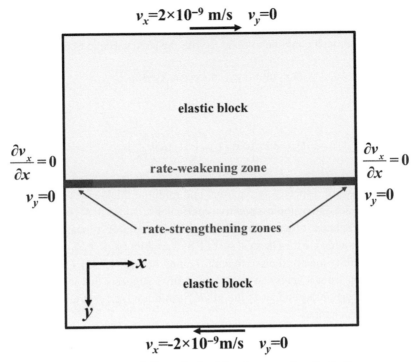

Fig. 15.2. Numerical setup for modelling of seismicity along a strike-slip fault.

pressure. Modify the visco-elasto-plastic Picard iteration on Eulerian basic nodes with Eqs. (15.23)–(15.27). Implement adaptive time stepping with Eqs. (15.14)–(15.22). Set a tolerance level of 10 Pa for $\Delta\sigma_{yield(nodal)}$ (Eq. 13.55). Run numerical experiments and investigate the dynamics of seismic events by plotting maximal values of $V_{p(i,j)}$ in the model (Eq. 15.23) as a function of time. Try to vary RSF model parameters and see how this will affect the behaviour of the fault. Please note that this behaviour is strongly dependent (e.g. Liu and Rice, 2007; Herrendörfer et al., 2018) on the characteristic slip distance L and the ratio between the length of the rate-weakening section of the fault W and the *nucleation length* h^{*} defined as (Rubin and Ampuero, 2005)

$$h^{*} = \frac{2\mu bL}{\pi(b-a)^{2}P(1-v)},\qquad(15.33)$$

where $0.5 < a/b < 1$.

Different regimes of the slip spectrum on the fault (Fig. 15.3) may include stable sliding, aseismic slip transients and seismic slip. Try to obtain these regimes by varying L and the initial condition for P in accordance with the diagram shown in Fig. 15.3. An example is in **i2elvisSTM.m**.

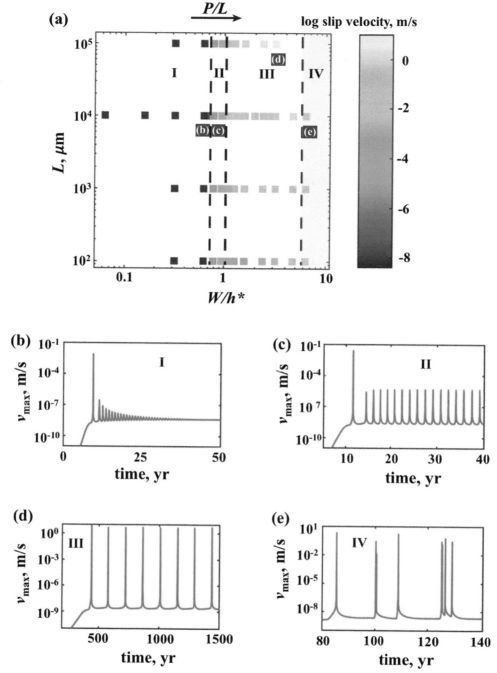

Fig 15.3. Slip spectrum along a strike-slip fault as a function of nucleation size h^* and characteristic slip distance L (Herrendörfer et al., 2018). (a) Slip spectrum is expressed as slip velocity (colour code), as a function of W/h^* (W is the length of the rate-weakening zone of the fault, Fig. 15.2) and L. Four different regimes are identified: I decaying oscillations towards stable sliding, II periodic aseismic slip transients, III periodic seismic slip and IV aperiodic seismic slip affected by seismic waves reflected from model boundaries. (b)–(e) Evolution of the maximum slip velocity V_p in four reference modes representing the different regimes: (b) I, (c) II, (d) III, (e) IV.

16

Hydro-thermomechanical modelling

> **Theory:** What is hydro-thermomechanical modelling? Fluid percolation
> processes. Darcy law and its derivation from simple principles. Permeability
> and its dependence on porosity. Governing equations for modelling coupled
> hydro-thermomechanical visco-elasto-plastic systems. Visco-elasto-plastic
> compaction. Mass transfer. Organization of a hydro-thermomechanical code.
> Hydro-thermomechanical iterations. Toy melting and dehydration models.
> **Exercises:** Programming of 2D hydro-thermomechanical codes.

16.1 What is hydro-thermomechanical modelling?

Hydro-thermomechanical (*HTM*) modelling is an important branch of geodynamic
modelling, which investigates evolution of *coupled fluid-solid systems*, where the
fluid is a low viscosity phase, which could be gas, liquid, melt and their mixtures.
In the geodynamic community, this class of numerical modelling problems is often
referred to as *two-phase flow problems*, where the two phases are solid and fluid.
Such coupled systems commonly form during geodynamic processes and require
special analytical and numerical treatment due to intrinsic coupling between fluid
percolation and deformation of the solid matrix. In this chapter, I deliberately
combine modelling of various two-phase flow processes (aqueous fluid release and
percolation, melt generation and transport, etc.) under the common name *hydro*
(taken from *hydrodynamics* – the study of liquids in motion) since the physics of
these processes is similar in many respects, which leads to similarities in their
mathematical description. We will discuss how fluid-solid coupling could be implemented self-consistently into *thermomechanical* (*TM*) geodynamic codes, thereby
making them capable of modelling hydro-thermomechanical processes. Please note
that conservation equations given in this chapter *look more complicated* than the
ones used before due to the various thermomechanical and chemical interactions
between the solid and fluid phases. Therefore it is essential to carefully repeat

derivations and transformations of these new equations, thereby understanding their structure, logic and beauty ☺. The content of this chapter is somewhat *dense and tough but affordable*, please have a pen and paper handy when going through it.

16.2 Fluid percolation processes and the Darcy law

Percolation is a differential motion of the *fluid* through a *porous solid matrix*. The intensity of this motion depends on the fluid viscosity η^f and pressure gradients $\partial P^f/\partial x_i$ and on the solid matrix permeability k^ϕ (m^2), which can be formulated in the form of the *Darcy law*

$$q_i^D = -\frac{k^\phi}{\eta^f}\left(\frac{\partial P^f}{\partial x_i} - \rho^f g_i\right),$$ (16.1a)

$$q_i^D = \phi\left(v_i^f - v_i^s\right),$$ (16.1b)

where q_i^D are components of the *Darcy flux* (or *Darcy velocity*) vector $\vec{q}^D$, which is the *fluid percolation volume flux relative to the moving solid phase* (m^3/m^2/s = m/s, Chapter 1); ϕ is the volume fraction of fluid (also referred to as *porosity* of the solid matrix); ρ^f is fluid density; v_i^f and v_i^s are fluid and solid velocity, respectively.

The Darcy law could be derived, for example, from a simplified analysis of the laminar incompressible fluid flow through multiple parallel planar vertical channels separated by solid impermeable, rigid walls moving with the same constant vertical velocity v_y^s (Fig. 16.1a). The width of the channels (d_c) and walls (d_w) is assumed to be constant, as well as the viscosity, density, and vertical pressure gradients in the fluid. Flow in each channel (Figs. 5.2, 16.1b) is then described by Eq. (5.29) with $L-d_c$, $v_y = v_y^f$ and $v_{y0} = v_{y1} = v_y^s$

$$v_y^f = v_y^s + \frac{1}{2\eta^f}\left(\frac{\partial P^f}{\partial y} - \rho^f g_y\right)\left(x^2 - xd_c\right).$$ (16.2)

The horizontally averaged relative fluid velocity in each channel is (derive as an exercise)

$$\left(v_y^f - v_y^s\right)_{average} = \frac{1}{d_c}\int_{x=0}^{x=d_c}\left(v_y^f - v_y^s\right)dx,$$ (16.3a)

$$\left(v_y^f - v_y^s\right)_{average} = -\frac{d_c^2}{12\eta^f}\left(\frac{\partial P^f}{\partial y} - \rho^f g_y\right).$$ (16.3b)

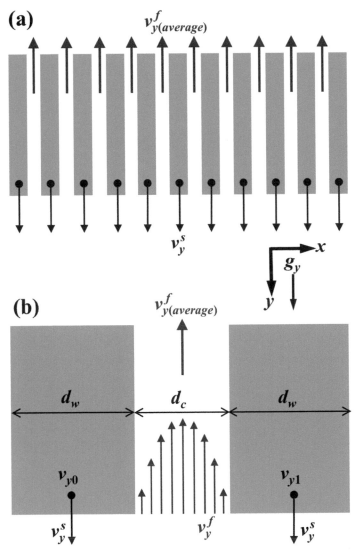

Fig. 16.1. Percolation of fluid through multiple planar channels separated by moving rigid imperme-able walls: (a) general view, (b) zoom-in of a single channel. $v^f_{y(average)}$ corresponds to the average fluid velocity in the channel.

Taking into account that there is no fluid flow inside channel walls and that the numbers of walls and channels are equal (Fig. 16.1a), the average relative fluid volume flux (Darcy flux) is

$$q_y^D = \phi \left(v_y^f - v_y^s \right)_{average} = -\frac{\phi d_c^2}{12\eta^f} \left(\frac{\partial P^f}{\partial y} - \rho^f g_y \right), \qquad (16.4)$$

$$\phi = \frac{d_c}{d_c + d_w}, \qquad (16.5)$$

where ϕ is the volume fraction of fluid in the system (i.e., 'porosity'). By expressing d_c through d_w and ϕ we obtain

$$d_c = \frac{\phi d_w}{(1 - \phi)}, \qquad (16.6)$$

$$q_y^D = -\frac{k^\phi}{\eta^f} \left(\frac{\partial P^f}{\partial y} - \rho^f g_y \right), \qquad (16.7)$$

$$k^\phi = \frac{\phi^3 d_w^2}{12(1 - \phi)^2}, \qquad (16.8)$$

where the permeability k^ϕ depends on the fluid fraction ϕ and the thickness of the walls d_w. The latter could be considered as a proxy for the grain size h (if fluid channels form between aligned grains of the solid).

This simplified analysis gives a helpful illustration for the derivation of various macroscopic *mixture models* or *continuum multi-phase models* used in HTM modelling. The key is that the different material phases (e.g., solid and fluid) are not resolved in their microscopic arrangement, but rather we find governing equations that are based on a spatially averaged representation of the phase mixture over a control volume that defines a point in the continuum field. That way, we end up describing phase properties (pressure, velocity, etc.) as a set of interpenetrating and interacting continuum fields coexisting in each location. This issue of *scale separation* or *spatial averaging* is really at the core of HTM models and needs to be understood in order to appreciate the strengths and limitations of these models.

16.3 Derivation of mass and momentum conservation equations

In the last four decades, several different formulations of conservation equations for modelling coupled fluid-solid systems have been proposed (e.g. Sleep, 1974; Stevenson, 1980; Scott and Stevenson, 1984, 1986; Fowler, 1984, 1985; McKenzie, 1984; Stevenson and Scott, 1991; Spiegelman, 1993; Connolly and Podladchikov, 1998, 2000; Schmeling, 2000; Bercovici et al., 2001, Morency et al., 2007; Rozhko et al., 2007; Simpson et al., 2010a,b; Dymkova and Gerya, 2013; Keller et al., 2013; Yarushina and Podladchikov, 2015). The relatively simple cases concern purely viscous systems with compacting/decompacting viscous solid (e.g. Spiegelman, 1993; Schmeling, 2000; Morency et al., 2007) whereas other formulations combine compressible low viscosity fluid with visco-elasto-plastic compressible compacting/decompacting solid matrix (Yarushina and Podladchikov, 2015; Omlin et al., 2018).

Different formulations also differ in the logic of their numerical implementation (e.g. Dymkova and Gerya, 2013) and here we will choose one that is both robust and requires relatively small modifications to the visco-elasto-plastic codes programmed in previous chapters. Such a formulation has recently been developed by Yarushina and Podladchikov (2015), who derived their system of mass and momentum conservation equations based on principles of irreversible thermodynamics formulated for a two-phase fluid-solid system. This formulation is also consistent with Biot's poroelasticity theory (Biot, 1941).

The mass and momentum conservation equations for fluid and solid are based on the following general *Eulerian* relationship (Yarushina and Podladchikov, 2015; Yarushina, personal communication):

mass conservation
for solid

$$\frac{\partial\left(\rho^s(1-\phi)\right)}{\partial t} + \text{div}\left(\rho^s(1-\phi)\vec{v}^s\right) = -\Gamma^{mass},\tag{16.9}$$

for fluid

$$\frac{\partial\left(\rho^f\phi\right)}{\partial t} + \text{div}\left(\rho^f\phi\vec{v}^f\right) = \Gamma^{mass},\tag{16.10}$$

momentum conservation
for solid

$$\frac{\partial\left(\rho^s(1-\phi)v_i^s\right)}{\partial t} + \text{div}\left(\rho^s(1-\phi)v_i^s\vec{v}^s\right) - \frac{\partial\left((1-\phi)\sigma_{ij}^s\right)}{\partial x_j} - \rho^s(1-\phi)g_i = -\Gamma_i^{mom},\tag{16.11}$$

for fluid

$$\frac{\partial\left(\rho^f\phi v_i^f\right)}{\partial t} + \text{div}\left(\rho^f\phi v_i^f\vec{v}^f\right) - \frac{\partial\left(\phi\sigma_{ij}^f\right)}{\partial x_j} - \rho^f\phi g_i = \Gamma_i^{mom},\tag{16.12}$$

where g_i is the ith component of the gravity vector $\vec{g}$, Γ^{mass} is the *rate of mass transfer* (per unit system volume) from solid matrix to fluid (kg/m^3/s), Γ_i^{mom} is the rate of ith momentum component transfer from solid matrix to fluid (N/m^3). Similarly to the usual momentum equation (Chapter 5), Eqs. (16.11) and (16.12) should be written separately for each velocity component v_i. The conservation equations (16.9)–(16.12) can also be rewritten in the *Lagrangian* form (please derive as an exercise):

mass conservation
for solid

$$\frac{D^s\left(\rho^s(1-\phi)\right)}{Dt}+\rho^s(1-\phi)\mathrm{div}(\vec{v}^s)=-\Gamma^{mass},\tag{16.13}$$

for fluid

$$\frac{D^f\left(\rho^f\phi\right)}{Dt}+\rho^f\phi\,\mathrm{div}\left(\vec{v}^f\right)=\Gamma^{mass},\tag{16.14}$$

momentum conservation
for solid

$$\frac{D^s\left(\rho^s(1-\phi)v_i^s\right)}{Dt}+\rho^s(1-\phi)v_i^s\mathrm{div}(\vec{v}^s)-\frac{\partial\left((1-\phi)\sigma_{ij}^s\right)}{\partial x_j}-\rho^s(1-\phi)g_i=-\Gamma_i^{mom},\tag{16.15}$$

for fluid

$$\frac{D^f\left(\rho^f\phi v_i^f\right)}{Dt}+\rho^f\phi v_i^f\mathrm{div}\left(\vec{v}^f\right)-\frac{\partial\left(\phi\sigma_{ij}^f\right)}{\partial x_j}-\rho^f\phi g_i=\Gamma_i^{mom},\tag{16.16}$$

where D^s/Dt and D^f/Dt denote Lagrangian time derivatives in respectively solid and fluid velocity frames. For any scalar property A, these derivatives can be formulated as

$$\frac{D^sA}{Dt}=\frac{\partial A}{\partial t}+\vec{v}^s\mathrm{grad}(A)=\frac{D^fA}{Dt}-\left(\vec{v}^f-\vec{v}^s\right)\mathrm{grad}(A),\tag{16.17a}$$

$$\frac{D^fA}{Dt}=\frac{\partial A}{\partial t}+\vec{v}^f\mathrm{grad}(A)=\frac{D^sA}{Dt}+\left(\vec{v}^f-\vec{v}^s\right)\mathrm{grad}(A),\tag{16.17b}$$

$$\frac{D^fA}{Dt}-\frac{D^sA}{Dt}=\left(\vec{v}^f-\vec{v}^s\right)\mathrm{grad}(A)=\frac{1}{\phi}\vec{q}_{Darcy}\mathrm{grad}(A),\tag{16.18}$$

where $\partial A/\partial t$ is the Eulerian time derivative. Please note that the solid and fluid Lagrangian reference frames are different (because of differences of fluid and solid velocities) and are used for the physical properties of solid matrix and pore fluid, respectively. In particular, we will consider the fluid fraction ϕ (porosity) as the property of the solid matrix and therefore compute its Lagrangian time derivative in the solid velocity frame. It is also important to mention that the velocity vectors $\vec{v}^s$ and $\vec{v}^f$ are macroscopic and represent averaged (unknown) microscopic velocity fields of solid and fluid (cf. Eqs. 16.2–16.4).

Equations (16.9)–(16.16) can be combined and converted to Lagrangian forms with deviatoric stresses and pressure (derive as an exercise) that are more suitable to implement numerically on the basis of our previous programming exercises

mass conservation
for solid

$$\frac{D^s \ln\rho^s}{Dt} + \frac{D^s \ln(1-\phi)}{Dt} + \operatorname{div}(\vec{v}^s) = -\frac{\Gamma^{mass}}{\rho^s(1-\phi)}, \tag{16.19}$$

for fluid

$$\phi\left(\frac{D^f \ln\rho^f}{Dt} - \frac{D^s \ln\rho^s}{Dt}\right) - \frac{D^s \ln(1-\phi)}{Dt} + \operatorname{div}(\vec{q}^D) = \frac{\Gamma^{mass}\rho^t}{\rho^f\rho^s(1-\phi)}, \tag{16.20}$$

momentum conservation
for bulk (solid + fluid, i.e. by summing up Eqs. 16.15, 16.16)

$$\frac{\partial\sigma'^t_{ij}}{\partial x_j} - \frac{\partial P^t}{\partial x_i} + \rho^t g_i = \rho^f\phi\frac{D^f v^f_i}{Dt} + \rho^s(1-\phi)\frac{D^s v^s_i}{Dt} + \left(v^f_i - v^s_i\right)\Gamma_{mass}, \tag{16.21}$$

for fluid

$$\rho^f\phi\frac{D^f v^f_i}{Dt} + v^f_i\Gamma^{mass} - \frac{\partial\left(\phi\sigma'^f_{ij}\right)}{\partial x_j} + \frac{\partial(\phi P^f)}{\partial x_i} - \rho^f\phi g_i = \Gamma^{mom}_i, \tag{16.22}$$

$$\sigma'^t_{ij} = \phi\sigma'^f_{ij} + (1-\phi)\sigma'^s_{ij}, \tag{16.23}$$

$$P^t = \phi P^f + (1-\phi)P^s, \tag{16.24}$$

$$\rho^t = \phi\rho^f + (1-\phi)\rho^s, \tag{16.25}$$

where P^s, P^f and P^t are respectively solid, fluid and total pressure, ρ^s, ρ^f and ρ^t are respectively solid, fluid and total density, σ'^s_{ij}, σ'^f_{ij} and σ'^t_{ij} are respectively solid, fluid and total deviatoric stress.

Fluid deviatoric stress is negligible due to very low fluid viscosity so that we can assume

$$\sigma'^f_{ij} = 0. \tag{16.26}$$

As a result, total deviatoric stresses are formulated in the *solid strain rate frame* on the basis of the *Maxwell visco-elasto-plastic constitutive relationship* that we studied in Chapter 12 (Eqs. 12.54–12.62)

$$\dot{\varepsilon}^{\prime s}_{ij} = \frac{\sigma^{\prime t}_{ij}}{2\eta^t_{vp}} + \frac{1}{2\mu^t}\frac{D^s \sigma^{\prime t}_{ij}}{Dt}, \tag{16.27}$$

$$\dot{\varepsilon}^{\prime s}_{ij} = \dot{\varepsilon}^s_{ij} - \frac{1}{3}\delta_{ij}\dot{\varepsilon}^s_{kk}, \tag{16.28}$$

$$\dot{\varepsilon}^s_{ij} = \frac{1}{2}\left(\frac{\partial v^s_i}{\partial x_j} + \frac{\partial v^s_j}{\partial x_i}\right), \tag{16.29}$$

$$\dot{\varepsilon}^s_{kk} = \dot{\varepsilon}^s_{xx} + \dot{\varepsilon}^s_{yy} + \dot{\varepsilon}^s_{zz} = \frac{\partial v^s_i}{\partial x_i} = \operatorname{div}(\vec{v}^s), \tag{16.30}$$

where η^t_{vp} and μ^t are the visco-plastic viscosity (Chapter 13, Eq. 13.4) and shear modulus of the bulk (i.e., solid + fluid) material deformed in the solid strain rate frame (which is used for rheological laboratory measurements). The properties of the bulk material η^t_{vp} and μ^t are not equivalent to the respective properties of the pure solid and may also depend on the fluid fraction (porosity).

The momentum transfer term is formulated in a manner that allows recovery of the Darcy equation (Eq. 16.1) from Eq. (16.22) under the condition of negligible mass transfer, inertial forces and deviatoric fluid stress (verify as an exercise)

$$\Gamma^{mom}_i = -\frac{\phi\eta^f}{k^\phi}q^D_i + P^f\frac{\partial\phi}{\partial x_i}. \tag{16.31}$$

Equation (16.31) is however incomplete and omits inertial interchange due to one phase accelerating through the other (e.g. Drew 1971, 1983; Drew and Segel, 1971; Nigmatulin, 1991; Zhang and Prosperetti, 1994; Drew and Passman, 1999; Bercovici and Michaut, 2010; Yarushina et al., 2015). The most common result of such interactions is the *virtual mass* (or *added mass*) *effect* wherein a small particle accelerating through a fluid gains extra inertia equal to 1/2 the displaced mass of fluid. For example, an accelerating parcel of magma with density ρ^{magma}, volume δV and mass $\delta M = \rho^{magma}\delta V$ moving through gas with density ρ^{gas} gets an effective mass $\delta M_{eff} = (\rho^{magma} + \frac{1}{2}\rho^{gas})\delta V$ (Bercovici and Michaut, 2010). Here, we ignore these effects by assuming that for the problems of our interest concerning relatively slow percolation of fluid through the solid rock matrix, differences in the acceleration of solid and fluid can be neglected.

Using Eqs. (16.22), (16.26) and (16.31), the fluid momentum conservation equation becomes (derive as an exercise)

$$q_i^D = -\frac{k^\phi}{\eta^f}\left(\frac{\partial P^f}{\partial x_j} - \rho^f g_i + \rho^f \frac{D^f v_i^f}{Dt} + \frac{v_i^f}{\phi}\Gamma^{mass}\right). \tag{16.32}$$

The mass conservation equations (16.19), (16.20) can be further modified by adding the following *closure relations* for porosity and density changes consistent with Biot's poroelasticity theory and Gassmann's relation (Biot, 1941; Gassmann, 1951; Yarushina and Podladchikov, 2015; Yarushina, personal communication)

$$\frac{D^s ln(1-\phi)}{Dt} = \frac{\beta^\phi}{(1-\phi)}\left(\frac{D^s P^t}{Dt} - \frac{D^f P^f}{Dt}\right) - \frac{\alpha^\phi}{(1-\phi)}\frac{D^s T}{Dt} + \frac{P^t - P^f}{(1-\phi)\eta^\phi} - \Gamma^{mass}A^\phi, \tag{16.33}$$

$$\frac{D^s ln\rho^s}{Dt} = \frac{\beta^s}{1-\phi}\left(\frac{D^s P^t}{Dt} - \phi\frac{D^f P^f}{Dt}\right) - \alpha^s\frac{D^s T}{Dt} + \Gamma^{mass}A^s, \tag{16.34}$$

$$\frac{D^f ln\rho^f}{Dt} = \beta^f\frac{D^f P^f}{Dt} - \alpha^f\frac{D^f T}{Dt} + \Gamma^{mass}A^f, \tag{16.35}$$

where the temperature T is assumed to be the same for both fluid and solid, β^s and β^f are compressibility of solid and fluid, respectively, α^s and α^f are thermal expansion of solid and fluid, respectively, α^ϕ is effective thermal expansion of pores, β^ϕ and η^ϕ are respectively effective compressibility of pores and effective *compaction viscosity* (also often called *bulk viscosity*), which characterize solid matrix resistance to reversible (i.e., elastic, β^ϕ) and irreversible (i.e., visco-plastic, η^ϕ) pore compaction/decompaction, A^ϕ, A^s and A^f are coefficients relating mass transfer to the time derivatives of porosity, solid density and fluid density, respectively.

Expressions for A^ϕ, A^s and A^f are obtained by analysing mass transfer from the solid to the fluid. Such transfer often involves complex processes resulting from several multi-component chemical reactions (melting/solidification, hydration/dehydration, dissolution/precipitation, etc.). However, here we will characterize reactive mass transfer simply by considering a net mass transfer (ΔM) from the solid to the fluid during a time increment Δt: positive ΔM values correspond to mass transfer from the solid to the fluid (dehydration, melting, dissolution, etc.), whereas negative ΔM values imply mass transfer from the fluid to the solid (hydration, solidification, precipitation, etc.). The transferred mass is formally described as a single chemically complex component of the solid and fluid (we will call it the *C component*) that has different density in its solid and fluid states (ρ_C^s and ρ_C^f, respectively), which can also differ from the bulk density of the solid (ρ^s) and fluid (ρ^f). As the result of the mass transfer, the initial volume V^o of the Lagrangian solid+fluid system can change during the time increment Δt to the new value V

$$V = V^o + \Delta V, \tag{16.36}$$

$$\Delta V = \Delta V^f - \Delta V^s, \tag{16.37}$$

$$\Delta V^s = \frac{\Delta M}{\rho_C^s}, \tag{16.38}$$

$$\Delta V^f = \frac{\Delta M}{\rho_C^f}, \tag{16.39}$$

where ΔV is the system volume change, ΔV^s is volume decrease of the solid, ΔV^f is volume increase of the fluid. The change of the solid fraction $\Delta(1-\phi)$ is given by (derive as an exercise)

$$
\begin{aligned}
\Delta(1-\phi) &= (1-\phi) - (1-\phi^o) = \frac{V^s}{V} - (1-\phi^o) = \frac{V^{so} - \Delta V^s}{V} - (1-\phi^o) \\
&= \frac{(1-\phi^o)V^o - \Delta V^s}{V} - (1-\phi^o) = \frac{(1-\phi^o)(V^o - V) - \Delta V^s}{V} \\
&= -\frac{(1-\phi^o)\Delta V + \Delta V^s}{V} = -\frac{(1-\phi^o)(\Delta V^f - \Delta V^s) + \Delta V^s}{V} \\
&= -\frac{(1-\phi^o)\Delta V^f + \phi^o \Delta V^s}{V},
\end{aligned}
\tag{16.40}
$$

$$\frac{1}{(1-\phi)}\frac{\Delta(1-\phi)}{\Delta t} = -\left(\frac{\Delta M}{V\Delta t}\right)\frac{\rho_C^f \phi^o + \rho_C^s(1-\phi^o)}{(1-\phi)\rho_C^s \rho_C^f}, \tag{16.41}$$

where ϕ^o and $\phi = V^s/V$ are the old and new porosity, respectively; $V^{so} = (1-\phi^o)V^o$ and $V^s = V^{so} - \Delta V^s$ are the old and new volume of solid, respectively. Under the condition of Δt, ΔM, $\Delta(1-\phi)$ and ΔV tending to zero, ϕ tending to ϕ^o and $\Delta M/(V\Delta t)$ tending to $\Gamma^{mass} = (1/V)(D^sM/Dt)$ Eq. (16.41) becomes

$$\frac{D\ln(1-\phi)}{Dt} = -\Gamma^{mass}\frac{\rho_C^f \phi + \rho_C^s(1-\phi)}{(1-\phi)\rho_C^s \rho_C^f}, \tag{16.42}$$

which implies

$$A^\phi = \frac{1}{\rho_C^s}\left(\frac{\phi}{1-\phi}\right) + \frac{1}{\rho_C^f}. \tag{16.43}$$

Similarly, changes of the solid and fluid density ($\Delta\rho^s$ and $\Delta\rho^f$, respectively) are given by

$$\Delta\rho^s = \frac{(1-\phi^o)V^o\rho^s - \Delta M}{(1-\phi^o)V^o - \Delta V^s} - \rho^s, \tag{16.44}$$

$$\Delta\rho^f = \frac{\phi^o V^o\rho^f + \Delta M}{\phi^o V^o + \Delta V^f} - \rho^f, \tag{16.45}$$

which lead to the following expressions for A^s and A^f (derive as an exercise)

$$A^s = \frac{1}{(1-\phi)} \left(\frac{1}{\rho_C^s} - \frac{1}{\rho^s} \right), \qquad (16.46)$$

$$A^f = \frac{1}{\phi} \left(\frac{1}{\rho^f} - \frac{1}{\rho_C^f} \right). \qquad (16.47)$$

It should be noted that the *C component approach* presented above by Eqs. (16.36)–(16.47) is not a unique (and not the most common) way of considering the mass transfer. It is 'invented' for this textbook in order to derive conservation equations without considering thermodynamically any chemical reactions responsible for the mass transfer. Conventional treatment of mass transfer typically involves multi-component thermodynamic derivations (e.g. Katz, 2008; Rudge et al., 2011; Keller and Katz, 2016; Plümper et al., 2017). It will be demonstrated below in toy melting and dehydration models how the quantities involved in the mass transfer (Γ^{mass}, ρ_C^s, ρ_C^f, ρ^s, ρ^f) could be obtained from thermodynamic calculations.

By combining Eqs. (16.19), (16.20) and (16.33)–(16.35), (16.43), (16.46), (16.47) the following mass conservation equations can be obtained (derive as an exercise):

for solid (porous matrix divergence)

$$\text{div}(\vec{v}^s) + \beta_d \left(\frac{D^s P^t}{Dt} - K_{BW} \frac{D^f P^f}{Dt} \right) + \frac{P^t - P^f}{(1-\phi)\eta^\phi} = \left(\alpha^s + \frac{\alpha^\phi}{(1-\phi)} \right) \frac{D^s T}{Dt} - \Gamma_{mass} \left(\frac{1}{\rho^s(1-\phi)} - A^\phi + A^s \right),$$

$$(16.48)$$

or by finding out that

$$\frac{1}{\rho^s(1-\phi)} - A^\phi + A^s = \frac{1}{\rho_C^s} - \frac{1}{\rho_C^f}$$

we obtain

$$\text{div}(\vec{v}^s) + \beta_d \left(\frac{D^s P^t}{Dt} - K_{BW} \frac{D^f P^f}{Dt} \right) + \frac{P^t - P^f}{(1-\phi)\eta^\phi} = \left(\alpha^s + \frac{\alpha^\phi}{(1-\phi)} \right) \frac{D^s T}{Dt} + \Gamma_{mass} \left(\frac{1}{\rho_C^f} - \frac{1}{\rho_C^s} \right),$$

$$(16.49)$$

for fluid (Darcy flux divergence)

$$\text{div}(\vec{q}^D) - K_{BW}\beta_d \left(\frac{D^s P^t}{Dt} - \frac{1}{K_{Sk}} \frac{D^f P^f}{Dt} \right) - \frac{P^t - P^f}{(1-\phi)\eta^\phi} =$$

$$= \phi \left[\alpha^f \frac{D^f T}{Dt} - \left(\alpha^s + \frac{\alpha^\phi}{(1-\phi)\phi} \right) \frac{D^s T}{Dt} \right] + \Gamma^{mass} \left(\frac{\rho^t}{\rho^f \rho^s (1-\phi)} - A^\phi + \phi A^s - \phi A^f \right),$$

$$(16.50a)$$

or by finding out that

$$\frac{\rho_t}{\rho^f \rho^s (1 - \phi)} - A_\phi + \phi A^s - \phi A^f = 0$$

we obtain

$$\operatorname{div}(\vec{q}^D) - K_{BW}\beta_d \left(\frac{D^s P^t}{Dt} - \frac{1}{K_{Sk}} \frac{D^f P^f}{Dt} \right) - \frac{P^t - P^f}{(1 - \phi)\eta^\phi} = \phi \left[\alpha^f \frac{D^f T}{Dt} - \left(\alpha^s + \frac{\alpha^\phi}{(1 - \phi)\phi} \right) \frac{D^s T}{Dt} \right],$$

$$\text{(16.50b)}$$

$$\beta_d = \frac{\beta^\phi + \beta^s}{1 - \phi}, \tag{16.51}$$

$$K_{BW} = 1 - \frac{\beta^s}{\beta_d}, \tag{16.52}$$

$$K_{Sk} = \frac{\beta_d - \beta^s}{\beta_d - \beta^s + \phi(\beta^f - \beta^s)}, \tag{16.53}$$

where β_d is drained compressibility, K_{BW} is the Biot–Willis coefficient and K_{Sk} is Skempton's coefficient (Yarushina and Podladchikov, 2015). The Biot–Willis and Skempton coefficients have different standard notations (α and B, respectively) in poromechanics and the geophysical literature, which we do not use here to avoid confusion with respectively the thermal expansion coefficient and the bulk modulus. Equations (16.50a), (16.50b) imply that there is no direct effect of mass transfer rate Γ^{mass} on the Darcy flux divergence. Indirect effects can however be present due to changes of porosity (Eq. 16.33). Equations (16.32) and (16.50) are sometimes combined together to eliminate Darcy fluxes from the list of coupled unknown parameters and thereby from the global matrix (e.g. Dymkova and Gerya, 2013; Omlin et al., 2017).

16.4 Energy conservation equation

The energy conservation equation for a coupled deforming visco-elasto-plastic fluid-solid system should take into account differences in movement of solid and fluid, heat and mass transfer between them and various heat dissipation sources present in the system (e.g. McKenzie, 1984; Katz, 2008; Rudge et al., 2011; Yarushina and Podladchikov, 2015; Keller and Katz, 2016; Schmeling et al., 2018). We focus on a relatively simple case when temperature of the fluid and solid does not differ ($T^f = T^s = T$) and local thermodynamic equilibrium exists in between these two phases. In this case, a single energy conservation equation can be formulated for the bulk material as (e.g. McKenzie, 1984; Yarushina and Podladchikov, 2015; Keller and Katz, 2016)

$$(1 - \phi)\rho^s C_P^s \frac{D^s T}{Dt} + \phi \rho^f C_P^f \frac{D^f T}{Dt} = -\frac{\partial q_i^t}{\partial x_i} + H_r^t + H_a^t + H_s^t + H_L^t, \tag{16.54}$$

$$q_i^t = -k^t \frac{\partial T}{\partial x_i}, \tag{16.55}$$

$$H_r^t = (1 - \phi)H_r^s + \phi H_r^f, \tag{16.56}$$

$$H_a^t = T\left[\alpha^s\left(\frac{D^s P^t}{Dt} - \phi \frac{D^f P^f}{Dt}\right) + \phi \alpha^f \frac{D^f P^f}{Dt} + \alpha^\phi\left(\frac{D^s P^t}{Dt} - \frac{D^f P^f}{Dt}\right)\right]$$
$$= T\left[(\alpha^s + \alpha^\phi)\frac{D^s P^t}{Dt} + (\phi \alpha^f - \phi \alpha^s - \alpha^\phi)\frac{D^f P^f}{Dt}\right], \tag{16.57}$$

$$H_s^t = \frac{\left(\sigma'^t_{ij}\right)^2}{2\eta^t_{vp}} + \frac{\left(P^t - P^f\right)^2}{(1 - \phi)\eta^\phi} + \frac{\eta^f\left(q_i^D\right)^2}{k^\phi}, \tag{16.58}$$

$$H_L^t = -\Delta H^{f-s} \Gamma^{mass}, \tag{16.59}$$

where $\rho^s C_P^s$ and $\rho^f C_P^f$ are the volumetric isobaric heat capacity of solid and fluid respectively (as before, we treat the ρC_P product as a *single variable*, Chapter 9), q_i^t are components of the total (i.e., solid+fluid) heat flux vector, H_r^t, H_a^t, H_s^t and H_L^t are volumetric radioactive, adiabatic, shear and latent heat production, respectively (W/m^3), ΔH^{f-s} is the enthalpy difference between the fluid and the solid (J/kg), k^t is the total thermal conductivity. One possibility to compute k^t is to use the formalism developed by Budiansky (1970)

$$\frac{\phi k^t}{2k^t + k^f} + \frac{(1 - \phi)k^t}{2k^t + k^s} = \frac{1}{3}, \tag{16.60}$$

or by expressing k^t (derive as an exercise)

$$k^t = \sqrt{\frac{k^f k^s}{2} + \frac{[k^s(3\phi - 2) + k^f(1 - 3\phi)]^2}{16}} - \frac{k^s(3\phi - 2) + k^f(1 - 3\phi)}{4}, \tag{16.61}$$

where k^s and k^f are the thermal conductivity of solid and fluid, respectively.

The energy conservation equation (16.54) can be modified into a different form written in solid velocity frame (derive as an exercise)

$$\rho^t C_P^t \frac{D^s T}{Dt} + \phi \rho^f C_P^f \left(\frac{D^f T}{Dt} - \frac{D^s T}{Dt}\right) = -\frac{\partial q_i^t}{\partial x_i} + H_r^t + H_a^t + H_s^t + H_L^t, \tag{16.62a}$$

or by using Eqs. (16.1b), (16.18)

$$\rho^t C_P^t \frac{D^s T}{Dt} + \rho^f C_P^f \vec{q}^D \mathrm{grad}\,(T) = -\frac{\partial q_i^t}{\partial x_i} + H_r^t + H_a^t + H_s^t + H_L^t, \tag{16.62b}$$

where $\rho^t C_P^t$ is the volumetric isobaric heat capacity of the bulk material

$$\rho^t C_P^t = (1 - \phi)\rho^s C_P^s + \phi\rho^f C_P^f. \tag{16.63}$$

16.5 Influence of porosity and fluid pressure on material properties

Equation (16.8) in particular shows that the permeability is a non-linear function of porosity. The exact expression depends on the internal geometry of pores and related fluid pathways. Theoretically, functions of permeability depending on porosity can vary from quadratic to cubic (Gueguen and Dienes, 1989), while the value suggested for natural pore distribution is around three (Zhu et al., 1995; Connolly and Podladchikov, 2000; Morency et al., 2007). However, there are a number of experimental works that show much higher values for the *permeability exponent*, which can vary from 5–25 in sandstones to 25–55 in shales (e.g. David et al., 1994; Dong et al., 2010; Yarushina et al., 2013 and references therein). It is often assumed that at relatively low porosity the permeability could be expressed with an empirical power law (e.g. Dymkova and Gerya, 2013; Keller et al., 2013; Yarushina et al., 2013)

$$k^\phi = k_r^\phi \left(\frac{\phi}{\phi_r}\right)^m \left(\frac{1 - \phi}{1 - \phi_r}\right)^n, \tag{16.64}$$

where k_r^ϕ is the permeability at a reference porosity ϕ_r and m and n are empirical power law exponents ($m = 3$ and $n = 0$ values are often used).

Expressions for the effective compaction viscosity (also often called bulk viscosity) and compressibility of pores can be introduced on the basis of analyses of microscopic deformation in simplified porous systems (e.g. Simpson et al., 2010a; Schmeling et al., 2012; Yarushina and Podladchikov, 2015)

$$\beta^\phi = \frac{\phi}{K_p\mu^t}, \tag{16.65}$$

$$\eta^\phi = K_p \frac{\eta_{vp}^t}{\phi}, \tag{16.66}$$

where K_p is a scaling coefficient depending on pore geometry. This coefficient is 1 and 4/3 for cylindrical and spherical pores, respectively, and goes down to almost zero for very narrow elliptical pores/cracks with large aspect ratio (e.g. Yarushina et al., 2013). The effective thermal expansion of pores α^ϕ may also show some correlation with porosity, which can be either positive or negative depending on the coupled effect of the thermal expansions and bulk moduli of the solid constituents (e.g. Ghabezloo, 2010, 2012; Zeng et al., 2012).

At low melt/fluid content ($\phi<0.2$), the dependence of the effective ductile viscosity of the bulk material (solid+fluid) on porosity is often expressed as

$$\eta^t = \eta^s \exp(-\alpha_\eta \phi), \tag{16.67}$$

where η^s is the effective shear viscosity of the solid and $\alpha_\eta = 28\pm3$ is an experimentally derived melt-weakening coefficient (e.g. Kelemen et al., 1995; Mei et al., 2002; Katz et al., 2006). At larger variations of fluid/melt fraction, more complicated expressions are needed to compute η^t that take into account a strong decrease in viscosity after a certain (*disaggregation*) threshold ϕ_d of the melt/fluid fraction is reached, for example (Caricci et al., 2007; Costa et al., 2009)

$$\eta^t = \eta^f \left(1 + \left(\frac{1-\phi}{1-\phi_d}\right)^\delta\right) \left\{1 - (1-\xi)\mathrm{erf}\left[\frac{\sqrt{\pi}(1-\phi)}{2(1-\xi)(1-\phi_d)}\left(1 + \left(\frac{1-\phi}{1-\phi_d}\right)^\lambda\right)\right]\right\}^{-B_E(1-\phi_d)}, \tag{16.68}$$

where $B_E=2.5$ is the Einstein coefficient (Einstein, 1906), ϕ_d, δ, λ, $\xi \ll 1$ are empirical parameters such that $\eta^t = \eta^f$ at $\phi = 1$ and $\eta^t = \eta^s$ at $\phi = 0$. It should however be understood very clearly that in the case when a melt fraction increase leads to the disaggregation of the rock matrix producing *melt-crystal suspensions* and fully molten magma bodies, the assumptions used for the derivation of our conservation equations need to be revisited to ensure their validity at high melt fractions (e.g. Keller et al., 2013).

As we already discussed in Chapter 12 (Eq. 12.51), the plastic strength σ_{yield} of a rock can be strongly affected by changes in the melt/fluid pressure (P^f) relative to the total pressure (P^t). The difference between the total and fluid pressures is often defined as the *effective pressure* (P_{eff}), which is used in particular in the brittle/plastic yielding condition

$$P_{eff} = P^t - P^f, \tag{16.69}$$

$$\sigma_{yield} = \max[0, \min(\sigma_c + \gamma_{int}P_{eff}, \sigma_t + P_{eff})], \tag{16.70}$$

where γ_{int} is the internal friction coefficient, σ_c and σ_t are respectively *compressive strength* and *tensile strength* (Griffith, 1924; Cai, 2010; Keller et al., 2013), which should typically decrease with increasing porosity. It is however obvious that the 'wet' yielding condition (16.70) should not apply for nearly dry rocks with very low fluid/melt fraction, in which only a small amount of isolated fluid/melt-filled pores is present. Fluid pressure in these pores will be close to the total pressure (i.e., $P_{eff} = 0$) but this will not affect friction along the predominantly dry grain boundaries and high tensile strength of the solid matrix. In this case, the usual dry brittle/plastic yielding condition (Eq. 12.50) should be applied

$$\sigma_{yield} = \max[0, \min(\sigma_c + \gamma_{int}P^t, \sigma_t + P^t)]. \tag{16.71}$$

The transition between the wet (Eq. 16.70) and dry (Eq. 16.71) yielding conditions could be defined either abruptly or gradually at a certain small ($\phi \ll 0.1$) porosity threshold (e.g. Dymkova and Gerya, 2013; Keller et al., 2013). More accurate treatment of this transition requires analysis of various visco-elasto-plastic hydro-mechanical microporosity models (e.g. Yarushina and Podladchikov, 2015).

16.6 Simplified conservation equations

The derived general conservation equations (16.21), (16.32), (16.49), and (16.50b) look quite complicated. In some cases, they could be simplified but we should carefully analyse reasons for such simplification and their consequences. For example, in the case of hydration/dehydration reactions we can assume that the density of the transferred *C component* (crystalline water, which is much less dense than the silicate component of the hydrated solid phase) in its solid and fluid state does not differ from the bulk fluid density (i.e., $\rho_C^s = \rho_C^f = \rho^f$) (e.g. Omlin et al. 2017). In this case, coefficients

$$A^\phi = \frac{1}{(1-\phi)\rho^f} \ , \ A^s = \frac{1}{(1-\phi)}\left(\frac{1}{\rho^f} - \frac{1}{\rho^s}\right), \ A^f = 0$$

and mass conservation equations are given by
for solid

$$\text{div}(\vec{v}^s) + \beta_d\left(\frac{D^sP^t}{Dt} - K_{BW}\frac{D^fP^f}{Dt}\right) + \frac{P^t - P^f}{(1-\phi)\eta^\phi} = \left(\alpha^s + \frac{\alpha^\phi}{(1-\phi)}\right)\frac{D^sT}{Dt}, \qquad (16.72)$$

for fluid

$$\text{div}(\vec{q}^D) - K_{BW}\beta_d\left(\frac{D^sP^t}{Dt} - \frac{1}{K_{Sk}}\frac{D^fP^f}{Dt}\right) - \frac{P^t - P^f}{(1-\phi)\eta^\phi} = \phi\left(\alpha^f\frac{D^fT}{Dt} - \left(\alpha^s + \frac{\alpha^\phi}{(1-\phi)\phi}\right)\frac{D^sT}{Dt}\right),$$
$$(16.73)$$

which imply that there will be no direct effect of fluid release/consumption on the divergence of neither the solid velocity nor the Darcy flux. Indirect effects can however be present due to changes in porosity (Eq. 16.33)

$$\frac{D^s\ln(1-\phi)}{Dt} = \frac{\beta^\phi}{(1-\phi)}\left(\frac{D^sP^t}{Dt} - \frac{D^fP^f}{Dt}\right) - \frac{\alpha^\phi}{(1-\phi)}\frac{D^sT}{Dt} + \frac{P^t - P^f}{(1-\phi)\eta^\phi} - \frac{\Gamma^{mass}}{(1-\phi)\rho^f},$$
$$(16.74)$$

which will in turn affect the permeability and viscosity of the matrix and thereby both fluid percolation and rock deformation.

In the case of melting/solidification, we can assume that the density of the transferred *C component* in its solid and fluid state does not differ from respectively the bulk solid (i.e., $\rho_C^s = \rho^s$) and bulk melt (i.e., $\rho_C^f = \rho^f$) (e.g. Schmeling, 2000). In this case, coefficients

$$A^\phi = \frac{1}{\rho^s}\left(\frac{\phi}{1-\phi}\right) + \frac{1}{\rho^f}, \ A^f = A^s = 0$$

and the mass conservation equation for solid simplifies to

$$\text{div}(\vec{v}^s) + \beta_d\left(\frac{D^s P^t}{Dt} - K_{BW}\frac{D^f P^f}{Dt}\right) + \frac{P^t - P^f}{(1-\phi)\eta^\phi} = \left(\alpha^s + \frac{\alpha^\phi}{(1-\phi)}\right)\frac{D^s T}{Dt} + \Gamma^{mass}\left(\frac{1}{\rho^f} - \frac{1}{\rho^s}\right),$$

(16.75)

whereas the mass conservation equation for fluid is given by Eq. (16.73). The porosity evolution equation (16.33) is then given by

$$\frac{D^s\ln(1-\phi)}{Dt} = \frac{\beta^\phi}{(1-\phi)}\left(\frac{D^s P^t}{Dt} - \frac{D^f P^f}{Dt}\right) - \frac{\alpha^\phi}{(1-\phi)}\frac{D^s T}{Dt} + \frac{P^t - P^f}{(1-\phi)\eta^\phi} - \Gamma^{mass}\left(\frac{1}{\rho^f} + \left(\frac{\phi}{1-\phi}\right)\frac{1}{\rho^s}\right).$$

(16.76)

In the case of modelling melting dynamics in the mantle, it is also often assumed that deformation is very slow and purely viscous and both fluid and solid can be considered as incompressible materials (e.g. McKenzie, 1984; Schmeling, 2000; Katz, 2008). In this case mass (Eqs. 16.75, 16.50b) and momentum (Eqs. 16.21, 16.32) conservation equations simplify respectively to

mass conservation
for solid

$$\text{div}(\vec{v}^s) + \frac{P^t - P^f}{(1-\phi)\eta^\phi} = \Gamma^{mass}\left(\frac{1}{\rho^f} - \frac{1}{\rho^s}\right),$$

(16.77)

for fluid

$$\text{div}(\vec{q}^D) - \frac{P^t - P^f}{(1-\phi)\eta^\phi} = 0,$$

(16.78)

momentum conservation
for bulk (solid + fluid)

$$\frac{\partial\left(2\eta^t\dot{\varepsilon}_{ij}^{'s}\right)}{\partial x_j} - \frac{\partial P^t}{\partial x_i} + \rho^t g_i = 0,$$

(16.79)

for fluid

$$q_i^D = -\frac{k^\phi}{\eta^f}\left(\frac{\partial P^f}{\partial x_j} - \rho^f g_i\right),$$
(16.80)

where η^f is the effective shear viscosity of the bulk material, which depends on stress, pressure, temperature, composition and melt content. The porosity evolution equation (16.76) simplifies to

$$\frac{D^s\ln(1-\phi)}{Dt} = \frac{P^t - P^f}{(1-\phi)\eta^\phi} - \Gamma^{mass}\left(\frac{1}{\rho^f} + \left(\frac{\phi}{1-\phi}\right)\frac{1}{\mu^s}\right).$$
(16.81)

It should be noted that by using purely viscous Eqs. (16.77)–(16.81) we neglect brittle/plastic deformation of the rock matrix. Such deformation may however be present in the melt-bearing mantle even at very high pressure and temperatures due to the reduced difference between the total pressure and melt pressure ($P^t - P^f \approx 0$), which will strongly reduce the brittle/plastic strength of the solid matrix (Eq. 16.70).

Another set of simplified equations can be derived for the condition of slow (non-inertial) visco-elasto-plastic deformation and no mass transfer between the solid and fluid ($\Gamma^{mass} = 0$), which can apply for some crustal fluid-rock settings (e.g., sediment compaction, geothermal systems, Yarushina and Podladchikov, 2015; Omlin et al., 2018). In this case mass conservation equations are again given by Eqs. (16.72), (16.73), whereas momentum conservation equations become

for bulk (solid + fluid)

$$\frac{\partial \sigma_{ij}^{'t}}{\partial x_j} - \frac{\partial P^t}{\partial x_i} + \rho^t g_i = 0,$$
(16.82)

for fluid

$$q_i^D = -\frac{k^\phi}{\eta^f}\left(\frac{\partial P^f}{\partial x_j} - \rho^f g_i\right),$$
(16.83)

where the deviatoric stress $\sigma_{ij}^{'t}$ is given by the Maxwell relationship (16.27)–(16.30). The porosity evolution equation (16.33) does not contain the mass transfer term and goes to

$$\frac{D^s\ln(1-\phi)}{Dt} = \frac{\beta^\phi}{(1-\phi)}\left(\frac{D^sP^t}{Dt} - \frac{D^fP^f}{Dt}\right) - \frac{\alpha^\phi}{(1-\phi)}\frac{D^sT}{Dt} + \frac{P^t - P^f}{(1-\phi)\eta^\phi}.$$
(16.84)

In the case of long-term and large-scale geodynamic processes, the mass conservation equations (16.72), (16.73) can be further simplified by assuming incompressibility of the fluid and solid $(\alpha^s = \alpha^f = \alpha^\phi = 0,\ \beta^s = \beta^f = 0,\ \beta_d = \beta^\phi/(1-\phi),\ K_{BW} = K_{Sk} = 1)$ which gives

mass conservation
for solid

$$\mathrm{div}(\vec{v}^s) + \frac{\beta^\phi}{1-\phi}\left(\frac{D^s P^t}{Dt} - \frac{D^f P^f}{Dt}\right) + \frac{P^t - P^f}{(1-\phi)\eta^\phi} = 0, \qquad (16.85)$$

for fluid

$$\mathrm{div}(\vec{q}^D) - \frac{\beta^\phi}{1-\phi}\left(\frac{D^s P^t}{Dt} - \frac{D^f P^f}{Dt}\right) - \frac{P^t - P^f}{(1-\phi)\eta^\phi} = 0. \qquad (16.86)$$

The porosity evolution equation (16.84) further simplifies to

$$\frac{D^s \ln(1-\phi)}{Dt} = \frac{\beta^\phi}{(1-\phi)}\left(\frac{D^s P^t}{Dt} - \frac{D^f P^f}{Dt}\right) + \frac{P^t - P^f}{(1-\phi)\eta^\phi}. \qquad (16.87)$$

Equations (16.82), (16.83), (16.85)–(16.87) could be seen as a 'minimum complexity' model for long-term deformation of non-reacting fluid-bearing visco-elasto-plastic rocks.

16.7 Structure of hydro-thermomechanical code

As usually seen in this book, the algorithms developed in the previous chapters for modelling of visco-elasto-plastic deformation can be relatively easily adapted to account for solid-fluid coupling (Fig. 16.2). Specifically for this reason, we formulated our momentum conservation equation (16.21) for the bulk material (solid+fluid) rather than for the solid matrix alone, which made this equation rather similar to our 'single phase' Stokes and Navier–Stokes equations (Chapter 5). The previously used staggered grid structure also remains optimal for discretization of the derived conservation equations (Fig. 16.3). Three main modifications are needed to adapt the previously developed thermomechanical codes for coupled fluid-solid problems. The first one consists in changing the mechanical solver by adding to it the conservation equations for fluid (Eqs. 16.32, 16.50b) and modifying the conservation equations for solid (Eqs. 16.21, 16.49). The second modification concerns introducing fluid pressure into the brittle/plastic yielding condition (Eqs. 16.69, 16.70) while performing visco-elasto-plastic Picard iteration. The third modification concerns update and advection of porosity, temperature, solid and fluid pressure in respectively solid and fluid velocity frames. Markers represent the solid matrix and are generally advected along the solid velocity field (Step 12 in Fig. 16.2).

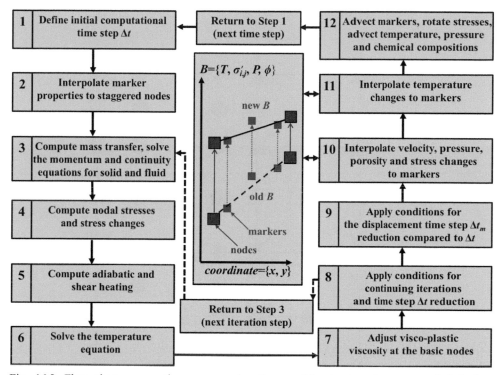

Fig. 16.2. Flow chart representing an example of a possible structure for a numerical hydro-thermomechanical 2D code, which uses staggered finite differences and marker-in-cell technique (SFD-MIC) for solving mass, momentum and energy conservation equations for a coupled solid-fluid system.

16.8 Discretization of the conservation equations

We will discuss discretization of the conservation equations for the relatively simple 2D case of incompressible visco-elasto-plastic solid and incompressible fluid with no mass transfer between the two phases (Eqs. 16.54, 16.82, 16.83, 16.85–16.87). Expressions for other cases can be obtained by analogy.

The expressions for deviatoric stresses of the bulk material are in fact analogous to the equations written for the compressible visco-elasto-plastic solid (Eqs. 13.11–13.14)

$$\sigma_{xx}^{\prime t} = 2\eta_{vp}^{t}\dot{\varepsilon}_{xx}^{\prime s}\frac{\mu^{t}\Delta t}{\mu^{t}\Delta t + \eta_{vp}^{t}} + \sigma_{xx}^{\prime to}\frac{\eta_{vp}^{t}}{\mu^{t}\Delta t + \eta_{vp}^{t}} = -\sigma_{yy}^{\prime t}, \qquad (16.88)$$

$$\sigma_{xy}^{\prime t} = 2\eta_{vp}^{t}\dot{\varepsilon}_{xy}^{\prime s}\frac{\mu^{t}\Delta t}{\mu^{t}\Delta t + \eta_{vp}^{t}} + \sigma_{xy}^{\prime to}\frac{\eta_{vp}^{t}}{\mu^{t}\Delta t + \eta_{vp}^{t}} = \sigma_{yx}^{\prime t}, \qquad (16.89)$$

$$\dot{\varepsilon}_{xx}^{\prime s} = \frac{1}{2}\left(\frac{\partial v_{x}^{s}}{\partial x} - \frac{\partial v_{y}^{s}}{\partial y}\right) = -\dot{\varepsilon}_{yy}^{\prime s}, \qquad (16.90)$$

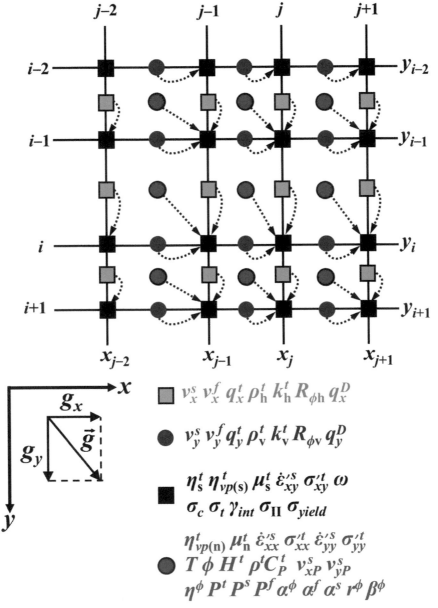

Fig. 16.3. Staggered 2D irregularly spaced numerical grid corresponding to the hydro-thermomechanical code. The algorithm for this code is presented in Fig. 16.2.

$$\dot{\varepsilon}_{xy}^{\prime s} = \frac{1}{2}\left(\frac{\partial v_x^s}{\partial y} + \frac{\partial v_y^s}{\partial x}\right) = \dot{\varepsilon}_{yx}^{\prime s}. \tag{16.91}$$

Equation (16.90) assumes that $\dot{\varepsilon}_{zz}^s = (\dot{\varepsilon}_{xx}^s + \dot{\varepsilon}_{yy}^s)/2$ and thus $\dot{\varepsilon}_{kk}^s = 3(\dot{\varepsilon}_{xx}^s + \dot{\varepsilon}_{yy}^s)/2$, $\dot{\varepsilon}_{zz}^{\prime s} = 0$ (Eqs. 4.11, 4.12).

The conservative FD representation of the x momentum conservation equation for the bulk material (Eq. 16.82) with free surface stabilization (Chapter 8) and visco-elasto-plastic rheology (Eqs. 16.88–16.91) is as follows (see Fig. 16.3 for the indexing of the grid points):

$$
\frac{4\Delta t}{x_{j+1}-x_{j-1}}\left(\frac{\eta^t_{vp(n)(i,j+1)}\dot\varepsilon'^s_{xx(i,j+1)}\mu^t_{n(i,j+1)}}{\mu^t_{n(i,j+1)}\Delta t+\eta^t_{vp(n)(i,j+1)}}-\frac{\eta^t_{vp(n)(i,j)}\dot\varepsilon'^s_{xx(i,j)}\mu^t_{n(i,j)}}{\mu^t_{n(i,j)}\Delta t+\eta^t_{vp(n)(i,j)}}\right)
$$

$$
+\frac{2\Delta t}{y_i-y_{i-1}}\left(\frac{\eta^t_{vp(s)(i,j)}\dot\varepsilon'^s_{xy(i,j)}\mu^t_{s(i,j)}}{\mu^t_{s(i,j)}\Delta t+\eta^t_{vp(s)(i,j)}}-\frac{\eta^t_{vp(s)(i-1,j)}\dot\varepsilon'^s_{xy(i-1,j)}\mu^t_{s(i-1,j)}}{\mu^t_{s(i-1,j)}\Delta t+\eta^t_{vp(s)(i-1,j)}}\right)
$$

$$
-2\frac{P^t_{i,j+1}-P^t_{i,j}}{x_{j+1}-x_{j-1}}-g_x\Delta t v^s_{x(i,j)}\frac{\rho^t_{h(i,j+1)}-\rho^t_{h(i,j-1)}}{x_{j+1}-x_{j-1}}
$$

$$
-g_x\Delta t\left(v^s_{y(i-1,j)}+v^s_{y(i-1,j+1)}+v^s_{y(i,j)}+v^s_{y(i,j+1)}\right)\frac{\rho^t_{h(i+1,j)}-\rho^t_{h(i-1,j)}}{2(y_{i+1}+y_i-y_{i-1}-y_{i-2})}
$$

$$
=-\rho^t_{h(i,j)}g_x-\frac{2}{x_{j+1}-x_{j-1}}\left(\frac{\sigma'^{to}_{xx(i,j+1)}\eta^t_{vp(n)(i,j+1)}}{\mu^t_{n(i,j+1)}\Delta t+\eta^t_{vp(n)(i,j+1)}}-\frac{\sigma'^{to}_{xx(i,j)}\eta^t_{vp(n)(i,j)}}{\mu^t_{n(i,j)}\Delta t+\eta^t_{vp(n)(i,j)}}\right)
$$

$$
-\frac{1}{y_i-y_{i-1}}\left(\frac{\sigma'^{to}_{xy(i,j)}\eta^t_{vp(s)(i,j)}}{\mu^t_{s(i,j)}\Delta t+\eta^t_{vp(s)(i,j)}}-\frac{\sigma'^{to}_{xy(i-1,j)}\eta^t_{vp(s)(i-1,j)}}{\mu^t_{s(i-1,j)}\Delta t+\eta^t_{vp(s)(i-1,j)}}\right),
$$

$$\tag{16.92}$$

$$
\dot\varepsilon'^s_{xx(i,j)}=\frac{1}{2}\left(\frac{v^s_{x(i,j)}-v^s_{x(i,j-1)}}{x_j-x_{j-1}}-\frac{v^s_{y(i,j)}-v^s_{y(i-1,j)}}{y_i-y_{i-1}}\right),
$$

$$
\dot\varepsilon'^s_{xx(i,j+1)}=\frac{1}{2}\left(\frac{v^s_{x(i,j+1)}-v^s_{x(i,j)}}{x_{j+1}-x_j}-\frac{v^s_{y(i,j+1)}-v^s_{y(i-1,j+1)}}{y_i-y_{i-1}}\right),
$$

$$
\dot\varepsilon'^s_{xy(i-1,j)}=\frac{v^s_{x(i,j)}-v^s_{x(i-1,j)}}{y_i-y_{i-2}}+\frac{v^s_{y(i-1,j+1)}-v^s_{y(i-1,j)}}{x_{j+1}-x_{j-1}},
$$

$$
\dot\varepsilon'^s_{xy(i,j)}=\frac{v^s_{x(i+1,j)}-v^s_{x(i,j)}}{y_{i+1}-y_{i-1}}+\frac{v^s_{y(i,j+1)}-v^s_{y(i,j)}}{x_{j+1}-x_{j-1}},
$$

where σ'^{to}_{xy} and σ'^{to}_{xx} are the deviatoric stress tensor components from the previous time step (corrected for advection and rotation), which are interpolated from markers; Δt is the computational time step. Discretization of the y momentum equation is analogous to Eq. (16.92) (derive as an exercise).

The FD representation of the x momentum conservation equation for the fluid (Eq. 16.83) is as follows (see Fig. 16.3 for the indexing of the grid points):

$$R_{\phi h(i,j)} q^D_{x(i,j)} + 2 \frac{P^f_{x(i,j+1)} - P^f_{x(i,j)}}{x_{j+1} - x_{j-1}} = \rho^f_{h(i,j)} g_x, \qquad (16.93)$$

where $R_{\phi h}$ corresponds to the η^f / k_ϕ ratio interpolated from markers to v_x velocity points. The form of the discretized y momentum conservation equation for the fluid is analogous to Eq. (16.93) (derive as an exercise).

The time-dependent mass conservation equations for the solid (Eq. 16.85) and fluid (Eq. 16.86) are discretized respectively as (see Fig. 16.3 for the indexing of the grid points)

$$\frac{v^s_{x(i,j)} - v^s_{x(i,j-1)}}{x_j - x_{j-1}} + \frac{v^s_{y(i,j)} - v^s_{y(i-1,j)}}{y_i - y_{i-1}} + \left(\frac{\beta^\phi_{x(i,j)}}{\Delta t} + \frac{1}{\eta^\phi_{i,j}} \right) \frac{P^t_{i,j} - P^f_{i,j}}{1 - \phi_{i,j}} = \frac{\beta^\phi_{x(i,j)}}{\Delta t (1 - \phi_{i,j})} \left(P^{to}_{i,j} - P^{fo}_{i,j} \right), \quad (16.94)$$

$$\frac{q^D_{x(i,j)} - q^D_{x(i,j-1)}}{x_j - x_{j-1}} + \frac{q^D_{y(i,j)} - q^D_{y(i-1,j)}}{y_i - y_{i-1}} - \left(\frac{\beta^\phi_{x(i,j)}}{\Delta t} + \frac{1}{\eta^\phi_{i,j}} \right) \frac{P^t_{i,j} - P^f_{i,j}}{1 - \phi_{i,j}} = \frac{-\beta^\phi_{x(i,j)}}{\Delta t (1 - \phi_{i,j})} \left(P^{to}_{i,j} - P^{fo}_{i,j} \right),$$

$$(16.95)$$

where $P^{ot}_{i,j}$ and $P^{of}_{i,j}$ are respectively total and fluid pressure from the previous time step corrected for advection (e.g., by using *back-tracing* of Lagrangian points overlapping with Eulerian pressure points along both fluid and solid velocity fields; this is a semi-Lagrangian scheme that we discussed in Chapter 15).

Indexing of variables in the global matrix composed for solving of the mass and momentum conservation equations should take into account six unknowns per node $(v^s_x, v^s_y, P^t, q^D_x, q^D_y, P^f)$ indexed respectively as

$$in_{vxs} = 6[(j-1)(N_y+1) + (i-1)] + 1, \ in_{vys} = in_{vxs} + 1, \ in_{Pt} = in_{vxs} + 2, \ in_{qxD}$$
$$= in_{vxs} + 3, \ in_{qyD} = in_{vxs} + 4, \ in_{Pf} = in_{vxs} + 5,$$

$$(16.96)$$

where i and j are the indices of the horizontal and vertical gridlines which are intersecting at the basic node (cf. Fig. 16.3 for the connectivity of the unknowns to the basic nodes), and N_y is the vertical resolution of the model.

Boundary conditions for solid velocity are identical to our usual velocity boundary conditions (Chapter 7). Boundary conditions for Darcy flux are only needed for components that are normal to boundaries. We could, however, additionally define boundary conditions for components that are parallel to the boundaries (cf. external v_x and v_y boundary nodes in Fig. 7.15), which we can then use for fluid velocity calculation and interpolation to markers and other geometrical points located close to the boundaries. The boundary condition for *either total or fluid pressure* should be defined for one grid cell. The complementary pressure value for this cell is obtained by solving one of the mass conservation equations (Eq. 16.94 for P^t or Eq. 16.95 for P^f).

Numerical techniques for discretizing and solving the energy conservation equation (*the temperature equation*) are very similar to those used in the thermomechanical codes of Chapters 11, 13, 14 (Eqs. 11.8, 13.31–13.34, 14.20). Equation (16.54) is represented with conservative FD as (see Fig. 16.3 for the indexing of the grid points)

$$\rho^t C_{P(i,j)}^t \frac{T_{i,j}}{\Delta t} + \frac{q_{x(i,j)}^t - q_{x(i,j-1)}^t}{x_j - x_{j-1}} + \frac{q_{y(i,j)}^t - q_{y(i-1,j)}^t}{y_i - y_{i-1}} = \rho^t C_{P(i,j)}^t \frac{T_{i,j}^o}{\Delta t} + H_{r(i,j)}^t + H_{a(i,j)}^t + H_{s(i,j)}^t, \quad (16.97)$$

$$q_{x(i,j-1)}^t = -2k_{h(i,j-1)}^t \frac{T_{i,j} - T_{i,j-1}}{x_j - x_{j-2}},$$

$$q_{x(i,j)}^t = -2k_{h(i,j)}^t \frac{T_{i,j+1} - T_{i,j}}{x_{j+1} - x_{j-1}}$$

$$q_{y(i-1,j)}^t = -2k_{v(i-1,j)}^t \frac{T_{i,j} - T_{i-1,j}}{y_i - y_{i-2}},$$

$$q_{y(i,j)}^t = -2k_{v(i,j)}^t \frac{T_{i+1,j} - T_{i,j}}{y_{i+1} - y_{i-1}},$$

$$H_{a(i,j)}^t = T_{i,j}^o \left[\left(\alpha_{i,j}^s + \alpha_{x(i,j)}^\phi \right) \frac{P_{i,j}^t - P_{i,j}^{to}}{\Delta t} + \left(\phi_{i,j} \alpha_{i,j}^f - \phi_{i,j} \alpha_{i,j}^s - \alpha_{i,j}^\phi \right) \frac{P_{i,j}^f - P_{i,j}^{fo}}{\Delta t} \right],$$

$$H_{s(i,j)}^t = \frac{\left(\sigma_{xx(i,j)}'^t \right)^2}{\eta_{vp(n)(i,j)}^t} + \frac{1}{4} \left(\frac{\left(\sigma_{xy(i-1,j-1)}'^t \right)^2}{\eta_{vp(s)(i-1,j-1)}^t} + \frac{\left(\sigma_{xy(i,j-1)}'^t \right)^2}{\eta_{vp(s)(i,j-1)}^t} + \frac{\left(\sigma_{xy(i-1,j)}'^t \right)^2}{\eta_{vp(s)(i-1,j)}^t} + \frac{\left(\sigma_{xy(i,j)}'^t \right)^2}{\eta_{vp(s)(i,j)}^t} \right)$$
$$+ \frac{1}{2} \left(R_{\phi x(i,j-1)} \left(q_{x(i,j-1)}^D \right)^2 + R_{\phi x(i,j)} \left(q_{x(i,j)}^D \right)^2 \right) + \frac{1}{2} \left(R_{\phi y(i-1,j)} \left(q_{y(i-1,j)}^D \right)^2 + R_{\phi v(i,j)} \left(q_{y(i,j)}^D \right)^2 \right)$$
$$+ \frac{\left(P_{i,j}^t - P_{i,j}^f \right)^2}{(1 - \phi_{i,j}) \eta_{i,j}^\phi},$$

where T^o is temperature from the previous time step corrected for different advection paths of solid and fluid. This correction can be done for each marker with

$$T_{m(corrected)}^o = \frac{(1 - \phi_m) \rho^s C_{P(m)}^s T_m^o + \phi_m \rho^f C_{P(m)}^f \left(T_m^o + \Delta T_{m(f-s)}^o \right)}{(1 - \phi_m) \rho^s C_{P(m)}^s + \phi_m \rho^f C_{P(m)}^s}, \quad (16.98a)$$

$$\Delta T_{m(f-s)}^o = T_{m(nodal)}^f - T_{m(nodal)}^s, \quad (16.98b)$$

where $\Delta T_{m(f-s)}^o$ is the temperature difference between the fluid and the solid related to their different advection trajectories, $T_{m(nodal)}^s$ is the nodal temperature interpolated to the marker before any advection, $T_{m(nodal)}^f$ is the nodal temperature interpolated to the marker after its forward advection along the solid velocity field followed by the backward advection (i.e., back-tracing) along the fluid velocity field (example is provided in **i2visHTM.m**). This is a semi-Lagrangian scheme (Chapter 15), which can also be used for other parameters advected with a fluid velocity field (e.g., for fluid composition, an example is provided in **i2visHTM_melting.m**).

Porosity changes are computed for the pressure nodes by using the following equation derived from Eq. (16.87)

$$\frac{D^s \ln\left(\frac{1-\phi}{\phi}\right)}{Dt} = \frac{\beta^\phi}{(1-\phi)\phi}\left(\frac{D^s P^t}{Dt} - \frac{D^f P^f}{Dt}\right) + \frac{P^t - Pf}{(1-\phi)\phi\eta^\phi}. \tag{16.99}$$

Equation (16.99) can be used for the following porosity update at the pressure nodes (Fig. 16.3)

$$\phi_{i,j}^{new} = \frac{\phi_{i,j}}{(1-\phi_{i,j})\exp\left(r_{\phi(i,j)}\Delta t\right) + \phi_{i,j}}, \tag{16.100}$$

$$r_{i,j}^\phi = \frac{\beta_{i,j}^\phi\left(P_{i,j}^t - P_{i,j}^{to} - P_{i,j}^f + P_{i,j}^{fo}\right)}{\Delta t(1-\phi_{i,j})\phi_{i,j}} + \frac{P_{i,j}^t - P_{i,j}^f}{(1-\phi_{i,j})\phi_{i,j}\eta_{i,j}^\phi}, \tag{16.101}$$

where

$$r_{i,j}^\phi = \frac{D^s \ln\left(\frac{1-\phi}{\phi}\right)}{Dt}$$

is the logarithmic rate of the porosity ratio change defined at the pressure points (Fig. 16.3). Equations (16.99)–(16.101) guarantee that porosity will always remain within physically meaningful limits ($0 < \phi < 1$). We can also use Eq. (16.100) to update porosity on markers by using r_m^ϕ values interpolated from the pressure points

$$\phi_m^{new} = \frac{\phi_m}{(1-\phi_m)\exp\left(r_m^\phi\Delta t\right) + \phi_m}. \tag{16.102}$$

Porosity changes per time step should be limited to relatively small values to ensure accuracy of hydro-thermomechanical calculations.

16.9 Computing the mass transfer rate

The mass transfer rate from solid to liquid (Γ^{mass}) and related quantities ($\rho_C^s, \rho_C^f, A^\phi, A^s, A^f$, Eqs. 16.43, 16.46, 16.47) can be computed based on the assumption of local chemical reactions between the fluid/melt and the solid using various thermodynamic computational approaches ranging from relatively simple calculations involving melting/solidification curves (e.g. Schmeling, 2000; Katz, 2008) to a Gibbs energy minimization approach based on internally consistent thermodynamic databases for minerals, fluid and melt (e.g. Connolly, 2005; Chapter 2). It is then possible to compute new (reacted) composition, enthalpy, density and fractions (including ϕ) of the fluid and solid phases based on the local composition of the bulk (solid + fluid) material, pressure and temperature.

In this case, Γ_{mass}, A^ϕ, A^s and A^f can be formulated locally as a function of *six independent quantities*: porosity and density of the solid and fluid before (i.e., for the beginning of the time step, $\phi^o, \rho^{so}, \rho^{fo}$) and after (i.e., for the end of the time step, ϕ, ρ^s, ρ^f) chemical reactions. We start derivations from considering the mass transfer Γ_{mass} during the time step Δt

$$\Gamma^{mass} = \frac{1}{V}\frac{\Delta M}{\Delta t}, \tag{16.103}$$

$$\Delta M = M^{so} - M^s = \rho^{so}V^o(1-\phi^o) - \rho^s V(1-\phi) = M^f - M^{fo} = \rho^f V\phi - \rho^{fo}V^o\phi^o, \tag{16.104}$$

where ΔM is the mass transferred from the solid to the fluid, V^o and V are the local system volume before and after the reactions, respectively. The local system volume change (Eqs. 16.36, 16.37) can be computed from the mass balance condition (16.104)

$$V\left(\rho^s(1-\phi) + \rho^f\phi\right) = V^o\left(\rho^{so}(1-\phi^o) + \rho^{fo}\phi^o\right), \tag{16.105}$$

$$R_V = \frac{V^o}{V} = \frac{\rho^s(1-\phi) + \rho^f\phi}{\rho^{so}(1-\phi^o) + \rho^{fo}\phi^o}, \tag{16.106}$$

$$\Delta V = V - V^o = V - VR_V = V(1-R_V) = V\left[1 - \frac{\rho^s(1-\phi) + \rho^f\phi}{\rho^{so}(1-\phi^o) + \rho^{fo}\phi^o}\right], \tag{16.107}$$

where R_V is the initial to final system volume ratio.

Similarly, volume and density of the transferred *C component* (with mass ΔM) in its solid and fluid state can be evaluated as (Eqs. 16.38, 16.39)

$$\Delta V^s_C = V^{so} - V^s = V^o(1-\phi^o) - V(1-\phi), \tag{16.108}$$

$$\rho^s_C = \frac{\Delta M}{\Delta V^s_C} = \frac{R_V\rho^{so}(1-\phi^o) - \rho^s(1-\phi)}{R_V(1-\phi^o) - (1-\phi)}, \tag{16.109}$$

$$\Delta V^f_C = V^f - V^{fo} = V\phi - V^o\phi^o, \tag{16.110}$$

$$\rho_C^f = \frac{\Delta M}{\Delta V_C^f} = \frac{\rho^f \phi - R_V \rho^{fo} \phi^o}{\phi - R_V \phi^o}, \qquad (16.111)$$

which can then be used to compute A^ϕ, A^s, A^f from Eqs. (16.43), (16.46), (16.47), respectively. Finally, the mass transfer rate and related terms can be evaluated as

$$\Gamma^{mass} = \frac{1}{V}\frac{\Delta M}{\Delta t} = \frac{R_V \rho^{so}(1 - \phi^o) - \rho^s (1 - \phi)}{\Delta t} = \frac{\rho^f \phi - R_V \rho^{fo}\phi^o}{\Delta t}, \qquad (16.112a)$$

$$\Gamma^{mass} A^\phi = \frac{R_V (\phi - \phi^o)}{\Delta t (1 - \phi)} \,(\text{cf. Eq. } 16.33), \qquad (16.112b)$$

$$\Gamma^{mass} A^s = \frac{R_V}{\Delta t}\left(\frac{1 - \phi^o}{1 - \phi}\right)\left(1 - \frac{\rho^{so}}{\rho^s}\right) \,(\text{cf. Eq. } 16.34), \qquad (16.112c)$$

$$\Gamma^{mass} A^f = \frac{R_V \phi^o}{\Delta t \phi}\left(1 - \frac{\rho^{fo}}{\rho^f}\right) \,(\text{cf. Eq. } 16.35), \qquad (16.112d)$$

$$\Gamma_{mass}\left(\frac{1}{\rho_C^f} - \frac{1}{\rho_C^s}\right) = \frac{1 - R_V}{\Delta t} \,(\text{cf. Eq. } 16.49). \qquad (16.112e)$$

As already mentioned above, Eqs. (16.106), (16.112a)–(16.112e) contains only six independent parameters ($\phi^o, \rho^{so}, \rho^{fo}$ and ϕ, ρ^s, ρ^f). They are more convenient to use numerically than directly computing ρ_C^s, ρ_C^f with Eqs. (16.109), (16.111), which may cause numerical problems in the case of negligible mass transfer ($\Delta M \approx 0$).

Latent heating can be then computed as

$$H_L = \Gamma^{mass}(H^t - H^{to}), \qquad (16.113)$$

where H^{to} and H^t are the initial and final enthalpy of the bulk (solid+fluid) system (per unit mass) computed thermodynamically.

Equations (16.106), (16.112), (16.113) can be implemented on either nodal points or markers and require hydro-thermomechanical iteration (e.g., Picard iteration, Chapter 14) to compute equilibrium values of ρ^f, ρ^s, ϕ and H^t as a function of new (yet unknown) values of pressure, temperature, composition and reactions rates. Below we will discuss such numerical implementation on the basis of toy melting and dehydration models.

16.10 Toy thermodynamic model of mantle melting

Numerical modelling of mantle convection processes with melting is challenging and involves thermodynamic calculations describing various melting reactions between mantle minerals and moving melt (e.g. Schmeling, 2000; Katz, 2008). Here, we will discuss a *toy thermodynamic model of mantle melting*, which assumes *ideal binary mixing* of basaltic (*B*) and lherzolitic (L) components in both solid and liquid state. This toy model neglects the actual mineralogy of the solid mantle and treats it as a single phase (an ideal basalt-lherzolite *solid solution*), which is obviously incorrect but will give us a 'flavour' of thermodynamic calculations (e.g. Gerya et al., 2004d; Connolly, 2005), which can be involved in hydro-thermomechanical modelling. Two end-member melting reactions will then characterize the melting process:

$$\text{melting of the basaltic component } B^s = B^f, \tag{16.114}$$

$$\text{melting of the lherzolitic component } L^s = L^f, \tag{16.115}$$

which imply that the *C component* transferred from the solid to the fluid is a mixture of basaltic and lherzolitic components.

Thermodynamic equilibrium conditions for these two reactions can be formulated as

$$\Delta G_{B(P^f,T)} = 0 = \Delta H_B - T\Delta S_B + P^f \Delta V_B + RT ln \left(\frac{X_B^f}{X_B^s} \right), \tag{16.116}$$

$$\Delta G_{L(P^f,T)} = 0 = \Delta H_L - T\Delta S_L + P^f \Delta V_L + RT ln \left(\frac{X_L^f}{X_L^s} \right), \tag{16.117}$$

$$X_B^s + X_L^s = 1, \tag{16.118}$$

$$X_B^f + X_L^f = 1, \tag{16.119}$$

where X_B^s, X_L^s and X_B^f, X_L^f are molar fractions of respective components in the solid and melt, respectively, $R = 8.314$ J/(K mol) is the gas constant, ΔG, ΔH, ΔS and ΔV are molar Gibbs free energy, enthalpy, entropy and volume change in respective reactions. Characteristic melting relationships for this toy basalt-lherzolite system are shown in Fig. 16.4.

Additional mass balance constraints are given by the known local composition of the system X_B^t, X_L^t

$$X_B^t = X_B^f X^f + X_B^s X^s, \tag{16.120}$$

$$X_L^t = X_L^f X^f + X_L^s X^s, \tag{16.121}$$

$$X_B^t + X_L^t = 1, \tag{16.122}$$

$$X^s + X^f = 1, \tag{16.123}$$

where X^s and X^f are equilibrium molar fractions of solid and fluid in the system. Unknown quantities can then be computed from Eqs. (16.116)–(16.123) as (please derive as an exercise)

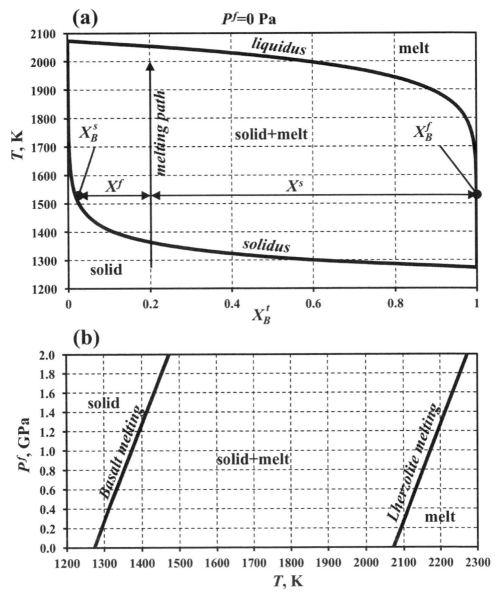

Fig. 16.4. Melting relationships for the toy mantle melting model discussed in the text. Model parameters are as follows: $M_B = M_L = 1$ kg/mol, $\rho_B^f = \rho_L^f = 3100$ kg/m^3, and $\rho_B^s = \rho_L^s = 3305$ kg/m^3, $\Delta S_B = \Delta S_L = 200$ J/(K mol), $\Delta H_B = 254600$ J/mol, $\Delta H_L = 414600$ J/mol.

$$K_B = \frac{X_B^f}{X_B^s} = \exp\left(-\frac{\Delta H_B - T\Delta S_B + P^f\Delta V_B - \Delta G_B}{RT}\right), \tag{16.124}$$

$$K_L = \frac{X_L^f}{X_L^s} = \exp\left(-\frac{\Delta H_L - T\Delta S_L + P^f \Delta V_L - \Delta G_L}{RT}\right),$$ (16.125)

$$X_B^s = \frac{1 - K_L}{K_B - K_L},$$ (16.126)

$$X_L^s = 1 - X_B^s,$$ (16.127)

$$X_B^f = K_B X_B^s,$$ (16.128)

$$X_L^f = 1 - X_B^f,$$ (16.129)

$$X^f = \frac{X_B^t - X_B^s}{X_B^f - X_B^s},$$ (16.130)

$$X^s = 1 - X^f,$$ (16.131)

where K_L and K_B are two *reaction constants*, ΔG_B and ΔG_L characterize the *affinity* (i.e., deviation from the thermodynamic equilibrium) of the reactions (16.114) and (16.115) ($\Delta G_B = \Delta G_L = 0$ in the case of complete local thermodynamic equilibration of the solid and melt).

The calculated molar quantities can then be used to compute the required new (i.e., reacted) porosity and density of the solid and melt (ϕ, ρ^s, ρ^f)

$$\phi = \frac{X^f \left(X_B^f V_B^f + X_L^f V_L^f\right)}{X^f \left(X_B^f V_B^f + X_L^f V_L^f\right) + X^s \left(X_B^s V_B^s + X_L^s V_L^s\right)},$$ (16.132)

$$V_B^f = \frac{M_B}{\rho_B^f},$$ (16.133)

$$V_L^f = \frac{M_L}{\rho_L^f},$$ (16.134)

$$V_B^s = \frac{M_B}{\rho_B^s},$$ (16.135)

$$V_L^s = \frac{M_L}{\rho_L^s},$$ (16.136)

$$\Delta V_B = V_B^f - V_B^s,$$ (16.137)

$$\Delta V_L = V_L^f - V_L^s, \tag{16.138}$$

where M_B and M_L are molar masses of respective components, V_B^f, V_L^f, and V_B^s, V_L^s are molar volumes of respective components in the solid and melt, respectively, ρ_B^f, ρ_L^f and ρ_B^s, ρ_L^s are densities of these components. Solid and fluid densities are then given by

$$\rho^s = \frac{M_B X_B^s + M_L X_L^s}{V_B^s X_B^s + V_L^s X_L^s}, \tag{16.139}$$

$$\rho^f = \frac{M_B X_B^f + M_L X_L^f}{V_B^f X_B^f + V_L^f X_L^f}. \tag{16.140}$$

The relative enthalpy of the system (per unit mass) can be computed as

$$H^t = X^f \left(X_B^f \frac{\Delta H_B}{M_B} + X_L^f \frac{\Delta H_L}{M_L} \right). \tag{16.141}$$

A further useful equation derived from Eq. (16.132) allows computation of the old (non-reacted) fluid fraction X^{fo} from known (old) porosity ϕ^o and compositions of coexisting solid and fluid phases

$$X^{fo} = \frac{\phi^o \left(X_B^{so} V_B^s + X_L^{so} V_L^s \right)}{(1 - \phi^o)\left(X_B^{fo} V_B^f + X_L^{fo} V_L^f \right) + \phi^o \left(X_B^{so} V_B^s + X_L^{so} V_L^s \right)}, \tag{16.142}$$

where X_B^{fo}, $X_L^{fo} = 1 - X_B^{fo}$, X_B^{so}, $X_L^{so} = 1 - X_B^{so}$, ϕ^o are taken from the previous time step.

We can also assume a local solid-melt equilibration time $\Delta t_{reaction}$ to compute ΔG_B and ΔG_L for Eqs. (16.124), (16.125) as

$$\Delta G_B = \Delta G_L = 0 \; if \; \Delta t \geq \Delta t_{reaction}, \tag{16.143a}$$

$$\Delta G_B = \left(1 - \frac{\Delta t}{\Delta t_{reaction}} \right) \left[\Delta H_B - T\Delta S_B + P^f \Delta V_B + RT\ln\left(\frac{X_B^{fo}}{X_B^{so}}\right) \right] \; if \; \Delta t < \Delta t_{reaction},$$
$$\tag{16.143b}$$

$$\Delta G_L = \left(1 - \frac{\Delta t}{\Delta t_{reaction}} \right) \left[\Delta H_L - T\Delta S_L + P^f \Delta V_L + RT\ln\left(\frac{X_L^{fo}}{X_L^{so}}\right) \right] \; if \; \Delta t < \Delta t_{reaction}.$$
$$\tag{16.143c}$$

This approach allows the avoidance of numerical instability due to an increase of mass transfer rates at small time step Δt (Eq. 16.112).

Further calculations of the mass transfer and latent heat effects can be done with Eqs. (16.106), (16.112), (16.113). Porosity evolution can be computed for markers in two steps: (1) adopting new (i.e. reacted) porosity values computed with Eq. (16.132) and then (2) updating these values with Eqs. (16.101), (16.102) by using r_m^ϕ values interpolated from the pressure points. The composition of the system (X_B^t, X_L^t) can also be computed for markers based on the known old porosity and compositions of solid and melt from the previous time step (Eqs. 16.142, 16.120–16.123). These solid and melt compositions and porosity $(X_B^{fo}, X_B^{so}, \phi^o)$ can be advected with markers taking into account different advection paths of solid and melt (e.g., by marker back-tracing, see explanations to Eq. 16.98). It should be noted that numerical implementation of the thermodynamic calculations (Eqs. 16.114–16.143) will typically require global hydro-thermomechanical iteration due to the dependence of melt and solid fractions and compositions on the yet unknown values of temperature and fluid pressure affected by the local mass transfer rate.

One possible way of organizing the iteration is shown in Fig. 16.2, where thermodynamic calculations are done at every iteration step before solving momentum and continuity equations. Mass transfer related terms are calculated with markers as follows:

- compute the old fluid fraction X^{fo} from Eq. (16.142) based on $X_B^{fo}, X_B^{so}, \phi^o$;
- compute the system composition X_B^t, X_L^t from Eqs. (16.120)–(16.123) based on $X^{fo}, X_B^{fo}, X_B^{so}$;
- interpolate the current nodal temperature T and fluid pressure P^f to the marker;
- compute ΔG_B and ΔG_L from Eq. (16.143) based on $X_B^{fo}, X_L^{fo} = 1 - X_B^{fo}$ and $X_B^{so}, X_L^{so} = 1 - X_B^{so}$;
- compute and save new compositions of solid (X_B^s) and melt (X_B^f) from Eqs. (16.124)–(16.129), (16.133)–(16.138);
- compute and save the new porosity ϕ from Eqs. (16.132)–(16.136);
- compute old (ρ^{so}, ρ^{fo}) and new (ρ^s, ρ^f) densities of solid and melt from Eqs. (16.139), (16.140);
- compute the system volume ratio R_V with Eq. (16.106);
- compute old (H^{to}) and new (H^t) enthalpies of the system from Eq. (16.141) using $X^{fo}, X_B^{fo}, X_L^{fo}$ and X^f, X_B^f, X_L^f, respectively;
- compute the mass transfer term $\Gamma_{mass}\left(1/\rho_C^f - 1/\rho_C^s\right)$ with Eq. (16.112e) and latent heating H_L with Eqs. (16.112a), (16.113), interpolate these quantities to pressure points and use them in the mass conservation equation for solid and the temperature equation, respectively.

A programming example of such an iterative numerical implementation for the toy melting model (Fig. 16.4) is given in **i2visHTM_melting.m** which is a toy mantle plume model with melting ☺.

16.11 Toy thermodynamic model of rocks hydration/dehydration

Numerical modelling of rock hydration and dehydration is frequent in geodynamic model-ling, especially in numerical simulations of subduction processes (e.g. Gerya, 2011 and references therein) and models of sedimentary basins (e.g. Omlin et al., 2017). There are many hydration/dehydration reactions, which depend on rock composition and *P-T* conditions and can follow each other during metamorphic processes associated with tectonic evolution of rocks in different geodynamic settings. Here, we will discuss a *toy thermodynamic model of rock hydration and dehydration*, which assumes *ideal binary mixing* of dry ($D_{silicate}$) and wet ($W_{silicate}$) silicate components in the solid whereas *fluid is represented by pure water* (H_2O). As in the case of the toy melting model, the toy hydration/dehydration model neglects actual mineralogy of solid rock and treats it as a single phase (an *ideal solid solution* of dry and wet silicate end-members). A single dehydration reaction will then characterize the water liberation process

$$W_{silicate} = D_{silicate} + H_2O, \qquad (16.144)$$

which implies that the *C component* transferred from the solid to the fluid is water.

The thermodynamic equilibrium condition for this reaction can be formulated as

$$\Delta G_{WD(P^f,T)} = 0 = \Delta H_{WD} - T\Delta S_{WD} + P^f \Delta V_{WD} + RT\ln\left(\frac{X_D^s}{X_W^s}\right), \qquad (16.145)$$

$$X_D^s + X_W^s = 1, \qquad (16.146)$$

where X_D^s and X_W^s are molar fractions of respective components in the solid, ΔG_{WD}, ΔH_{WD}, ΔS_{WD} and ΔV_{WD} are molar Gibbs free energy, enthalpy, entropy and volume change in reaction (16.144). It should be noted that reaction (16.144) does not necessarily have to correspond to the condition of no system volume change (cf. derivation of Eqs. 16.72, 16.73), which is only satisfied when $\Delta V_{WD} = 0$. Characteristic water content changes for this toy wet-dry rock system are shown in Fig. 16.5.

Mass balance constraints are given by the local composition of the system expressed in terms of molar fractions of water ($X_{H_2O}^t$) and dry silicate (X_D^t), which take into account that wet silicate can be represented by the sum of dry silicate and water (Eq. 14.144)

$$X_{H_2O}^t = \frac{X_W^s X^s + X^f}{1 + X_W^s X^s}, \qquad (16.147)$$

$$X_D^t = \frac{X^s}{1 + X_W^s X^s}, \qquad (16.148)$$

$$X_{H_2O}^t + X_D^t = 1, \qquad (16.149)$$

$$X^s + X^f = 1, \qquad (16.150)$$

where X^s and X^f are equilibrium molar fractions of solid and fluid in the system. Unknown equilibrium quantities can then be computed as (please derive as an exercise)

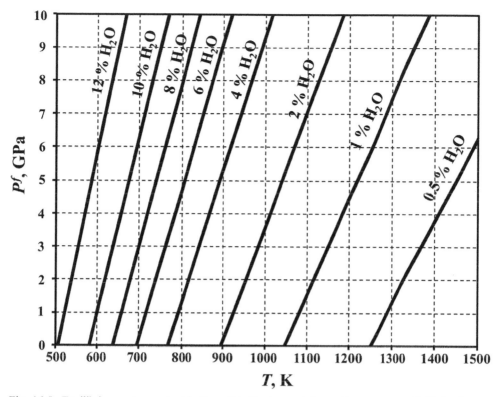

Fig. 16.5. Equilibrium water content in the solid silicate coexisting with an aqueous fluid for the toy hydration/dehydration model discussed in the text. Model parameters are as follows: $M_{H_2O} = 0.018$ kg/mol, $M_D = 0.120$ kg/mol, $\rho_D^s — 3300$ kg/m^3, $\rho_W^s = 2600$ kg/m^3, $\rho_{H_2O}^f = 1000$ kg/m^3, $\Delta S_{WD} = 60$ J/(K mol), $\Delta H_{WD} = 40\,000$ J/mol.

$$K_{WD} = \frac{X_D^s}{X_W^s} = \exp\left(-\frac{\Delta H_{WD} - T\Delta S_{WD} + P^f \Delta V_{WD} - \Delta G_{WD}}{RT}\right),\qquad(16.151)$$

$$X_W^s = \frac{1}{K_{WD}+1},\qquad(16.152)$$

$$X_D^s = 1 - X_W^s,\qquad(16.153)$$

$$X^s = \frac{X_D^t}{1 - X_D^t X_W^s},\qquad(16.154)$$

$$X^f = 1 - X^s,\qquad(16.155)$$

where K_{WD} is the dehydration reaction constant, ΔG_{WD} characterizes the affinity of reaction (16.144) ($\Delta G_{WD} = 0$ in the case of complete local thermodynamic equilibration of the solid and fluid).

The calculated molar quantities can then be used to compute equilibrium porosity and density of the fluid and solid phases

$$\phi = \frac{X^f V_{H_2O}^f}{X^f V_{H_2O}^f + X^s \left(X_W^s V_W^s + X_D^s V_D^s \right)}, \tag{16.156}$$

$$V_{H_2O}^f = \frac{M_{H_2O}}{\rho_{H_2O}^f}, \tag{16.157}$$

$$V_D^s = \frac{M_D}{\rho_D^s}, \tag{16.158}$$

$$V_W^s = \frac{M_D + M_{H_2O}}{\rho_W^s}, \tag{16.159}$$

$$\Delta V_{WD} = V_D^s + V_{H_2O}^f - V_W^s, \tag{16.160}$$

where M_D and M_{H_2O} are molar masses of respective components, V_D^s, V_W^s and $V_{H_2O}^f$ are molar volumes of respective components in the solid and fluid, respectively, ρ_D^s, ρ_W^s and $\rho_{H_2O}^f$ are densities of these components. Solid and fluid densities are given by

$$\rho^s = \frac{M_D + M_{H_2O} X_W^s}{V_D^s X_D^s + V_W^s X_W^s}, \tag{16.161}$$

$$\rho^f = \rho_{H_2O}^f. \tag{16.162}$$

Relative enthalpy of the system (per unit mass) can be computed as

$$H^t = -X^s X_W^s \frac{\Delta H_{WD}}{M_D + M_{H_2O}}. \tag{16.163}$$

A further useful equation derived from Eq. (16.156) allows computation of the old (non-reacted) fluid fraction X^{fo} from known (old) porosity ϕ^o and composition of the solid phase

$$X^{fo} = \frac{\phi^o \left(X_W^{so} V_W^s + X_D^{so} V_D^s \right)}{(1 - \phi^o) V_{H_2O}^f + \phi^o \left(X_W^{so} V_W^s + X_D^{so} V_D^s \right)}, \tag{16.164}$$

where X_W^{so}, $X_D^{so} = 1 - X_W^{so}$, ϕ^o are taken from the previous time step.

Similarly to melting, we can also assume a local equilibrium time $\Delta t_{reaction}$ to compute ΔG_{WD} for Eq. (16.145) as

$$\Delta G_{WD} = 0 \ \textit{if} \ \Delta t \geq \Delta t_{reaction}, \tag{16.165a}$$

$$\Delta G_{WD} = \left(1 - \frac{\Delta t}{\Delta t_{reaction}}\right)\left[\Delta H_{WD} - T\Delta S_{WD} + P^f \Delta V_{WD} + RTln\left(\frac{X_D^{so}}{X_W^{so}}\right)\right] \text{ if } \Delta t < \Delta t_{reaction},$$

$$(16.165b)$$

where X_D^{so}, $X_W^{so} = 1 - X_D^{so}$ are taken from the previous time step.

Further calculations of the mass transfer and latent heat effects can be done with Eqs. (16.106), (16.112), (16.113). Porosity evolution can be computed for markers in two steps: (1) adopting new (i.e. reacted) porosity values computed with Eq. (16.156) and then (2) updating these values with Eqs. (16.101), (16.102) by using r_m^ϕ values interpolated from the pressure points. Composition of the system ($X_{H_2O}^t$, X_D^t) can also be computed for markers based on the known porosity (ϕ^o) and composition of the solid ($X_D^{so}, X_W^{so} = 1 - X_D^{so}$) from the previous time step (Eqs. 16.164, 16.147–16.150). This solid composition should be advected with markers using the solid velocity field. Similarly to the melting model, numerical solution of Eqs. (16.144)–(16.165) will require global hydro-thermomechanical iteration due to the dependence of fluid and solid fractions and solid composition on the yet unknown values of temperature and fluid pressure.

Mass transfer related terms can be calculated iteratively (Fig. 16.2) with markers similarly to the toy melting model (outline an algorithm as an exercise and compare to the provided program example **i2visHTM_hydration.m** which is a toy subduction model with crust dehydration ☺).

Programming exercises

Exercise 16.1.

Using viscous thermomechanical code from Exercise 11.1, program a viscous hydro-thermomechanical (HTM) code (Figs. 16.2, 16.3). Use a regularly spaced staggered grid with 51×51 basic nodes (4 × 4 markers per cell). Discretize Eqs. (16.62), (16.77)–(16.80) (with no mass transfer, $\Gamma^{mass}=0$) similarly to Eqs. (16.92)–(16.102). Modify the model setup with the mantle plume (ρ^s=3250 kg/m³, η^s=10¹⁹ Pa s) by prescribing a 70 km thick lithosphere (ρ^s=3350 kg/m³, η^s=10²³ Pa s, other properties are the same as for the mantle) with linear vertical temperature gradient (T=273–1500 K) on top of the 380 km thick mantle (ρ^s=3300 kg/m³, η^s=10¹⁹ Pa s). Use Eq. (16.64) with $k_r^\phi = 10^{-14}$ m², ϕ_r=0.01, m=3 and n=2 for all materials. Use Eq. (16.67) with α_η=28 to compute η^t for markers. Set the initial melt content for the plume ϕ=0.03, use ϕ=0.001 for other materials. Use the following properties of the melt: ρ^f=2700 kg/m³, $\rho^f C_P^f = 3\times10^6$ J/(m³ K), α^f=3×10⁻⁵ 1/K, $H_r^f = 2 \times 10^{-7}$ W/m³ η^f =10 Pa s. Use Eq. (16.61) to compute the thermal conductivity k^t. Compute the effective bulk viscosity on pressure nodes as $\eta_{i,j}^\phi = \eta_{vp(n)(i,j)}^t/\phi$, where $\eta_{vp(n)(i,j)}^t$ is the effective shear viscosity η^t interpolated from markers. In the end of each time step, program back-tracing of displaced solid markers along the fluid velocity field ($v_x^f = v_x^s + q_x^D/\phi, v_y^f = v_y^s + q_y^D/\phi$) and evaluate the fluid-solid temperature difference $\Delta T_{m(f-s)}^o$ for the markers to re-compute their temperature with Eq. (16.98). Restrict the time step by limiting the maximal marker displacement by 0.5 of the grid step per time step along both fluid and solid velocity fields. Compute the logarithmic rate of the porosity ratio change $r^{\phi}_{(i,j)}$ at the pressure nodes using Eq. (16.101) with $\beta^{\phi} = 0$. Additionally, restrict the time step by using $max|r^{\phi}_{(i,j)}|\Delta t < 0.1$ criterion to limit porosity changes per time step. Update porosity on markers with Eq. (16.102) using r^{ϕ}_m interpolated from the pressure nodes. Modify permeability by varying k^{ϕ}_r $(10^{-18}-10^{-14}$ m$^2)$ and check its effect on the model evolution. An example is provided in **i2visHTM.m**.

Exercise 16.2.

Using visco-elasto-plastic thermomechanical code from Exercise 13.3, program a visco-elasto-plastic hydro-thermomechanical (HTM) code (Figs. 16.2, 16.3). Use the same model setup (Fig. 13.4) with the regularly spaced staggered grid (101×101 basic nodes, 4 × 4 markers per cell). Use fluid-pressure dependent yielding condition (Eqs. 16.69, 16.70) in the global Picard iterations. Use the following material parameters: block (100×60 km^2), $\mu^s = 10^{11}$ Pa, $\eta^s=10^{23}$ Pa s, $\sigma_c=10^8$ Pa, $\sigma_t=3\times10^7$ Pa, $\gamma_{int}=0.6$; 10×10 km^2 squared inclusion and weak medium, $\mu^s =10^{11}$ Pa, $\eta^s=10^{17}$ Pa s, $\sigma_c =10^7$ Pa, $\sigma_t=3\times10^6$ Pa, $\gamma_{int}=0$. Discretize Eqs. (16.54), (16.82), (16.83) and (16.85)–(16.87) using Eqs. (16.88)–(16.102). Use Eq. (16.64) with $k^{\phi}_r=10^{-14}$ m^2, $\phi_r=0.01$, $m=3$ and $n=2$ for all materials. Use Eq. (16.67) with α_{η} =28 to compute η^t. Set the initial fluid content for all materials to $\phi=0.01$. Use the following properties of the fluid: $\rho^f=1000$ kg/m^3, $\rho^f C^f_P=3\times10^6$ J/(m^3 K), $\alpha^f=3\times10^{-5}$ 1/K, $H^f_r=0$, η^f =0.001 Pa s. Use Eq. (16.61) to compute the thermal conductivity k^t. Evaluate $\Delta T^o_{m(f-s)}$ and correct the temperature for markers similarly to Exercise 16.1. Trace positions of material points from pressure nodes back along both the solid and fluid velocity fields. Use these traced points to interpolate $P^{to}_{i,j}$ (for points traced back along the solid velocity field) and $P^{fo}_{i,j}$ (for points traced back along the fluid velocity field), which will be needed at the next time step. Compute the logarithmic rate of the porosity ratio change $r^{\phi}_{(i,j)}$ at the pressure nodes using Eq. (16.101) and update the porosity on markers with Eq. (16.102) using r^{ϕ}_m interpolated from the pressure nodes. Modify the permeability by varying k^{ϕ}_r $(10^{-18}-10^{-14}$ m$^2)$ and check its effect on the characteristic fault angle (cf. Fig. 13.3). An example is in **i2elvisHTM.m**.

17

Adaptive mesh refinement

> **Theory:** What is AMR? Mesh refinement with staggered finite differences. 'Swiss cross' approach. Block-structured AMR approach. Conditions for 'hanging nodes'. Refinement criteria. Convergence of the numerical solution. AMR for different conservation equations.
> **Exercises:** Programming of AMR code.

17.1 What is AMR?

AMR stands for adaptive mesh refinement – the method of numerical discretization that allows higher numerical resolution in the area(s) of interest and lower resolution elsewhere. AMR thus allows computational costs to be minimized without decreasing the accuracy of calculations in the area(s) of interest. This approach is often essential for problems with localization of deformation, which develops spontaneously in regional or global thermome-chanical models (e.g. Stadler et al., 2010; Gerya, 2010b, 2013), for example in models of global mantle convection with narrow plate boundaries, requiring very high resolution (Stadler et al., 2010). The use of AMR is gradually becoming more widespread in geody-namics (e.g. Albers, 2000; Stadler et al., 2010; Kronbichler et al., 2012; Gerya et al., 2013 and references therein; Rudi et al., 2015; Heister et al., 2017) and this trend is dictated by the fact that it is not feasible and useful to resolve the entire Earth at uniform high resolution as needed for example at narrow plate boundaries (Stadler et al., 2010). It is also apparent that the Earth's coupled geodynamic system is driven by inherently multiscale processes. Considering only the compositional layering, we observe a wide range of relevant length scales. For example, sedimentary and volcanic processes lead to the formation of centimetric to kilometric lithological units, localization of deformation can occur from the millimetric up to the kilometric scale, topographic variations occur on the kilometric scale, tectonic motions involve plates of several thousands of kilometres separated by quasi-discrete plate bound-aries, the length scale of mantle heterogeneities may be on the order of thousands of kilometres, and processes such as core formation take place at planetary scale. These multi-scale features make the use of AMR nearly unavoidable and it is therefore good to have some

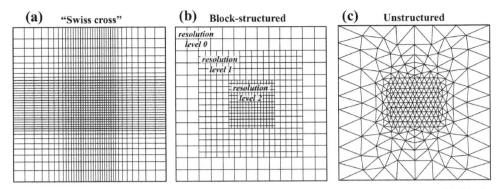

Fig. 17.1. Different types of adaptive mesh refinement (AMR): (a) 'Swiss cross', (b) block-structured, (c) unstructured. See text for discussion.

knowledge about it. There are several types of AMR, which could be applied to geodynamic problems. They differ mainly in the geometry of the grid and related differences in the discretization of the governing equations. Three very popular types are (Fig. 17.1) the 'Swiss cross' grid refinement (Fig. 17.1a), block-structured grid refinement (Fig. 17.1b) and unstructured grid refinement (Fig. 17.1c). Below, we will discuss only the first two approaches since numerical implementation of unstructured grid refinement is more natural with the finite element method (FEM) and is rather complicated with staggered finite differences.

17.2 'Swiss cross' grid refinement

The most simple AMR approach, often called the *'Swiss cross'* grid refinement because of the peculiar cross-like geometry of the grid (Fig. 17.1a), consists in gradual densification of the grid in both horizontal and vertical directions toward the area of interest. The advantage of this grid refinement method is that the discretization of the conservation equations does not change, except for using locally variable rather than constant grid steps. The global indexing of unknowns also remains the same as in the case of a constant grid step. The disadvantage is in the appearance of cells with large aspect ratio and in the limited geometrical flexibility of this AMR method, which is only well suited for refining the grid in a single rectangular area. Indeed the 'Swiss cross' method is widely used in computational geodynamics, especially in regional-scale models (e.g. Gerya et al., 2004a; Gerya and Burg, 2007; Chapter 21).

In order to implement the 'Swiss cross' refinement, the grid spacing increases gradually away from the area of high resolution by a constant factor F at every nodal point. The grid step change should typically not exceed 20% per grid step ($1 < F < 1.2$). In order to compute the incremental factor F, the following formula can be applied, which is solved iteratively

$$F = \left(1 + \frac{D}{b}\left(1 - \frac{1}{F}\right)\right)^{1/N}, \tag{17.1}$$

where D is the distance that should be covered by N non-uniform grid steps (cells) and b is the grid spacing in the high-resolution area (i.e. grid spacing from which the incremental increase should start). In the case of moving high-resolution regions, the grid modification should be done at each time step upon changes of the model geometry. Re-meshing has no major effect on the computational algorithm since in our staggered finite difference marker-in-cell (SFD+MIC) approach, relative positions of markers and nodes change at each time step anyway. Nodal values of physical parameters (including temperature changes) are only used for updating properties at the moving markers. Therefore, an Eulerian node has no 'memory' and can be shifted between two time steps.

During interpolation *from markers to nodes*, one can also take into account the variable size of the cell surface area (Fig. 8.8)

$$B_{i,j} = \frac{\sum_m B_m w_{m(i,j)}}{\sum_m w_{m(i,j)}}, \tag{17.2}$$

$$w_{m(i,j)} = \frac{1}{\Delta x \times \Delta y}\left(1 - \frac{\Delta x_m}{\Delta x}\right) \times \left(1 - \frac{\Delta y_m}{\Delta y}\right), \tag{17.3}$$

where $\Delta x \times \Delta y$ is the surface area of the cell in which marker m is located. An example of the 'Swiss cross' grid refinement is given in **i2vis_Swiss_cross.m**.

17.3 Block-structured grid refinement

The block-structured grid refinement is more complex than the 'Swiss cross' approach and requires changes in the discretization of governing equations and global indexing of unknowns. This approach consists in defining several different *resolution levels*, which are gradually introduced in the region(s) of interest based on some local grid refinement criteria. Advantages of this method are in the good geometrical flexibility of the grid adaptation, uniform aspect ratio of computational cells and good convergence of the numerical solution with increasing resolution (Gerya et al., 2013). Among the disadvantages are abrupt changes of grid step between two resolution levels (typically by a factor of 2), which creates some numerical inaccuracy (e.g. Gerya et al., 2013), the necessity of dealing with *'hanging' (slave)* velocity nodes (Fig. 17.2), for which some additional constraints need to be introduced, and more complex stencils for governing equations when they are discretized at the boundary between the two different resolution levels. Below we will discuss an example of implementation of the block-structured AMR for the case of viscous incompressible Stokes flow based on the paper of Gerya et al. (2013).

Figure 17.2 shows the geometry of an adaptive staggered grid used for the formulation of the momentum, continuity, temperature and Poisson equations in 2D. The main principle of constructing this type of adaptive grid consists in (i) uniform structure of computational cells at all levels of resolution and (ii) using conservative finite differences both within and

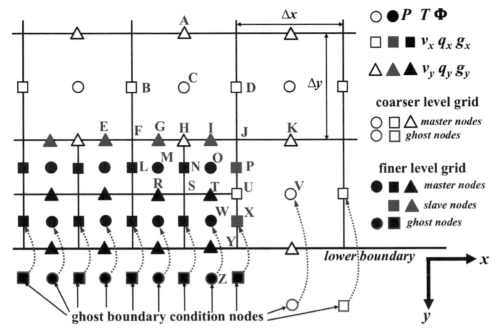

Fig. 17.2. Geometry of a block-structured, adaptive staggered grid for solving thermomechanical problems with self-gravitation (Gerya et al., 2013). Dotted arrows show the relation of external ghost boundary condition nodes to internal nodes.

between resolution levels. Each computational cell has five staggered nodal points (two v_x nodes, two v_y nodes and one P node) where various types of equations are formulated.

The 2D conservation of mass is approximated by the incompressible continuity equation

$$\frac{\partial v_x}{\partial x} + \frac{\partial v_y}{\partial y} = 0,$$ (17.4)

where x, y are coordinates, v_x and v_y are respectively horizontal and vertical velocity components. This equation is discretized in the P node of each cell according to the standard finite difference formula (Fig. 17.2):

$$\text{for coarser level cell C} \quad \frac{v_{xD} - v_{xB}}{\Delta x} + \frac{v_{yH} - v_{yA}}{\Delta y} = 0,$$ (17.5a)

$$\text{for finer level cell M} \quad \frac{v_{xN} - v_{xL}}{\frac{1}{2}\Delta x} + \frac{v_{yR} - v_{yG}}{\frac{1}{2}\Delta y} = 0,$$ (17.5b)

$$\text{for finer level cell O} \quad \frac{v_{xP} - v_{xN}}{\frac{1}{2}\Delta x} + \frac{v_{yT} - v_{yI}}{\frac{1}{2}\Delta y} = 0,$$ (17.5c)

where letters at v_x and v_y denote respective geometrical points (see blue letters in Fig. 17.2). *The 2D Stokes equations* for the viscous incompressible flow (Chapter 5) take the form

$$\frac{\partial \sigma'_{xx}}{\partial x} + \frac{\partial \sigma'_{xy}}{\partial y} - \frac{\partial P}{\partial x} = -\rho(x,y)g_x, \tag{17.6}$$

$$\frac{\partial \sigma'_{xy}}{\partial x} + \frac{\partial \sigma'_{yy}}{\partial y} - \frac{\partial P}{\partial y} = -\rho(x,y)g_y, \tag{17.7}$$

$$\sigma'_{xy} = 2\eta \dot{\varepsilon}_{xy},$$

$$\sigma'_{xx} = 2\eta \dot{\varepsilon}_{xx},$$

$$\sigma'_{yy} = 2\eta \dot{\varepsilon}_{yy},$$

$$\dot{\varepsilon}_{xy} = \frac{1}{2}\left(\frac{\partial v_x}{\partial y} + \frac{\partial v_y}{\partial x}\right),$$

$$\dot{\varepsilon}_{xx} = \frac{\partial v_x}{\partial x},$$

$$\dot{\varepsilon}_{yy} = \frac{\partial v_y}{\partial y}$$

where σ'_{ij} are components of the deviatoric stress tensor, $\dot{\varepsilon}_{ij}$ are components of the strain rate tensor, P is pressure, $\rho(x, y)$ is density which depends on coordinates and g_x and g_y are gravitational acceleration components, which may also change with coordinates. Equations (17.6) and (17.7) are formulated in v_x and v_y master nodes, respectively. Discretization of these equations within the same resolution level is based on stress conservative finite differences suggested for the uniform rectangular staggered grid (Chapter 7). In contrast, different treatment is used between resolution levels: Eqs. (17.6), (17.7) are discretized for coarser level (*master*) nodes only (Fig. 17.2), whereas velocity at finer level (*slave*) nodes is computed differently. The following example explains the logic of an across-level stress conservative discretization of Eq. (17.7) for the coarser level v_y node H (see blue letter H in Fig. 17.2):

$$\frac{\sigma'_{xyJ} - \sigma'_{xyF}}{\Delta x} + \frac{\sigma'_{yyN} - \sigma'_{yyC}}{\frac{3}{4}\Delta y} - \frac{P_N - P_C}{\frac{3}{4}\Delta y} = -g_{yH}\frac{\rho_J + \rho_F}{2}, \tag{17.8}$$

$$\sigma'_{yyN} = \frac{\sigma'_{yyM} + \sigma'_{yyO}}{2}$$

$$P_N = \frac{P_M + P_O}{2}$$

$$\sigma'_{xyF} = 2\eta_F \dot{\varepsilon}_{xyF},$$

$$\sigma'_{xyJ} = 2\eta_J \dot{\varepsilon}_{xyJ},$$

$$\sigma'_{yyC} = 2\eta_C \dot{\varepsilon}_{yyC}$$

$$\sigma'_{yyM} = 2\eta_M \dot{\varepsilon}_{yyM}$$

$$\sigma'_{yyO} = 2\eta_O \dot{\varepsilon}_{yyO}$$

$$\dot{\varepsilon}_{xyF} = \frac{1}{2}\left(\frac{v_{xL} - v_{xB}}{\frac{3}{4}\Delta y} + \frac{v_{yG} - v_{yE}}{\frac{1}{2}\Delta x}\right),$$

$$\dot{\varepsilon}_{xyJ} = \frac{1}{2}\left(\frac{v_{xP} - v_{xD}}{\frac{3}{4}\Delta y} + \frac{v_{yK} - v_{yI}}{\frac{3}{4}\Delta x}\right),$$

$$\dot{\varepsilon}_{yyC} = \frac{v_{yH} - v_{yA}}{\Delta y},$$

$$\dot{\varepsilon}_{yyM} = \frac{v_{yR} - v_{yG}}{\frac{1}{2}\Delta y},$$

$$\dot{\varepsilon}_{yyO} = \frac{v_{yT} - v_{yI}}{\frac{1}{2}\Delta y},$$

where letters at variables denote respective geometrical points (see blue letters in Fig. 17.2). This discretization essentially means that both σ'_{yy} and P for the point N are obtained by averaging (interpolating) these parameters from the points M and O where they are naturally formulated using standard staggered grid schemes (Chapter 7). Also note that σ'_{xy} shear stress component discretization is naturally based on both coarser and finer level velocity nodes. Similarly, discretization of Eq. (17.6) for finer level v_x node N (see blue letter N in Fig. 17.2) is the following (derive as an exercise)

$$\frac{\sigma'_{xyS} - \sigma'_{xyH}}{\frac{1}{2}\Delta y} + \frac{\sigma'_{xxO} - \sigma'_{xxM}}{\frac{1}{2}\Delta x} - \frac{P_O - P_M}{\frac{1}{2}\Delta x} = -g_{xN}\frac{\rho_S + \rho_H}{2}, \qquad (17.9)$$

$$\sigma'_{xyH} = \left(\sigma'_{xyF} + \sigma'_{xyJ}\right)/2,$$

$$\sigma'_{xxM} = 2\eta_M \dot{\varepsilon}_{xxM},$$

$$\sigma'_{xxO} = 2\eta_O \dot{\varepsilon}_{xxO},$$

$$\dot{\varepsilon}_{xx\mathrm{M}} = \frac{v_{x\mathrm{N}} - v_{x\mathrm{L}}}{\frac{1}{2}\Delta x},$$

$$\dot{\varepsilon}_{xx\mathrm{O}} = \frac{v_{x\mathrm{P}} - v_{x\mathrm{N}}}{\frac{1}{2}\Delta x},$$

where the σ'_{xy} shear stress component for the point H is again obtained by averaging (interpolating) from the points J and F where σ'_{xy} is naturally formulated using standard staggered grid schemes (Chapter 7). It should also be mentioned that the suggested discretization schemes require that viscosity η and density ρ are defined in four corners of each cell and that viscosity is also defined in the centre of each cell.

In order to obtain velocity in finer level *slave nodes* (or '*hanging nodes*') two principles should be used: (i) conservation of volume flux across resolution levels and (ii) stress-based interpolation of velocity gradients between σ'_{xy} points. Conservation of volume flux across resolution boundaries implies that volume flux (i.e., average velocity across the coarser cell boundary) should not change between resolution levels, for example

$$v_{y\mathrm{H}} = \frac{1}{2}\left(v_{y\mathrm{G}} + v_{y\mathrm{I}}\right). \qquad (17.10)$$

Stress-based interpolation of velocity gradients implies that the velocity gradient between two slave nodes on a coarser cell boundary should be inversely proportional to the local viscosity, similarly to the velocity gradients composing σ'_{xy} stress components (see e.g. Eq. 17.8), for example

$$\eta_{\mathrm{H}}\left(\frac{\partial v_y}{\partial x}\right)_{\mathrm{H}} = \frac{1}{2}\left[\eta_{\mathrm{F}}\left(\frac{\partial v_y}{\partial x}\right)_{\mathrm{F}} + \eta_{\mathrm{J}}\left(\frac{\partial v_y}{\partial x}\right)_{\mathrm{J}}\right], \qquad (17.11)$$

$$\left(\frac{\partial v_y}{\partial x}\right)_{\mathrm{F}} = \frac{v_{y\mathrm{G}} - v_{y\mathrm{E}}}{\frac{1}{2}\Delta x},$$

$$\left(\frac{\partial v_y}{\partial x}\right)_{\mathrm{H}} = \frac{v_{y\mathrm{I}} - v_{y\mathrm{G}}}{\frac{1}{2}\Delta x},$$

$$\left(\frac{\partial v_y}{\partial x}\right)_{\mathrm{J}} = \frac{v_{y\mathrm{K}} - v_{y\mathrm{I}}}{\frac{3}{4}\Delta x},$$

$$\eta_{\mathrm{H}} = \frac{1}{2}(\eta_{\mathrm{F}} + \eta_{\mathrm{J}})$$

where viscosity in the point H is computed as an arithmetic average of viscosity for the points F and J, where σ'_{xy} stress components are naturally formulated. Combining Eqs. (17.10) and (17.11) gives the following finite difference formulas for the 'hanging' (slave) v_y nodes G and I (derive as an exercise)

$$(\eta_F + \eta_J)\frac{v_{yH} - v_{yG}}{\frac{1}{4}\Delta x} = \eta_F\frac{v_{yG} - v_{yE}}{\frac{1}{2}\Delta x} + \eta_J\frac{v_{yK} - v_{yI}}{\frac{3}{4}\Delta x}, \tag{17.12}$$

$$(\eta_F + \eta_J)\frac{v_{yI} - v_{yH}}{\frac{1}{4}\Delta x} = \eta_F\frac{v_{yG} - v_{yE}}{\frac{1}{2}\Delta x} + \eta_J\frac{v_{yK} - v_{yI}}{\frac{3}{4}\Delta x}. \tag{17.13}$$

The 2D temperature equation (Chapter 9) takes the form

$$\rho C_P\frac{DT}{Dt} = -\frac{\partial q_x}{\partial x} - \frac{\partial q_y}{\partial y} + H_r + H_s + H_a, \tag{17.14}$$

$$q_x = -k\frac{\partial T}{\partial x},$$

$$q_y = -k\frac{\partial T}{\partial y},$$

$$H_s = \frac{\sigma'^2_{xx}}{\eta} + \frac{\sigma'^2_{xy}}{\eta},$$

$$H_a = T\alpha\rho\Big(g_x v_x + g_y v_y\Big),$$

where q_x and q_y are respectively horizontal and vertical components of the heat flux vector, T is temperature, C_P is isobaric heat capacity, k is thermal conductivity, H_r, H_s and H_a are respectively radioactive, sear and adiabatic heating contributions, α is the thermal expansion coefficient. Discretization of Eq. (17.14) within the same resolution level is based on heat flux-conservative finite differences suggested for the uniform rectangular staggered grid (Chapter 10). In contrast, at the boundary between resolution levels, heat fluxes in master and slave nodes are used for discretizing respectively coarser and finer level temperature equations (Fig. 17.2)

$$\text{for coarser level } \left(-\frac{\partial q_x}{\partial x} - \frac{\partial q_y}{\partial y}\right)_C = -\frac{q_{xD} - q_{xB}}{\Delta x} - \frac{q_{yH} - q_{yA}}{\Delta y}, \tag{17.15}$$

$$\text{for finer level } \left(-\frac{\partial q_x}{\partial x} - \frac{\partial q_y}{\partial y}\right)_O = -\frac{q_{xP} - q_{xN}}{\frac{1}{2}\Delta x} - \frac{q_{yT} - q_{yI}}{\frac{1}{2}\Delta y}. \tag{17.16}$$

Heat flux components in the master nodes are obtained by averaging of temperatures defined at the finer level, for example (Fig. 17.2)

$$q_{yH} = -(k_F + k_J)\frac{T_M + T_O - 2T_C}{3\Delta y}, \tag{17.17}$$

$$q_{xU} = (k_J + k_Y) \frac{T_O + T_W - 2T_V}{3\Delta x}. \tag{17.18}$$

Heat flux components in the slave nodes are obtained based on conservation of heat flux across resolution boundaries, for example (Fig. 17.2)

$$q_{yI} = q_{yH} = q_{yG}, \tag{17.19}$$

$$q_{xP} = q_{xX} = q_{xU}. \tag{17.20}$$

Based on Eqs. (17.15)–(17.20) *implicit* discretization of the temperature equation in the point O can be formulated as (derive as an exercise)

$$\rho C_{PO} \frac{T_O - T_O^o}{\Delta t} = -\frac{q_{xP} - q_{xN}}{\frac{1}{2}\Delta x} - \frac{q_{yT} - q_{yI}}{\frac{1}{2}\Delta y} + H_{rO} + H_{sO} + H_{aO}, \tag{17.21}$$

$$q_{xN} = -(k_H + k_S) \frac{T_O - T_M}{\Delta x},$$

$$q_{yT} = -(k_S + k_U) \frac{T_W - T_O}{\Delta y},$$

$$H_{sO} = \frac{\sigma'^2_{xxO}}{\eta_O} + \frac{1}{4} \left(\frac{\sigma'^2_{xyII}}{\eta_H} + \frac{\sigma'^2_{xyJ}}{\eta_J} + \frac{\sigma'^2_{xyS}}{\eta_S} + \frac{\sigma'^2_{xyU}}{\eta_U} \right),$$

$$H_{aO} = \frac{1}{4} T_O^o \alpha_O \rho_O [(g_{xN} + g_{xP})(v_{xN} + v_{xP}) + (g_{yI} + g_{yT})(v_{yI} + v_{yT})],$$

where T_O^o corresponds to temperature in the point O at the previous time step corrected for advection and T_O is the new (unknown) temperature in the same node (Chapter 10)

The 2D Poisson equation (Chapter 2) takes the form

$$\frac{\partial^2 \Phi}{\partial x^2} + \frac{\partial^2 \Phi}{\partial y^2} = -\frac{\partial g_x}{\partial x} - \frac{\partial g_y}{\partial y} = 4K\pi G\rho(x,y), \tag{17.22}$$

$$g_x = -\frac{\partial \Phi}{\partial x},$$

$$g_y = -\frac{\partial \Phi}{\partial y},$$

where G is the gravitational constant and K depends on the 3D geometry of a self-gravitating body modelled in 2D ($K=1$ and $K=2/3$ stand for cylindrical and spherical geometries, respectively, Chapter 11). Discretization of Eq. (17.22) follows the same logic as used for the discretization of the temperature equation, for example (Fig. 17.2)

$$-\frac{g_{xP} - g_{xN}}{\frac{1}{2}\Delta x} - \frac{g_{yT} - g_{yI}}{\frac{1}{2}\Delta y} = 4KG\pi\rho_O, \tag{17.23}$$

$$g_{xP} = g_{xX} = g_{xU} = \frac{\Phi_O + \Phi_W - 2\Phi_V}{\frac{3}{2}\Delta x},$$

$$g_{xN} = -\frac{\Phi_O - \Phi_M}{\frac{1}{2}\Delta x},$$

$$g_{yT} = -\frac{\Phi_W - \Phi_O}{\frac{1}{2}\Delta y},$$

$$g_{yI} = g_{yG} = g_{yH} = -\frac{\Phi_M + \Phi_O - 2\Phi_C}{\frac{3}{2}\Delta y},$$

where the principle of gravity force conservation across the resolution boundary is used.

Since all velocity, temperature and gravitational potential points are connected to specific cells, there are no external boundary condition nodes, in which boundary conditions for these parameters can be prescribed (e.g., external boundary condition nodes in Fig. 7.15). Therefore, boundary conditions should be used directly in discretization stencils for respective equations using relations between *ghost boundary condition nodes* (Chapter 7, Fig. 7.16) and real (ordinary) nodal points, for example (Fig. 17.2)

$$\text{constant potential condition} \quad \frac{\Phi_Z + \Phi_W}{2} = \Phi_{lower}$$

$$\left(\frac{\partial^2 \Phi}{\partial y^2}\right)_W = \frac{\Phi_T - 2\Phi_W + \Phi_Z}{\frac{1}{4}\Delta y^2} = \frac{\Phi_T - 3\Phi_W + 2\Phi_{lower}}{\frac{1}{4}\Delta y^2}, \tag{17.24}$$

$$\text{symmetry condition} \quad \Phi_Z = \Phi_W$$

$$\left(\frac{\partial^2 \Phi}{\partial y^2}\right)_W = \frac{\Phi_T - 2\Phi_W + \Phi_Z}{\frac{1}{4}\Delta y^2} = \frac{\Phi_T - \Phi_W}{\frac{1}{4}\Delta y^2}, \tag{17.25}$$

where Φ_{lower} is the prescribed constant value for the gravitational potential at the lower boundary, Φ_Z is the gravitational potential value in the ghost boundary condition node Z.

17.4 Indexing of unknowns and connectivity within the block-structured grid

Compared to the simple staggered grid discussed in Chapters 7 and 11, indexing of the block-structured grid is more challenging, since the number of nodes, cells and unknowns can change at each time step and they are not ordered in space in a simple manner. How should we deal with such a 'messy', complex, evolving grid? And how do we define global indexes of unknowns when composing global matrices for different conservation equations presented above? In order to deal with these problems we could use a few simple observations (Fig. 17.2):

- each grid cell has 3 unknowns located in the cell centre (P, T, Φ), which are not shared with any other cells;
- each grid cell has 4 velocity unknowns in v_x and v_y nodes located at the cell boundaries; each of these unknowns can be shared with one neighbouring cell; the unknown is not shared if the neighbouring cell belongs to a different resolution level;
- each grid cell has 4 corners, which are basic nodes of the grid; each of these basic nodes can be surrounded by up to 4 cells.

Based on these observations we could compose global matrices cell-wise given that the following conditions are fulfilled:

(i) each cell has a unique index from 1 to N_{cells},
(ii) each basic node has a unique index from 1 to N_{nodes},
(iii) each basic node 'knows' indices of (up to) 4 surrounding cells,
(iv) each unknown has a unique index in respective global matrices (indexes vary from 1 to N_{Stokes} in the Stokes+continuity equations matrix and from 1 to N_{cells} in the temperature and Poisson equation matrices),
(v) each cell 'knows' its resolution level, indexes of 4 corners (basic nodes) and indexes of 7 unknowns (P, T, Φ, two v_x and two v_y) in respective global matrices.

Our task is thus to create these conditions in the beginning of calculations and maintain them upon changes in the structure of the grid.

Why are these conditions important? Let us do one simple exercise using cell O in Fig. 17.2. We start by composing the continuity equation (17.5c). For this equation we need global indices of v_{xN}, v_{xP}, v_{yI}, v_{yT} and P_O (index of pressure in the cell; this is also the index of the continuity equation itself), which should be thus listed for this cell. Then we proceed with composing up to two x Stokes (Eq. 17.6) and up to two y Stokes (Eq. 17.7) equations (given that they have not yet been composed by processing of neighbouring cells, which should be indicated for respective unknowns). For that we will not only need global indexes of pressure and velocity unknowns for the cell O but also indexes of other unknowns (e.g., P_C, P_M, v_{xD}, v_{yA} etc.), which we can obtain by defining neighbouring cell indexes through cell-node and then node-cell connectivity. We should also take into account that the resolution levels of the neighbouring cells may be different, which will strongly affect both the geometry of the local stencil and the finite difference representation of Stokes, temperature and Poisson equations. It is therefore quite obvious that the composing of matrices for different conservation equations within the block-structured grid will require many logical operators, which will make the resulting code notably longer compared to our usual thermomechanical codes. An example of the discussed logic and procedures is given in **AMR_mechanical.m**.

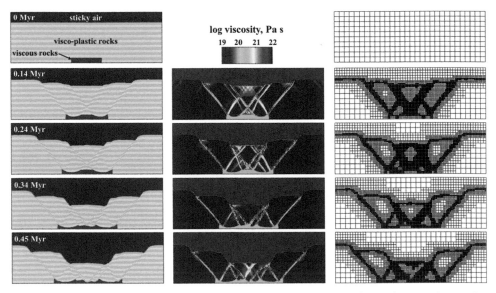

Fig. 17.3. Initial conditions and results of a 2D thermomechanical experiment for crustal extension (Gerya et al., 2013). Evolution of the (left) lithological structure, (middle) viscosity field and (right) grid structure. The model is computed with the code **AMR_markers.m** associated with this chapter.

17.5 Block-structured grid refinement and quad trees

Refinement of the block-structured grid is done by splitting a cell of resolution level n into four cells (in 2D) of resolution level $n + 1$. The refinement should be indicated by saving information about indexes of four daughter cells for the refined parent cell as well as by saving information about the parent cell index for each of the four daughter cells. Thereby a so-called *quad tree* is built, which can be followed in either refinement or coarsening direction and defines local connectivity between different levels of resolution.

Criteria for the refinement can be based on variations of different physical parameters (viscosity, density, composition, velocity etc.) in the cell, among which viscosity contrast within the cell seems to be the most important (e.g. Gerya et al., 2013). It is also important to make sure that at any resolution boundary the grid does not refine for more than one level and thus the width of the zone of resolution n located in between areas of resolution $n - 1$ and $n + 1$ is at least 1–2 cells wide (Fig. 17.1b).

In addition, in dynamically evolving models, it is also important to implement grid cell coarsening, which will allow high resolution to be removed in the areas where it is not needed anymore, thereby minimizing the use of computational resourses (Fig. 17.3).

17.6 Interpolation between marker and nodes

Interpolation of marker properties to the grid should naturally use the quad-tree structure formed during the grid refinement, which establishes the relationship between the parent and daughter cells. The first cell index for a given marker should be defined at the resolution level 0 and then the index of the cell is refined by following the local quad tree (starting from the initial level 0 cell). After reaching the finest level of resolution, marker properties are interpolated to the centre of the cell and four cell edges (i.e., to basic nodes indexed for this cell) based on a distance-dependent weighted average that takes into account the resolution level n of the cell (Fig. 8.8)

$$B_i = \frac{\sum_m B_m w_{m(i)}}{\sum_m w_{m(i)}}, \tag{17.26}$$

$$w_{m(i)} = 4^n \left(1 - \frac{\Delta x_m}{\frac{1}{2^n} \Delta x}\right) \times \left(1 - \frac{\Delta y_m}{\frac{1}{2^n} \Delta y}\right), \tag{17.27}$$

where m is the index of the marker and i is the index of the cell/node, Δx_m and Δy_m are respectively horizontal and vertical distances from the marker m to the node/cell centre, Δx and Δy are respectively horizontal and vertical grid steps for the resolution level 0.

In dynamic problems with markers, the grid refinement works most naturally when a uniformly spaced grid of resolution level 0 is defined before calculations. This *preliminary grid* is then adapted iteratively to the initial model geometry represented by markers before going to the first time step. During the calculations, grid adaptation could be performed in the beginning of each time step iteratively, by performing interpolation of marker properties to the grid and then modifying it based on the adopted grid refinement and coarsening criteria (Figs. 17.3, 17.4).

17.7 Convergence properties of AMR

Compared to a regular staggered grid, AMR allows faster convergence of the numerical solution to a true solution with increasing number of cells and nodal points and thereby in the number of solved equations and respective unknowns (i.e., in the *number of degrees of freedom* of the numerical system) (Fig. 17.5). Indeed some of the error norms converge better than others. For example, the mean error (L_1 *norm*) and mean square error (L_2 *norm*) converge notably better than the maximal absolute error (*L-infinity norm*). For mechanical problems with large abrupt viscosity contrast, the *L-infinity* error norm for pressure converges only in the case of the unstructured mesh refinement, in which boundaries of the grid cells can follow accurately the viscosity contrast boundaries (*shape fitting approach*). In contrast, 'Swiss cross' and block-structured AMR approaches do not demonstrate convergence of the maximal absolute pressure error for problems in which high viscosity contrast boundaries go across the orthogonal boundaries of computational cells (Fig. 17.5c, code example **AMR_mechanical.m**), which is one of the important weaknesses of these two relatively simple methods.

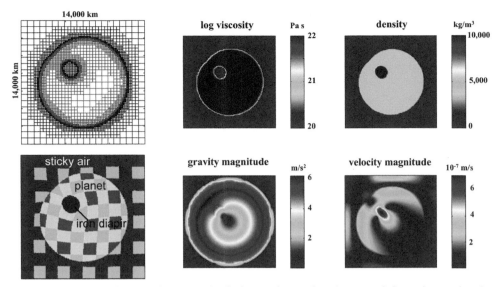

Fig. 17.4. Results of a 2D thermomechanical experiment for planetary deformation under the condition of self-gravitation (Gerya et al., 2013). The model is computed with the code **AMR_planet.m** associated with this chapter.

Programming exercises

Exercise 17.1.

Program a 'Swiss cross' AMR for the case of a plume moving in the mantle. Use the thermomechanical code and model setup programmed in Exercise 11.1 but without the 'sticky air' layer. Take into account variable grid spacing in both the discretization of the conservation equations (Chapters 7 and 10) and interpolation between markers and nodes (Chapter 8). In particular, use Eqs. (17.2), (17.3) for interpolation from markers to nodes (Fig. 8.8). Program the 'Swiss cross' grid (Fig. 17.1a) by defining a 20×20 cell high-resolution area around the moving marker from the plume centre. In this area, use a constant grid step, which is twice as high as the average grid step in the model. Compute the grid step in the remaining model area using Eq. (17.1). Adjust the 'Swiss cross' grid to the marker movement in the beginning of each time step. Do not allow a grid step smaller than in the high-resolution area around the plume. Change plume movements by varying gravitational acceleration components and observe variations in the grid structure following the moving plume. An example is provided in **AMR_Swiss_cross.m**.

Exercise 17.2.

Compose a flow chart (e.g. Fig. 11.1) for the mechanical code with block-structured AMR without markers for the case of pre-defined non-changing viscosity and density structure (e.g., circular high viscosity inclusion in low viscosity matrix subjected to pure shear deformation, Fig. 17.5c). Compare your flow chart to the provided code example **AMR_mechanical.m**.

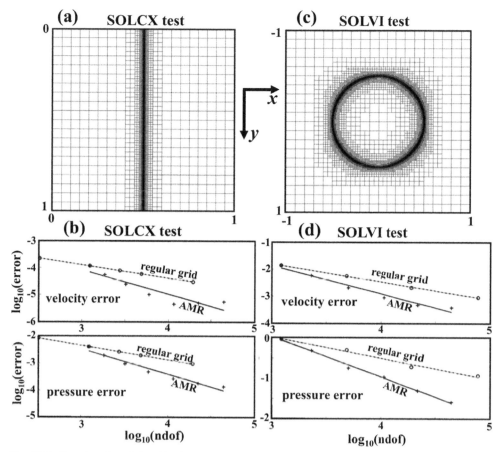

Fig. 17.5. Results of applying block-structured AMR for (a),(b) SOLCX and (c),(d) SOLVI numerical tests (Gerya et al., 2013). (a),(c) The mesh structure obtained using viscosity contrast based grid refinement. (b),(d) The decrease of pressure and velocity error (L_1 norm) with increasing *number of degrees of freedom* (ndof, number of unknowns in the global matrix). The tests are computed with the code **AMR_mechanical.m** associated with this chapter.

Exercise 17.3.

Compose a flow chart (e.g. Fig. 11.1) for the mechanical code with block-structured AMR with markers for the case of an arbitrary changing velocity and density structure (e.g., extensional deformation of the visco-plastic crust in the vertical gravity field, Fig. 17.3). Compare your flow chart to the provided code example **AMR_markers.m**.

Exercise 17.4.

Compose a flow chart (e.g. Fig. 11.3) for the thermomechanical spherical-Cartesian code with block-structured AMR with markers for the case of a deforming viscous self-gravitating planet (e.g., Fig. 17.4). Compare your flow chart to the provided code example **AMR_planet_T.m**.

18

The multigrid method

> **Theory:** Principles of the multigrid method. Multigrid method for solving the Poisson equation in 2D. Coupled solving of momentum and continuity equations in 2D with the multigrid method for cases with constant and variable viscosity.
>
> **Exercises:** Programming of multigrid methods for solving the Poisson equation and coupled solving of the momentum and continuity equations in 2D.

18.1 Multigrid – what is it?

The use of direct methods to solve the system of equations places a strong limitation on the maximum possible number of nodal points within the numerical grid due to limitations in computer memory and computational speed. This limitation is particularly critical in 3D where, to reach the same resolution as in 2D (tens and hundreds of grid poins in each direction), the number of linear equations to be solved increases by at least 2 orders of magnitude and the number of computational operations required to solve the same equations grows by at least 3 orders of magnitude. Try to increase the resolution in your visco-elasto-plastic code (Exercise 13.2) by 2–3 times in each direction and you will see how much slower it will run . . .

Therefore, for very high resolution in 2D and for any reasonable resolution in 3D, we are forced to use iterative methods, which do not have such strong memory limitations. However, many iterative methods also have the problem that they require an increasing number of iterations with an increasing number of grid points (Fig. 18.1). Try to increase the resolution in your code which solves the Poisson equation via Gauss–Seidel iteration (Exercise 3.3) by 2–3 times in each direction and you will see how many more iterations will be needed with this simple iterative method to obtain an accurate solution . . .

What can we do to overcome these problems? One possible way is by using a *multigrid method*. Multigrid is a relatively new type of numerical algorithm explicitly formulated for the first time in 1964 (Fedorenko, 1964) and has been actively developed since the 1980s (a good introduction to multigrid methods can be found in the book by Wesseling, 1992). This

51×51 nodes   **401×401 nodes**

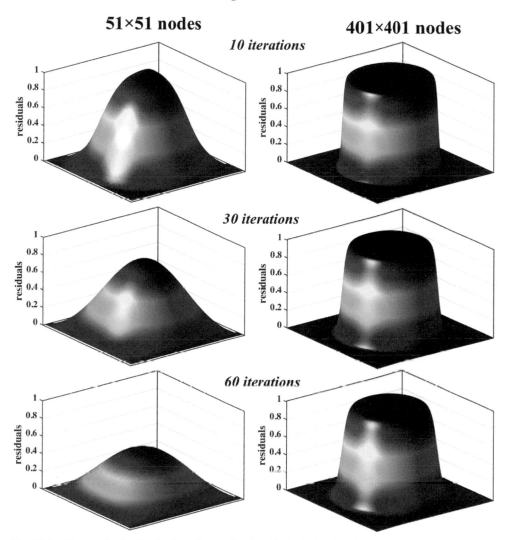

Fig. 18.1. Changes in the distribution of normalized residuals during iterative solution of the 2D Poisson equation for the model with density field corresponding to a circular body (planet) embedded in a massless medium (space). Note that for the model with higher resolution (right column), residuals decay much slower than for the model with lower resolution (left column) with the same number of Gauss–Seidel iterations. The model is computed with the code **Gauss_Seidel_iterations_Poisson_planet.m** associated with this chapter.

method greatly speeds up the convergence of iterations and makes the number of iteration cycles independent of the number of grid points. How does it do it? By solving the same equations in parallel on several grids (typically having different resolution) and by exchanging information between these grids. This is why it is called MULTI-grid.

Multigrid is based on the simple idea that any linear equation

$$C_1 x_1 + C_2 x_2 + \cdots + C_n x_n = R, \qquad (18.1)$$

where C_1, C_2, ..., C_n are coefficients, x_1, x_2, ..., x_n are unknowns and R is the right hand side, can be represented in *additive* form as

$$C_1\left(x_1' + \Delta x_1\right) + C_2\left(x_2' + \Delta x_2\right) + \cdots + C_n\left(x_n' + \Delta x_n\right) = R, \qquad (18.2)$$

$$x_1 = x_1' + \Delta x_1,$$

$$x_2 = x_2' + \Delta x_2,$$

$$x_n = x_n' + \Delta x_n,$$

where x_1', x_2', ..., x_n' are the current (known) approximations of x_1, x_2, ..., x_n and Δx_1, Δx_2, ..., Δx_n are unknown *corrections* needed to satisfy Eq. (18.1). Equation (18.2) can be further transformed to

$$C_1 \Delta x_1 + C_2 \Delta x_2 + \cdots + C_n \Delta x_n = \Delta R, \qquad (18.3)$$

$$\Delta R = R - \left(C_1 x_1' + C_2 x_2' + \cdots + C_n x_n'\right),$$

where ΔR is the current residual of Eq. (18.1) (see Chapter 3).

When some correction approximations are known, Eq. (18.3) can also be represented in the additive form

$$C_1\left(\Delta x_1' + \Delta\Delta x_1\right) + C_2\left(\Delta x_2' + \Delta\Delta x_2\right) + \cdots + C_n\left(\Delta x_n' + \Delta\Delta x_n\right) = \Delta R, \qquad (18.4)$$

$$\Delta x_1 = \Delta x_1' + \Delta\Delta x_1,$$

$$\Delta x_2 = \Delta x_2' + \Delta\Delta x_2,$$

$$\Delta x_n = \Delta x_n' + \Delta\Delta x_n,$$

where $\Delta x_1'$, $\Delta x_2'$, ..., $\Delta x_n'$ are current (known) approximations of Δx_1, Δx_2, ..., Δx_n and $\Delta\Delta x_1$, $\Delta\Delta x_2$, ..., $\Delta\Delta x_n$ are unknown *corrections to corrections* needed to satisfy Eqs. (18.1) and (18.3). Equation (18.4) can be further transformed to

$$C_1 \Delta\Delta x_1 + C_2 \Delta\Delta x_2 + \cdots + C_n \Delta\Delta x_n = \Delta\Delta R, \qquad (18.5)$$

$$\Delta\Delta R = \Delta R - \left(C_1 \Delta x_1' + C_2 \Delta x_2' + \cdots + C_n \Delta x_n'\right),$$

where $\Delta\Delta R$ is the current residual of Eq. (18.3); see Chapter 3. Obviously, the additive representation of Eq. (18.5) can also be done, etc. any desirable amount of times.

The multigrid method implies that we use *different numerical grids* to formulate the complementary equations (18.1), (18.3), (18.5), etc. for the same numerical model. Please note that the coefficients $C_1, C_2, \ldots, C_n$ in Eqs. (18.1), (18.3) and (18.5) are identical and only the right hand sides of the equations are different. In the case of numerical solutions, this means that the discretization scheme for the governing equations will always be the same, independent of whether we formulate (i) equations for unknown $x_1, x_2, \ldots, x_n$ or (ii) equations for corrections to these unknowns $\Delta x_1, \Delta x_2, \ldots, \Delta x_n$ or (iii) equations for corrections to these corrections $\Delta\Delta x_1, \Delta\Delta x_2, \ldots, \Delta\Delta x_n$ or (iv) etc. Another important point to note is that for such a hierarchical additive representation, residuals of *approximated* equations go into the right hand side *of correcting* equations (i.e. residuals of Eq. 18.1 go to the right hand side of Eq. 18.3, residuals of Eq. 18.3 go to the right hand side of Eq. 18.5 etc.).

How does multigrid help us? What is required to rapidly obtain a global accurate solution for the time-independent (steady-state) equations (like Poisson, Stokes and incompressible continuity equations) is that the '*numerical information*' propagates quickly across the entire model. During one iteration cycle, the information about updates of unknowns propagates to (or *is felt by*) only neighbouring grid points. Therefore, the finer the grid resolution, the shorter the physical distance over which information propagates during one iteration step. *Residuals with short wavelengths* (in terms of number of grid points) decay relatively fast (within a few iterations), while *residuals with longer wavelength* decay much more slowly (Fig. 18.2). Therefore, any increase in resolution produces even longer wavelengths in the residual distribution, which will thus require more iterations to decay.

Multigrid resolves this problem by performing additional iterations on several *hierarchically arranged coarser grids* (*resolution levels*, Chapter 17), which propagate the solution over larger distances and rapidly smooth out longer wavelength residuals. In this manner, residuals of all wavelengths decay with the same (small) number of iterations, which results in a solution convergence that is independent of the grid resolution. A typical way to program the multigrid method is to use several grids whose resolution increases by a fixed factor (e.g. a factor of 2, see Fig. 18.3). The finest grid (Level 1) is the *principal* one, on which an accurate solution is obtained, and the coarser grids are used to compute corrections for solutions on finer grids (cf. Eqs. 18.1–18.5). The coarser grid that is one level above the finest one will always compute corrections to the real solution, while the other grids will typically compute *corrections to corrections to the real solution, corrections to corrections to corrections to the real solution*, etc. (continue as an exercise ☺).

The equations that are formulated on the various grids (including boundary conditions) are identical, with the exception of the right hand sides of the equations; on coarser grids this is substituted by residuals that are interpolated (typically) from the nearest finer grid (cf. Eqs. 18.1–18.5). Material properties such as viscosity, shear modulus etc. necessary to formulate the equations are also interpolated from the finer grid. These properties could also be defined in some different manner (e.g., interpolated directly from markers) but we will

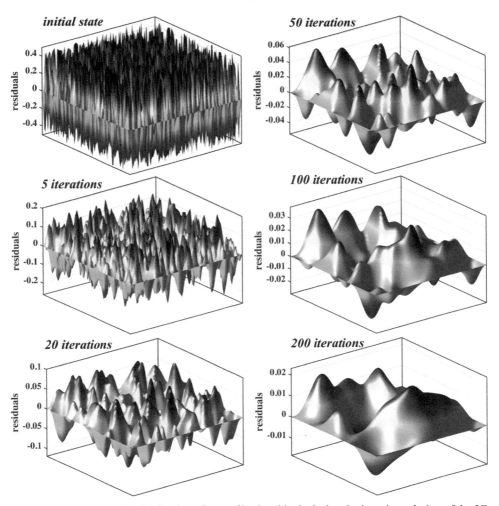

Fig. 18.2. Changes in the distribution of normalized residuals during the iterative solution of the 2D Poisson equation for a model with initially random distribution of the density field. Note that long wavelength residuals decay much slower then short wavelength ones. The model is computed with the code **Gauss_Seidel_iterations_Poisson.m** associated with this chapter. Model resolution is 100×100 grid points.

not do this here. As a result, the solution obtained on the coarser grid (e.g. pressure and velocity values) is in itself a *correction* (small addition) to the solution on the finer grid. It can then be used to update the solution on the finer grid by interpolating the corrections. To sum up: residuals and material properties are interpolated from finer to coarser levels (*restriction operation*) while computed corrections are interpolated back from coarser to finer levels (*prolongation operation*). During one iteration cycle, an accurate solution should only be obtained on the *coarsest* (last) grid where many iterations or a direct matrix inversion can be employed (which can be done at low computational cost since the

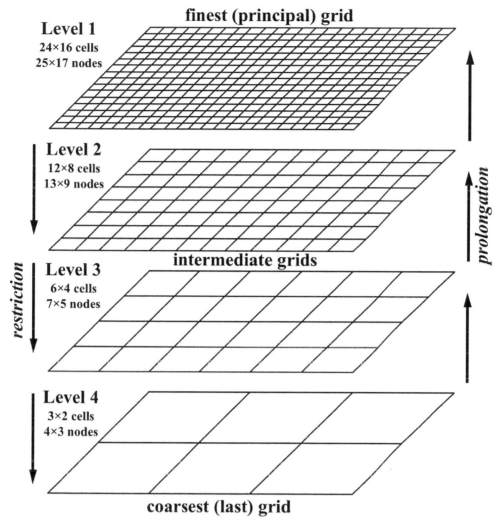

finest (principal) grid

Level 1
24×16 cells
25×17 nodes

Level 2
12×8 cells
13×9 nodes

intermediate grids

restriction

Level 3
6×4 cells
7×5 nodes

Level 4
3×2 cells
4×3 nodes

prolongation

coarsest (last) grid

Fig. 18.3. Multigrid structure for a uniformly spaced, 2D rectangular non-staggered grid with 4 levels of resolution. Resolution of the grid between two nearest levels changes by a factor of 2 (see changes in the number of cells). Coarser levels of resolution are responsible for the decay of larger residual wavelengths (see Fig. 18.2).

resolution of this grid is low). On other grids, some *limited number of iterations* (typically increasing by some factor with increasing grid level) should be performed in order to propagate information about the solution update and to compute new residuals. This process is called a *smoothing operation* (the reason for using this term is obvious from Fig. 18.2).

It should be pointed out that an increase in the resolution by an integer factor is not a strict requirement for multigrid. Generally, grid structures on different levels could be made in

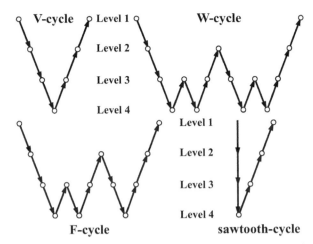

Fig. 18.4. Various *multigrid schedules* (or *cycles*) shown for the case of a 4-level multigrid algorithm (Fig. 18.3). Circles denote smoothing operation, arrows show restriction (downward) and prolongation (upward) operations.

a fully independent manner (e.g. by using independent irregularly spaced meshes) and the only requirement is that coarser grids should efficiently smooth residuals with larger wavelengths (Fig. 18.2). In particular, the multigrid approach could naturally combine with an adaptive block-structured grid (Fig. 17.1b) by removing the finest level of resolution when moving from a finer to coarser level grid structure. In the case of complex grid structures, special care should be taken when organizing the restriction and prolongation operations between grids since the nodal points of the coarser grid should not necessarily overlap those of the finer grid, even in the case of a regular rectangular non-staggered grid. This situation is also very common for staggered grids (even in the case when the resolution increases by an even factor) – non-overlap of points between different levels for a staggered grid is unavoidable and, therefore, the resolution at different levels can be chosen rather independently. Indeed, the choice of grid resolution and structure at various levels may notably affect the convergence of the solution. This choice can thus be different for different numerical problems and can be optimized empirically. Examples of programming such 'arbitrary resolution multigrids' for the Poisson equation as well as for the momentum and continuity equations are given respectively in the programs **Poisson_Multigrid_planet_arbitrary.m** and **Variable_viscosity_Multigrid_arbitrary.m** associated with this chapter.

The order in which grids are visited is called the *multigrid schedule* or *cycle*. Several standard schedules exist (Fig. 18.4): the V-cycle, the W-cycle, the F-cycle and the sawtooth-cycle. The *V-cycle* is the simplest and most commonly used multigrid schedule (Fig. 18.4): restriction+smoothing go uniformly from the finest level to the coarsest level through all intermediate levels, then prolongation+smoothing go uniformly from the coarsest level to the finest level, also through all intermediate levels. The order of operations for W- and F-cycles is more complicated and contains several cycles of restriction+smoothing followed by prolongation+smoothing between coarser levels (Fig. 18.4) before returning corrections to the *principal* (finest) Level 1. The sawtooth-cycle is, in a way, similar to the V-cycle but the smoothing operations are omitted during interpolation of residuals from finer to coarser levels. Thus, this cycle first solves all equations on the coarsest grid and then gradually 'refines' the solution toward the principal level by applying prolongation+smoothing operations. The sawtooth-cycles are typically applied when no good initial approximation of the solution exists on the principal level, which is a common situation for the beginning of a numerical experiment.

Interpolation of residuals and material properties *from finer to coarser grid* (i.e. restriction operation) can be made in the same manner as the interpolation of various parameters *from markers to nodes* (Eq. 8.18, Fig. 8.8). Consequently, the interpolation of corrections *from coarser to finer grid* (i.e. prolongation operation) can be organized by analogy of interpolation *from nodes to markers* (Eq. 8.19, Fig. 8.9). Since interpolation between markers and nodes has already been extensively discussed in Chapter 8, programming of restriction and prolongation operations is rather straightforward, as exemplified by several written MATLAB functions (e.g. **Poisson_restriction_planet**, **Poisson_ prolongation_planet**, **Stokes_Continuity_prolongation**, **Viscosity_restriction**, **Stokes_Continuity_viscous_restriction**) used in the codes associated with this chapter. There are also more sophisticated schemes of organizing restriction and prolongation operations which give a higher multigrid performance in specific cases (e.g. Wesseling, 1992).

It should also be mentioned that the method described in this chapter is called *geometric multigrid*, which requires the definition of several grids for the same model, formulation of the same differential equations separately for each grid, and storing material properties, solutions and corrections for all grids. Computational and memory costs for the geometric multigrid are relatively small since coarser grids have much fewer nodal points then the principal one. For example, in the case of grid coarsening by a factor of two, all coarser grids will have in 2D and 3D less than 50% and 25% of grid points, respectively, compared to the finest grid.

However, there is also a class of more sophisticated multigrid approaches called *algebraic multigrid* (*AMG*) which do not require the explicit definition of the coarser grids, but rather use algebraic operations based on multigrid principles to process and solve a global matrix constructed for the finest (principal) grid. In an algebraic multigrid scheme, the coarse level equations are generated from finer level equations without the use of any geometry or re-discretization on the coarse levels. This has the advantage that no coarse level grid has to be generated or stored, and no flux or source term needs be calculated on the coarse levels. This feature makes AMG particularly important for use on unstructured meshes (e.g. Geenen et al., 2009).

How efficient is multigrid? It is extremely efficient for simple cases like solving the Poisson equation on a regular grid (Fig. 18.2) and speeds up convergence by several orders of magnitude (Fig. 18.5). In more complex, thermomechanical modelling cases, it is typically less efficient, particularly when physical phenomena (such as e.g. localization of

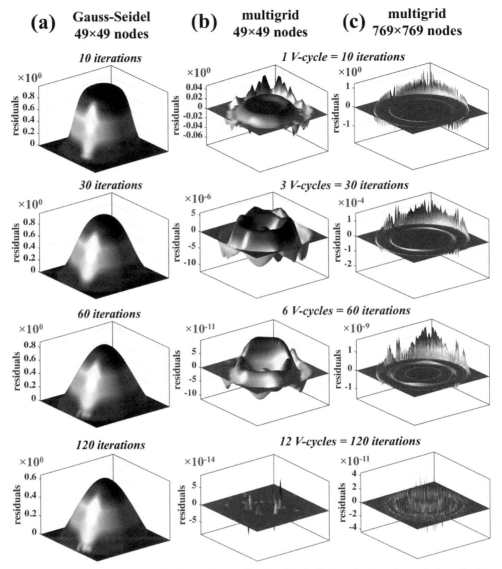

Fig. 18.5. Changes in the distribution of normalized residuals during the iterative solution of a 2D Poisson equation for a model with a density field that corresponds to a circular body (planet) embedded in a massless medium (space). (a) Gauss–Seidel iterations at low grid resolutions. (b), (c) Multigrid iterations at both low (b) and high (c) grid resolution, The multigrid cycle corresponds to a V-cycle with 5+5=10 Gauss–Seidel iterations on the finest grid per cycle; 4 and 7 levels of resolution are used for (b) and (c) respectively. Note that the convergence of the numerical solution (decay of residuals) in the multigrid case is several orders of magnitude faster (see bold numbers above vertical axes defining the order of magnitude for residuals) than for the same number of simple Gauss–Seidel iterations performed on the finest level. Also, in contrast to simple Gauss–Seidel iterations, the convergence of the multigrid solutions is independent of grid resolution (compare (b) and (c) with Fig. 18.1). The models are computed with the code **Poisson_Multigrid.m**.

5-point cross

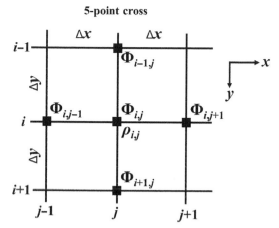

Fig. 18.6. Stencil of the regular rectangular grid used for discretization of the Poisson equation for the iterative Gauss–Seidel smoother used with multigrid.

deformation) are not properly reproduced on the coarser levels. For many geodynamic applications, the multigrid provides one of the best options to build efficient and robust codes and is therefore widely used in 3D numerical modelling of mantle convection and plate tectonic processes (e.g. Tackley, 2000, 2008; Gerya, 2010b; Burov and Gerya, 2014; Gerya et al., 2015). Efficiency of the multigrid also greatly depends on the ability of the *smoother* (i.e. an iterative approach used for performing the *smoothing operation*, Chapter 3) to transfer information efficiently through the grid. For the numerical applications discussed in this chapter, we will always use *Gauss–Seidel smoothers*, which seem to be much more efficient than *Jacobi smoothers* for numerical problems of our interest (Chapter 3).

18.2 Solving the Poisson equation with multigrid

Implementation of a *multigrid solver* for the Poisson equation is relatively simple: a standard *Gauss–Seidel iteration* with a relatively high relaxation parameter $\theta_{relaxation}^{Poisson}$ (up to 1.75 uniformly applied for all nodes in all grids) can be used as an efficient smoother and the interpolation of residuals (restriction operation) and corrections (prolongation operation) is very straightforward, particularly when the resolution of a regular grid between adjacent levels increases by an integer factor (2, 3 etc.) and grid lines of the coarser grid overlap with grid lines of the finer grid (Fig. 18.3). A 5-point stencil in 2D for the discretization of the Poisson equation on such a regular grid is shown in Fig. 18.6 and the following iterative FD representation is used for updating the solution with a Gauss–Seidel smoother

$$\Delta R_{i,j} = R_{i,j} - \left(\frac{\partial^2 \Phi}{\partial x^2}\right)_{i,j} - \left(\frac{\partial^2 \Phi}{\partial y^2}\right)_{i,j}, \tag{18.6}$$

$$\Phi_{i,j}^{new} = \Phi_{i,j} + \frac{\Delta R_{i,j}}{C_{i,j}} \theta_{relaxation}^{Poisson}, \tag{18.7}$$

$$\left(\frac{\partial^2 \Phi}{\partial x^2}\right)_{i,j} = \frac{\Phi_{i,j-1} - 2\Phi_{i,j} + \Phi_{i,j+1}}{\Delta x^2}, \tag{18.8}$$

$$\left(\frac{\partial^2 \Phi}{\partial y^2}\right)_{i,j} = \frac{\Phi_{i-1,j} - 2\Phi_{i,j} + \Phi_{i+1,j}}{\Delta y^2}, \tag{18.9}$$

$$C_{i,j} = -\frac{2}{\Delta x^2} - \frac{2}{\Delta y^2}, \tag{18.10}$$

where $\Phi_{i,j-1}$, $\Phi_{i-1,j}$, $\Phi_{i,j}$, $\Phi_{i+1,j}$, $\Phi_{i,j+1}$ are either the current values of gravity potential (at finest level) or corrections for this potential (at coarser levels) in respective nodal points, $C_{i,j}$ is the coefficient at $\Phi_{i,j}$ in the discretized Poisson equation, $\Delta R_{i,j}$ is the current residual and $R_{i,j}$ is the right hand side of the Poisson equation. On the principal level (finest grid), the right hand side is computed from the standard equation

$$R_{i,j} = 4K\pi G \rho_{i,j}, \tag{18.11}$$

where G is the gravitational constant and K depends on the geometry of the self-gravitating body modelled in 2D ($K = 1$ and $K = 2/3$ stand for cylindrical and spherical geometry, respectively, Chapter 11). For coarser levels, $R_{i,j}$ is composed by residuals interpolated from finer levels. Obviously, grid steps Δx and Δy are also different for different levels of resolution. In a standard case, the simplest possible boundary condition equation $\Phi_{i,j} = 0$ is used for all marginal nodes on all grids, which also poses no difficulty for programming.

A peculiar case occurs when an internal boundary is present within the model on which a boundary condition to solve the Poisson equation has to be defined (Fig. 18.7). For example, this is the case when we want to compute the gravity potential inside and around a planet, which is a component of a spherical-Cartesian approach for modelling self-gravitating bodies on a rectangular Cartesian grid (Fig. 11.5). In order to force the planet to remain in the centre of the grid and obtain a natural distribution of the gravitational acceleration vector *inside the planet* (Chapter 11), a constant gravity potential boundary condition ($\Phi = \Phi_b$) in 2D can be defined on a circle located at a distance from the planetary surface (Fig. 18.7). In this case, the Poisson equation is solved only for the nodes located inside the circle (see solid squares in Fig. 18.7), whereas the boundary condition $\Phi = \Phi_b$ is applied for all other nodes of the grid. In order to have consistent solutions for all levels of resolution, the FD representation of the Poisson equation should be modified for the nodes located immediately next to the internal boundary with the use of a *ghost boundary condition nodes approach* (Chapter 17, Fig. 17.2). In this case, the derivative of the gravity potential on the side of a 5-point cross which crosses the internal boundary should be defined in such a manner that it satisfies the boundary condition $\Phi = \Phi_b$. This situation is

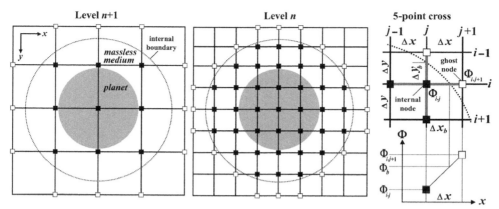

Fig. 18.7. Rectangular grids for two levels of resolution for solving the Poisson equation for a 2D numerical model with an internal boundary based on multigrid. The Poisson equation is solved for internal nodes of the grids located inside the boundary (solid squares). Ghost nodes (open squares) located immediately outside the boundary are used to formulate internal boundary conditions when discretizing the Poisson equation for the nearest internal nodes.

shown in Fig. 18.7 (right hand side), where the horizontal derivative of the gravity potential to the right of the internal *ij*th node should satisfy the boundary condition $\Phi=\Phi_b$ and the following FD equation can be formulated

$$\frac{\partial \Phi}{\partial x} = \frac{\Phi_{i,j+1} - \Phi_{i,j}}{\Delta x} = \frac{\Phi_b - \Phi_{i,j}}{\Delta x_b}, \tag{18.12}$$

where Δx is the horizontal grid step, Δx_b is the distance from the *ij*th node to the circular boundary and $\Phi_{i,j+1}$ is the gravity potential for an imaginary (ghost) node located at the distance Δx from the *ij*th node. The value of gravity potential for the ghost node can then be computed as

$$\Phi_{(i,j+1)} = \Phi_b \frac{\Delta x}{\Delta x_b} - \Phi_{(i,j)} \frac{\Delta x - \Delta x_b}{\Delta x_b}. \tag{18.13}$$

Equation (18.8) for the second *x* derivative of gravity potential can then be reformulated by excluding the ghost node value $\Phi_{i,j+1}$ (derive as an exercise)

$$\left(\frac{\partial^2 \Phi}{\partial x^2}\right)_{i,j} = \Phi_{i,j-1} \frac{1}{\Delta x^2} - \Phi_{i,j} \frac{\Delta x_b + \Delta x}{\Delta x^2 \Delta x_b} + \frac{\Phi_b}{\Delta x \Delta x_b}. \tag{18.14}$$

A similar transformation can be done for Eq. (18.9) by excluding the ghost node value $\Phi_{i-1,j}$ (Fig. 18.7, derive as an exercise):

$$\left(\frac{\partial^2 \Phi}{\partial y^2}\right)_{i,j} = \Phi_{i,j-1}\frac{1}{\Delta y^2} - \Phi_{i,j}\frac{\Delta y_b + \Delta y}{\Delta y^2 \Delta y_b} + \frac{\Phi_b}{\Delta y \Delta y_b}. \qquad (18.15)$$

Consequently, the coefficient $C_{i,j}$ in Eq. (18.10) will change to

$$C_{i,j} = -\frac{\Delta x_b + \Delta x}{\Delta x^2 \Delta x_b} - \frac{\Delta y_b + \Delta y}{\Delta y^2 \Delta y_b}. \qquad (18.16)$$

Ghost nodes are thus only used for reformulating the Poisson equation in the nearest internal nodes, and values of the gravity potential in the ghost nodes are not computed explicitly. Moreover, the 'implied' gravity potential value (Eq. 18.13) in a ghost node is generally different when the same ghost node is used in different Poisson equations formulated for different internal nodes. This is because the gravity potential has a 'kink' on the circular boundary and it is taken to be constant ($\Phi = \Phi_b$) for all nodes outside this boundary, including all ghost nodes in the final solution. However, the uniform use of Eqs. (18.14)–(18.16) to reformulate the Poisson equation on different resolution levels is important and ensures the *geometrical compatibility* of solutions between all multigrid levels. This is because the boundary conditions for all grids are formulated on the same internal boundary, irrespective of the resolution and actual positions of the nodal points relative to this boundary. Note that the value of Φ_b should be set to zero for all levels of resolution with the exception of the finest (principal) level since coarser levels are used for computing corrections to the solution and these corrections should tend to zero (and not to Φ_b) with an increasing number of iterations. An example of a multigrid implementation with the circular internal boundary for gravity potential is given in the program **Poisson_Multigrid_planet.m** associated with this chapter.

Solving the Poisson equation on an irregularly spaced grid is not very different from the above procedures. Modifications only concern the manner of computing second derivatives of gravity potential in Eq. (18.6), the $C_{i,j}$ coefficient in Eq. (18.7), and the way of finding a correspondence between nodal points of coarser and finer grids when programming restriction and prolongation operations. The necessary modifications of the Poisson equation can be made on the basis of respective FD equations given in Chapter 11 (Fig. 11.4, Eq. 11.11). Finding a correspondence between nodal points is analogous to that between Lagrangian markers (equivalent to nodes of finer grid) and Eulerian grid points (equivalent to nodes of coarser grid) (Figs. 8.8, 8.9, Eqs. 8.18, 8.19) and can be based on the bisection procedure presented in Chapter 8 (Fig. 8.10).

18.3 Solving Stokes and continuity equations with multigrid

The main challenge for solving coupled momentum and continuity equations with a multigrid method consists of programming a robust smoother that uses a *primitive variable* (*pressure-velocity*) formulation. In the case of constant viscosity, this challenge can be avoided by using a stream function formulation that requires double solving of the Poisson equation (Chapter 5). However, we are more interested in creating a *Gauss–Seidel pressure-velocity smoother* that allows explicitly computation of the pressure distribution in the model and is further applicable (with some modifications) for the variable viscosity case, which is much more relevant for geodynamic applications. The main obstacle to building such a pressure-velocity smoother comes from solving the incompressible continuity equation $\text{div}(\vec{v}) = 0$, which does not contain pressure and therefore (without modifications) cannot be converted into a pressure solution update procedure as for example the Stokes equation (for velocity) or Poisson equation (for gravity potential, see Eqs. 18.6, 18.7). This problem can be overcome with the *computational compressibility approach* according to which an iterative pressure update in a specific location is made proportional to the current residual of the continuity equation computed for the same location

$$\Delta R_{i,j}^{continuity} = R_{i,j}^{continuity} - \text{div}(\vec{v})_{i,j}, \tag{18.17}$$

$$P_{i,j}^{new} = P_{i,j} + \frac{\Delta R_{i,j}^{continuity}}{\beta_{i,j}^{computational}} \theta_{relaxation}^{continuity}, \tag{18.18}$$

where $R_{i,j}^{continuity}$ is the right hand side of the continuity equation (it is zero on the finest level and is made of residuals for coarser levels), $\beta_{i,j}^{computational}$ is computational compressibility and $\theta_{relaxation}^{continuity}$ is the relaxation coefficient used for the continuity equation. Despite the artificial origin of this scheme for the incompressible viscous medium, a surprisingly efficient choice of $\beta_{i,j}^{computational}$ can be made for all levels of resolution on the basis of the simple relation

$$\beta_{i,j}^{computational} = \frac{1}{\eta_{i,j}}, \tag{18.19}$$

where $\eta_{i,j}$ is the local viscosity. Equations (18.17)–(18.19) are applicable for both variable and constant viscosity cases (in the constant viscosity case $\eta_{i,j}$ is obviously equal to the global viscosity η for the entire model) and give stable convergence of solutions for coupled momentum and continuity equations if the relaxation coefficient $\theta_{relaxation}^{continuity}$ is chosen to be 0.1–0.3. An explanation for this surprising efficiency is that the computational compressibility provides a natural way of coupling between pressure and velocity equations: in places where for the current iteration step the fluid converges and $\beta_{i,j}^{computational} = 1/\eta_{i,j}$, Eqs. (18.17)–(18.19) produce an increase in pressure, thus creating outward directed pressure gradients that force (through the Stokes equation) divergence of velocity and improve the solution of the continuity equation; in places where fluid diverges and $\text{div}(\vec{v}) > 0$, the reaction is opposite.

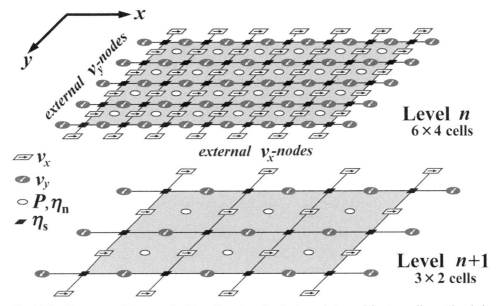

Fig. 18.8. Geometry of staggered grids with external velocity points used for two adjacent levels in the case of coupled solving of momentum and continuity equations. The internal (working) part of the grid is shaded grey. Note that distances from the grid boundaries to the external velocity nodes are different for different levels of resolution.

An efficient numerical representation of the momentum and continuity equations for the multigrid can be based on a staggered grid with external velocity points (Fig. 18.8), as discussed in Chapter 7. Since the global numbering of unknowns is not needed in the case of iterative methods, the indexing of arrays for different parameters can be done separately and these arrays will also have different dimensions (Fig. 18.8): $N_x \times N_y$ for ρ and η_s (viscosity used for shear stress formulation) located in basic nodes, $N_x \times (N_y+1)$ for v_x, $(N_x+1) \times N_y$ for v_y and $(N_x-1) \times (N_y-1)$ for P and η_n (the viscosity used for normal deviatoric stress components formulation), where N_x and N_y are respectively the horizontal and vertical resolutions of the basic grid at a given multigrid level (see black rectangles in Fig. 18.8).

Constant viscosity case

A simple pressure-velocity update scheme based on the Gauss–Seidel iteration and a computational compressibility approach (Eqs. 18.17–18.19) can be constructed on a regular grid for the case of constant viscosity (Fig. 18.9):

$$\Delta R_{i,j}^{x\text{-}Stokes} = R_{i,j}^{x\text{-}Stokes} - \eta \left(\frac{\partial^2 v_x}{\partial x^2}\right)_{i,j} - \eta \left(\frac{\partial^2 v_x}{\partial y^2}\right)_{i,j} + \left(\frac{\partial P}{\partial x}\right)_{i,j}, \tag{18.20}$$

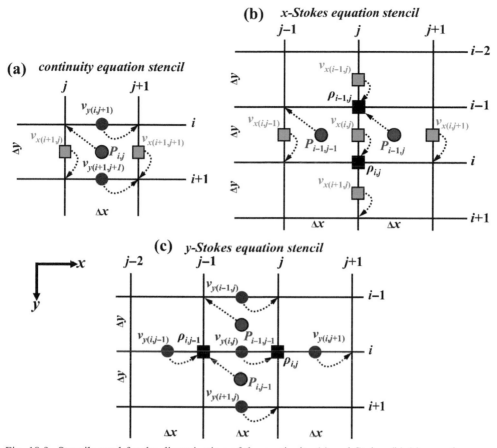

Fig. 18.9. Stencils used for the discretization of the continuity (a) and Stokes (b),(c) equations on a 2D regular staggered grid (Fig. 18.8) for models with constant viscosity. Indexing of grid lines corresponds to basic (density) nodal points. Indexing of different unknowns (dotted arrows) is made separately depending on the number of respective nodal points in the staggered grid (Fig. 18.8).

$$v_{x(i,j)}^{new} = v_{x(i,j)} + \frac{\Delta R_{i,j}^{x\text{-}Stokes}}{C_{v_x(i,j)}} \theta_{relaxation}^{Stokes}, \qquad (18.21)$$

$$\Delta R_{i,j}^{y\text{-}Stokes} = R_{i,j}^{y\text{-}Stokes} - \eta \left(\frac{\partial^2 v_y}{\partial x^2}\right)_{i,j} - \eta \left(\frac{\partial^2 v_y}{\partial y^2}\right)_{i,j} + \left(\frac{\partial P}{\partial y}\right)_{i,j}, \qquad (18.22)$$

$$v_{y(i,j)}^{new} = v_{y(i,j)} + \frac{\Delta R_{i,j}^{y\text{-}Stokes}}{C_{v_y(i,j)}} \theta_{relaxation}^{Stokes}, \qquad (18.23)$$

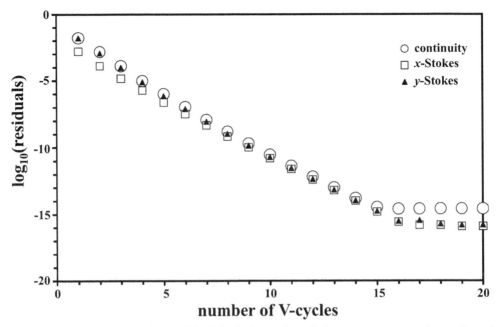

Fig. 18.10. Decay of normalized residuals for Stokes and continuity equations versus the number of multigrid V-cycles for a constant viscosity model. Residuals stabilize at computer accuracy level. A 4-level multigrid with resolution 49×49 points on the finest level is used with relaxation parameters $\theta_{relaxation}^{continuity} = 0.3$ and $\theta_{relaxation}^{Stokes} = 0.9$. Numerical setup: rectangular block having higher density sinks in lower density fluid. Iterations start from a hydrostatic pressure field and zero velocities. Results are obtained with the program **Constant_Viscosity_Multigrid_ghost.m**.

$$\Delta R_{i,j}^{continuity} = R_{i,j}^{continuity} - \left(\frac{\partial v_x}{\partial x}\right)_{i,j} - \left(\frac{\partial v_y}{\partial y}\right)_{i,j}, \tag{18.24}$$

$$P_{i,j}^{new} = P_{i,j} + \eta \Delta R_{i,j}^{continuity} \theta_{relaxation}^{continuity}, \tag{18.25}$$

$$\left(\frac{\partial^2 v_x}{\partial x^2}\right)_{i,j} = \frac{v_{x(i,j-1)} - 2v_{x(i,j)} + v_{x(i,j+1)}}{\Delta x^2}, \text{ (Fig. 18.9b)} \tag{18.26}$$

$$\left(\frac{\partial^2 v_x}{\partial y^2}\right)_{i,j} = \frac{v_{x(i-1,j)} - 2v_{x(i,j)} + v_{x(i+1,j)}}{\Delta y^2}, \text{ (Fig. 18.9b)} \tag{18.27}$$

$$\left(\frac{\partial^2 v_y}{\partial x^2}\right)_{i,j} = \frac{v_{y(i,j-1)} - 2v_{y(i,j)} + v_{y(i,j+1)}}{\Delta x^2}, \text{ (Fig. 18.9c)} \tag{18.28}$$

$$\left(\frac{\partial^2 v_y}{\partial y^2}\right)_{i,j} = \frac{v_{y(i-1,j)} - 2v_{y(i,j)} + v_{y(i+1,j)}}{\Delta y^2}, \text{ (Fig. 18.9c)} \tag{18.29}$$

$$\left(\frac{\partial P}{\partial x}\right)_{i,j} = \frac{P_{i-1,j} - P_{i-1,j-1}}{\Delta x}, \quad \text{(Fig. 18.9b)} \tag{18.30}$$

$$\left(\frac{\partial P}{\partial y}\right)_{i,j} = \frac{P_{i,j-1} - P_{i-1,j-1}}{\Delta y}, \quad \text{(Fig. 18.9c)} \tag{18.31}$$

$$\left(\frac{\partial v_x}{\partial x}\right)_{i,j} = \frac{v_{x(i+1,j+1)} - v_{x(i+1,j)}}{\Delta x}, \quad \text{(Fig. 18.9a)} \tag{18.32}$$

$$\left(\frac{\partial v_y}{\partial y}\right)_{i,j} = \frac{v_{y(i+1,j+1)} - v_{y(i,j+1)}}{\Delta y}, \quad \text{(Fig. 18.9a)} \tag{18.33}$$

$$C_{v_x(i,j)} = -\frac{2\eta}{\Delta x^2} - \frac{2\eta}{\Delta y^2}, \tag{18.34}$$

$$C_{v_y(i,j)} = -\frac{2\eta}{\Delta x^2} - \frac{2\eta}{\Delta y^2}, \tag{18.35}$$

where $\theta_{relaxation}^{Stokes}$ is a relaxation parameter for the Stokes equations: $v_{x(i,j)}$, $v_{y\,(i,j)}$, $P_{i,j}$ etc. are either current values of velocity components and pressure (at finest level) or corrections for these values (at coarser levels) at respective nodal points; $C_{v_x(i,j)}$ and $C_{v_y(i,j)}$ are coefficients at respectively $v_{x(i,j)}$ and $v_{y(i,j)}$ in the discretized x and y Stokes equations, respectively; $\Delta R_{i,j}^{x\text{-}Stokes}$, $\Delta R_{i,j}^{y\text{-}Stokes}$, $\Delta R_{i,j}^{continuity}$ and $R_{i,j}^{x\text{-}Stokes}$, $R_{i,j}^{y\text{-}Stokes}$, $R_{i,j}^{continuity}$ are current residuals, and right hand side for the momentum and continuity equations, respectively. On the finest level of resolution, these right hand side contributions are computed from the standard equations

$$R_{i,j}^{x\text{-}Stokes} = -g_x \frac{\rho_{i,j} + \rho_{i-1,j}}{2}, \quad \text{(Fig. 18.9b)} \tag{18.36}$$

$$R_{i,j}^{y\text{-}Stokes} = -g_y \frac{\rho_{i,j-1} + \rho_{i,j}}{2}, \quad \text{(Fig. 18.9c)} \tag{18.37}$$

$$R_{i,j}^{continuity} = 0, \tag{18.38}$$

where g_x and g_y are respective components of the gravitational acceleration vector. At coarser levels, $R_{i,j}^{x\text{-}Stokes}$, $R_{i,j}^{y\text{-}Stokes}$ and $R_{i,j}^{continuity}$ are composed of respective residuals interpolated from finer levels. Obviously, grid steps Δx and Δy are different for each multigrid level. Standard boundary condition equations for no slip and free slip conditions are always applied to the same external boundaries of the basic grid (see grey areas in Fig. 18.8) and can be formulated uniformly for all levels as follows:

upper boundary

$$v_{y(i=1,j)} = 0,$$

$$v_{x(i=1,j)} = v_{x(i=2,j)} \text{ for free slip, i.e. } \frac{\partial v_x}{\partial y} = 0 \text{ across the boundary,}$$

$$v_{x(i=1,j)} = -v_{x(i=2,j)} \text{ for no slip, i.e. } v_x = 0 \text{ on the boundary;}$$

left boundary

$$v_{x(i,j=1)} = 0,$$

$$v_{y(i,j=1)} = v_{y(i,j=2)} \text{ for free slip, i.e. } \frac{\partial v_y}{\partial x} = 0 \text{ across the boundary,}$$

$$v_{y(i,j=1)} = -v_{y(i,j=2)} \text{ for no slip, i.e. } v_y = 0 \text{ on the boundary.}$$

Conditions for lower and right boundaries are applied similarly. These boundary condition equations can either be called directly in the Gauss–Seidel iteration cycle or (which often gives better convergence of the solution) be implemented within the x and y Stokes equations (Eqs. 18.20–18.35) discretized for the nearest internal velocity nodes (*ghost boundary condition nodes approach*). For example, a free slip condition at the upper boundary can be implemented as

$$\left(\frac{\partial^2 v_x}{\partial y^2}\right)_{i=2,j} = \frac{-v_{x(i=2,j)} + v_{x(i=3,j)}}{\Delta y^2} \text{ (Fig. 18.9b compare with Eq. 18.27),} \qquad (18.39)$$

$$C_{v_x(i=2,j)} = -\frac{\eta}{\Delta x^2} - \frac{2\eta}{\Delta y^2} \text{ (compare with Eq. 18.34),} \qquad (18.40)$$

$$\left(\frac{\partial^2 v_y}{\partial y^2}\right)_{i=2,j} = \frac{-2v_{y(i=2,j)} + v_{y(i=3,j)}}{\Delta y^2} \text{ (Fig. 18.9c, compare with Eq. 18.29),} \qquad (18.41)$$

$$\left(\frac{\partial v_y}{\partial y}\right)_{i=1,j} = \frac{v_{y(i=2,j+1)}}{\Delta y} \text{ (Fig. 18.9a, compare with Eq. 18.33).} \qquad (18.42)$$

Velocity conditions for other boundaries can be implemented in a similar way. An example of such boundary condition implementation is given in the MATLAB function **Stokes_Continuity_smoother_ghost.m**.

In order to be able to compute pressure fields, we also need to prescribe the pressure value *in one selected cell*. This can be done on the finest level at the end of each smoothing cycle by subtracting uniformly from all pressure nodes an estimated current difference between required and actual pressure values in the selected cell. This operation does not change the pressure gradients in the Stokes equations and thus does not affect the accuracy of the solution. During the smoothing procedure, an iterative pressure update is done uniformly with Eq. (18.25) for all pressure nodes, including the selected one, as such uniformity typically gives better solution convergence. Also, faster convergence is often obtained when the hydrostatic pressure distribution is initially defined in the computational domain. This distribution can be computed from the density field and gravity vector as follows:

- a pressure value is first defined in a first cell $P_{i=1,j=1}$ and then computed in the first row of cells based on density and the horizontal (g_x) component of the gravity vector

$$P_{i=1,j} = P_{i=1,j-1} + g_x \frac{\rho_{i=1,j} + \rho_{i=2,j}}{2} \Delta x \text{ (Fig. 18.9b),} \qquad (18.43)$$

- pressure in the remaining cells is computed by columns based on density and the vertical (g_y) component of the gravity vector

$$P_{i,j-1} = P_{i-1,j-1} + g_y \frac{\rho_{i,j} + \rho_{i,j-1}}{2} \Delta y \quad \text{(Fig. 18.9c).} \qquad (18.44)$$

A multigrid solver based on the described procedures is very efficient for the case of constant viscosity, and residuals of both Stokes and continuity equations decay very rapidly (up to an order of magnitude per V-cycle, Fig. 18.10) to computer accuracy in 15–20 cycles. Examples of the described multigrid implementation for the case of constant viscosity are given in the programs **Stokes_Continuity_Multigrid.m** (boundary condition equations called directly in the Gauss–Seidel iteration cycle) and **Constant_Viscosity_ Multigrid_ghost.m** (boundary condition equations implemented to momentum and continuity equations).

Adding variable viscosity

A variable viscosity multigrid solver is based essentially on the same principles as for the constant viscosity but conservative finite differences discussed in Chapter 7 should be used to re-formulate the Stokes equations. Pressure-velocity update schemes based on Gauss–Seidel iteration in the case of variable viscosity and regular grids (Fig. 18.11) can be written as follows (only equations which are different from Eqs. 18.20–18.33 are shown)

$$\Delta R_{i,j}^{x\text{-}Stokes} = R_{i,j}^{x\text{-}Stokes} - \left(\frac{\partial \sigma'_{xx}}{\partial x}\right)_{i,j} - \left(\frac{\partial \sigma'_{xy}}{\partial y}\right)_{i,j} + \left(\frac{\partial P}{\partial x}\right)_{i,j}, \qquad (18.45)$$

$$\Delta R_{i,j}^{y\text{-}Stokes} = R_{i,j}^{y\text{-}Stokes} - \left(\frac{\partial \sigma'_{yy}}{\partial y}\right)_{i,j} - \left(\frac{\partial \sigma'_{yx}}{\partial x}\right)_{i,j} + \left(\frac{\partial P}{\partial y}\right)_{i,j}, \qquad (18.46)$$

$$P_{i,j}^{new} = P_{i,j} + \eta_{n(i,j)} \Delta R_{i,j}^{continuity} \theta_{relaxation}^{continuity}, \text{ (Fig. 18.11a)} \qquad (18.47)$$

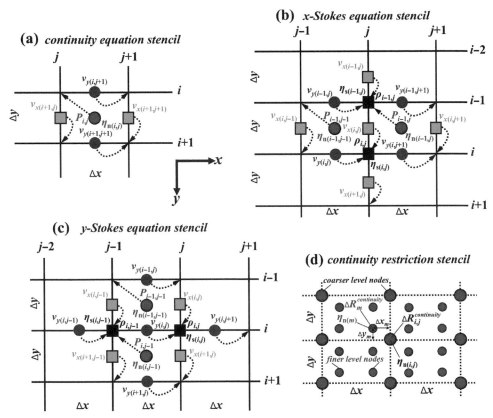

Fig. 18.11. Stencils used for discretization of the continuity (a) and Stokes (b),(c) equations and for restriction of continuity residuals (d) on a 2D regular staggered grid (Fig. 18.8) for models with variable viscosity. Indexing of solid grid lines corresponds to basic (density) nodal points. Indexing of different unknowns is done separately depending on the number of respective nodal points in the staggered grid (Fig. 18.8).

$$\left(\frac{\partial \sigma'_{xx}}{\partial x}\right)_{i,j} = 2\eta_{n(i-1,j)}\frac{v_{x(i,j+1)} - v_{x(i,j)}}{\Delta x^2} - 2\eta_{n(i-1,j-1)}\frac{v_{x(i,j)} - v_{x(i,j-1)}}{\Delta x^2}, \quad \text{(Fig. 18.11b)}$$

$$(18.48)$$

$$\left(\frac{\partial \sigma'_{xy}}{\partial y}\right)_{i,j} = \eta_{s(i,j)}\left(\frac{v_{x(i+1,j)} - v_{x(i,j)}}{\Delta y^2} + \frac{v_{y(i,j+1)} - v_{y(i,j)}}{\Delta x \Delta y}\right) -$$
$$-\eta_{s(i-1,j)}\left(\frac{v_{x(i,j)} - v_{x(i-1,j)}}{\Delta y^2} + \frac{v_{y(i-1,j+1)} - v_{y(i-1,j)}}{\Delta x \Delta y}\right) \quad , \quad \text{(Fig. 18.11b)} \quad (18.49)$$

$$\left(\frac{\partial \sigma'_{yy}}{\partial y}\right)_{i,j} = 2\eta_{n(i,j-1)}\frac{v_{y(i+1,j)} - v_{y(i,j)}}{\Delta y^2} - 2\eta_{n(i-1,j-1)}\frac{v_{y(i,j)} - v_{y(i-1,j)}}{\Delta y^2}, \quad \text{(Fig. 18.11c)}$$

$$(18.50)$$

$$\left(\frac{\partial \sigma'_{yx}}{\partial x}\right)_{i,j} = \eta_{s(i,j)}\left(\frac{v_{y(i,j+1)} - v_{y(i,j)}}{\Delta x^2} + \frac{v_{x(i+1,j)} - v_{x(i,j)}}{\Delta x \Delta y}\right) -$$
$$-\eta_{s(i,j-1)}\left(\frac{v_{y(i,j)} - v_{y(i,j-1)}}{\Delta x^2} + \frac{v_{x(i+1,j-1)} - v_{x(i,j-1)}}{\Delta x \Delta y}\right) \qquad \text{, (Fig. 18.11c)} \qquad (18.51)$$

$$C_{v_x(i,j)} = -2\frac{\eta_{n(i-1,j)} + \eta_{n(i-1,j-1)}}{\Delta x^2} - \frac{\eta_{s(i,j)} + \eta_{s(i-1,j)}}{\Delta y^2}, \qquad (18.52)$$

$$C_{v_y(i,j)} = -2\frac{\eta_{n(i,j-1)} + \eta_{n(i-1,j-1)}}{\Delta y^2} - \frac{\eta_{s(i,j)} + \eta_{s(i,j-1)}}{\Delta x^2}. \qquad (18.53)$$

The viscosity is defined (Fig. 18.11) both in the cell centres (η_n) and in the basic nodes (η_s), which are used to formulate normal (σ'_{xx}, σ'_{yy}) and shear ($\sigma'_{xy} = \sigma'_{yx}$) deviatoric stress components, respectively. These two types of viscosity should be interpolated from finer to coarser levels (i.e. *viscosity restriction*) before starting any multigrid iteration. Note that continuity equation residuals are multiplied to a *local viscosity value* η_n defined in the centre of the respective cell when computing pressure updates for this cell (Eq. 18.47). Moreover, in the case of large viscosity contrasts, convergence is notably improved when continuity residuals are rescaled based on local viscosity values at both finer and coarser levels, during restriction operations. The following first order of accuracy bilinear scheme is used to calculate the right hand side of the continuity equation $R_{i,j}^{continuity}$ for the *ij*th pressure node at a coarser level based on the continuity equation residuals $\Delta R_m^{continuity}$ computed for finer level nodes located within one grid step distance around the coarser level node (Fig. 18.11d):

$$R_{i,j}^{continuity} = \frac{\sum\limits_m \eta_{n(m)} \Delta R_m^{continuity} w_{m(i,j)}}{\eta_{n(i,j)} \sum\limits_m w_{m(i,j)}}, \qquad (18.54)$$

$$w_{m(i,j)} = \left(1 - \frac{\Delta x_m}{\Delta x}\right) \times \left(1 - \frac{\Delta y_m}{\Delta y}\right), \qquad (18.55)$$

where $w_{m(i,j)}$ represents a statistical weight of the *m*th finer level node at the *ij*th coarser level node; Δx_m and Δy_m are distances from *m*th node to *ij*th node. Similarly to interpolation from markers to nodes, Eq. (18.55) only accounts for finer level nodes located within a limited (one coarser grid step) distance around the coarser level node. Equation (18.54) guarantees that pressure corrections computed at coarser levels will always be proportional to the product of continuity residuals and local viscosities at the finest (principal) level as required by Eq. (18.47). Obviously, in the constant viscosity case, Eq. (18.54) turns into a standard bilinear interpolation scheme (Chapter 8, Eq. 8.18, Fig. 8.8). It should also be mentioned that more sophisticated pressure update and restriction/prolongation schemes (Tackley, 2008) are based on computational compressibility (Eq. 18.18) defined as the local pressure derivative of velocity divergence

$$\beta_{i,j}^{computational} = \left(\frac{\partial \mathrm{div}(\vec{v})}{\partial P}\right)_{i,j}. \tag{18.56}$$

This derivative can be computed numerically in the centre of a cell by using the discretized Stokes equations for four surrounding velocity nodes (Fig. 18.11a) that contain the pressure value for this specific cell (Eqs. 18.45, 18.46). Indeed, Eq. (18.56) always predicts an inverse proportionality between $\beta_{i,j}^{computational}$ and local viscosity $\eta_{n(i,j)}$ (Tackley, 2008), which explains why the simplified update scheme of Eq. (18.47) is sufficiently robust.

One more modification to the multigrid solution algorithm, which helps in obtaining an accurate solution for the first time step, is a gradual increase in the viscosity contrast. When we initialize a computation that has a large viscosity contrast ($>10^3$), we typically do not have any good initial approximation of the velocity field (since we cannot use the velocity field from the previous time step). If we start from zero velocity and a hydrostatic pressure field, the convergence of the solution can be very slow and the velocity field may remain unrealistic for many iterations. This is particularly the case when the velocity field is defined *by the weakest, rather than by the strongest* medium. This happens, for example, in the case of a hard Stokes sphere/cylinder that passes through a low viscosity fluid (cf. Stokes cylinder test, Popov and Sobolev, 2008; Schmeling et al., 2008) or in the case of a rigid *isolated* dense slab/block sinking in a weak medium (cf. falling block test, Gerya and Yuen, 2003a). Model setups with isolated rigid objects are in strong contrast with *Rayleigh–Taylor instability* models (Chapter 20) where a strong layer is attached to the model boundaries and the velocity field is therefore dictated by the rate of its internal deformation. In the latter case, the multigrid solution converges rapidly even for large viscosity contrasts. In the former case, a gradual increase of the *computational viscosity contrast* may indeed notably improve and speed up the solution (Fig. 18.12). Initially (in the beginning of multigrid cycles) the viscosity field is rescaled to a low/no viscosity contrast for which an accurate pressure-velocity solution can be rapidly computed, starting from a hydrostatic pressure and zero velocity fields (Fig. 18.10). Then, after either a limited number of iterations or after reaching some level of accuracy, the computational viscosity contrast is gradually increased by a certain factor (1.5 to 10) and the original viscosity field is rescaled to this new contrast. The operations are repeated until the *original viscosity contrast* of the model is recovered. Rescaling of viscosity for the model can be made on the basis of the following formula

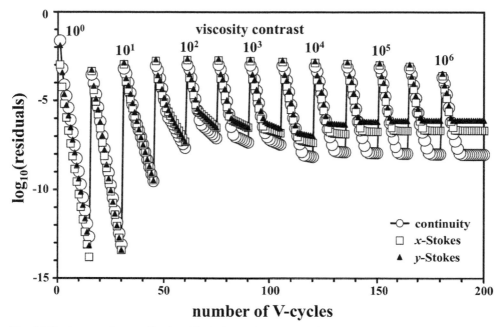

Fig. 18.12. Decay of normalized residuals for Stokes and continuity equations with the number of multigrid V-cycles for a model with variable viscosity. Residuals stabilize above the computer accuracy level. A 4-level multigrid with resolution 49×49 points on the finest level is used with relaxation parameters $\theta_{relaxation}^{continuity} - 0.3$ and $\theta_{relaxation}^{Stokes} - 0.9$. Numerical setup: rectangular block having higher density and viscosity (by a factor of 10^6) sinks in lower density and viscosity fluid. Iterations start from a hydrostatic pressure field and zero velocities and no viscosity contrast. Spikes in the solutions are caused by an increase in viscosity contrast by a factor of 3.333 every 15 multigrid cycles. Results are obtained with the program **Variable_viscosity_Multigrid_arbitrary.m**.

$$\eta_{i,j} = \eta_{min}^{computational} \exp\left[\frac{\ln\left(\eta_{max}^{computational}/\eta_{min}^{computational}\right)}{\ln\left(\eta_{max}^{original}/\eta_{min}^{original}\right)}\ln\left(\frac{\eta_{i,j}}{\eta_{min}^{original}}\right)\right], \tag{18.57}$$

where $\eta_{min}^{original}$, $\eta_{max}^{original}$ and $\eta_{min}^{computational}$, $\eta_{max}^{computational}$, are respectively the original and computational minimal and maximal viscosity for the model. An example of using such an algorithm (Fig. 18.12) is given in the program **Variable_viscosity_Multigrid_arbitrary. m**. It should however be mentioned that at large ($\gg 10^3$) and sharp (on one/few nodal points) viscosity contrasts, the accuracy of the multigrid solution is typically reduced compared to cases with lower viscosity contrast (see decreasing depth of residual minimization 'spikes' with increasing viscosity contrast in Fig. 18.12). A reasonably high level of accuracy (10^{-4}–10^{-7}) for such sharply inhomogeneous models can indeed be reached and the use of more complex multigrid schedules such as F- and W-cycles can also improve convergence. Time steps following the first time step typically do not require viscosity

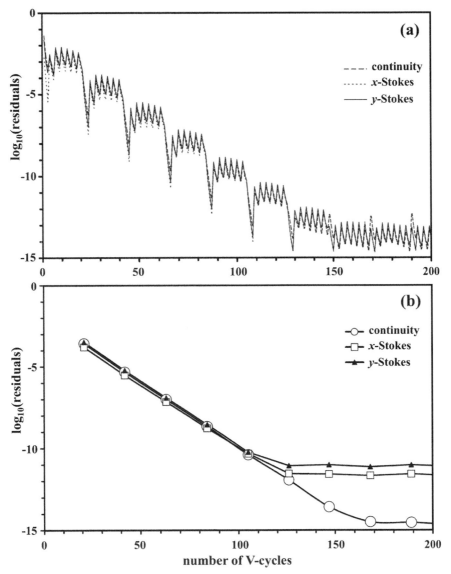

Fig. 18.13. Decay of normalized residuals for Stokes and continuity equations with the number of multigrid V-cycles for a model with variable viscosity in the case of a 'multi-multigrid' approach, which uses repetitive cycles of gradual increase in computational viscosity contrast. Residuals stabilize at the computer accuracy level. A 5-level multigrid with resolution 49×49 points on the finest level is used with relaxation parameters $\theta_{relaxation}^{continuity} = 0.3$ and $\theta_{relaxation}^{Stokes} = 0.9$. Numerical setup: rectangular block having higher density and viscosity (by a factor of 10^6) sinks into a lower density and viscosity fluid. Iterations start from a hydrostatic pressure field and zero velocities. (a) Decay of local residuals computed with current corrections and right hand side *within* each cycle (steps) of gradual viscosity contrast increase (spikes). (b) Decay of global residuals computed *after* each cycle of gradual viscosity contrast increase. Spikes in (a) are caused by an increase in viscosity contrast by a factor of 10 every 3 multigrid cycles. Results are obtained with the program **Variable_viscosity_MultiMultigrid_arbitrary.m**.

rescaling, as they have a much better initial guess for pressure and velocity. In addition, as discussed in Chapter 13, the numerical viscosity contrast can be efficiently decreased by using visco-elastic rheological models in which the upper limit of the *numerical viscosity* decreases proportionally with a decreasing computational time step (see Eqs. 13.5–13.9).

Another efficient possibility for improving convergence in the case of large viscosity contrasts is to use *repetitive cycles of gradual increase in a computational viscosity contrast*. In the beginning of each cycle (the first one excepted) residuals obtained for the finest grid level in the end of the previous cycle are assigned to the right hand side of respective equations *at the same finest level*. At the end of the cycle, corrections computed at the finest level are added to the global solution and new residuals are then computed at the finest level to be used in the next cycle.

This method is an art of '*multi-multigrid*' approach which uses a hierarchical representation of governing equations on the same numerical grid, which is analogous to the derivation of Eqs. (18.1)–(18.5). In many cases, this approach allows a computer accuracy solution to be reached within a finite number of iterations, even for very large viscosity contrasts (Fig. 18.13). An example of using such an algorithm (Fig. 18.13) is given in the program **Variable_viscosity_MultiMultigrid_arbitrary.m**.

Finally, it is also important to mention that besides large viscosity contrasts, further convergence problems can be caused by strongly irregular grid spacing, by significant differences in grid spacing used for different dimensions, by strong (e.g. plastic) localization of deformation characterizing the mechanical solution at the finest grid which is not captured on coarser levels etc. Therefore, be prepared that some of your thermomechanical models which utilize multigrid will be 'demanding', and will require special efforts in tuning and adjusting the iteration procedures.

Programming exercises

Exercise 18.1.
Program the multigrid solution based on a V-cycle for solving the Poisson equation in 2D for the case of a circular planetary body embedded in a massless-like medium (Eqs. 18.6–18.16, Figs. 18.6, 18.7). Use a ghost node approach to define the boundary conditions $\Phi=0$, along a circular boundary located at a distance from the planet. Program a Poisson equation smoother based on Gauss–Seidel iteration (Exercise 3.3 of Chapter 3) as an external MATLAB function and call it for different levels of resolution. Program external functions for restriction (Eq. 8.18, Fig. 8.8) and prolongation (Eq. 8.19, Fig. 8.9) operations to be called for different multigrid levels. Model parameters: radius of the planet 6000 km, density of the planet 6000 kg/m^3, model size 18 000×18 000 km^2, radius for gravity potential boundary 8999 km, number of resolution levels is 4, resolution on the coarsest (last) grid 7×7 nodal points, factor of increase in resolution between the levels is 2, relaxation coefficient for Gauss–Seidel iterations $\theta_{relaxation}^{Poisson} = 1.5$, number of smoothing iterations on the finest level is 5, factor of increase in the number of iterations with the level coarsening is 2. An example is provided in **Poisson_Multigrid_planet.m**.

Exercise 18.2.
Program a multigrid solution for solving the Stokes and continuity equations in 2D for a constant viscosity case using a pressure-velocity formulation and a ghost node approach (Eqs. 18.20–18.44, Figs. 18.8, 18.9). The setup corresponds to a dense rectangular block sinking in a lower density medium. Model parameters: model size 100×100 km^2, block size 20×20 km^2 (located in the middle of the model), block density 3100 kg/m^3, medium density 3000 kg/m^3, acceleration of gravity (vertical) 9.81 m/s^2, model viscosity 10^{20} Pa s, boundary conditions are free slip at all boundaries, number of resolution levels is 4, resolution of the basic grid on the coarsest (last) level 7×7 nodal points, factor of increase in resolution between the levels is 2, relaxation coefficients for Gauss–Seidel iterations $\theta^{stokes}_{relaxation} = 1.2$ and $\theta^{continuity}_{relaxation} = 0.3$, number of smoothing iterations on the finest (basic) level is 5, factor of increase in the number of iterations with the level coarsening is 2. An example is provided in **Constant_Viscosity_Multigrid_ghost.m**.

Exercise 18.3.
Modify the previous example to include a variable viscosity (Eqs. 18.45–18.55, Fig. 18.11). Use a high viscosity for the block (10^{23} Pa s) in comparison to the surrounding medium. Use $\theta^{Stokes}_{relaxation} = 1.0$ and $\theta^{continuity}_{relaxation} = 0.3$, and program a gradual increase in the viscosity contrast by a factor of 10$^{1/2}$ (Eq. 18.57) to reach an accurate solution (Fig. 18.12). An example is provided in **Variable_viscosity_Multigrid_arbitrary.m**.

19

Programming of 3D problems

> **Theory:** Formulation of thermomechanical problems in 3D and their numerical implementation. Methods for solving temperature, Poisson, momentum and continuity equations in 3D.
> **Programming:** Programming of numerical solutions for temperature and Poisson equations and coupled solving of momentum and continuity equations in 3D.

19.1 Why not simply always 3D?

We know very well that the Earth is a three-dimensional (3D), nearly spherical object and all dynamic processes inside our planet are thus intrinsically three dimensional. Therefore, it is very logical to assume that realistic geodynamic modelling should always be done in 3D. When you talk to geoscientists studying various natural geological objects, you are frequently told that such objects can only be modelled in 3D. This is a normal expectation since they are perfectly aware of the spatial 2D variability of geological structures on the Earth's surface and they thus know that a similar variability also exists in depth. Therefore, 3D modelling appears to be the natural choice for 'observers'. What about 'modellers'? Why don't they always use 3D modelling? What is wrong with it? The 'uncensored' truth about 3D modelling is the following.

- 3D thermomechanical modelling is quite easy from a methodological point of view – it is fairly straightforward to formulate and discretize the governing equations in a 3D Cartesian geometry for both simple viscous and more realistic visco-elasto-plastic rheologies (especially using the same relatively simple finite differences and marker-in-cell techniques that we have discussed extensively in this book).
- 3D modelling is much more difficult from a technical point of view. This mainly concerns the coupled solving of momentum and continuity equations. Highly accurate direct solvers that are applicable in 2D are too slow and consume too much memory to yield the same spatial resolution of numerical grids in 3D.

Iterative solvers, on the other hand, are very efficient for simple cases (such as constant viscosity problems) but do not always converge well for realistic geodynamic problems, which involve strain localization and large viscosity contrasts on sharp boundaries.

It is obvious that when changing from 2D to 3D models, *we do not want to reduce the numerical resolution*. This requirement, however, immediately implies a several orders of magnitude increase in the number of grid points and markers and, thus, in the number of equations that have to be solved.

- A 100×100 2D grid with 5×5 markers per cell implies around $3 \times 100 \times 100 = 30\,000$ momentum and continuity equations to be solved and $5 \times 5 \times 100 \times 100 = 250\,000$ markers to be followed at each time step.
- A $100 \times 100 \times 100$ 3D grid with $5 \times 5 \times 5$ markers per cell involves about $4 \times 100 \times 100 \times 100 = 4\,000\,000$ momentum and continuity equations to be solved (i.e. 130 times more than in 2D) and $5 \times 5 \times 5 \times 100 \times 100 \times 100 = 125\,000\,000$ markers to be followed at each time step (i.e. 500 times more than in 2D).

This is the main reason why modellers prefer to apply 2D rather then 3D approaches, where geodynamic problems can be justifiably simplified to lower dimensions. Logically, adaptive mesh refinement methods (AMR, Chapter 17) can also play an increasingly important role in 3D numerical geodynamic modelling (e.g. Albers, 2000; Stadler et al., 2010; Kronbichler et al., 2012; Rudi et al., 2015; Heister et al., 2017 and references therein). This field is developing very actively and significant progress has already been achieved in the field of mantle convection in both Cartesian and spherical geometry, in the large scale modelling of plate tectonics processes and in some other directions. Several groups are working on the development of more efficient and universal all-in-one 3D numerical geodynamic codes and 3D thermomechanical modelling has already become a standard tool in all fields of computational geodynamics.

This chapter gives a practical summary that allows a relatively simple implementation of 3D thermomechanical modelling, based on conservative finite differences and marker-in-cell techniques combined with iterative *multigrid* solvers similar to those discussed in Chapter 18 for 2D problems.

19.2 3D staggered grid and discretization of momentum, continuity, temperature and Poisson equations

Let us first discuss the discretization of various equations in 3D. Figure 19.1 shows an elementary volume (cell) of a 3D staggered grid that can be used for discretization of momentum, continuity, Poisson and temperature equations in the case of viscous flow with variable viscosity and variable thermal conductivity. The grid is constructed in a specific

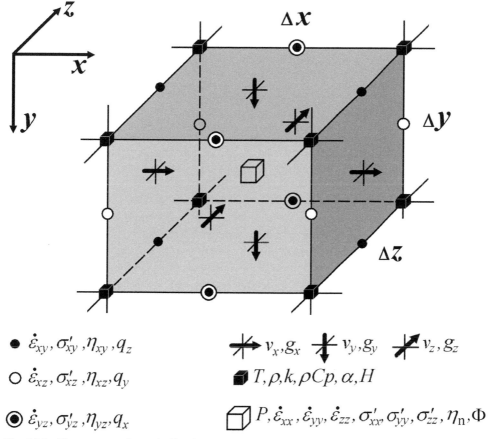

$$\bullet \quad \dot{\varepsilon}_{xy}, \sigma'_{xy}, \eta_{xy}, q_z$$

$$\circ \quad \dot{\varepsilon}_{xz}, \sigma'_{xz}, \eta_{xz}, q_y$$

$$\odot \quad \dot{\varepsilon}_{yz}, \sigma'_{yz}, \eta_{yz}, q_x$$

$$v_x, g_x \quad v_y, g_y \quad v_z, g_z$$

$$T, \rho, k, \rho Cp, \alpha, H$$

$$P, \dot{\varepsilon}_{xx}, \dot{\varepsilon}_{yy}, \dot{\varepsilon}_{zz}, \sigma'_{xx}, \sigma'_{yy}, \sigma'_{zz}, \eta_n, \Phi$$

Fig. 19.1. Elementary volume (cell) of a 3D staggered grid used for discretization of momentum, continuity, Poisson and temperature equations in the case of incompressible viscous flow with variable viscosity and thermal conductivity.

way that allows a natural representation of all governing equations with conservative finite differences:

- pressure, deviatoric normal stresses and strain rates and gravity potential (when needed) are located at the centre of the cell,
- components of the velocity vector v_x, v_y and v_z and variable gravitational acceleration vector g_x, g_y and g_z (when needed) are located in the middle of the faces orthogonal to the x, y and z axes, respectively,
- shear stresses and strain rates are located in the middle of the edges formed by the intersection of faces containing respective velocity components, i.e. by intersection of v_x and v_y faces in the case of σ'_{xy}, $\dot{\varepsilon}_{xy}$ etc,

- viscosity is defined in four different places corresponding to positions of normal (η_n, in the centre of the cell) and shear (η_{xy}, η_{xz}, η_{yz}, in the middle of respective edges) stress components,
- heat fluxes q_x, q_y and q_z are located in the middle of edges parallel to the x, y and z axes, respectively,
- other material properties and temperature are located at the cell corners, which are the basic nodes of the grid.

Before discretizing the governing equations on a 3D staggered grid, an important step is (although it might sound really boring . . .) *to properly understand the indexing* of different field variables located around a grid cell (Fig. 19.1):

- Indexing of variables located in the basic nodes (T, ρ, α, ρC_P etc.) of the grid is simple since the respective arrays have dimension of $N_x \times N_y \times N_z$, where N_x, N_y and N_z are the numbers of nodes of the basic grid in the respective directions.
- Arrays for the variables located in cell centres (P, σ'_{xx}, η_n, Φ, etc.) will be $(N_x - 1) \times (N_y - 1) \times (N_z - 1)$.
- Arrays for various shear stresses, strain rates and respective viscosity values located on cell edges will be $N_x \times N_y \times (N_z - 1)$ for σ'_{xy}, $\dot\varepsilon_{xy}$ and η_{xy}, $N_x \times (N_y - 1) \times N_z$ for σ'_{xz}, $\dot\varepsilon_{xz}$ and η_{xz}, $(N_x - 1) \times N_y \times N_z$ for σ'_{yz}, $\dot\varepsilon_{yz}$ and η_{yz}.
- Finally, the indexing of the velocity nodes should take into account nodes located outside the basic grid, which are used for formulating boundary conditions and interpolation of velocity components to markers (similarly to ones that we discussed in Chapter 18 for 2D grids, Fig. 18.8). Consequently, velocity arrays will be larger in two directions compared to the basic grid resolution (Fig. 19.2): $N_x \times (N_y + 1) \times (N_z + 1)$ for v_x, $(N_x + 1) \times N_y \times (N_z + 1)$ for v_y and $(N_x + 1) \times (N_y + 1) \times N_z$ for v_z.

Based on these considerations, we can now understand the logic of indexing for various grid points (Fig. 19.3), which will then be used to construct conservative 3D finite difference schemes for the momentum, continuity, Poisson and temperature equations.

After extensive discussions on composing conservative FD schemes in 1D and 2D, the discretization of various equations on a 3D staggered grid shown in Figs. 19.1–19.3 is quite straightforward and therefore we only discuss it briefly.

The representation of the incompressible 3D continuity equation on a stencil with six velocity nodes around a cell (Fig. 19.4) is

$$\frac{v_{x(i+1,j+1,l+1)} - v_{x(i+1,j,l+1)}}{\Delta x_{j+1/2}} + \frac{v_{y(i+1,j+1,l+1)} - v_{y(i,j+1,l+1)}}{\Delta y_{i+1/2}} + \frac{v_{z(i+1,j+1,l+1)} - v_{z(i+1,j+1,l)}}{\Delta z_{l+1/2}} = 0,$$

$$(19.1)$$

where i, j and l are indexes in respectively the y, x and z directions.

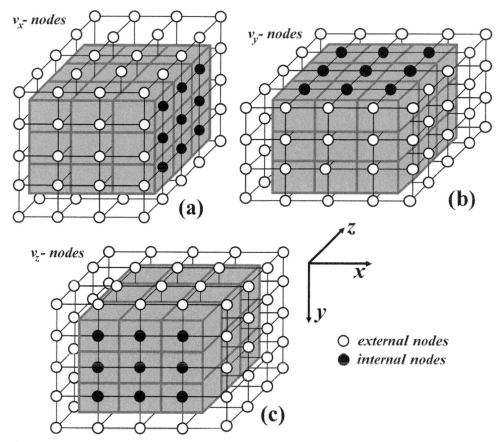

Fig. 19.2. Distribution of various velocity nodal points in 3D in the case when external nodes (open circles) are used to formulate boundary conditions for v_x (a), v_y (b) and v_z (c) velocity components. The basic grid of the model (see solid lines in Fig. 19.1) is shown in grey.

Discretization of the *3D Stokes equation* for an *incompressible* fluid with variable viscosity uses a stencil containing 15 velocity nodes and 2 pressure nodes. An example of this stencil is shown in Fig. 19.5 for the x Stokes equation

$$\frac{\partial \sigma'_{xx}}{\partial x} + \frac{\partial \sigma'_{xy}}{\partial y} + \frac{\partial \sigma'_{xz}}{\partial z} - 2\frac{P_{i-1,j,l-1} - P_{i-1,j-1,l-1}}{\Delta x_{j-1/2} + \Delta x_{j+1/2}} = \frac{1}{4}\left(\rho_{i-1,j,l-1} + \rho_{i,j,l-1} + \rho_{i-1,j,l} + \rho_{i,j,l}\right)g_x,$$

$$(19.2)$$

$$\frac{\partial \sigma'_{xx}}{\partial x} = 4\eta_{n(i-1,j,l-1)}\frac{v_{x(i,j+1,l)} - v_{x(i,j,l)}}{\Delta x_{j+1/2}\left(\Delta x_{j-1/2} + \Delta x_{j+1/2}\right)} - 4\eta_{n(i-1,j-1,l-1)}\frac{v_{x(i,j,l)} - v_{x(i,j-1,l)}}{\Delta x_{j-1/2}\left(\Delta x_{j-1/2} + \Delta x_{j+1/2}\right)},$$

$$(19.3)$$

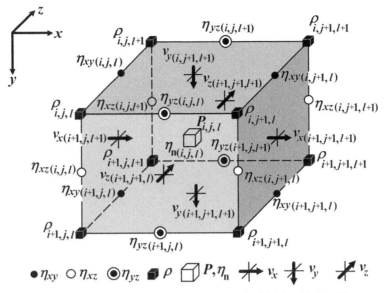

Fig. 19.3. Indexing of different variables for a 3D staggered grid (Fig. 19.1) with external velocity nodes (Fig. 19.2).

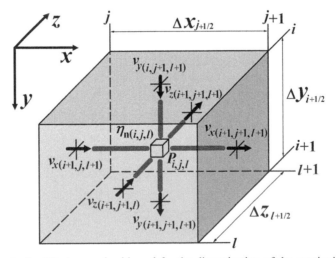

Fig. 19.4. Stencil of a 3D staggered grid used for the discretization of the continuity equation for iterative solution. The small open cube in the centre corresponds to the pressure node at which the continuity equation is formulated. Notation of different nodal points is as in Fig. 19.1. Indexing of different variables corresponds to a 3D staggered grid with external velocity nodes (Fig. 19.3).

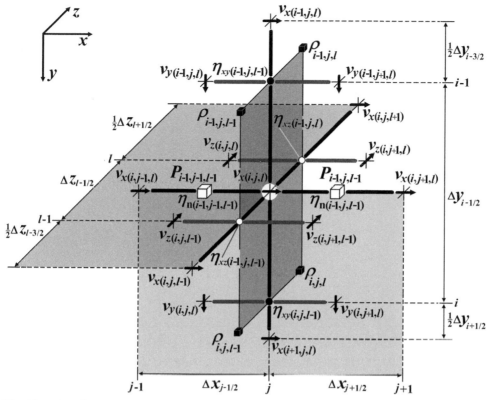

Fig. 19.5. Stencil of a 3D staggered grid used for the discretization of the x Stokes equations with variable viscosity. The white circle in the centre corresponds to a horizontal velocity node for which the x Stokes equation is formulated. The notation of different nodal points is as in Fig. 19.1. Indexing of different variables corresponds to a 3D staggered grid with external velocity nodes (Fig. 19.3).

$$\frac{\partial \sigma'_{xy}}{\partial y} = 2\eta_{xy(i,j,l-1)}\left(\frac{v_{x(i+1,j,l)} - v_{x(i,j,l)}}{\Delta y_{i-1/2}\left(\Delta y_{i-1/2} + \Delta y_{i+1/2}\right)} + \frac{v_{y(i,j+1,l)} - v_{y(i,j,l)}}{\Delta y_{i-1/2}\left(\Delta x_{j-1/2} + \Delta x_{j+1/2}\right)} \right)$$
$$-2\eta_{xy(i-1,j,l-1)}\left(\frac{v_{x(i,j,l)} - v_{x(i-1,j,l)}}{\Delta y_{i-1/2}\left(\Delta y_{i-3/2} + \Delta y_{i-1/2}\right)} + \frac{v_{y(i-1,j+1,l)} - v_{y(i-1,j,l)}}{\Delta y_{i-1/2}\left(\Delta x_{j-1/2} + \Delta x_{j+1/2}\right)} \right). \tag{19.4}$$

$$\frac{\partial \sigma'_{xz}}{\partial z} = 2\eta_{xz(i-1,j,l)}\left(\frac{v_{x(i,j,l+1)} - v_{x(i,j,l)}}{\Delta z_{l-1/2}\left(\Delta z_{l-1/2} + \Delta z_{l+1/2}\right)} + \frac{v_{z(i,j+1,l)} - v_{z(i,j,l)}}{\Delta z_{l-1/2}\left(\Delta x_{j-1/2} + \Delta x_{j+1/2}\right)} \right)$$
$$-2\eta_{xz(i-1,j,l-1)}\left(\frac{v_{x(i,j,l)} - v_{x(i,j,l-1)}}{\Delta z_{l-1/2}\left(\Delta z_{l-3/2} + \Delta z_{l-1/2}\right)} + \frac{v_{z(i,j+1,l-1)} - v_{z(i,j,l-1)}}{\Delta z_{l-1/2}\left(\Delta x_{j-1/2} + \Delta x_{j+1/2}\right)} \right). \tag{19.5}$$

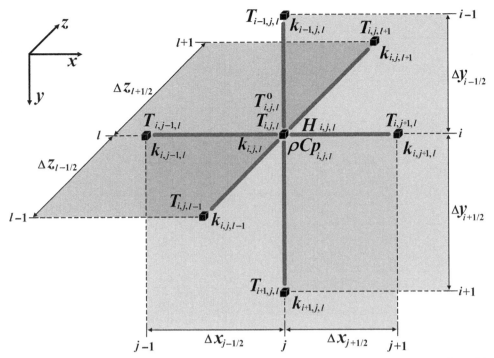

Fig. 19.6. Stencil of a 3D staggered grid used for discretization of the temperature equation with variable thermal conductivity. The temperature equation is formulated for the central node $T_{i,j,l}$ which is one of the basic nodes of the 3D staggered grid (Fig. 19.1). The notation of different nodal points is the same as in Fig. 19.1. Indexing of different variables corresponds to a 3D staggered grid with external velocity nodes (Fig. 19.3).

Discretization of the y Stokes and z Stokes equations is rather obvious and the respective conservative FD schemes can be constructed and indexed by analogy to Fig. 19.5 and Eqs. (19.2)–(19.5) (derive as an exercise).

Given that temperature advection is solved with markers (see Chapter 11), discretization of the *3D temperature equation* with a variable thermal conductivity can be done in a simple Lagrangian form, which does not include advective terms and uses a stencil with 7 temperature nodes (7-point cross, Fig. 19.6). In implicit form, the conservative FD can be written as follows

$$\rho Cp_{i,j,l}\frac{T_{i,j,l} - T^o_{i,j,l}}{\Delta t} + \frac{\partial q_x}{\partial x} + \frac{\partial q_y}{\partial y} + \frac{\partial q_z}{\partial z} = H_{i,j,l}, \tag{19.6}$$

$$\frac{\partial q_x}{\partial x} = \frac{\left(k_{i,j-1,l} + k_{i,j,l}\right)\left(T_{i,j,l} - T_{(i,j-1,l)}\right)}{\Delta x_{j-1/2}\left(\Delta x_{j-1/2} + \Delta x_{j+1/2}\right)} - \frac{\left(k_{i,j,l} + k_{i,j+1,l}\right)\left(T_{i,j+1,l} - T_{i,j,l}\right)}{\Delta x_{j+1/2}\left(\Delta x_{j-1/2} + \Delta x_{j+1/2}\right)}, \tag{19.7}$$

$$\frac{\partial q_y}{\partial y} = \frac{(k_{i-1,j,l} + k_{i,j,l})(T_{i,j,l} - T_{i-1,j,l})}{\Delta y_{i-1/2}(\Delta y_{i-1/2} + \Delta y_{i+1/2})} - \frac{(k_{i,j,l} + k_{i+1,j,l})(T_{i+1,j,l} - T_{i,j,l})}{\Delta y_{i+1/2}(\Delta y_{i-1/2} + \Delta y_{i+1/2})}, \tag{19.8}$$

$$\frac{\partial q_z}{\partial z} = \frac{(k_{i,j,l-1} + k_{i,j,l})(T_{i,j,l} - T_{i,j,l-1})}{\Delta z_{l-1/2}(\Delta z_{l-1/2} + \Delta z_{l+1/2})} - \frac{(k_{i,j,l} + k_{i,j,l+1})(T_{i,j,l+1} - T_{i,j,l})}{\Delta z_{l+1/2}(\Delta z_{l-1/2} + \Delta z_{l+1/2})}, \tag{19.9}$$

where $T^o_{i,j,l}$ is temperature for the current moment of time (t) in the central *ijl*th node of the cross and $T_{i,j,l-1}$, $T_{i,j,l+1}$, etc. are temperatures in seven nodal points for the next moment of time ($t + \Delta t$), $k_{i,j,l-1}$, $k_{i,j,l+1}$, etc. stand for thermal conductivity that can be different at different nodes, $\rho Cp_{i,j,l-1}$, and $H_{i,j,l-1}$ are volumetric isobaric heat capacity and heat production values for the central node of the cross, respectively. It should be noted that temperature points could alternatively be placed in centres of cells (i.e., similarly to pressure points, Chapter 11). In this case, heat flux components will be located in respective velocity nodes and external temperature points located outside of the grid will be needed to formulate thermal boundary conditions.

As for the temperature equation, the discretization of the *3D Poisson equation* is also based on a 7-point cross (Fig. 19.7)

$$\frac{\partial^2 \Phi}{\partial x^2} + \frac{\partial^2 \Phi}{\partial y^2} + \frac{\partial^2 \Phi}{\partial z^2}$$
$$= \pi G\Big(\rho_{i,j,l} + \rho_{i+1,j,l} + \rho_{i,j+1,l} + \rho_{i+1,j+1,l} + \rho_{i,j,l+1} + \rho_{i+1,j,l+1} + \rho_{i,j+1,l+1} + \rho_{i+1,j+1,l+1}\Big)/2, \tag{19.10}$$

$$\frac{\partial^2 \Phi}{\partial x^2} = 2\frac{\Phi_{i,j+1,l} - \Phi_{i,j,l}}{\Delta x_{j+1}(\Delta x_j + \Delta x_{j+1})} - 2\frac{\Phi_{i,j,l} - \Phi_{i,j-1,l}}{\Delta x_j(\Delta x_j + \Delta x_{j+1})}, \tag{19.11}$$

$$\frac{\partial^2 \Phi}{\partial y^2} = 2\frac{\Phi_{i+1,j,l)} - \Phi_{i,j,l}}{\Delta y_{i+1}(\Delta y_i + \Delta y_{i+1})} - 2\frac{\Phi_{i,j,l} - \Phi_{i-1,j,l}}{\Delta y_i(\Delta y_i + \Delta y_{i+1})}, \tag{19.12}$$

$$\frac{\partial^2 \Phi}{\partial z^2} = 2\frac{\Phi_{i,j,l+1} - \Phi_{i,j,l}}{\Delta z_{l+1}(\Delta z_l + \Delta z_{l+1})} - 2\frac{\Phi_{i,j,l} - \Phi_{i,j,l-1}}{\Delta z_l(\Delta z_l + \Delta z_{l+1})}. \tag{19.13}$$

The density at the central node of the cross in Eq. (19.10) is computed as an arithmetic average from eight surrounding basic nodes. Alternatively, additional density points can be defined in the same gravity potential nodes (i.e. in the centre of cells, Fig. 19.1) and respective density values can be separately interpolated from markers when deformation of a self-gravitating body is modelled in 3D.

Finally, *interpolation between markers and nodes in 3D* is also based on the same principles as those discussed for 2D interpolation in Chapter 10. It can be done with the following standard first order of accuracy *trilinear* interpolation schemes (Fig. 19.8):

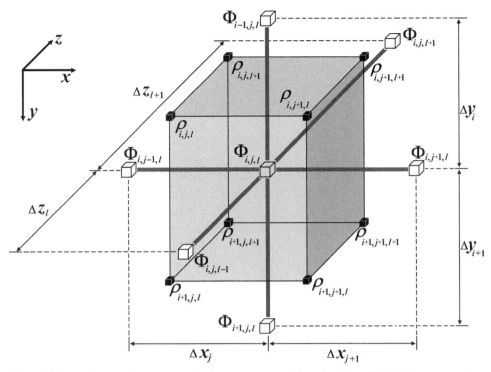

Fig. 19.7. Stencil of a 3D staggered grid used for the discretization of the Poisson equation. The Poisson equation is formulated for the central node $\Phi_{i,j,l}$ located in one of the cell centres (pressure nodes) of the 3D staggered grid (Fig. 19.1). The notation of different nodal points is the same as in Fig. 19.1. Indexing of different variables corresponds to a 3D staggered grid with external velocity nodes (Fig. 19.3).

interpolation to the *ijl*th node from markers found in 8 cells surrounding this node

$$B_{(i,j,l)} = \frac{\sum_m B_m w_{m(i,j,l)}}{\sum_m w_{m(i,j,l)}}, \tag{19.14}$$

interpolation to a marker in a cell from 8 nodes surrounding the cell

$$B_m = B_{i,j,l} w_{m(i,j,l)} + B_{i-1,j,l} w_{m(i-1,j,l)} + B_{i,j-1,l} w_{m(i,j-1,l)} + B_{i-1,j-1,l} w_{m(i-1,j-1,l)}$$
$$+ B_{i,j,l-1} w_{m(i,j,l-1)} + B_{i-1,j,l-1} w_{m(i-1,j,l-1)} + B_{i,j-1,l} w_{m(i,j-1,l-1)}$$
$$+ B_{i-1,j-1,l-1} w_{m(i-1,j-1,l-1)} . \tag{19.15}$$

The statistical weight of the *m*th marker for the *ijl*th node depends on Δx_m, Δy_m, Δz_m distances to this node as

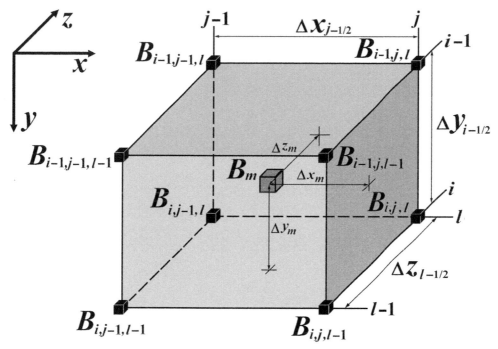

Fig. 19.8. Stencil of a 3D grid (8-node cell) used for the interpolation between marker (grey cube) and nodes (black cubes).

$$w_{m(i,j,l)} = \left(1 - \frac{\Lambda x_m}{\Delta x_{j-1/2}}\right) \times \left(1 - \frac{\Delta y_m}{\Delta y_{i-1/2}}\right) \times \left(1 - \frac{\Delta z_m}{\Delta z_{l-1/2}}\right). \tag{19.16}$$

19.3 Solving discretized 3D equations

After discussing discretization of various equations required for 3D thermomechanical geodynamic modelling, let us consider methods for solving these equations. All approaches considered here are iterative and based on the *Gauss–Seidel method*, which is more efficient than Jacobi iteration for the numerical problems of our interest (Chapter 3).

Temperature equation. Efficient solving of the time-dependent non-steady-state 3D temperature equation (19.6) does not require a multigrid since, in contrast to steady-state Stokes and Poisson equations, the time-dependent solutions for temperature are controlled locally by local heat fluxes rather than globally and can be implemented on the basis of a Gauss–Seidel iteration with a relatively large relaxation parameter $\theta_{relaxation}^{temperature}$ that ranges from 0.5 to 1.5. The respective iterative temperature update scheme *for a regularly spaced grid* based on Eqs. (19.6)–(19.9) is as follows:

$$T_{i,j,l}^{new} = T_{i,j,l} + \frac{\Delta R_{i,j,l}^{temperature}}{C_{T(i,j,l)}} \theta_{relaxation}^{temperature} , \qquad (19.17)$$

$$\Delta R_{i,j,l}^{temperature} = H_{i,j,l} - \rho C_{Pi,j,l} \frac{T_{i,j,l} - T_{i,j,l}^{o}}{\Delta t} - \frac{\partial q_x}{\partial x} - \frac{\partial q_y}{\partial y} - \frac{\partial q_z}{\partial z} , \qquad (19.18)$$

$$\frac{\partial q_x}{\partial x} = \frac{\left(k_{i,j,l} + k_{i,j-1,l}\right)\left(T_{i,j,l} - T_{i,j-1,l}\right) - \left(k_{i,j+1,l} + k_{i,j,l}\right)\left(T_{i,j+1,l} - T_{i,j,l}\right)}{2\Delta x^2} , \qquad (19.19)$$

$$\frac{\partial q_y}{\partial y} = \frac{\left(k_{i,j,l} + k_{i-1,j,l}\right)\left(T_{i,j,l} - T_{i-1,j,l}\right) - \left(k_{i+1,j,l} + k_{i,j,l}\right)\left(T_{i+1,j,l} - T_{i,j,l}\right)}{2\Delta y^2} , \qquad (19.20)$$

$$\frac{\partial q_z}{\partial z} = \frac{\left(k_{i,j,l} + k_{i,j,l-1}\right)\left(T_{i,j,l} - T_{i,j,l-1}\right) - \left(k_{i,j,l+1} + k_{i,j,l}\right)\left(T_{i,j,l+1} - T_{i,j,l}\right)}{2\Delta z^2} , \qquad (19.21)$$

$$C_{T(i,j,l)} = \frac{\rho C p_{i,j,l}}{\Delta t} + \frac{k_{i,j-1,l} + 2k_{i,j,l} + k_{i,j+1,l}}{2\Delta x^2} + \frac{k_{i-1,j,l} + 2k_{i,j,l} + k_{i+1,j,l}}{2\Delta y^2} + \frac{k_{i,j,l-1} + 2k_{i,j,l} + k_{i,j,l+1}}{2\Delta z^2} , \qquad (19.22)$$

where Δx, Δy, and Δz are regular grid steps in respective directions, $T_{i,j,l}^{o}$ is the temperature at the ijlth node for the current moment of time (t), $T_{i,j,l}$ and $T_{i,j,l}^{new}$ are old and new (updated) values of temperature at the ijlth node for the next moment of time ($t + \Delta t$). To satisfy the boundary condition equations, these equations are called for marginal temperature nodes in the same Gauss–Seidel iteration cycle

upper boundary ($i = 1$)

> no vertical heat flux $T_{i=1,j,l} = T_{i=2,j,l}$
>
> constant temperature $T_{i=1,j,l} = T_{top}$

left boundary ($j = 1$)

> no horizontal heat flux $T_{i,j=1,l} = T_{i,j=2,l}$
>
> constant temperature $T_{i,j=1,l} = T_{left}$

front boundary ($l = 1$)

> no lateral heat flux $T_{i,j,l=1} = T_{i,j,l=2}$
>
> constant temperature $T_{i,j,l=1} = T_{front}$

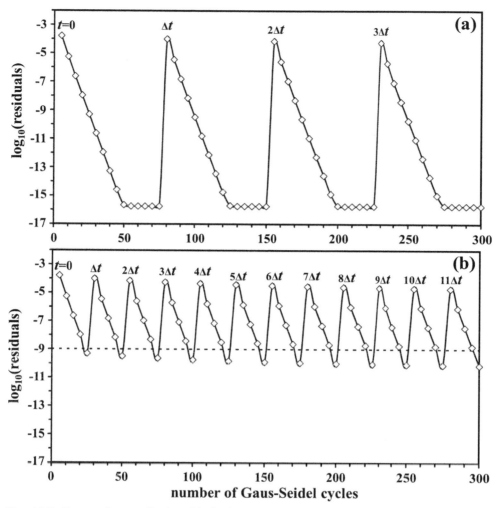

Fig. 19.9. Decay of normalized residuals for the temperature equations with the number of Gauss–Seidel cycles for a 3D model with variable thermal conductivity. Residuals in (a) stabilize at computer accuracy level (10^{-16}) after around 50 iterations (per time step). Iterations in (b) are terminated after reaching a given level of tolerance (10^{-9}) that takes around 25 iterations per time step. Model resolution is $51 \times 51 \times 51$ points with regularly spaced grid. The relaxation parameter $\theta_{relaxation}^{temperature} = 1.25$ is used (in Eq. 19.17). Numerical setup: rectangular block having a higher temperature is placed in a lower temperature surrounding. Results are obtained with the program **Temperature3D_Gauss_Seidel.m**.

and similarly for other boundaries. An example implementation of the above algorithm is given in the program **Temperature3D_Gauss_Seidel.m**. As can be seen from Fig. 19.9, Gauss–Seidel iterations allow high accuracy solutions to be obtained with 10–20 iterations per time step, whereas a computer accuracy solution is obtained with a few tens of iterations.

Poisson equation. As in 2D (Chapter 18), an efficient solution of the 3D Poisson equation (19.10) can be based on a multigrid approach, that can again be implemented on the basis of Gauss–Seidel iterations with a relatively high relaxation parameter $\theta_{relaxation}^{Poisson}$ ranging from 0.5 to 1.5. The respective iterative gravity potential update scheme *for a regularly spaced grid* based on Eqs. (19.10)–(19.13) is as follows:

$$\Phi_{i,j,l}^{new} = \Phi_{i,j,l} + \frac{\Delta R_{i,j,l}}{C_{\Phi(i,j,l)}} \theta_{relaxation}^{Poisson}, \tag{19.23}$$

$$\Delta R_{i,j,l}^{Poisson} = R_{i,j,l}^{Poisson} - \frac{\partial^2 \Phi}{\partial x^2} - \frac{\partial^2 \Phi}{\partial y^2} - \frac{\partial^2 \Phi}{\partial z^2}, \tag{19.24}$$

$$\frac{\partial^2 \Phi}{\partial x^2} = \frac{\Phi_{i,j-1,l} - 2\Phi_{i,j,l} + \Phi_{i,j+1,l}}{\Delta x^2}, \tag{19.25}$$

$$\frac{\partial^2 \Phi}{\partial y^2} = \frac{\Phi_{i-1,j,l} - 2\Phi_{i,j,l} + \Phi_{i+1,j,l}}{\Delta y^2}, \tag{19.26}$$

$$\frac{\partial^2 \Phi}{\partial z^2} = \frac{\Phi_{i,j,l-1} - 2\Phi_{i,j,l} + \Phi_{i,j,l+1}}{\Delta z^2}. \tag{19.27}$$

$$C_{\Phi(i,j,l)} = -\frac{2}{\Delta x^2} - \frac{2}{\Delta y^2} - \frac{2}{\Delta z^2}. \tag{19.28}$$

$\Phi_{i,j-1,l}$, $\Phi_{i-1,j,l}$, etc. are current values of either gravity potential (at finest level) or corrections for this potential (at coarser levels) at respective nodal points, $C_{\Phi(i,j,l)}$ is the coefficient at $\Phi_{i,j,l}$ in the discretized Poisson equation, $\Delta R_{i,j,l}$ is the current residual and $R_{i,j,l}$ is the right hand side of the Poisson equation. On the finest principal level of resolution, the right hand side is computed from the standard equation

$$R_{i,j,l} = 4\pi G \rho_{\Phi(i,j,l)}, \tag{19.29}$$

where G is the gravitational constant and $\rho_{\Phi(i,j,l)}$ is the density defined at the same location as $\Phi_{i,j,l}$ (alternatively use Eq. 19.10 when the gravity potential and density are defined at different points). For coarser levels, $R_{i,j,l}$ is composed of residuals interpolated from finer levels. Obviously, grid steps Δx, Δy and Δz are different at different levels of resolution. Various boundary conditions are defined as in 2D (see Fig. 18.7, Eqs. 18.12–18.16). An example implementation of the above algorithm, for the case of a self-gravitating planet and gravity potential boundary condition defined on the internal spherical surface inside the grid, is given in the program **Poisson3D_Multigrid_planet_arbitrary.m**. As can be seen from Fig. 19.10, multigrid allows a computer accuracy solution to be obtained within around 10 V-cycles, independent of the model resolution.

Momentum and continuity equations. As in 2D (Chapter 18), the efficient solution of the 3D Stokes and continuity equations (19.1–19.5) can be based on a multigrid approach that can be implemented on the basis of Gauss–Seidel iterations with pressure updates

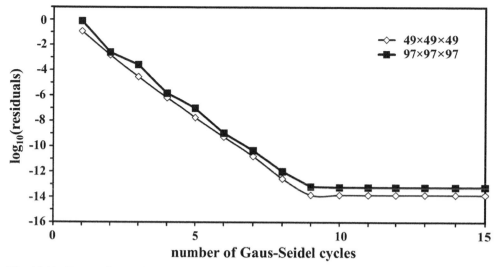

Fig. 19.10. Decay of normalized residuals for the 3D Poisson equation at various resolutions versus number of multigrid V-cycles for a model with a spherical planet embedded in a massless medium (space). Residuals stabilize at computer accuracy level. 5- and 6-level multigrids with resolutions respectively of $49 \times 49 \times 49$ (open diamonds) and $97 \times 97 \times 97$ (solid squares) nodes on the finest level are used with a relaxation parameter $\theta^{Poisson}_{relaxation} = 1.5$. Results are obtained with the program **Poisson3D_Multigrid_planet_arbitrary.m**.

computed from local divergence scaled to local viscosity. The respective iterative pressure and velocity update schemes *for a regularly spaced staggered grid* can be derived on the basis of Eqs. (19.1)–(19.5):

$$P^{new}_{i,j,l} = P_{i,j,l} + \eta_{n(i,j,l)}\Delta R^{continuity}_{i,j,l} \theta^{continuity}_{relaxation}, \tag{19.30}$$

$$\Delta R^{continuity}_{i,j,l} = R^{continuity}_{i,j,l} - \frac{\partial v_x}{\partial x} - \frac{\partial v_y}{\partial y} - \frac{\partial v_z}{\partial z}, \tag{19.31}$$

$$v^{new}_{x(i,j,l)} = v_{x(i,j,l)} + \frac{\Delta R^{x-Stokes}_{i,j,l}}{C_{v_x(i,j,l)}} \theta^{Stokes}_{relaxation}, \tag{19.32}$$

$$v^{new}_{y(i,j,l)} = v_{y(i,j,l)} + \frac{\Delta R^{y-Stokes}_{i,j,l}}{C_{v_y(i,j,l)}} \theta^{Stokes}_{relaxation}, \tag{19.33}$$

$$v^{new}_{z(i,j,l)} = v_{y(i,j,l)} + \frac{\Delta R^{z-Stokes}_{i,j,l}}{C_{v_z(i,j,l)}} \theta^{Stokes}_{relaxation}. \tag{19.34}$$

For models with constant viscosity η, the respective residuals and coefficients in Eqs. (19.30)–(19.34) become

$$\Delta R_{i,j,l}^{x-Stokes} = R_{i,j,l}^{x-Stokes} - \eta \left(\frac{\partial^2 v_x}{\partial x^2} + \frac{\partial^2 v_x}{\partial y^2} + \frac{\partial^2 v_x}{\partial z^2} \right) + \frac{\partial P}{\partial x}, \qquad (19.35)$$

$$\Delta R_{i,j,l}^{y-Stokes} = R_{i,j,l}^{y-Stokes} - \eta \left(\frac{\partial^2 v_y}{\partial x^2} + \frac{\partial^2 v_y}{\partial y^2} + \frac{\partial^2 v_y}{\partial z^2} \right) + \frac{\partial P}{\partial y}, \qquad (19.36)$$

$$\Delta R_{i,j,l}^{z-Stokes} = R_{i,j,l}^{z-Stokes} - \eta \left(\frac{\partial^2 v_z}{\partial x^2} + \frac{\partial^2 v_z}{\partial y^2} + \frac{\partial^2 v_z}{\partial z^2} \right) + \frac{\partial P}{\partial z}, \qquad (19.37)$$

$$\eta_{n(i,j,l)} = \eta, \qquad (19.38)$$

$$\frac{\partial^2 v_x}{\partial x^2} = \frac{v_{x(i,j-1,l)} - 2v_{x(i,j,l)} + v_{x(i,j+1,l)}}{\Delta x^2}, \ (\text{Fig. 19.5}) \qquad (19.39)$$

$$\frac{\partial^2 v_x}{\partial y^2} = \frac{v_{x(i-1,j,l)} - 2v_{x(i,j,l)} + v_{x(i+1,j,l)}}{\Delta y^2}, \ (\text{Fig. 19.5}) \qquad (19.40)$$

$$\frac{\partial^2 v_x}{\partial z^2} = \frac{v_{x(i,j,l-1)} - 2v_{x(i,j,l)} + v_{x(i,j,l+1)}}{\Delta z^2}, \ (\text{Fig. 19.5}) \qquad (19.41)$$

$$\frac{\partial P}{\partial x} = \frac{P_{i-1,j,l-1} - P_{i-1,j-1,l-1}}{\Delta x}, \ (\text{Fig. 19.5}) \qquad (19.42)$$

$$C_{v_x(i,j,l)} = -\frac{2\eta}{\Delta x^2} - \frac{2\eta}{\Delta y^2} - \frac{2\eta}{\Delta z^2}, \qquad (19.43)$$

and other terms can be derived in a similar manner (derive as an exercise).

For models with variable viscosity, the respective residuals and coefficients in Eqs. (19.30)–(19.34) become

$$\Delta R_{i,j,l}^{x-Stokes} = R_{i,j,l}^{x-Stokes} - \frac{\partial \sigma_{xx}'}{\partial x} - \frac{\partial \sigma_{xy}'}{\partial y} - \frac{\partial \sigma_{xz}'}{\partial z} + \frac{\partial P}{\partial x}, \qquad (19.44)$$

$$\Delta R_{i,j,l}^{y-Stokes} = R_{i,j,l}^{y-Stokes} - \frac{\partial \sigma_{yy}'}{\partial y} - \frac{\partial \sigma_{yx}'}{\partial x} - \frac{\partial \sigma_{yz}'}{\partial z} + \frac{\partial P}{\partial y}, \qquad (19.45)$$

$$\Delta R_{i,j,l}^{z-Stokes} = R_{i,j,l}^{z-Stokes} - \frac{\partial \sigma_{zz}'}{\partial z} - \frac{\partial \sigma_{zx}'}{\partial x} - \frac{\partial \sigma_{zy}'}{\partial y} + \frac{\partial P}{\partial z}, \qquad (19.46)$$

$$\frac{\partial \sigma_{xx}'}{\partial x} = 2\eta_{n(i-1,j,l-1)} \frac{v_{x(i,j+1,l)} - v_{x(i,j,l)}}{\Delta x^2} - 2\eta_{n(i-1,j-1,l-1)} \frac{v_{x(i,j,l)} - v_{x(i,j-1,l)}}{\Delta x^2} \ (\text{Fig. 19.5}),$$
$$(19.47)$$

$$\frac{\partial \sigma'_{xy}}{\partial y} = \eta_{xy(i,j,l-1)} \left(\frac{v_{x(i+1,j,l)} - v_{x(i,j,l)}}{\Delta y^2} + \frac{v_{y(i,j+1,l)} - v_{y(i,j,l)}}{\Delta y \Delta x} \right)$$
$$- \eta_{xy(i-1,j,l-1)} \left(\frac{v_{x(i,j,l)} - v_{x(i-1,j,l)}}{\Delta y^2} + \frac{v_{y(i-1,j+1,l)} - v_{y(i-1,j,l)}}{\Delta y \Delta x} \right), \quad \text{(Fig. 19.5)} \tag{19.48}$$

$$\frac{\partial \sigma'_{xz}}{\partial z} = \eta_{xz(i-1,j,l)} \left(\frac{v_{x(i,j,l+1)} - v_{x(i,j,l)}}{\Delta z^2} + \frac{v_{z(i,j+1,l)} - v_{z(i,j,l)}}{\Delta z \Delta x} \right) -$$
$$\eta_{xz(i-1,j,l-1)} \left(\frac{v_{x(i,j,l)} - v_{x(i,j,l-1)}}{\Delta z^2} + \frac{v_{z(i,j+1,l-1)} - v_{z(i,j,l-1)}}{\Delta z \Delta x} \right) \quad \text{(Fig. 19.5)}, \tag{19.49}$$

$$C_{v_x(i,j,l)} = -2 \frac{\eta_{n(i-1,j,l-1)} + \eta_{n(i-1,j-1,l-1)}}{\Delta x^2} - \frac{\eta_{xy(i,j,l-1)} + \eta_{xy(i-1,j,l-1)}}{\Delta y^2} - \frac{\eta_{xz(i-1,j,l)} + \eta_{xz(i-1,j,l-1)}}{\Delta z^2}, \tag{19.50}$$

and other terms can be derived similarly (derive as an exercise).

The methodology of using a multigrid approach for 3D models is the same as in 2D cases, with the single difference that trilinear (Eqs. 19.14–19.16) and not bilinear interpolation schemes should be used to construct the restriction and prolongation operations. Example implementations of the above algorithms for the case of constant and variable viscosity are given respectively in the programs **Stokes_Continuity3D_Multigrid.m**, **Variable_viscosity3D_Multigrid.m** and **Variable_viscosity3D_MultiMultigrid.m** associated with this chapter. As can be seen from Figs. 19.11, 19.12, multigrid allows high accuracy 3D mechanical solutions to be obtained at various viscosity contrasts. In the case of large viscosity contrasts, a computer accuracy solution can often be obtained with a 'multi-multigrid' approach (Fig. 19.12) using repetitive cycles of gradual increase in a computational viscosity contrast, as described in Chapter 18.

Elastic stress rotation. 3D numerical models including elasticity should take into account the elastic stress rotation (Chapter 12). One possible way of doing this is to use the general form of *Jaumann stress rate* (see Eqs. 12.28–12.36), which allows the rate of change caused by rotation to be computed for various stress components

$$\dot{\sigma}'_{ij(Jaumann)} = \omega_{ik} \sigma'_{kj} - \sigma'_{ik} \omega_{kj}, \tag{19.51}$$

where $\dot{\sigma}'_{ij(Jaumann)}$ is the rate of change for the σ'_{ij} deviatoric stress components, the repeated index k indicates summation and ω_{kj}, ω_{ik} are components of the anti-symmetric rotation rate tensor (Eq. 12.28). Using Eq. (19.51) in 3D for σ'_{xx} gives (see Eq. 12.31)

$$\dot{\sigma}'_{xx(Jaumann)} = 2\sigma'_{xy} \omega_{xy} + 2\sigma'_{xz} \omega_{xz}. \tag{19.52}$$

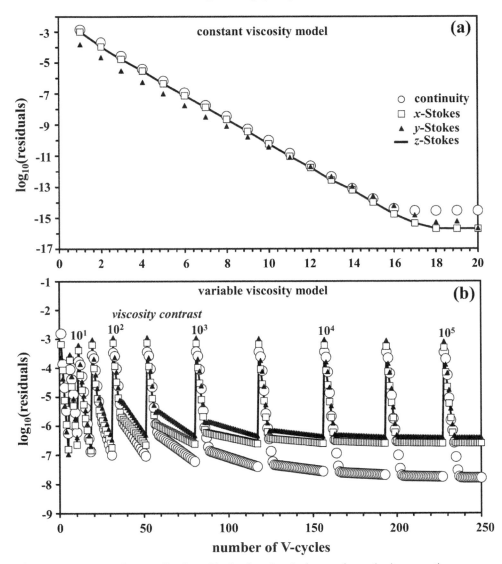

Fig. 19.11. Decay of normalized residuals for the Stokes and continuity equations versus number of multigrid V-cycles for 3D models with constant (a) and variable (b) viscosity. A 5-level multigrid with resolution 49×49×49 nodes on the finest level is used with relaxation parameters $\theta_{relaxation}^{continuity} = 0.3$ and $\theta_{relaxation}^{Stokes} = 0.9$. Numerical setup: rectangular block having higher density (and viscosity in (b) by factor 10^5) sinks in a lower density fluid. Iterations start from a hydrostatic pressure field and zero velocities and no viscosity contrast. Residuals in (a) stabilize at computer accuracy. Spikes in the solutions in (b) are caused by an increase in viscosity contrast by a factor of 3.333 after reaching a given level of tolerance (10^{-6}) for the sum of residuals. Results are obtained with the programs **Stokes_Continuity3D_Multigrid.m** (a) and **Variable_viscosity3D_Multigrid.m** (b).

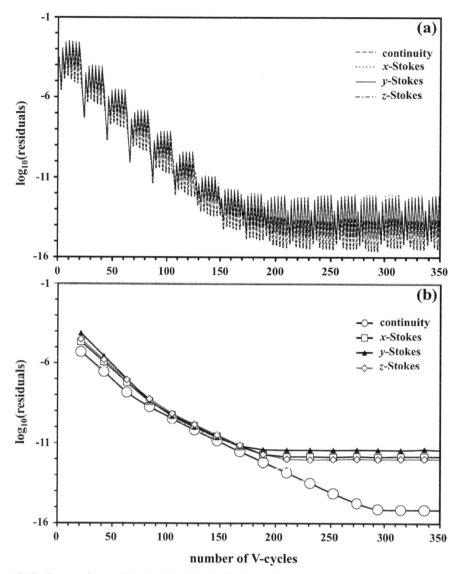

Fig. 19.12. Decay of normalized residuals for the Stokes and continuity equations with the number of multigrid V-cycles for a 3D model with variable viscosity in the case of a 'multi-multigrid' approach using repetitive cycles of gradual increase in computational viscosity contrast. Residuals stabilize at computer accuracy level. A 5-level multigrid with resolution 49×49×49 points on the finest level is used with relaxation parameters $\theta_{relaxation}^{continuity} = 0.75$ and $\theta_{relaxation}^{Stokes} = 1.25$. Numerical setup: rectangular block having higher density and viscosity (by a factor of 10^6) sinks in fluid with a lower density and viscosity. Iterations start from a hydrostatic pressure field and zero velocities. (a) Decay of local residuals computed with current corrections and right hand side *within* each cycle (steps) of gradual viscosity contrast increase (spikes). (b) Decay of global residuals computed *after* each cycle of gradual viscosity contrast increase. Spikes in (a) are caused by an increase in viscosity contrast by the factor of 10 every 3 multigrid cycles. Results are obtained with the program **Variable_viscosity3D_MultiMultigrid.m**.

Similar derivations can be made for other stress components (see Eqs. 12.32–12.36). The numerical implementation of stress rotation is done by re-computing elastic stress components stored at markers according to the fourth-order (*in stress space*) Runge-Kutta scheme (Popov et al., 2014a, personal communication, Chapter 12, code example **Stress_rotation_markers.m**)

$$\sigma'_{ij(rotated)} = \sigma'_{ij(m)} + \Delta t_m \times \dot{\sigma}'_{ij(Jaumann)(eff)}, \tag{19.53a}$$

$$\dot{\sigma}'_{ij(Jaumann)(eff)} = \frac{1}{6}\left(\dot{\sigma}'_{ij(Jaumann)(A)} + 2\dot{\sigma}'_{ij(Jaumann)(B)} + 2\dot{\sigma}'_{ij(Jaumann)(C)} + \dot{\sigma}'_{ij(Jaumann)(D)}\right), \tag{19.53b}$$

$$\dot{\sigma}'_{ij(Jaumann)(A)} = \omega_{ik}\sigma'_{kj(A)} - \sigma'_{ik(A)}\omega_{kj}, \tag{19.53c}$$

$$\dot{\sigma}'_{ij(Jaumann)(B)} = \omega_{ik}\sigma'_{kj(B)} - \sigma'_{ik(B)}\omega_{kj}, \tag{19.53d}$$

$$\dot{\sigma}'_{ij(Jaumann)(C)} = \omega_{ik}\sigma'_{kj(C)} - \sigma'_{ik(C)}\omega_{kj}, \tag{19.53e}$$

$$\dot{\sigma}'_{ij(Jaumann)(D)} = \omega_{ik}\sigma'_{kj(D)} - \sigma'_{ik(D)}\omega_{kj}, \tag{19.53f}$$

where the coordinates of points *A, B, C* and *D* in stress space are computed as

$$\sigma'_{ij(A)} = \sigma'_{ij(m)}, \tag{19.53g}$$

$$\sigma'_{ij(B)} = \sigma'_{ij(A)} + \frac{\Delta t_m}{2} \times \dot{\sigma}'_{ij(Jaumann)(A)}, \tag{19.53h}$$

$$\sigma'_{ij(C)} = \sigma'_{ij(A)} + \frac{\Delta t_m}{2} \times \dot{\sigma}'_{ij(Jaumann)(B)}, \tag{19.53i}$$

$$\sigma'_{ij(D)} = \sigma'_{ij(A)} + \Delta t_m \times \dot{\sigma}'_{ij(Jaumann)(C)}, \tag{19.53j}$$

where $\sigma'_{ij(m)}$ are the deviatoric stress components for a given marker, Δt_m is the marker displacement time step (Chapter 13) and ω_{kj} are the rotation rate components that are defined at the same nodal points and computed with similar FD schemes as the respective $\dot{\varepsilon}_{kj} = \frac{1}{2}\left(\partial v_k/\partial x_j + \partial v_j/\partial x_k\right)$ strain rate components (see Figs. 19.1, 19.5) and then interpolated to markers using standard interpolation formulas (Eqs. 19.15, 19.16, Fig. 19.8). Another efficient way of calculating 3D stress rotation is to use the 3D finite angle stress rotation approach developed by Anton Popov (Popov et al., 2014a, personal communication, Chapter 12, Eqs. 12.37–12.42, code example **Stress_rotation_markers.m**).

Numerical algorithms. Numerical algorithms for the thermomechanical codes in 3D do not differ from the algorithms described in Chapters 11 and 13 for 2D codes, with the exception that direct solvers for various equations should rather be substituted by the iterative ones described above.

Programming exercises

Exercise 19.1.
Program solution of the 3D temperature equation on a regular $51 \times 51 \times 51$ grid based on Gauss–Seidel iteration (Eqs. 19.17–19.22). The model setup corresponds to a hot rectangular block ($20 \times 20 \times 20$ km^3, $T = 1500$ K, $\rho = 3100$ kg/m^3, $C_P = 1500$ J/(K kg), $k = 1$ W/(m K)) which is located in a colder medium ($T = 1000$ K, $\rho = 3000$ kg/m^3, $C_P = 1000$ J/(K kg), $k = 3$ W/(m K)). The block is located in the middle of the model, which is $100 \times 100 \times 100$ km^3 in size. Boundary conditions are insulating at all boundaries. Use the relaxation parameter $\theta_{relaxation}^{temperature} = 1.25$ in Eq. (19.17). An example is provided in **Temperature3D_Gauss_Seidel.m**.

Exercise 19.2.
Generalize Exercise 18.1 to 3D. Solve the Poisson equation for the case of a spherical planetary body embedded in a massless-like medium (Eqs. 19.23–19.28, Fig. 19.7). The number of resolution levels is 4, the resolution on the coarsest (last) grid is $7 \times 7 \times 7$ nodal points, and the factor of increase in resolution between the levels is 2. All other model parameters and material properties are the same as in the 2D exercise. Use a ghost node approach along the spherical boundary surface for the gravity potential in the same way as in 2D (Fig. 18.7, Eqs. 18.12–18.16), i.e. independently in the x, y and z directions. An example is provided in **Poisson3D_Multigrid_planet_arbitrary.m**.

Exercise 19.3.
Generalize Exercises 18.2 and 18.3 to 3D. Solve the Stokes and continuity equations for both constant and variable viscosity cases using a pressure-velocity formulation and ghost node approach (Eqs. 19.30–19.50, Figs. 19.1–19.5). Model parameters: model size $100 \times 100 \times 100$ km^3, block size $20 \times 20 \times 20$ km^3. Material properties are the same as in 2D cases. Boundary conditions are free slip at all boundaries, number of resolution levels is 4, resolution of the basic grid on the coarsest (last) level $4 \times 4 \times 4$ nodal points, factor of increase in resolution between the levels is 2, relaxation coefficients for Gauss–Seidel iterations $\theta_{relaxation}^{Stokes} = 0.9$ and $\theta_{relaxation}^{Continuity} = 0.3$, number of smoothing iterations on the finest (basic) level is 5, factor of increase in the number of iterations with the level coarsening is 2. For the variable viscosity case, use a gradual increase in the viscosity contrast by a factor of $10^{1/2}$ (Eq. 18.57), to reach an accurate solution (Fig. 19.11). An example is in **Variable_viscosity3D_Multigrid.m**.

20

Numerical benchmarks

Theory: Numerical benchmarks: testing of numerical codes for various problems. Examples of thermomechanical benchmarks. Manufactured solutions.
Exercises: Programming of models for various numerical benchmarks.

20.1 Code benchmarking: why should we spend time on it?

Benchmarking of a numerical code means comparing the numerical solution obtained from solving the system of linear equations with (i) analytical solutions, (ii) results of physical (analogue) experiments, (iii) numerical results from other (well established) codes and (iv) general physical considerations. Benchmarking of newly created numerical tools is sometimes very tedious, but is an absolutely necessary stage of code development as its purpose is to test the robustness of the code in a broad range of situations relevant to geodynamic modelling applications. For instance, if you plan to model shear heating processes in deforming rocks – make sure that your code provides the correct temperature changes related to mechanical energy dissipation; if you model subduction – make sure that your code handles correctly large viscosity contrasts and has no notable numerical diffusion of the temperature field and composition; if you intend to model self-gravitating planetary bodies – make sure that your code computes the correct gravity field etc. We should not be lazy and limit ourselves to one or two common benchmarks, such as the Rayleigh–Taylor instability and convection with constant viscosity, hoping that everything else will work automatically. No, it will not! Therefore, test your code on a broad range of challenging cases (several of which are discussed below), make it 'screaming' to explore its limitations and be creative in inventing and calibrating new numerical benchmarks. Then, in the end, you will be really proud of your 'numerical child'. Many of the analytical solutions that can be used for testing of thermomechanical codes *for geodynamically relevant situations* can be taken from the textbook of Turcotte and Shubert (2002, 2014), which we will also use for constraining some of our numerical benchmarks.

Below we discuss details of several important benchmarks, which test various aspects of thermomechanical codes. These calibrating tests aim to verify the efficacy of numerical solutions for a variety of circumstances relevant to geodynamics. These include the following:

(a) sharply discontinuous viscosity distribution (tests 1 and 2);
(b) strain rate dependent viscosity (test 3);
(c) non-steady-state development of temperature field (test 4);
(d) shear heating for temperature-dependent viscosity (test 5);
(e) advection of a sharp temperature front (test 6);
(f) heat conduction for temperature-dependent thermal conductivity (test 7);
(g) thermal convection with constant and variable viscosity (test 8);
(h) elastic stress buildup and advection (tests 9 and 10);
(i) localization of visco-elasto-plastic deformation (test 11);
(j) kinetic to potential energy conversion during impact (test 12).

Of course this list is incomplete and many additional benchmarks exists, or can be invented. Yet, on performing the discussed benchmarks, we will at least get some confidence that we have created a state of the art numerical geodynamic modelling tool which correctly reproduces a number of challenging geodynamic models.

20.2 Test 1: Rayleigh–Taylor instability benchmark

This is a typical analytical solution based benchmark. To test the correctness of the numerical velocity solution for gravity driven flows, in the case of sharply heterogeneous density and viscosity fields, one can use a two-layer *Rayleigh–Taylor instability* model in a purely vertical gravity field (e.g. Ramberg, 1968) with a no-slip condition on the top and at the bottom and symmetry conditions along the vertical walls (Fig. 20.1a). An initial sinusoidal disturbance of the boundary between the upper (η_1, ρ_1,) and the lower (η_2, ρ_2) layers of thicknesses h_1 and h_2, respectively, has a small initial amplitude (ΔA) and a wavelength (λ). Under this condition, the velocity of the diapiric growth (v_y) is given by the relation (Ramberg, 1968)

$$\frac{v_y}{\Delta A} = -K\frac{\rho_1 - \rho_2}{2\eta_2}h_2 g_y,$$ (20.1)

$$K = \frac{-d_{12}}{c_{11}j_{22} - d_{12}i_{21}},$$

$$c_{11} = \frac{\eta_1 2\phi_1^2}{\eta_2\left(\cosh 2\phi_1 - 1 - 2\phi_1^2\right)} - \frac{2\phi_2^2}{\cosh 2\phi_2 - 1 - 2\phi_2^2}$$

$$d_{12} = \frac{\eta_1\left(\sinh 2\phi_1 - 2\phi_1\right)}{\eta_2\left(\cosh 2\phi_1 - 1 - 2\phi_1^2\right)} + \frac{\sinh 2\phi_2 - 2\phi_2}{\cosh 2\phi_2 - 1 - 2\phi_2^2},$$

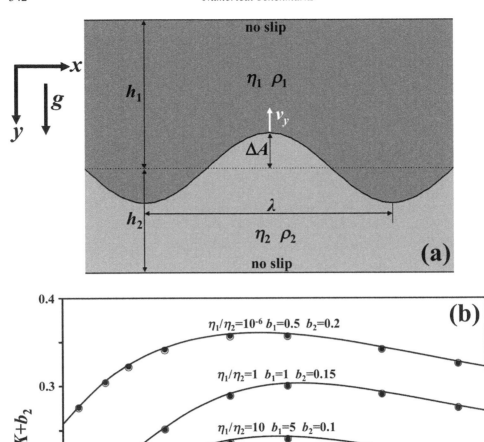

Fig. 20.1. Rayleigh–Taylor instability benchmark. (a) Initial setup. (b) Comparison of numerical (symbols) and analytical (lines, Eq. 16.1) solutions for the case of two layers with equal thicknesses ($h_1 = h_2$). Arbitrary scaling coefficients b_1 and b_2 are used in (b) for plotting results computed at variable viscosity contrasts on the same diagram. The growth factor K for numerical cases is computed from the velocity field based on Eq. (20.1). Numerical results are calculated at resolutions of 51×51 for nodes and 250×250 for markers with the code **Variable_viscosity_Ramberg.m**.

$$i_{21} = \frac{\eta_1 \phi_2 (\sinh 2\phi_1 + 2\phi_1)}{\eta_2 (\cosh 2\phi_1 - 1 - 2\phi_1^2)} + \frac{\phi_2 (\sinh 2\phi_2 + 2\phi_2)}{\cosh 2\phi_2 - 1 - 2\phi_2^2},$$

$$j_{22} = \frac{\eta_1 2\phi_1^2 \phi_2}{\eta_2 (\cosh 2\phi_1 - 1 - 2\phi_1^2)} - \frac{2\phi_2^3}{\cosh 2\phi_2 - 1 - 2\phi_2^2},$$

$$\phi_1 = \frac{2\pi h_1}{\lambda},$$

$$\phi_2 = \frac{2\pi h_2}{\lambda},$$

where g_y is the vertical component of the gravitational acceleration, K is a dimensionless growth factor.

With a marker-in-cell method, a small layer boundary perturbation (much smaller than one grid step) can easily be prescribed on a subgrid scale by small sinusoidal (vertical) displacements of markers that are initially distributed regularly inside the numerical grid

$$\Delta A_m = \cos\left(2\pi \frac{x_m - 0.5L}{\lambda}\right)\Delta A, \tag{20.2}$$

where x_m and ΔA_m are the horizontal coordinate and vertical displacement for a given marker m, and L is the horizontal width of the numerical model. For proper constraining of the numerical models, the relationship $L = 2\lambda$ can be used together with free slip (i.e. horizontal symmetry) conditions on two vertical boundaries.

Figure 20.1b compares numerical and analytical solutions for the growth rate of the instability estimated for two layers of equal thickness (i.e. $h_1 = h_2$) at different values of ΔA, λ and η_1/η_2. Good accuracy at large variations of the disturbance wavelength and layer viscosity contrasts ($\eta_1/\eta_2 - 10^{-6}$ to 5×10^{7}) suggests that the tested numerical code is capable of correctly modelling the velocity fields for gravity driven flows across a boundary with sharp changes in density and viscosity. Even very small subgrid perturbations of the horizontal boundary are properly captured by variations in relative position of markers via a bilinear density interpolation procedure from markers to nodes (Chapter 8). An example of the numerical setup for conducting the Rayleigh–Taylor instability benchmark is given by the code **Variable_viscosity_Ramberg.m**.

20.3 Test 2: falling block benchmark

This is a typical example of a benchmark which is based on general physical considerations. I personally like it very much, since it is simple to implement but creates challenging conditions to be handled numerically. In the case of an isolated rigid object, sinking in a low viscosity surrounding, the velocity of the object mainly depends on the viscosity of its surrounding (weakest medium). As was discussed in Chapter 18, this situation differs from

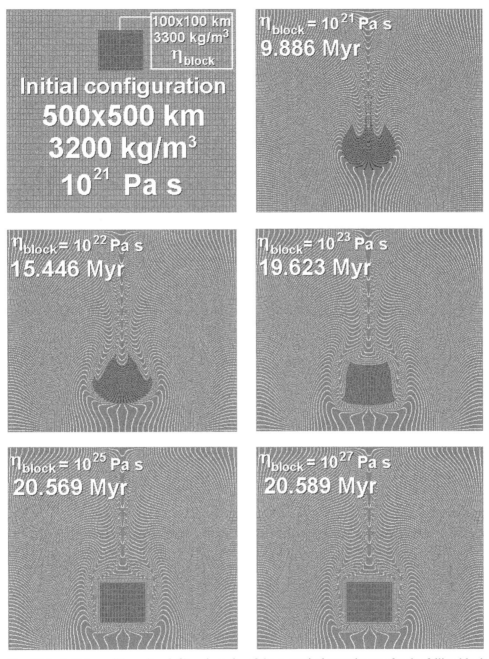

Fig. 20.2. Initial conditions (top left) and results of the numerical experiments for the falling block benchmark performed by Gerya and Yuen (2003a). Boundary conditions: free slip at all boundaries. Black and white dots represent positions of markers for the block and the medium, respectively. Grid resolution of the model is 51 × 51 nodes, 22 500 markers.

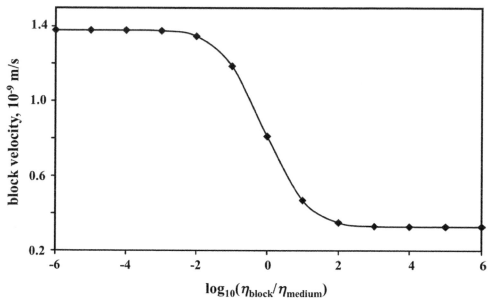

Fig. 20.3. Velocity of the rectangular block sinking in a viscous medium as a function of viscosity contrast between the block and the background medium. The model setup corresponds to Fig. 20.2 (top left). Numerical results are calculated at a resolution of 51×51 nodes and 250×250 markers with the code **Variable_viscosity_block.m**.

modelling the Rayleigh–Taylor instability where the strong layer is attached to the model boundaries and the velocity field is defined by the strongest medium. According to our physical intuition, (i) deformation of the block should vanish with increasing viscosity contrast (Fig. 20.2) and (ii) the sinking velocity at high viscosity contrasts should be independent of the absolute value of the viscosity of the block and should solely depend on that of the surrounding medium (Fig. 20.3). When using finite differences for solving the momentum equations, they should be formulated in a stress conservative manner (see Chapter 7). This test also proves the accurate conservation properties of a numerical procedure in terms of *preserving the block edges geometry* (Fig. 20.2) at large deformation and high (10^2–10^6) viscosity contrast between the stiffer block and the weak surroundings. An example of the numerical setup for conducting the falling block benchmark is given by the code **Variable_viscosity_block.m**.

20.4 Test 3: channel flow with a non-Newtonian rheology

This test can be conducted to check the numerical solution of the momentum and continuity equations for flows with a strongly strain-rate/stress-dependent rheology, which is characteristic of dislocation creep (Ranalli, 1995). The computation is carried out for vertical flow of a non-Newtonian (with a power law index n) viscous medium in a section of an infinite vertical channel (Fig. 5.2) of width L in the absence of gravity. Boundary conditions are taken as follows: a given constant vertical pressure gradient $\partial P/\partial y$ along the channel and no-slip conditions at the walls. The viscosity of the non-Newtonian flow is defined by the following rheological equation formulated in terms of second stress and strain rate invariants

$$2\dot{\varepsilon}_{\mathrm{II}} = C_1(\sigma_{\mathrm{II}})^n, \tag{20.3}$$

where C_1 is a material constant in units of $\mathrm{Pa}^{-n}{\cdot}\mathrm{s}^{-1}$. Equation (20.3) can be reformulated in terms of effective viscosity (Chapter 6) as a function of second strain rate invariant

$$\eta_{\mathit{eff}} = \frac{\sigma_{\mathrm{II}}}{2\varepsilon_{\mathrm{II}}} = C_1^{-1/n}(2\varepsilon_{\mathrm{II}})^{1/n-1}. \tag{20.4}$$

Analytical solutions for the velocity and viscosity profiles across the channel are given by (Turcotte and Schubert, 2002; Gerya and Yuen, 2003a)

$$v_y = \frac{C_1}{n+1}\left(-\frac{\partial P}{\partial y}\right)^n\left[\left(\frac{L}{2}\right)^{n+1} - \left(x - \frac{L}{2}\right)^{n+1}\right], \tag{20.5}$$

$$\eta_{\mathit{eff}} = \frac{\sigma_{\mathrm{II}}}{2\dot{\varepsilon}_{\mathrm{II}}} = \frac{\sigma'_{yx}}{2\dot{\varepsilon}_{yx}} = \frac{1}{C_1}\left(-\frac{\partial P}{\partial y}\right)^{1-n}\left(x - \frac{L}{2}\right)^{1-n}, \tag{20.6}$$

$$\frac{\partial P}{\partial y} = \frac{P_{\mathit{end}} - P_{\mathit{beg}}}{H}, \tag{20.7}$$

where P_{beg} and P_{end} are pressures at the beginning ($x = 0$) and at the end ($x = L$) of the channel section of height H, respectively. Figure 20.4 compares analytical and numerical (2D) solutions based on conservative finite differences with marker-in-cell techniques obtained with the code **Variable_viscosity_channel.m**. Numerical and analytical solutions overlap, implying high accuracy of the numerical method for modelling flows with strong lateral variations in viscosity caused by the non-Newtonian rheology. *Open channel boundary conditions* at the top and at the bottom imply an infinite vertical channel with constant vertical pressure gradients. These boundary conditions are programmed by defining P_{beg} and P_{end} in the first and the last row of pressure nodes, respectively, and prescribing $\partial v_x/\partial y = 0$, $\partial v_y/\partial y = 0$, at the upper and lower boundaries of the model. Note that the vertical length of the channel section H used in Eq. (20.7) for computing the pressure gradient corresponds to the distance between the first and the last row of pressure nodes and not to the vertical length of the 2D model.

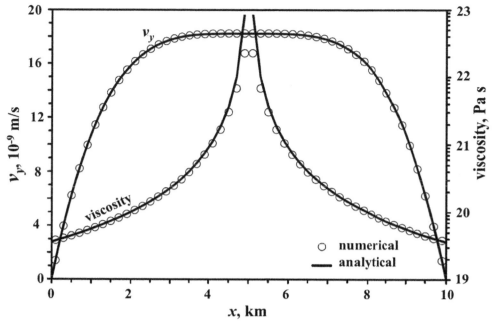

Fig. 20.4. Comparison of analytical and numerical solutions for the velocity and viscosity profiles across a channel with non-Newtonian flow rheology given by Eq. (20.3). Numerical results are calculated at the resolution of 51×21 nodes and 250×100 markers with the code **Variable_viscosity_channel.m** associated with this chapter. Model parameters: $L = 10$ km, $H = 9.5$ km, $n = 3$, $C_1 = 10^{-37}$ Pa$^{-3} \cdot$ s^{-1}, $P_{beg} = 10^9$ Pa, $P_{end} = 0$.

20.5 Test 4: non-steady-state temperature distribution in a Newtonian channel

Here we describe another channel flow based benchmark. This one can be performed to test the numerical accuracy of solution of the time-dependent (non-steady-state) temperature equations in cases when heat advection is coupled with heat diffusion. The model corresponds to the vertical flow of a heat-conductive medium of constant viscosity η in a channel, in the absence of gravity. Boundary conditions are: a given constant vertical pressure gradient, $\partial P/\partial y$ along the channel, non-slip conditions and $T = \text{const} = T_o(y)$ and $\partial T/\partial y = \text{const} = \partial T_o(y)/\partial y$ at the walls (Exercise 9.2). The initial conditions for the temperature distribution inside the model are $T = T_o(y)$, $\partial T/\partial y = \partial T_o(y)/\partial y$ and $\partial T/\partial x = 0$. The horizontal steady-state profile for vertical velocities, v_y, is defined by the equation which can be derived either from Eq. (20.5) with $n = 1$ and $C_1 = 1/\eta$ or from Eq. (5.30) with $g_y = 0$

$$v_y = -\frac{1}{2\eta}\left(\frac{\partial P}{\partial y}\right)\left(Lx - x^2\right). \qquad (20.8)$$

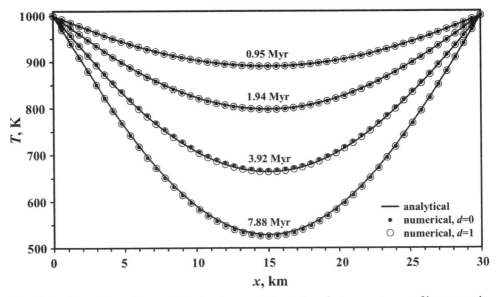

Fig. 20.5. Comparison of the analytical and numerical solutions for temperature profiles across the channel with constant viscosity. Numerical results are calculated at the resolution of 51×11 nodes and 250×50 markers with the code **Constant_viscosity_channel_T.m**. Model parameters: L=30 km, H = 11.25 km, $\eta = 10^{19}$ Pa s, $P_{beg} = 10^5$ Pa, $P_{end} = 0$, $\partial T/\partial y = 40$ K/km, $T(y = 0) = 1000$ K.

The corresponding temperature changes in the channel with time are then given by the following series expansion (Gerya and Yuen, 2003a, it is typically enough to compute the first 10 terms of the expansion for an accurate solution)

$$\Delta T(x, t) = \sum_{m=1}^{\infty} F_m E_{mt} \sin\left[\pi(2m - 1)\frac{x}{L}\right], \tag{20.9}$$

$$F_m = -8\xi \frac{L^2}{[\pi(2m - 1)]^3},$$

$$E_{mt} = L^2 \frac{1 - \exp\left\{-\frac{\kappa t}{L^2}[\pi(2m - 1)]^2\right\}}{\kappa[\pi(2m - 1)]^2},$$

$$\kappa = \frac{k}{\rho C_P},$$

$$\xi = -\frac{1}{2\eta}\left(\frac{\partial P}{\partial y}\right)\left(\frac{\partial T_o(y)}{\partial y}\right),$$

where $\Delta T(x, t)$ is the temperature change as a function of the horizontal coordinate x and time t; κ is a constant thermal diffusivity in units of m^2 s^{-1}. Equation (20.9) does not account for shear heating: in this numerical test it is considered negligible. Figure 20.5 compares the analytical solution from Eq. (20.9) with the numerical solution obtained with the 2D thermomechanical code **Constant_viscosity_channel_T.m**. Mechanical boundary conditions for the infinite vertical channel are the same as for the previous benchmark. A constant temperature gradient is used as a boundary condition for temperature nodes located at the upper and lower thermal boundaries, implying infinity of the thermal profile in the vertical direction. Figure 20.5 shows that numerical and analytical results coincide well for calculations performed both with ($d = 1$) and without ($d = 0$) numerical subgrid diffusion (Eqs. 10.15–10.19), implying robustness of the coupled thermomechanical solution for the case of non-steady-state heat conduction associated with heat advection.

20.6 Test 5: Couette flow with shear heating

This benchmark is designed to verify the numerical solution of the coupled momentum and temperature equations for flows with temperature-dependent rheology in the situation of strong shear heating (viscous dissipation). The analytical model setup corresponds to a vertical *Couette flow* (simple shear deformation in a laterally limited planar zone of width L) in the absence of gravity. Boundary conditions are taken as follows: zero vertical pressure gradient, $\partial P/\partial y = 0$ along the flow, $v_y = 0$, $T = T_0$ and $\sigma_{yx} = \text{const} = \sigma_{yx1}$, $\partial T/\partial x = 0$ at the left and right walls, respectively. Viscosity of the flow is given by the following rheological equation (Turcotte and Schubert, 2002)

$$\eta = A\exp\left[\frac{E_a}{RT_0}\left(1 - \frac{T - T_0}{T_0}\right)\right],$$

where E_a is the activation energy, R is the gas constant and A is a pre-exponential rheological constant, which depends on the material. The analytical solution for steady temperature distribution $T(x)$ inside the flow is given by the relation (Turcotte and Schubert, 2002)

$$x = \frac{L}{B}\ln\left[\frac{(D+B)(C-B)}{(D-B)(C+B)}\right], \tag{20.10}$$

$$B = \ln\left[\frac{1 + \left(1 - \frac{2Br}{B^2}\right)^2}{1 - \left(1 - \frac{2Br}{B^2}\right)^2}\right], \tag{20.11}$$

$$C = \left\{2[\phi_1 - \phi(x)]Br\right\}^{1/2}, \tag{20.12}$$

$$D = [2(\phi_1 - 1)Br]^{1/2}, \tag{20.13}$$

$$\phi(x) = \exp[\theta(x)], \tag{20.14}$$

$$\theta(x) = \frac{E_a[T(x) - T_0]}{RT_0^2}, \tag{20.15}$$

$$\phi_1 = \frac{B^2}{2Br}, \tag{20.16}$$

$$\phi_1 = \exp(\theta_1), \tag{20.17}$$

$$\theta_1 = \frac{E_a(T_1 - T_0)}{RT_0^2}, \tag{20.18}$$

$$Br = \frac{\left(\sigma'_{yx1}L\right)^2 E_a}{kART_0^2} \exp\left(-\frac{E_a}{RT_0}\right), \tag{20.19}$$

where Br is the non-dimensional Brinkman number, θ is the non-dimensional temperature change, σ'_{yx1} is the shear stress that remains constant within the flow, k is the thermal conductivity of the flow medium, T_1 is the temperature at the right wall of the flow (i.e. maximal temperature). Solving the non-linear equations (20.10)–(20.19) analytically for given values of k, L, A, E_a, T_0 and σ_{yx1} is non-trivial and the solution for T_1 is not unique at a given value of σ'_{yx1}. Rather than defining σ'_{yx1}, non-negative values of B can be chosen and then the Brinkman number and shear stress in the channel can be computed from Eq. (20.11) and Eq. (20.19), respectively, as

$$Br = \frac{B^2}{2}\left[1 - \left(\frac{\exp B - 1}{\exp B + 1}\right)^2\right], \tag{20.20}$$

$$\sigma'_{yx1} = \left[Br\frac{kART_0^2}{L^2 E_a}\exp\left(\frac{E_a}{RT_0}\right)\right]^{1/2}. \tag{20.21}$$

Other unknown parameters can be computed from B and Br by using Eqs. (20.10)–(20.18). Based on such calculations, the dependence of the maximal non-dimensional temperature change in the channel θ_1 on the Brinkman number Br can be computed (Fig. 20.6a).

To test the Couette flow solution numerically, the constant vertical velocity boundary condition $v_y = v_{y1}$ should be applied at the right boundary instead of $\sigma'_{yx} = \sigma'_{yx1}$ used in the analytical model. The upper and lower boundary conditions are the same as in tests 3 and 4 but with zero vertical pressure and temperature gradients ($P_{beg} = P_{end} = 0$ and $\partial T/\partial y = 0$). This modification will ensure uniqueness of the numerical thermomechanical solution that becomes steady state in a finite number of time steps. The value of the parameter B should be computed iteratively with Eqs. (20.16)–(20.18) and (20.20) from the steady-state temperature (T_1) at the right boundary (see the computing procedure in the end of the

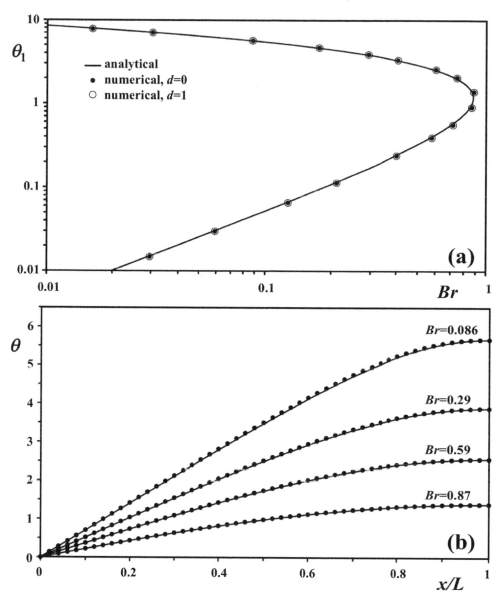

Fig. 20.6. Comparison of analytical and numerical results for a case of steady-state Couette flow with temperature-dependent viscosity (see Eq. 20.9) and shear heating. (a) Maximal temperature change within the flow θ_1 (Eq. 20.18) versus Brinkman number Br (Eq. 20.19). (b) Distribution of temperature changes θ (Eq. 20.15) across the flow at different values of Brinkman number. Numerical results are calculated at a resolution of 51 × 11 nodes and 250 × 50 markers with the code **Variable_viscosity_Couette_T.m** associated with this chapter. Model parameters: $L = 30$ km, $H = 11.25$ km, $A = 10^{15}$ Pa s, $E_a = 150$ kJ/mol, $k = 2$ W/(m K), $T_0 = 1000$ K.

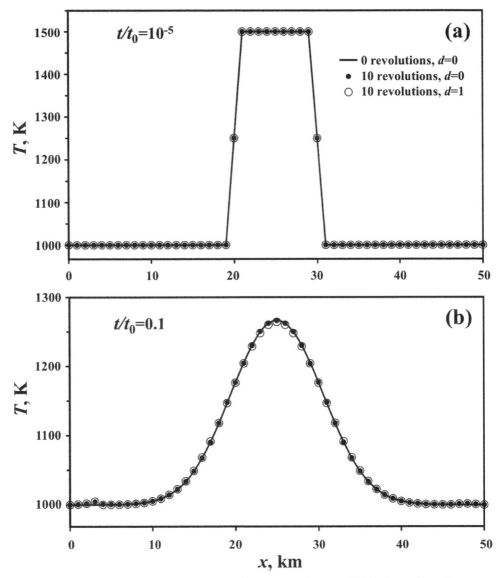

Fig. 20.7. The results of testing of the numerical solution for the solid body rotation of a square temperature wave. The figure shows the horizontal profiles across the wave at different times t after a given number of revolutions. (a) and (b) Results of numerical experiments at different characteristic thermal diffusion time scales $t_0 = \rho C_P L^2/k$, where $L = 10$ km is the initial length of the temperature wave. d is a numerical subgrid diffusion parameter (see Eq. 10.16 in Chapter 10). Numerical results are calculated at a resolution of 51×51 nodes and 250×250 markers with the code **Solid_Body_Rotation_T.m**. Model parameters: size 50×50 km^2, $C_P = 1000$ J/(kg K), $\rho = 3000$ kg/m^3, $k = 2 \times 10^{-6}$ and 0.02 W/(m K) for (a) and (b), respectively, $T_{medium} = 1000$ K, $T_{wave} = 1500$ K. The small temperature perturbation at the left boundary in (b) is a boundary effect (i.e. a trace of the rotating wave interacting with the model thermal boundary condition; run program **Solid_Body_Rotation_T.m** in order to see this trace in 2D).

program example **Variable_viscosity_Couette_T.m**). Other parameters can again be computed with B from Eqs. (20.10)–(20.21). Figure 20.6 shows that numerical and analytical results coincide well, implying that the numerical solution holds for thermomechanical effects of shear heating in the case of strongly variable temperature-dependent viscosity.

20.7 Test 6: advection of sharp temperature fronts

Verification of the ability to advect sharp temperature fronts is fundamental in numerical tests of various advection algorithms. The geodynamic relevance of this test is obvious when modelling rapidly moving subducting and detached slabs is envisaged. Numerical solutions for this type of benchmark (see e.g. Chapter 8) are typically calculated in 2D for the solid body rotation of a two-dimensional temperature wave of an arbitrary shape. One can, for example, perform such a test for a square wave with width L and thermal amplitude $\Delta T_o = 500$ K. The results of the test obtained with our finite difference and marker-in-cell techniques are shown in Fig. 20.7 for a regularly spaced grid of moderate resolution (51×51 nodes, 250×250 markers). If heat conduction is insignificant (Fig. 20.7a), the adopted marker-in-cell advection scheme is obviously not numerically diffusive, even for many revolutions, as far as the initial positions of markers (with the corresponding values of initially prescribed temperature field which is negligibly affected by the heat diffusion) are reproduced well with the classical fourth-order Runge–Kutta (Chapter 8) advection scheme (see code **Solid_Body_Rotation_T.m**). In the case of significant heat conduction (Fig. 20.7b), the final temperature distribution does not depend noticeably on the number of revolutions. This point suggests good conservation properties of the adopted numerical scheme when advecting diffusing temperature fronts. Introducing numerical subgrid diffusion (Chapter 10) only negligibly affects the temperature when heat conduction is significant (Fig. 20.7b). Obviously, this numerical diffusion, which gives a small addition to the physical diffusion, exerts little influence in the case of negligible heat conduction (Fig. 20.7a). Generally, the tested method of solving the temperature equation using markers works very well in the two distinct regimes of advection for both non-diffusive (Fig. 20.7a) and diffusive (Fig. 20.7b) sharp temperature fronts.

20.8 Test 7: channel flow with variable thermal conductivity

This analytical benchmark can be conducted to verify the accuracy of a thermomechanical code in the case of strong variations in temperature-dependent thermal conductivity, which are relevant to many geodynamic situations that involve large variations in temperature (mantle convection, lithospheric processes, etc.). For this purpose, we can again use vertical Newtonian channel flow (as in test 4 but without temperature gradients along the channel) with a velocity distribution defined by Eq. (20.8) and shear heating, which provides a strong heat source term in the temperature equation (Chapter 9). The thermal conductivity is taken

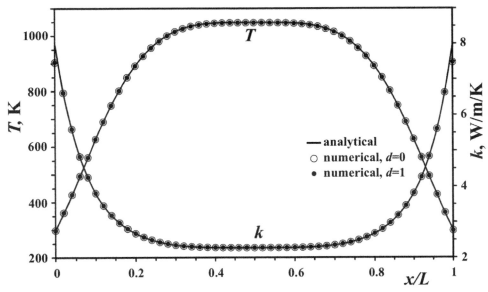

Fig. 20.8. Comparison of the analytical (Eq. 20.23) and numerical solutions for the steady temperature and thermal conductivity profiles across a channel with constant viscosity and strong shear
heating. Numerical results are calculated at a resolution of 51 × 11 nodes and 250 × 50 markers with
the code **Variable_conductivity_channel.m**. Model parameters: $L = 30$ km, $H = 11.25$ km, $\eta = 10^{19}$
Pa s, $P_{beg} = 3 \times 10^7$ Pa, $P_{end} = 0$, $T_0 = 298$ K, $k_0 = 8$ W/(m K), $b = 1$.

to be decreasing with temperature, which is very characteristic (e.g. Hofmeister, 1999) for
lattice conductivity defined by *phonons* in a crystal lattice

$$k = \frac{k_0}{1 + b(T - T_0)/T_0},$$
(20.22)

where T_0 is a constant temperature applied at the walls of the channel; k_0 is the thermal
conductivity at T_0; b is a dimensionless coefficient.

The steady temperature profiles across the channel $T(x)$ are then defined as (Gerya and
Yuen, 2003a)

$$T(x) = T_0 \frac{C(x) + b - 1}{b},$$
$$C(x) = \exp\left\{\frac{L^4 b}{192 k_0 T_0 \eta}\left(\frac{\partial P}{\partial y}\right)^2\left[1 - \left(2\frac{x}{L} - 1\right)^4\right]\right\},$$
(20.23)

where L is the channel width, η is a constant viscosity of the medium and $\partial P/\partial y$ is the
pressure gradient along the channel (Eq. 20.7).

Figure 20.8 compares the analytical solution for both temperature and thermal conductivity profiles with the numerical solutions obtained with the 2D thermomechanical code

Variable_conductivity_channel.m. This figure demonstrates the high accuracy of the numerical solution, suggesting that the adopted conservative FD scheme correctly computes heat transport in the case of strong variations in thermal conductivity (factor of 4 variation across the channel for the given case, Fig. 20.8).

20.9 Test 8: thermal convection with constant and variable viscosity

This benchmark can be conducted to test the ability of the code to model mantle convection. Blankenbach et al. (1989) tested several 2D mantle convection models with a broad variety of numerical techniques and reported steady-state values for a number of model parameters to which the numerical solution should converge with increasing grid resolution. Table 20.1 represents the physical parameters for five steady-state convection models with both constant (models 1a, 1b, 1c) and variable (models 2a, 2b) temperature- and depth-dependent viscosity. Convection is studied in a rectangular box of height H and width L ($H = L = 1000$ km for all models with the exception of model 2b). The boundary conditions are free slip along all boundaries, a specified temperature on the top (T_{top}) and at the bottom (T_{bottom}) and thermal insulation ($\partial T/\partial x = 0$) along the left and right walls. The difference between T_{top} and T_{bottom} in all experiments is 1000 K. The following formulation for temperature- and depth-dependent viscosity of the mantle is used:

$$\eta = \eta_0 \exp\left(b\,\frac{T - T_{top}}{T_{bottom} - T_{top}} + c\,\frac{y}{H} \right), \qquad (20.24)$$

where η_0 is viscosity at the top of the model (i.e. at $T = T_{top}$ and $y = 0$); b and c are coefficients establishing the dependences of viscosity with temperature and depth, respectively ($b = 0$ and $c = 0$ in constant viscosity tests 1a, 1b and 1c). Density in all models depends linearly on temperature

$$\rho = \rho_0[1 - \alpha(T - T_{top})],$$

where ρ_0=4000 kg/m^3 is the standard density and α=2.5×10^{-5} 1/K is the thermal expansion coefficient.

Despite this relatively simple setup, obtaining an accurate steady-state solution for mantle convection models is quite challenging. This is mainly due to (i) many (typically several thousand) time steps required to obtain a steady-state solution and (ii) a strong localization of thermal upwellings and downwellings along the walls (e.g. Fig. 20.9c) in models with low mantle viscosity (or more precisely with high *Rayleigh number* $Ra = \rho_0 \alpha (T_{bottom} - T_{top}) g_y H^3 C_P/(\eta k)$, where g_y is the vertical component of the gravitational acceleration, C_P is the isobaric heat capacity and k is the thermal conductivity. The problem of localization can be overcome either by using a high resolution for the entire model or (more efficiently) by using an irregularly spaced grid, which is denser at the model walls (Fig. 20.9a). The steady-state thermal structures computed for some of the

Table. 20.1 *Parameters of mantle convection benchmarks (Blankenbach et al., 1989)*

Test	1a	1b	1c	2a	2b
Gravitational acceleration, g (m/s^2)	10	10	10	10	10
Model height, H (km)	1000	1000	1000	1000	1000
Model width, L (km)	1000	1000	1000	1000	2500
Temperature at the top, T_{top} (K)	273	273	273	273	273
Temperature in the bottom, T_{bottom} (K)	1273	1273	1273	1273	1273
Thermal conductivity, k (W/(m K))	5	5	5	5	5
Heat capacity, C_P (J/kg)	1250	1250	1250	1250	1250
Standard density, ρ_0 (kg/m^3)	4000	4000	4000	4000	4000
Thermal expansion, α (1/K)	2.5×10^{-5}	2.5×10^{-5}	2.5×10^{-5}	2.5×10^{-5}	2.5×10^{-5}
Flow law parameters: η_0 (Pa s)	10^{23}	10^{22}	10^{21}	10^{23}	10^{23}
b	0	0	0	ln(1000)	ln(16384)
c	0	0	0	0	ln(64)
Nusselt number, $Nu = \frac{H}{T_{bottom}L} \int_{x=0}^{L} \left(\frac{\partial T}{\partial y}\right)_{top} dx$	4.8844	10.534	21.972	10.066	6.9229
Non-dimensional root mean square (rms) velocity, $v_{rms} = \frac{H\rho_0 C_P}{k} \sqrt{\frac{1}{HL} \int_{x=0}^{L}\int_{y=0}^{H} \left(v_x^2 + v_y^2\right) dy dx}$	42.865	193.21	833.99	480.43	171.76
Non-dimensional temperature gradients in the model corners, $q_{corner} = \frac{H}{T_{bottom}-T_{top}} \left(\frac{\partial T}{\partial y}\right)_{corner}$					
q_1 (top-left corner, above upwelling)	8.0593	19.079	45.964	17.531	18.484
q_2 (top-right corner, above downwelling)	0.5888	0.7228	0.8772	1.0085	0.1774
q_3 (bottom-right corner, below downwelling)	8.0593	19.079	45.964	26.809	14.168
q_4 (bottom-left corner, below upwelling)	0.5888	0.7228	0.8772	0.4974	0.6177
Local minimum along the central vertical temperature profile: $T_c = \frac{T-T_{top}}{T_{bottom}-T_{top}}$	0.4222	0.4284	0.4322	0.7405	0.3970
$Z_c = \frac{H-y}{H}$	0.2249	0.1118	0.0577	0.0623	0.1906
Local maximum along the central vertical temperature profile: $T_c = \frac{T-T_{top}}{T_{bottom}-T_{top}}$	0.5778	0.5716	0.5678	0.8323	0.5758
$Z_c = \frac{H-y}{H}$	0.7751	0.8882	0.9423	0.8243	0.7837

models of Table 20.1 are shown in Fig. 20.9b,c,d. Figure 20.10 presents the results of the mantle convection benchmark for these models obtained with the program **Variable_viscosity_convection_irregular_grid.m** associated with this chapter. As can be seen at the same model resolution of 51 × 51 nodes and 40 000 markers, models with an irregularly spaced grid show results that are much closer to the benchmark values.

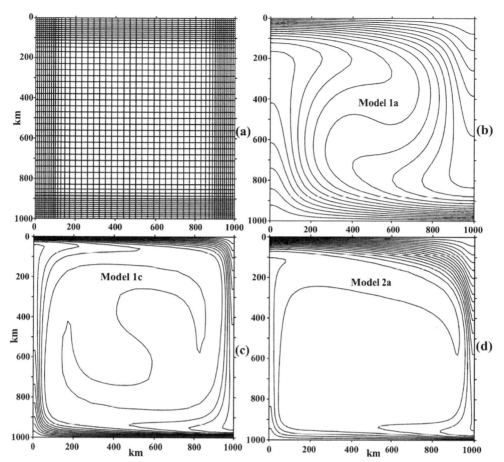

Fig. 20.9. Irregularly (10–30 km) spaced grid (a) and steady-state temperature structures (b)–(d) for the three mantle convection benchmarks from Table 20.1. Numerical results are computed at a resolution of 51×51 nodes and 200×200 randomly distributed markers with the code **Variable_viscosity_convection_irregular_grid.m**. Solid lines in (b)–(d) represent isotherms between T_{top} and T_{bottom} with an interval of 50 K.

Therefore the use of adaptive irregularly spaced grids (based on 'Swiss cross' AMR, Chapter 17), can in many cases significantly increase the accuracy of the numerical solution without a notable increase in computational cost.

20.10   Test 9: stress buildup in a visco-elastic Maxwell body

This test can be performed to verify the 2D numerical solutions for the case of a deforming visco-elastic Maxwell body (Exercise 12.1). In the case of uniform pure shear, deformation of an initially unstressed, incompressible visco-elastic medium with a constant strain rate $\dot{\varepsilon}_{xx}$, elastic deviatoric stress σ'_{xx}, grows with time t according to the equation

$$\sigma'_{xx} = 2\dot{\varepsilon}_{xx}\eta[1 - \exp(-t\mu/\eta)], \tag{20.25}$$

where t is the time from the beginning of deformation and η and μ are the constant viscosity and shear modulus of the medium, respectively. Based on Eq. (20.25), one can perform a numerical test of stress buildup shown in Fig. 20.11. The numerical experiment is designed on a rectangular model (cf. panel in Fig. 20.11) by prescribing constant outward directed velocity v_x along the vertical boundaries and inward directed velocity v_y for the horizontal boundaries of the model computed as

$$v_x = \tfrac{1}{2}\dot{\varepsilon}L_x,$$

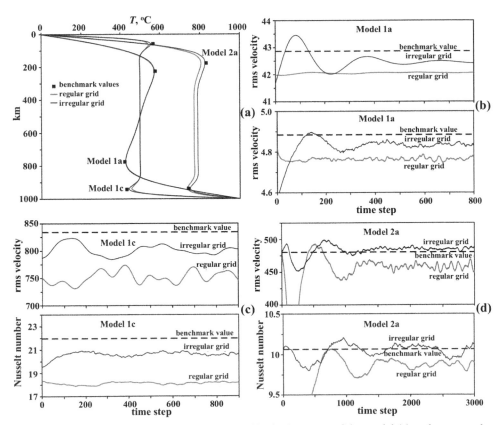

Fig. 20.10. Vertical steady-state temperature profiles in the centre of the model (a) and near-steady-state variations of root mean square (rms) velocity and Nusselt number (b)–(d) for the three mantle convection benchmarks from Table 20.1. Dashed lines in (b)–(d) show the benchmark values for respective parameters from Table 20.1. Solid lines show the numerical results calculated at a resolution of 51 × 51 nodes and 200 × 200 randomly distributed markers with the code **Variable_viscosity_convection_irregular_grid.m**.

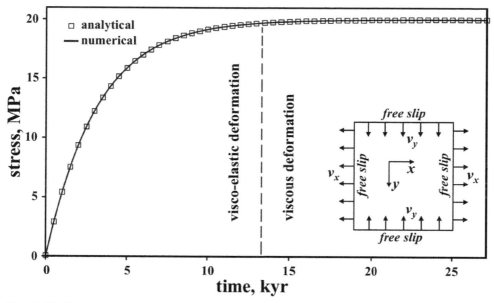

Fig. 20.11. Comparison of numerical (symbols) and analytical (solid line) solutions for the case of visco-elastic stress buildup due to pure shear (*x-y* direction) with constant normal strain rate and in the absence of gravity. Numerical and analytical (Eq. 20.25) solutions are compared for $\dot{\varepsilon}_{xx} = 10^{-14}$ s^{-1}, $\eta = 10^{21}$ Pa s and $\mu = 10^{10}$ Pa. The panel with numerical setup is shown in the right part of the diagram. Numerical results are calculated at a resolution of 51 × 51 nodes and 200 × 200 markers with the code **Stress_buildup.m**.

$$v_y = \tfrac{1}{2}\dot{\varepsilon}L_y,$$

where $\dot{c}$ is prescribed deviatoric strain rate, and L_x and L_y are respectively horizontal and vertical dimensions of the model. At each time step, all deviatoric stress components are interpolated from markers (either regularly or randomly distributed) to nodes, and stress increments are then interpolated back to markers (Fig. 13.1 in Chapter 13) after numerically solving the momentum and continuity equations for the entire model domain. Figure 20.11 is computed with the code **Stress_buildup.m** and demonstrates the high accuracy of the numerical solution, which overlaps with the analytical one, hence properly describing the transition from the dominant elastic regime to the prevailing viscous deformation.

20.11 Test 10: recovery of the original shape of an elastic slab

This benchmark can be performed to test the 2D visco-elastic numerical solutions in terms of proper advection and conservation of elastic stresses. Figure 20.12 shows the results of a numerical experiment for the recovery of the original shape of an elastic slab surrounded by a low density medium with much lower viscosity and much higher shear modulus than the slab.

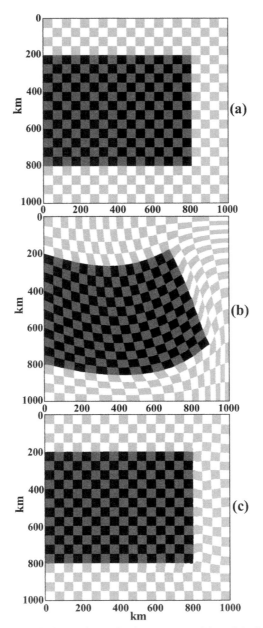

Fig. 20.12. Results of a numerical experiment for the recovery of the original shape of a visco-elastic slab (black, dark grey, $\rho = 4000$ kg/m^3, $\eta = 10^{27}$ Pa s and $\mu = 10^{10}$ Pa) embedded in a weak visco-elastic medium (light grey, white, $\rho = 1$ kg/m^3, $\eta = 10^{21}$ Pa s and $\mu = 10^{20}$ Pa). (a) Initial configuration. (b) Configuration after 20 kyr of deformation under constant vertical gravity field ($g_x = 0$, $g_y = 10$ m/s^2). (c) Configuration achieved within 9980 kyr of spontaneous deformation after switching off gravity (i.e. after the condition $g_x = g_y = 0$ is applied at 20 kyr). Boundary conditions: no slip at the left boundary and free slip at all other boundaries. Numerical results are calculated at a resolution of 51 × 51 nodes and 200 × 200 markers with the code **Slab_deformation.m** associated with this chapter. Note the irreversible viscous deformation of the weak surrounding medium, which is visible in its perturbed checkerboard structure close to slab corners in (c).

The initially unstressed slab is attached to the left wall of the box and is spontaneously deformed within 20 kyr under a purely vertical gravity field ($g_y = 10$ m/s^2, $g_x = 0$). The slab deformation is purely elastic due to the large Maxwell time (3 170 000 kyr) of slab material compared to the total deformation time (20 000 kyr). In contrast, the low viscosity medium is subjected to irreversible, purely viscous deformation since its Maxwell time (3.17×10^{-10} kyr) is negligible compared to the deformation time. The degree of elastic deformation in the slab is large (Fig. 20.12b) and the stresses stored on markers are, therefore, subjected to significant advection and rotation under both simple shear and pure shear deformation. After gravity is 'switched off' (i.e. after the condition $g_x = g_y = 0$ is set), the slab starts to unbend and finally fully recovers its original shape (Fig. 20.12c). In contrast, the low density medium does not recover its original configuration since the viscous deformation is irreversible (see perturbations of the checkerboard pattern in the weak medium around the slab corners).

For the model shown in Fig. 20.12, the deformation rate is independent of time step and is fully determined by the viscosity of the low density medium, which acts as a stronger material (note upbending of the lower right edge of the slab in response to the flow of the low density medium around the slab). This relationship is caused by the lower shear modulus of the slab (10×10^{10} Pa) compared to that of the low density medium (10×10^{20} Pa). In contrast, in Figure 12.2 from Chapter 12, another situation is shown (Gerya and Yuen, 2007) where shear moduli of both materials are the same and the low density medium acts as a weak material. The character of slab deformation changes correspondingly (dominant simple shear deformation and no significant upbending). In this case, however, the deformation rate is time step dependent, which does not preclude, indeed, testing the slab shape recovery (Fig. 12.2).

20.12 Test 11: numerical sandbox benchmark

Let us also consider the comparison of numerical results with physical (analogue) sandbox experiments. Numerical modelling of sandbox experiments poses significant computational challenges because the numerical code must be able (1) to calculate large strains along spontaneously forming narrow shear zones, (2) to represent complex boundary conditions, including frictional boundaries and free surfaces, and (3) to include a complex rheology involving both viscous and frictional/plastic materials. These challenges reflect directly the state of the art requirements for numerical modelling of large-scale tectonic processes. Several numerical sandbox benchmarks were described by Buiter et al. (2006, 2016), in which the results of analogue and numerical experiments for both shortening (Fig. 20.13) and extension settings were compared. One of the shortening experiments (Buiter et al., 2006) was conducted with the use of a mobile wall moving leftward at a velocity of 2.5 cm/hour (Fig. 20.13a). The original cross-section is composed of sand (density $\rho = 1560$ kg/m^3, cohesion $C = 10$ Pa, an initial internal friction angle of $\varphi_{initial} = 36°$ which changes linearly to the stable value of $\varphi_{stable} = 31°$ with strain increasing

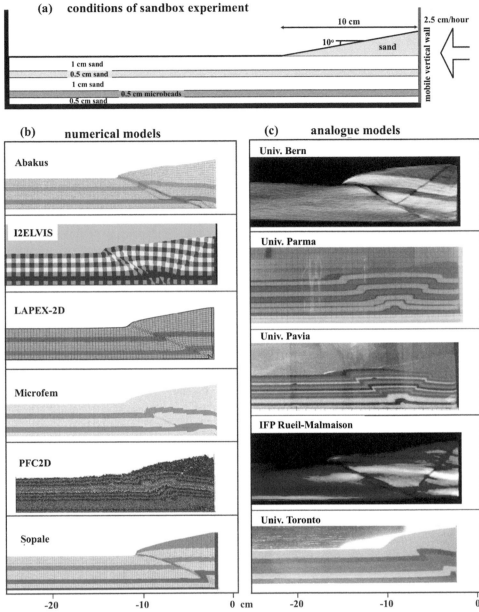

Fig. 20.13. Setup of a shortening experiment (a) and comparison of numerical (b) and analogue (c) models (at ~2 cm of shortening) performed by Buiter et al. (2006). (a) Horizontal layers of 'sand' (which have the same properties and differ in colour only) with an embedded layer of weaker 'microbeads' are shortened through a mobile wall on the right hand side which is pushed leftwards. (b),(c) Names of numerical codes (b) and analogue laboratories (c) are given in the respective model boxes.

from 0 to 1) and includes a 0.5 cm thick weak layer of microbeads ($\rho = 1480$ kg/m^3, $C = 10$ Pa, $\varphi_{initial} = 22°$, $\varphi_{stable} = 20°$). In the right part, the model includes a 10 cm wide surface wedge composed of sand. Boundary friction on all sandbox walls is lowered ($C = 0$, $\varphi_{initial} = 19°$, $\varphi_{stable} = 19°$). Boundary conditions corresponding to the mobile wall can be implemented in a number of ways. One option is to include a rigid (highly viscous) mobile wall and prescribe constant velocity conditions ($v_x = -2.5$ cm/hour, $v_y = 0$) on Eulerian nodes located inside this wall (Fig. 20.14a). This can be done in combination with a weak layer included in the model, which simulates air and shifts behind the wall as it moves. In order to ensure that the wall does not leave nodes with prescribed velocity, it can be thickened from behind, by accreting displaced air markers (Fig. 20.14b). It should be pointed out that the implementation of the mobile wall condition may notably affect the results of numerical experiments: for example, a backthrust that forms in most of the analogue experiments is absent in many numerical models where a mobile wall condition is implemented by prescribing a shortening velocity directly on the right model boundary (cf. Fig. 20.13b and 20.13c). The numerical and analogue models share many similarities (Buiter et al., 2006).

(1) Shortening is accommodated by an in-sequence forward propagation of thrusts (Fig. 20.14c, also see Fig. 12.6 in Chapter 12).
(2) The first-formed thrust roots at the base of the mobile wall (Fig. 20.14c).
(3) By 2 cm of displacement, an active thrust has formed in all models (Figs. 20.13b,c, 20.14c).
(4) The location where the first-formed forward thrust reaches the surface is influenced by the surface wedge in almost all of the experiments (Figs. 20.13b,c, 20.14c).

It should be pointed out, however, that details of shear zone patterns formed in individual analogue and numerical models are strongly variable. Such variations are an intrinsic feature of plastic deformation, and reproducing the exact pattern of shear zones should not be considered as the benchmarking goal. More importantly, with this benchmark a numerical code should rather demonstrate its ability to hold for large deformation, for strong strain localization along spontaneously forming narrow (1–2 grid cell wide) shear zones and for reproducing the general structural pattern of both forward and backward faults formed in analogue experiments. Figure 20.14 shows the results of the numerical sandbox experiments obtained with the code **Sandbox_shortening_ratio.m**. The differences between the numerical and analogue models are on the same order as the differences between analogue models from different laboratories (cf. Figs. 20.13b,c, 20.14b,c). The implemented numerical approach of plasticity treatment (Chapters 12 and 13) allows for spontaneous onset of narrow shear zones, which forms a sequence of forward and backward faults like in analogue experiments.

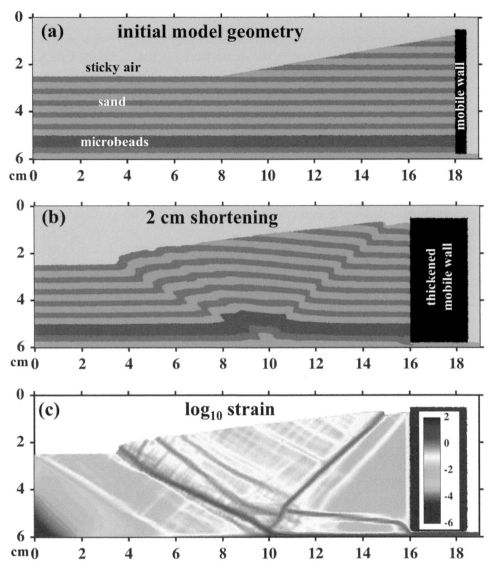

Fig. 20.14. Initial setup (a) and results (b),(c) of the numerical experiment for the shortening bench-mark (Fig. 20.13a). The numerical model employs a visco-elasto-plastic rheology with the following material properties: sand (light green, green), $\rho = 1560$ kg/m^3, $C = 10$ Pa, $\varphi_{initial} = 36°$, $\varphi_{stable} = 31°$, $\eta = 10^9$ Pa s, $\eta = 10^6$ Pa; microbeads (blue), $\rho = 1480$ kg/m^3, $C = 10$ Pa, $\varphi_{initial} = 22°$, $\varphi_{stable} = 20°$, $\eta = 10^9$ Pa s, $\mu = 10^6$ Pa; weak layer ('sticky air', very light green), $\rho = 1$ kg/m^3, $\eta = 10^2$ Pa s, $\mu = 10^6$ Pa; mobile wall (black), $\rho = 1520$ kg/m^3, $\eta = 10^{12}$ Pa s, $\mu = 10^{16}$ Pa. Boundary conditions: no slip at the left and bottom boundaries and free slip on all other boundaries. Boundary friction is implemented by prescribing $\varphi_{initial} = \varphi_{stable} = 19°$ for sand and microbeads located within 2 mm of the lower and left boundaries and of the mobile wall. The shortening condition ($v_x = -2.5$ cm/hour, $v_y=0$) is prescribed on the Eulerian nodes located inside the mobile wall. Note that the mobile wall is separated from the bottom by a 2 mm thick layer of sand and is thickening from the right by converting markers of the displaced 'sticky air'. Numerical results are calculated at a resolution of 191×61 nodes with 182 400 randomly distributed markers by using the code **Sandbox_shortening_ratio.m**.

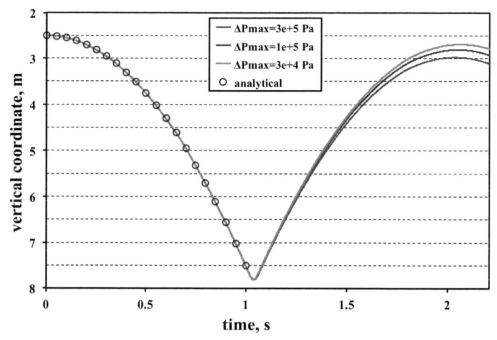

Fig. 20.15. Results of a numerical experiment for an elastic ball bouncing in a constant vertical gravity field ($g_x = 0$, $g_y = 10$ m/s², Fig. 14.3a). Material properties: ball, $\rho = 3000$ kg/m³, $\eta = 10^{22}$ Pa s, $\rho = 10^8$ Pa, $\beta = 10^{-8}$ 1/Pa, sticky air, $\rho = 1$ kg/m³, $\eta = 10^{-3}$ Pa s, $\mu = 10^8$ Pa, $\beta = 10^{-8}$ 1/Pa. Different curves show changes of vertical position of the ball centre with time for different time stepping conditions limited by the maximal allowed changes in deviatoric stresses and pressure per time step (ΔP_{max}). Numerical results are calculated at a resolution of 51×51 nodes and 200×200 markers with the code **Bouncing_Ball.m** associated with this chapter.

20.13 Test 12: bouncing ball benchmark

This simple elegant benchmark (Zhigadla, 2015) can be performed to test the 2D visco-elastic compressible numerical solutions with inertia (Chapter 14) in terms of proper conversion of kinetic energy into stresses and back. The numerical setup consists of a circular elastic object ('ball') surrounded by the sticky air (Fig. 14.3a, Exercise 14.2). The ball falls freely downward in the vertical gravity field until it collides with the lower model boundary and starts to deform elastically. Elastic deformation consumes kinetic energy of the ball by converting it into deviatoric stresses and pressure, thereby changing the velocity of the ball. As a result, the ball is reflected by the boundary and accelerates in the upward direction until all accumulated deviatoric stresses and pressure are released. Then the ball flies upward, being decelerated by the gravity field until its velocity goes to zero and it starts to fall back. If the deformation of the ball is purely elastic (i.e. fully reversible) and the numerical algorithm is robust, then the ball should keep bouncing without losing its potential energy and should always recover its original maximal altitude (or at least this altitude should decrease very slowly). Figure 20.15 shows the results of a numerical experiment for the recovery of the original altitude of the ball after the first bounce obtained with the code **Bouncing_Ball.m**. It can be seen that the quality of recovery mainly depends on the time stepping condition limited by the maximal allowed changes in deviatoric stresses and pressure per time step (Chapter 14). The smaller the allowed stress and pressure changes are the shorter the time step during the collision of the ball with the lower boundary and the more accurate the conversion of the kinetic energy into stresses and back. The results of this benchmark show that the SFD-MIC numerical algorithm discussed in Chapter 14 is robust and can be successfully used for accurate modelling of impact processes (Zhigadla, 2015).

20.14 Possible further benchmarks

Obviously, the potential number of benchmarks for testing numerical codes is infinite and not all of them are described in the present chapter. A few additional references for further numerical benchmarking problems are listed below:

- 2D analytical solutions for mantle thermal convection (Hager and O'Connell, 1981; Revenaugh and Parsons, 1987);
- 2D thermochemical convection (van Keken et al., 1997);
- 2D viscoplastic thermal convection (Tosi et al., 2015);
- 2D buoyancy-driven flows for strongly varying viscosity in the horizontal and vertical directions (Moresi et al., 1996; Zhong, 1996);
- 2D flow around deformable elliptic inclusions (Schmid and Podladchikov, 2003; Deubelbeiss and Kaus, 2008);
- 2D visco-elastic Rayleigh–Taylor instability (Kaus and Becker, 2007);
- 2D thermomechanical corner flows in subduction zones (van Keken et al., 2008);
- 2D spontaneous subduction with a free surface (Schmeling et al., 2008);
- 2D solitary waves in magma dynamics (Simpson and Spiegelman, 2011);
- 2D and 3D Stokes flow with variable viscosity (Popov et al., 2014b);
- 3D mantle convection in Cartesian geometry (Busse et al., 1994);
- 3D mantle convection in spherical geometry (Zhong et al., 2008);
- 3D infinitesimal and finite amplitude folding instability (Kaus and Schmalholz, 2006);
- 3D viscous incompressible Stokes flow in a spherical shell (Thieulot, 2017).

20.15 Method of manufactured solutions

One flexible and powerful approach to developing analytical solutions for geodynamic benchmarks is the *method of manufactured solutions* (MMS) (e.g. Roache, 1997, 1998; Dohrmann and Bochev, 2004; Galvan and Miller, 2013; May et al., 2013; Thieulot, 2014).

The MMS technique allows for the construction of arbitrarily complex solutions to any linear, or non-linear PDE. For Stokes flow problems, the MMS consists of prescribing some analytical functions (typically satisfying the incompressibility condition) for velocity components, viscosity and pressure to express their dependence on spatial coordinates (e.g. Dohrmann and Bochev, 2004). Based on these functions, boundary conditions and right parts of the conservation equations (e.g., buoyancy terms, ρg_i) are computed analytically, and the solutions are used for prescribing the internal model geometry. Accuracy of the numerical code is tested by comparing coupled numerical solutions of Stokes and continuity equations for the manufactured model with the originally prescribed analytical functions for velocity components and pressure. At present, the MMS is not commonly used in geodynamic modelling (e.g. Galvan and Miller, 2013; May et al., 2013; Thieulot, 2014), in stark contrast to the applied mathematics and engineering communities (e.g. Roache, 1997, 1998; Dohrmann and Bochev, 2004). Exploiting the MMS in computational geodynamics will help to establish a higher level of robustness and reliability in both the numerical methods used, and in the verification of their implementation (May et al., 2013).

Programming exercises

Exercise. 20.1.

Program an external MATLAB function for the 2D pressure-velocity Stokes+continuity variable viscosity solver for a regular staggered grid with external velocity nodes (Figs. 7.16, 18.8) based on the ghost node approach (Eqs. 18.39–18.42) and respective global indexing of unknowns as discussed in Chapter 7 (Fig. 7.16). Implement this solver in your viscous thermomechanical code (programming exercise 11.1 of Chapter 11). With this new viscous code, perform the falling block benchmark and compare the results with Figs. 20.2 and 20.3. The model setup corresponds to Fig. 20.2. An example is provided in **Variable_viscosity_block.m**.

Exercise 20.2.

Implement the ghost node based solver from the previous example into your visco-elastic thermomechanical code (Exercise 13.1). With this new code, perform the visco-elastic slab bending benchmark and compare the results with Fig. 20.12. The model setup corresponds to Fig. 20.12. An example is provided in **Slab_deformation.m**.

Exercise 20.3.

Modify Exercise 7.2 to build a 2D manufactured solution benchmark for isoviscous ($\eta = 10^{19}$ Pa s) incompressible Stokes flow with the following velocity and pressure fields defined analytically (cf. Exercise 1.2)

$$v_x = -v_{x0} \sin\left(2\pi \frac{x}{W}\right) \cos\left(\pi \frac{y}{H}\right),$$

$$v_y = v_{y0} \cos\left(2\pi \frac{x}{W}\right) \sin\left(\pi \frac{y}{H}\right),$$

$$P = P_0 + y\frac{\partial P_0}{\partial y} + \Delta P \cos\left(2\pi \frac{x}{W}\right) \sin\left(\pi \frac{y}{H}\right),$$

where x and y are respectively horizontal and vertical coordinates inside the box in metres; $W = 1\,000\,000$ m and $H = 1\,500\,000$ m are the width and height of the model, respectively; $v_{x0} = 1\times10^{-9}$ m/s and $v_{y0} = 3\times10^{-9}$ m/s are scaling values for respectively horizontal and vertical velocity components; $P_0 = 10^5$ Pa, $\partial P_0/\partial y = 33\,000$ Pa/m and $\Delta P = 10^6$ Pa are parameters of the pressure field. Derive analytical expressions for the right parts (i.e., ρg_x and ρg_y terms) of the following simplified x Stokes and y Stokes equations valid for incompressible constant viscosity flows (Chapter 5)

$$\eta\left(\frac{\partial^2 v_x}{\partial x^2} + \frac{\partial^2 v_x}{\partial y^2}\right) - \frac{\partial P}{\partial x} = -\rho g_x,$$

$$\eta\left(\frac{\partial^2 v_y}{\partial x^2} + \frac{\partial^2 v_y}{\partial y^2}\right) - \frac{\partial P}{\partial y} = -\rho g_y.$$

Use these expressions to compute the right parts of the discretized x Stokes and y Stokes equations using x and y coordinates of respective staggered velocity nodes. Compare the analytical and numerical solutions for velocity and pressure and visualize their differences. Check the influence of the model resolution on the numerical accuracy. An example is provided in **Manufactured.m**.

21

Design of 2D numerical geodynamic models

Theory: Warning message! What is numerical modelling all about? Rock properties for numerical geodynamic models. Design of numerical models for different geodynamic processes: visco-elasto-plastic slab bending, retreating subduction, lithospheric extension, collision, slab detachment, intrusion emplacement, mantle convection with phase changes, core formation.
Exercises: Designing a numerical model for studying extension of the continental lithosphere.

21.1 Warning message!

Several robust visco-elasto-plastic thermomechanical codes are provided with this chapter, which could be used as geodynamic modelling research tools. One can 'play' with these codes (*at one's own risk and responsibility, since I DO NOT PROVIDE ANY TECHNICAL SUPPORT for any use of these and other educational code examples*) by changing the model geometry and resolution, as well as the material properties and boundary conditions. There is nothing wrong with that and everyone is welcome to do it. However, be aware that numerical geodynamic modelling is not *pressing the button and automatically obtaining results* but *knowing in depth what you and your code is doing*. So, *don't play a lottery* by starting your numerical career by immediately using these codes as research tools. Before doing this, study this book carefully and make sure to correctly complete the exercises to learn about the *advantages* and *limitations* of the numerical modelling techniques used in the provided codes. Otherwise, there is a big risk that your 'automatically obtained results' appearing after 'pressing the button' will be EXTREMELY WRONG ...

21.2 What is numerical modelling all about?

Having continuously studied programming and numerical modelling techniques, we might get the impression that writing a good thermomechanical code is the main thing that guarantees success in numerical geodynamic modelling. Thinking that is a 'big mistake'!

Writing an extremely reliable code DOES NOT automatically imply that you will become a successful modeller ... We have to learn how to use our codes in the most efficient way, how to construct robust numerical models of various geodynamic, planetary and geotechnical processes, how to visualize and investigate these models and how to compare them to reality. In short: *designing thoughtful and realistic numerical models is at least as important as writing efficient numerical codes.* In order to help us with this issue, several examples of numerical models of various geodynamic and planetary processes are presented along with the design and technical details of the numerical experiments. The choice of these examples is, of course, subjective and mainly based on my scientific and aesthetic preferences, but this is what we have to live with.

What is numerical modelling about? Is it concerned with reproducing geological reality or investigating virtual ones? None of the two and both! We are not (either unfortunately or fortunately . . .) working in experimental physics where the conditions of experiments are 'relatively well' defined and known. Geological objects and systems are too complicated and their physical conditions are, in many cases, too poorly known to build fully deterministic numerical models. On the other hand, numerical modelling allows one to obtain some physical knowledge about such complex systems by studying, systematically, simpler end-member cases. And this is normal! Like in experimental petrology, rather simple systems such as MgO-Al_2O_3-SiO_2 (MAS) or CaO-FeO-MgO-Al_2O_3-SiO_2 (CFMAS) are often studied instead of natural rocks that are composed of at least 10–13 major oxides (Na_2O-K_2O-TiO_2-CaO-FeO-Fe_2O_3-MgO-MnO-Al_2O_3-SiO_2-H_2O-P_2O_5-CO_2). However, this simplification does not preclude the broad applicability of experimental results to natural rocks! Likewise, numerical modelling *is not a tool for fitting models to nature*, but instead *a research instrument to understand how nature works*.

21.3 Material properties

The choice of material properties in numerical models is very important. How models are set up, and which predictions they deliver, crucially depends on this choice. Indeed, there is a large uncertainty in material (*especially rheological*) properties of rocks that are strongly variable in nature. In addition, many physical properties of natural rocks (e.g., gross-scale effective rheology, Baumann and Kaus, 2015) are poorly constrained. Therefore, some subjectivity is always present in defining model parameters. Tables 21.1 and 21.2 summarize the material properties that are used in the following examples. The choice of properties is based on widely accepted geodynamic literature (such as Ranalli, 1995; Turcotte and Schubert, 2002) and (unavoidably) on the author's personal experience of building 'realistic geodynamic models' (to be honest this is a highly ambiguous term and the judgment between 'realistic' and 'non-realistic' is often affected by aesthetic preferences which are, in turn, strongly defined by cartoons provided in textbooks and geological literature . . .).

Table 21.1 *Rheological flow laws* used in numerical experiments (from a compilation by Ranalli, 1995)*

Material	A_D MPa^{-n} s^{-1}	n	E_a kJ mol^{-1}
olivine (dry)	2.5×10^4	3.5	532
olivine (wet)	2.0×10^3	4.0	471
rock salt	6.3	5.3	102
quartz	1.0×10^{-3}	2	167
plagioclase An$_{75}$	3.3×10^{-4}	3.2	238
orthopyroxene	3.2×10^{-1}	2.4	293
clinopyroxene	15.7	2.6	335
granite	1.8×10^{-9}	3.2	123
granite (wet)	2.0×10^{-4}	1.9	137
quartzite	6.7×10^{-6}	2.4	156
quartzite (wet)	3.2×10^{-4}	2.3	154
quartz diorite	1.3×10^{-3}	2.4	219
diabase	2.0×10^{-4}	3.4	260
anorthosite	3.2×10^{-4}	3.2	238
felsic granulite	8.0×10^{-3}	3.1	243
mafic granulite	1.4×10^4	4.2	445

* $\dot{\varepsilon}_{II} = A_D(\sigma_{II})^n \exp\left(-\frac{E_a}{RT}\right)$.

21.4 Visco-elasto-plastic slab bending

Modelling of slab bending is very important in geodynamics since this process is always associated with subduction and is related to the structural and seismic features in the trench area (e.g. Ranero et al., 2003, 2005). Of special interest is bending related faulting of the incoming plate, which creates a pervasive tectonic fabric that cuts across the crust, penetrating deep into the mantle (e.g. Ranero et al., 2003, 2005). Faulting is active across the entire oceanic trench slope, thereby promoting hydration of the cold crust and upper mantle surrounding these deep active faults, which may in turn control serpentinization, seismic anisotropy and seismicity of subducting slabs (Faccenda et al., 2008a, 2009, 2012; Dymkova and Gerya, 2013; Naliboff et al., 2013; van Dinther et al., 2014).

The numerical setup for bending of a subducting slab is rather simple (Fig. 21.1a,b), but requires relatively high resolution (≤2 km horizontal and vertical grid step in the slab bending area to adequately resolve the bending related normal faults) and realistic pressure, temperature- and stress-dependent visco-elasto-plastic rheology (Chapters 12, 13). One way to investigate spontaneous slab subduction and bending consists of using an initial setup for subduction initiation across a pre-existing transform fault (Hall et al., 2003; Gerya

Table 21.2 *Physical properties of rocks* used in numerical experiments*

Material	ρ_0 kg/m³	k W/(m K) (at T_K, P_{MPa})	$T_{solidus}$ K (at P_{MPa})	$T_{liquidus}$ K (at P_{MPa})	Q_L kJ/kg	H_r μW/m³	Flow law	μ GPa
Sediments	2700 (solid) 2400 (molten)	$0.64 + 807/(T + 77)$	$889 + 17900/(P + 54) + 20200/(P + 54)^2$ at $P < 1200$ MPa, $831 + 0.06P$ at $P > 1200$ MPa	$1262 + 0.09P$	300	0.5–5	wet quartzite	10
Upper continental crust	2700 (solid) 2400 (molten)	$0.64 + 807/(T + 77)$	$889 + 17900/(P + 54) + 20200/(P + 54)^2$ at $P < 1200$ MPa, $831 + 0.06P$ at $P > 1200$ MPa	$1262 + 0.09P$	300	0.5–5	wet quartzite	10
Lower continental crust	2800–3000 (solid) 2500–2700 (molten)	$1.18 + 474/(T + 77)$	$973 - 70400/(P + 354) + 77800000/(P + 354)^2$ at $P < 1600$ MPa, $935 + 0.0035P + 0.0000062P^2$ at $P > 1600$ MPa	$1423 + 0.105P$	380	0.25–0.5	plagioclase An$_{75}$	25
Upper oceanic crust (basalt)	3000–3500 (solid) 2700 (molten)	$1.18 + 474/(T + 77)$	$973 - 70400/(P + 354) + 77800000/(P + 354)^2$ at $P < 1600$ MPa, $935 + 0.0035P + 0.0000062P^2$ at $P > 1600$ MPa	$1423 + 0.105P$	380	0.25	plagioclase An$_{75}$	25
Lower oceanic crust (gabbro)	3000–3500 (solid) 2700 (molten)	$1.18 + 474/(T + 77)$	$973 - 70400/(P + 354) + 77800000/(P + 354)^2$ at $P < 1600$ MPa, $935 + 0.0035P + 0.0000062P^2$ at $P > 1600$ MPa	$1423 + 0.105P$	380	0.25	plagioclase An$_{75}$	25

Lithosphere–asthenosphere dry mantle	3300 (solid) 2700 (molten)	$[0.73 + 1293/(T + 77)] \times (1 + 0.00004P)$	$1394 + 0.132899P - 0.000005104P^2$ at $P < 10000$ MPa $2212 + 0.030819(P - 10000)$ at $P > 10000$ MPa	$2073 + 0.114P$	400	0.022	dry olivine	67
Lithosphere–asthenosphere hydrated mantle	3000–3300 (solid) 2700 (molten)	$[0.73 + 1293/(T + 77)] \times (1 + 0.00004P)$	$1240 + 49800/(P + 323)$ at $P < 2400$ MPa, $1266 - 0.0118P + 0.0000035P^2$ at $P > 2400$ MPa	$2073 + 0.114P$	400	0.022	wet olivine	67
References **	1, 2	3, 9	4, 5, 6, 7, 8	4	1, 2	1	Table 21.1	1

* Other properties (for all rock types): $C_P = 1000$ J kg^{-1} K^{-1}, $\alpha = 3 \times 10^{-5}$ K^{-1}, $\beta = 1 \times 10^{-11}$ Pa^{-1}.

** 1 Turcotte and Schubert, 2002; 2 Bittner and Schmeling, 1995; 3 Clauser and Huenges, 1995; 4 Schmidt and Poli, 1998; 5 Hess, 1989; 6 Hirschmann, 2000; 7 Johannes, 1985; 8 Poli and Schmidt, 2002; 9 Hofmeister, 1999.

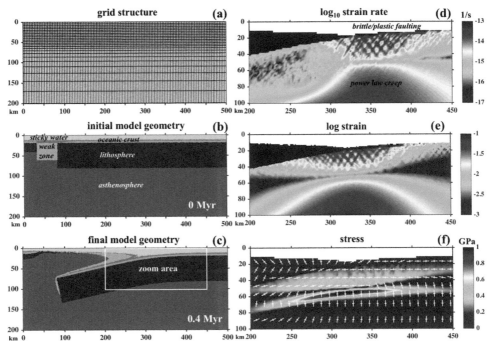

Fig. 21.1. Numerical grid (a), initial conditions (b), and results (c)–(f) of a numerical experiment for visco-elasto-plastic slab bending during spontaneously retreating subduction. Model resolution is 251 × 51 nodal points with 100 000 randomly distributed markers. Grid resolution (a) is non-uniform in the vertical direction (each second grid line is shown). Cooling ages of the left and right plates in (b) are 1 Myr and 70 Myr respectively. (d), (e) and (f) shows numerical results for the zoomed area outlined in (c). Note that plastic deformation along faults in (d) is deactivated in the subducted portion of the slab. White crosses in (f) show the orientation of the principal stress axes; long and short branches of the crosses show extension and shortening directions, respectively. Lithospheric and asthenospheric mantle in (b) and (c) do not differ in properties (dry mantle, Table 21.2), but different colours are used for better visualization of slab bending. Results are computed with the code **Subducting_slab_bending.m**.

et al., 2008a; Dymkova and Gerya, 2013). The experiment begins with two plates of different ages, juxtaposed along a transform fault (cf. light blue weak zone in Fig. 21.1b) with low plastic strength ($\gamma_{int} = 0$), which creates favourable conditions for spontaneous initiation of subduction and concurrent slab bending. The vertical thermal structure of the plates is computed according to the cooling of a semi-infinite half-space (Turcotte and Schubert, 2002)

$$T(d) = T_1 + (T_0 - T_1)\left(1 - \mathrm{erf}\left(\frac{d}{2\sqrt{\kappa\tau}}\right)\right), \tag{21.1}$$

where $T_0 = 273$ K is temperature at the surface for both plates, $T_1 = 1700$ K is temperature at the bottom of the model, d is depth in metres below the surface, κ is thermal diffusivity (10^{-6} m^2 s^{-1}) and τ is the age in seconds of the plates.

Bending is driven by strong negative buoyancy of the older plate while the weak fault allows initial displacement, which results in the spontaneous retreating subduction. To ensure self-sustaining, one-sided subduction, the weak, hydrated upper oceanic crust (basalts, sediments) is present atop the slab providing stable lubrication against the moving and cooling overriding plate (e.g. Sobolev and Babeyko, 2005; Gerya et al., 2008a; Dymkova and Gerya, 2013). On the other hand, a weak upper layer present above the crust ('sticky water', $\eta = 10^{18}$ Pa s, $\rho = 1000$ kg/m^3) provides a free-surface-like condition (Chapter 8), which is essential for natural slab bending to occur (e.g. Gerya et al., 2008a; Crameri et al., 2012a,b). The validity of the weak layer approach (Chapter 8) to approximate the free surface has recently been tested and proven (Schmeling et al., 2008; Crameri et al., 2012a) with the use of a large variety of numerical techniques (including our methodology based on conservative finite differences and marker-in-cell techniques) and comparison with analogue models. The thickness of the weak layer should be at least 4–5 grid cells and its viscosity should be at least 100 times less than that of the underlying lithosphere (e.g. Crameri et al., 2012a). For the geodynamic models presented here, we will use a 10–15 km weak layer with viscosity of 10^{18}–10^{19} Pa s (larger thickness and/or lower viscosity of this layer requires implementation of the free surface stabilization to avoid the 'drunken sailor' instability, Chapter 8).

Figure 21.1 shows the results of a numerical experiment for spontaneous bending of a retreating subducting plate obtained with the code **Subducting_slab_bending. m**. The deformation pattern in the bending slab is distinct (Fig. 21.1d): the top of the slab is subjected to intense plastic deformation with localized faults while the bottom of the slab deforms in a ductile way (i.e. by the temperature- and stress-activated dislocation creep, cf. Table 21.1) with enhancement of the deformation (see orange high strain rate zones inside the slab in Fig. 21.1d) due to high stresses (see yellow to red high stress zones in Fig. 21.1f) in the bending area. The plastic faulting and ductile deformation fields are characterized by extension and compression in the horizontal direction, respectively (see orientation of stress principal axes in Fig. 21.1f). These two fields are clearly separated by the narrow, non-deforming middle plane of the slab (see blue zone inside the slab in Fig. 21.1e), which is characterized by small deviatoric stresses (see blue zone inside the slab in Fig. 21.1f). The penetration depth of faults (10–50 km) (Fig. 21.1e) is in agreement with the observational constraints (e.g. Ranero et al., 2003, 2005). Results of the experiment show that a slab with a free upper surface can easily be bent by its own weight, thereby triggering spontaneous retreating subduction. Bending is facilitated by (i) lowered pressure in the extension region, which favours deep penetration of faults and (ii) large stresses in the compression region, which produces a local lowering of the slab viscosity due to the power-law nature of ductile creep.

21.5 Retreating oceanic subduction

This model is comparable to the previous model, but the model domain is much larger to allow a longer trench retreat and deeper slab penetration into the asthenospheric mantle. To avoid a significant increase in the number of grid points, we employ an adaptive non-uniformly spaced grid ('Swiss cross' AMR, Chapter 17) with a high-resolution area that moves together with the trench (Gerya et al., 2008a; Nikolaeva et al., 2008). The grid spacing increases gradually away from the area of high resolution by a constant factor F at every nodal point (cf. vertical resolution in the lower part of Fig. 21.2a). In order to compute this incremental factor, Eq. (17.1) is applied, which is solved iteratively in the beginning of each time step (Chapter 17). In addition, since the model is much deeper than the previous one and the pressure varies significantly from the top to the bottom, we should take into account the activation volume of dislocation creep.

The activation volume V_a for olivine creep (this creep is assumed to represent the mantle rheology, Tables 21.1, 21.2) varies from 0 (wet olivine) to 17 cm^3 (dry olivine) (e.g. Ranalli, 1995). Intermediate values of V_a are possible as the water content in the mantle varies. Subduction model development can be notably affected by this parameter (e.g. Mishin et al., 2008) and, therefore, investigation of a range of activation volume variations is required to understand how the model works. The same applies to the asthenospheric mantle temperature, the plastic strength and thickness of the upper oceanic crust (lubricating layer), plate ages, initial size of the weak zone and rheology etc. This generally means that *investigating the model parameter space* (to some degree, of course, since the parameter space of a model with 10 variable parameters has at least 2^{10} permutations and consequently at least 1024 numerical experiments have to be run to investigate it in a somewhat systematic manner . . .) is a necessary component of any numerical geodynamic study. Robust conclusions should only be based on the observations obtained from several/many models.

Figure 21.2 shows the evolving numerical grid and the results of a subduction experiment. The subducting slab spontaneously retreats toward the right hand side of the box and the high-resolution area of the grid follows the trench area (cf. change in position of the solid triangle in Fig. 21.2a and b). The slab steepens and the intensity of visco-elasto-plastic bending (Fig. 21.1) increases with time. The rate of trench retreat is very fast during the first 2 Myr (around 20 cm/year, Fig. 21.3) but it drops to a few cm/year as soon as the lower tip of the slab moves toward the lower boundary of the model box, and penetrates into the deeper mantle that has a larger effective viscosity (influence of the chosen $V_a = 12$ cm^3). Generally, in models like this one, trench retreat is mainly controlled by the density contrast between the slab and the asthenosphere and by the viscous resistance of the asthenospheric mantle, which in turn depends on its rheology and temperature. For example, the rate of trench retreat decreases if a higher activation volume and/or a lower temperature for the asthenosphere are used (try to experiment using the program **Subduction.m**).

The overall model behaviour is realistic and captures several important features of retreating intra-oceanic subduction zones. Slab bending is spontaneous and the slab deep

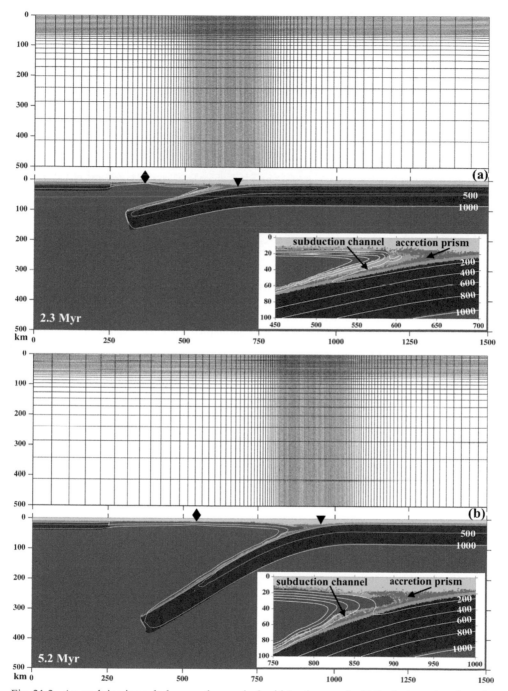

Fig. 21.2. An evolving irregularly spaced numerical grid (each second grid line is shown) and results (lithological field, see Fig. 21.1b for colour code) of a numerical experiment for spontaneous retreating subduction. The model resolution is 251 × 61 nodal points with 750 000 randomly

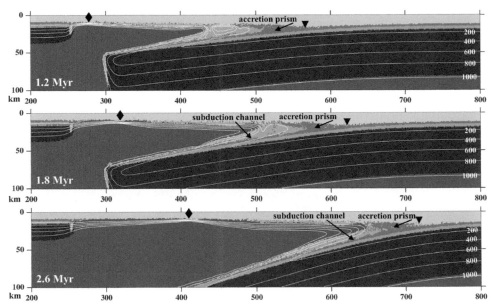

Fig. 21.3. Time evolution of the trench (triangle) and backarc spreading centre (diamond) associated with the growth and cooling of the overriding plate during a retreating subduction experiment (Fig. 21.2).

angle naturally increases with depth (Fig. 21.2b). An accretion prism and a subduction channel form and evolve spontaneously at the plate interface (cf. zoom-in in Fig. 21.2a,b). The upper (weak) part of the subducted oceanic crust (cf. dark green material in Fig. 21.2a, b) partly detaches from the slab and circulates in the subduction channel (Figs. 21.2, 21.3), which provides a pathway for the exhumation of high-pressure tectonic melanges to the surface (e.g. Cloos, 1982; Gerya et al., 2002; Gorczyk et al., 2007a). The overriding plate grows in length due to the trench retreat, and cools with time (cf. isotherms in Figs. 21.2 and 21.3). The plate growth process is accommodated by the spontaneous backarc spreading where the hot mantle approaches the surface (cf. diamonds in Fig. 21.3). The model topography also behaves in a realistic manner and shows a pronounced minimum in the

Caption for Fig. 21.2. *(cont.)*

distributed markers. Activation volume of mantle used in this experiment is 12 cm^3. The asthenospheric mantle temperature at the bottom of the model is 1850 K and decreases upward according to an adiabatic gradient of 0.5 K/km. The high-resolution area of the grid moves together with the retreating trench (black triangles). Inserts show a zoom-in of the moving trench area with a spontaneously evolving accretion prism and deep subduction channel. White labelled lines on the inserts are isotherms in °C. Black diamonds show the position of the backarc spreading centre. Results are computed with the code **Subduction.m**.

trench area as well as a visible maximum above the spreading centre (Fig. 21.3). Obviously, no topography maximum is formed in the middle of the overriding plate (i.e. in the area where a natural magmatic arc would grow) since our model does not account for magmatism and crustal growth. This aspect, however, can be improved by using more sophisticated and realistic models, which involve mantle wedge hydration and melting, and related melt extraction and crustal growth (e.g. Nikolaeva et al., 2008; Gerya and Meilick, 2011; Vogt et al., 2012).

21.6 Lithospheric extension

Lithospheric extension is an important geodynamic process, for example at mid-ocean ridges, backarc (Fig. 21.3) and intra-arc extension zones, passive and active rifting, continental break-up, formation of sedimentary basins, etc. (e.g. Turcotte and Shubert, 2002). Realistic modelling of such settings poses computational challenges (e.g. Burov and Poliakov, 2001; Gerya, 2010b, 2013; Huismans and Beaumont, 2011; Burov and Gerya, 2014) since extension of the lithosphere and the underlying mantle is associated with intense and simultaneous viscous and brittle/plastic (faulting) deformations, as well as with notable changes in topography. Modelling requires sufficiently high resolution (≤ 2 km horizontal and vertical grid step) in the extension area where strongly localized deformation along faults takes place. One way to address these challenges consists in using an evolving non-uniformly spaced numerical grid (as we used for modelling retreating subduction, Fig. 21.2) combined with a variable model size that evolves with time in response to imposed bulk extension. This approach suggests re-meshing at every time step, which can be easily done with our marker-in-cell algorithm (Chapter 17).

Figure 21.4 presents the results of modelling extension of a 70 Myr old oceanic lithosphere under an imposed constant extension rate ($v_{extension} = 2$ cm/year). The initial model is 400×300 km^2 in size and includes both lithospheric and asthenospheric domains as well as a weak top layer (sticky water) that allows a natural thermomechanical evolution of both upper and lower boundaries of the lithosphere.

As in previous examples, the lower lithosphere boundary is not prescribed and forms spontaneously as a rheological boundary between colder and stronger parts of the mantle and the underlying hotter and weaker (asthenospheric) region. Extension is prescribed symmetrically as constant horizontal outward velocity boundary conditions at two sides of the model

$$V_{outward} = \frac{1}{2} V_{extension.} \tag{21.2}$$

In order to ensure mass conservation in the computational model, a vertical inward velocity which changes at every time step is prescribed along the lower model boundary

$$v_{inward(t)} = \frac{H_{(t)}}{L_{(t)}} v_{extension} \tag{21.3}$$

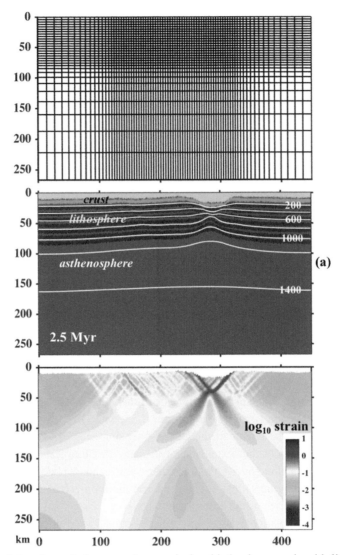

Fig. 21.4. Evolving irregularly spaced numerical grid (each second grid line is shown), lithological field (see Fig. 21.1b for colour code) and finite strain for two stages of an oceanic lithosphere extension experiment. Model resolution is 161 × 61 nodal points with 120 000 randomly distributed markers. Activation volume of the mantle used in this experiment is 10 cm³. No strain weakening is used for plastic deformation. Initial asthenospheric mantle temperature at the bottom of the model is 1750 K and decreases upward according to an adiabatic gradient of 0.5 K/km. Non-compositional layering of the mantle lithosphere is shown for visualizing deformation. White labelled lines are isotherms in °C. Results are computed with the code **Extension.m**.

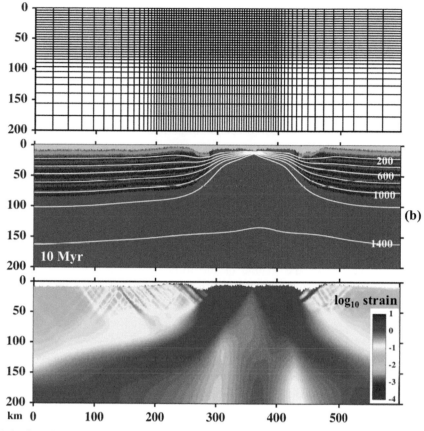

Fig. 21.4. *(cont.)*

where $L_{(t)}$ and $II_{(t)}$ are the width and height of the model, respectively. The model width and height change with time as

$$L_{(t)} = L_0 + t \times v_{extension} \tag{21.4}$$

$$H_{(t)} = L_0 \times H_0/L_{(t)} \tag{21.5}$$

where $L_0 = 400$ km is the initial model width, t is time from the beginning of the experiment and $H_0 = 300$ km is the initial model height. The resolution of the numerical grid is again based on the 'Swiss cross' AMR principle (Chapter 17) and contains 161×61 nodal points, with 120 000 randomly distributed markers. The dense 2×2 km^2 part of the grid covers the central 200×80 km^2 area of the model where the extension is localized (not necessarily exactly in the middle, of course). Resolution in the remaining grid changes (Eq. 17.1) at every time step in response to the model stretching.

Figure 21.4 presents results of the numerical experiment for the oceanic lithosphere extension computed with the code **Extension.m**. Two stages are clearly seen: (1) normal faulting and visco-plastic necking of the old oceanic lithosphere (Fig. 21.4a) and (2) growth of new oceanic lithosphere associated with the formation of a spreading centre (Fig. 21.4b). Extensional deformation associated with normal faulting (Fig. 21.4a) initially starts in a 200–300 km wide area of the lithosphere. It is followed by a spontaneous and gradual (within 1.5–2.5 Myr) focusing of the deformation in the necking area where a new spreading centre forms and the hot asthenospheric mantle comes close to the surface (see distribution of isotherms in Fig. 21.4b). The topography development is characteristic: the topographic low develops in the spreading centre and two elevated regions form on the rift flanks (e.g. Burov and Cloetingh, 1997; Burov and Poliakov, 2001). However, our model topography is assumed to be submarine and no erosion or sedimentation processes, which may notably affect lithospheric extension (e.g. Burov and Cloetingh, 1997; Burov and Poliakov, 2001), are taken into account. Like in the retreating subduction model, this experiment does not take into account decompression melting of the mantle or the formation of new oceanic crust in the spreading centre, which again points toward the need for developing more sophisticated and realistic models (e.g. Gerya, 2013).

21.7 Continental collision

Continental collision is another very 'popular' geodynamic setting that has been widely addressed by numerical modelling. In this case, much care should be taken to address erosion and sedimentation processes, which play major roles in orogeny associated with colliding continents (e.g. Willett, 1999; Beaumont et al., 2001; Ueda et al., 2015). According to numerical studies, variations in erosion/sedimentation rates during subduction and collision may significantly affect crustal mass flux and consequently alter crustal deformation and the behaviour of the crust-mantle interface (e.g. Willett, 1999; Beaumont et al., 2001; Pysklywec, 2006; Gerya et al., 2008b). One possibility for modelling these processes is to use *an internal evolving erosion/sedimentation surface* (Gerya and Yuen, 2003b) that separates the top boundary of the lithosphere from the overlaying sticky water/air layer with vertically stratified density (either 'air', 1 kg/m^3, for $y < y_{water}$ or 'water', 1000 kg/m^3, for $y > y_{water}$, where y_{water} is the water level adopted in the model). This surface evolves according to the transport equation solved at the Eulerian coordinates at each time step (Gerya and Yuen, 2003b):

$$\frac{\partial y_{es}}{\partial t} = v_y - v_x \frac{\partial y_{es}}{\partial x} + v_e - v_s, \tag{21.6}$$

where y_{es} is the vertical position of the surface as a function of the horizontal distance x; v_y and v_x are the vertical and horizontal components of the material velocity vector at the surface; v_s and v_e are sedimentation and erosion rates, respectively.

Erosion and sedimentation rates in Eq. (21.6) can be computed in various ways. The simplest is to use both slope and elevation independent *large-scale erosion and sedimentation rates* (e.g. Vance et al., 2003; Gerya and Yuen, 2003b; Gerya et al., 2008b), which correspond to the relations:

$$v_s = 0 \text{ mm/a}, \ v_e = v_{e0} \text{ when } y < y_{water} \text{ km,}$$

$$v_s = v_{s0} \text{ mm/a}, \ v_e = 0 \text{ when } y > y_{water} \text{ km,}$$

where v_{e0} and v_{s0} are imposed as constant erosion and sedimentation rates. Another possibility is to use more sophisticated models of surface processes that combine downhill diffusion erosion and fluvial erosion (e.g. Kooi and Beaumont, 1994; Braun and Sambridge, 1997; Burov and Cloetingh, 1997; Beaumont et al., 2001; Burov et al., 2001; Ueda et al., 2015). In the case of short-distance, downhill diffusion erosion (e.g. Kooi and Beaumont, 1994; Burov and Cloetingh, 1997), Eq. (21.6) can be modified to

$$\frac{\partial y_{es}}{\partial t} = v_y - v_x \frac{\partial y_{es}}{\partial x} + \frac{\partial}{\partial x}\left(K_s \frac{\partial y_{es}}{\partial x}\right), \tag{21.7}$$

where K_s is the effective 'topography diffusion' coefficient, which is highly variable (0–10^5 m^2/year, e.g. Kooi and Beaumont, 1994; Burov and Cloetingh, 1997). One possibility for solving the transport equations (21.6) and (21.7) consists in using a 1D Eulerian advection method such as upwind differences or the FCT method discussed in Chapter 8.

The oceanic lithosphere initially present between the two continental plates is another important component of continental collision models. Prescribing this lithosphere (e.g. Faccenda et al., 2008b; Gerya et al., 2008b; Warren et al., 2008; Duretz et al., 2011b; Sizova et al., 2012) allows a more natural beginning and more faithfully captures the initial stages of the collision process. During these stages, the model behaviour changes very rapidly due to the arrival of the positively buoyant continental crust in the subduction zone. Obviously, experiments of post-subduction collision should allow deep subduction of the oceanic slab and spontaneous bending of plates, which require the use of a sufficiently wide (>500 km) and deep (>200 km) model (e.g. Burov et al., 2001; Faccenda et al., 2008b; Gerya et al., 2008b; Warren et al., 2008; Duretz et al., 2011b; Sizova et al., 2012). One possibility to fulfil these requirements consists of using models with a variable size domain (e.g. Burov et al., 2001), like the one we explored for lithospheric extension. Another option is to use setups with a constant model size, internal plate velocity condition (as in the sandbox experiment, Chapter 20) and a permeable lower boundary (Burg and Gerya, 2005; Faccenda et al., 2008b; Gerya et al., 2008b) where infinity-like thermal and mechanical boundary conditions (Chapters 7 and 10) are prescribed.

Figure 21.5 shows the initial setup and results of a post-subduction collision experiment conducted with the code **Collision.m**. The 1000 × 300 km^2 model (Fig. 21.5a) uses a non-uniform 201 × 61 rectangular grid with a constant high resolution (2 × 2 km^2 grid cell size)

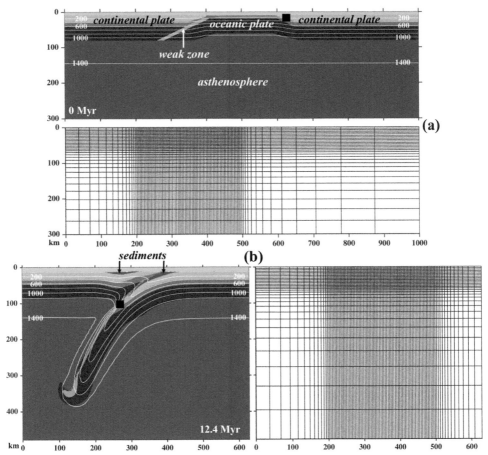

Fig. 21.5. Lithological field and evolving irregularly spaced numerical grid (each second grid line is shown) for the initial model setup (a) and culminating stage (b) of the continental collision experiment. Model resolution is 201 × 61 nodal points with 300 000 randomly distributed markers. Activation volume of the mantle used in this experiment is 10 cm^3. The initial asthenospheric mantle temperature at the bottom of the model is 1750 K and decreases upward according to an adiabatic gradient of 0.5 K/km. Convergence of 3 cm/yr is prescribed from the right. White labelled lines are isotherms in °C. The black square shows the position of a representative marker from the deeply subducting upper crustal rock unit for which the *P-T*-time path is shown in Fig. 21.8. Results are computed with the code **Collision.m**.

in the 300 × 60 km^2 area of the subduction/collision zone. Coarser resolution around this zone changes at every time step in response to the model shortening (leftward) and thickening (downward), which accommodates the convergence whose velocity is prescribed at the right model boundary. Boundary conditions are similar to those of the lithospheric extension example (Eqs. 21.3–21.5) but the model is shortening rather then extending and the left boundary does not change position with time. In order to compensate for the thickening of the sticky air/water layer on the top of the model, the water level changes at every time step as

$$y_{water(t)} = y_{water0} L_0 / L_{(t)}, \tag{21.8}$$

where y_{water0} = 7.5 km is the initial water level.

The initial material setup (Figure 21.5a) implies early oceanic-continental subduction with two continental sections (each 400 km wide) and a relatively short (200 km) intermediate oceanic plate, which is 40 Myr old. The continental crust is 35 km thick, with the upper and lower crustal layers (Table 21.2) of equal thickness. The nucleated subduction zone at the left ocean/continent boundary is prescribed as a 4–15 km wide weak zone cutting across the entire mantle lithosphere and reaching a depth of 90 km. The weak zone is prescribed as a wet brittle/plastic fault within mantle rocks, characterized by wet olivine rheological parameters and a low plastic strength of 1 MPa (i.e. assuming a high pore fluid pressure and γ_{int} = 0). During subduction, the pre-defined weak zone is spontaneously replaced by weak subducted crustal rocks, thereby preserving the decoupling along the interface. Obviously, the subduction zone is prescribed in a rather arbitrary way but this is what we have to live with since the issue of subduction initiation is controversial and over 10(!) conceptual models of subduction initiation have already been suggested in the literature (discussions by Ueda et al., 2008; Gerya, 2011; Stern and Gerya, 2018 and references therein).

The initial thermal structure of the continental lithosphere is laterally uniform and corresponds roughly to the usual continental geotherms (e.g. Turcotte and Schubert, 2002): 0°C at the surface linearly increasing to 1380°C at 100 km depth. The temperature structure of the oceanic plate corresponds to its age (Eq. 21.1). A gradual linear transition from the oceanic to the continental geotherm is prescribed within a 50 km wide area at the two ocean-continent boundaries. The initial temperature gradient in the asthenospheric mantle is 0.5°C/km. Due to thickening in response to shortening of the model with time, the temperature imposed for the lower boundary condition should increase with time according to the adiabatic gradient.

In this numerical model, the mechanisms driving subduction are a combined 'plate push' (prescribed constant convergence velocity at the right boundary) and 'slab pull' (temperature induced density contrast between the subducted oceanic lithosphere and surrounding mantle, see above retreating subduction model example). This type of boundary condition is widely applied in numerical models of subduction and collision (e.g. Burov et al., 2001; Burg and Gerya, 2005; Faccenda et al., 2008b; Warren et al., 2008), assuming that in the globally confined system of plates, the 'external push' imposed on a plate (coming from a different slab) can be significant. A spontaneously increasing slab pull mainly regulates slab bending dynamics and delamination of the slab from the overriding plate (e.g. Gerya et al., 2008b; Duretz et al., 2011b; Sizova et al., 2012; Ueda et al., 2012). Indeed, similar (and in relation) to the subduction initiation problem, the issue of choosing

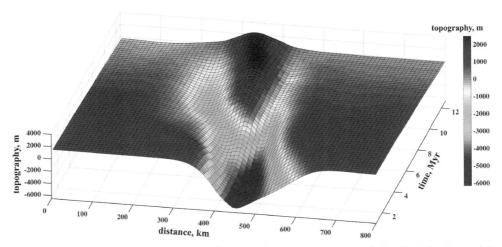

Fig. 21.6. Development of topography with time (relative to the water level, Eq. 21.8) for the model shown in Fig. 21.5. K_s = 0.00003 m²/s (1000 m²/year) is used in Eq. (21.7) which corresponds to approximately 1 mm/year gross-scale erosion/sedimentation rates. Results are computed with the code **Collision.m**.

proper convergence conditions is not fully resolved yet and the external push at the initial stages of convergence may be exaggerated compared to nature. Using spontaneously moving plates driven by the changing slab pull (e.g. Duretz et al., 2011b; Sizova et al., 2012) and mantle flow traction (e.g. Lu et al., 2015) can be noted as promising recent trends.

Figure 21.5b shows the final stage of an experiment for continental collision with a 200 km wide intermediate oceanic plate (30 Myr cooling age) moving leftward at a constant velocity of 3 cm/yr due to the model shortening imposed from the right boundary. Results are computed with the code **Collision.m**. The development of the continental collision zone is associated with deep (>100 km) subduction of the continental crust underneath the orogen (Fig. 21.5b). This feature appears in many numerical models of continental collision (e.g. Burov et al., 2001; Gerya et al., 2008b; Warren et al., 2008) including those driven by slab pull rather than by a prescribed convergence velocity (e.g. Faccenda et al., 2008b; Baumann et al., 2010; Duretz et al., 2011b; Sizova et al., 2012). This fits findings of *ultrahigh-pressure (UHP) rocks* (e.g. Chopin, 2003; Liou et al., 2004) that contain *metamorphic diamonds* (e.g. Rosen et al., 1972; Sobolev and Shatsky, 1990; Dobrzhinetskaya et al., 1995; Massonne, 1999) and coesite (e.g. Chopin, 1984; Smith, 1984) in Phanerozoic collision belts. The topography development (Fig. 21.6) predicted with our simplified model (Eq. 21.7) also appears realistic and demonstrates the growth of a positive topography (up to 2500 m above the water level) after the beginning of collision at around 5–7 Myr. Growth of the elevated region is associated with erosion and deposition of sediments on both sides of the 'orogen' (see brown material in Fig. 21.5b).

21.8 Slab breakoff

Slab breakoff (also called *slab detachment*) is an interesting 'hidden' process happening at depth within the mantle (ideal process for modellers indeed . . .). It was initially hypothesized on the basis of gaps in hypocentre distribution and tomographic images of subducted slabs (Isacks and Molnar, 1969; Barazangi et al., 1973; Pascal et al., 1973; Chung and Kanamori, 1976; Fuchs et al., 1979) and is supported by both theoretical considerations (Sacks and Secor, 1990; von Blanckenburg and Davies, 1995; Davies and von Blanckenburg, 1995) and detailed seismic tomography (Spakman et al., 1988; Wortel and Spakman, 1992, 2000; Xu et al., 2000; Levin et al., 2002). Slab detachment is often attributed to a decrease in subduction rate subsequent to continental collision (e.g. Davies and von Blanckenburg, 1995; Wong A Ton and Wortel, 1997), an effect caused by the buoyancy of the continental lithosphere introduced into the subduction zone (Fig. 21.5b). In addition to multiple geophysical, geological and geochemical investigations (cf. Andrews and Billen, 2009 and references therein), analytical models, laboratory and numerical experiments have been undertaken to characterize the breakoff processes (e.g. Davies and von Blanckenburg, 1995; Yoshioka and Wortel, 1995; Yoshioka et al., 1995; Wong A Ton and Wortel, 1997; Chemenda et al., 2000; Buiter et al., 2002; Gerya et al., 2004a; Faccenna et al., 2006; Faccenda et al., 2008b; Zlotnik et al., 2008; Andrews and Billen, 2009; Baumann et al., 2010; Burkett and Billen, 2010; Duretz et al., 2011b, 2014; van Hunen and Allen, 2011).

Recent thermomechanical models indicate two detachment modes (Andrews and Billen, 2009): (1) deep *viscous breakoff* which is characteristic of strong slabs and is controlled by thermal relaxation (heating) of the slab and subsequent thermomechanical necking in dislocation creep regime (Gerya et al., 2004a; Faccenda et al., 2008b; Zlotnik et al., 2008; Baumann et al., 2010; Duretz et al., 2011b), and (2) relatively fast, shallow *plastic breakoff* which is characteristic of weaker slabs and is controlled by plastic necking of the slab (Mishin et al., 2008; Ueda et al., 2008; Andrews and Billen, 2009; Duretz et al., 2011b). It was demonstrated that the time before the onset of viscous (but not plastic) detachment increases with the slab age, indicating that detachment time is controlled by the thickness and integrated stiffness of the thermally relaxing slabs (Gerya et al., 2004a; Andrews and Billen, 2009; Duretz et al., 2011b).

Breakoff can be modelled in a sufficiently self-consistent way starting, for example, from the configuration obtained in the continental collision experiment (Fig. 21.5b). We can essentially use the same code and stop convergence (either sharply or gradually) after the continental crust of the incoming plate reaches asthenospheric depths, assuming that the crustal buoyancy can potentially block further subduction. Even more consistent breakoff models use a spontaneous convergence of plates (driven by the slab pull) allowed after some period of forced convergence creating sufficient slab pull but before the actual beginning of collision (e.g. Faccenda et al, 2008b; Baumann et al., 2010; Duretz et al., 2011b; Sizova et al., 2012). In this case, plates should be detached from the model walls to permit horizontal movements.

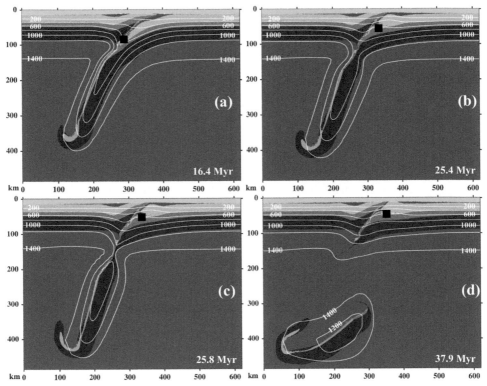

Fig. 21.7. Development of a viscous slab breakoff process. Initial conditions as well as model parameters correspond to Fig. 21.5b. Boundary conditions are free slip on all model boundaries. Black labelled lines are isotherms in °C. Four stages of breakoff are shown: (a) downward bending (steepening) and thermal relaxation of the slab, (b) beginning of the thermomechanical necking process, (c) slab detachment from the upper part of the plate, (d) rapid sinking and coherent rotation of the slab. The black square shows the position of a representative marker from deeply subducted upper crustal rock unit for which the *P-T*-time path is shown in Fig. 21.8. Time in figures is shown from the beginning of the collision experiment (Fig. 21.5). Results are computed with the code **Collision_and_breakoff.m**.

Figure 21.7 shows results of a breakoff experiment performed with the code **Collision_and_breakoff.m**. It uses the first approach and sharply stops model shortening (and obviously its thickening as well, otherwise the mass conservation condition in the model will be violated which would be really bad ...) after 12.7 Myr (i.e., shortly after reaching the mature continental collision configuration shown in Fig. 21.5b). The internal model geometry is then free to evolve spontaneously. In the beginning, dominating processes are downward bending (steepening) and thermal relaxation of the slab, as well as the buoyant escape of previously subducted continental crust toward shallower depths (see the movement of the black square in Fig. 21.7a,b,c and the respective *P-T*-time path in

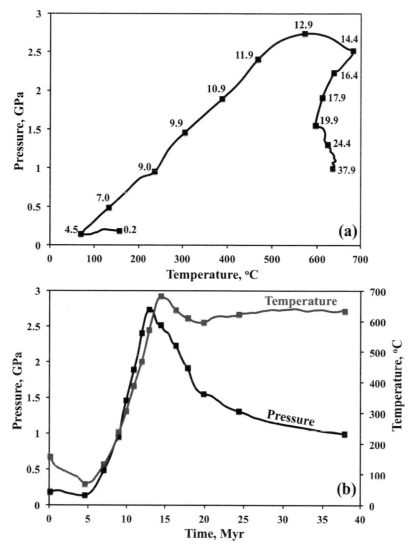

Fig. 21.8. Representative *P-T*-time path for the upper crustal rock unit deeply subducted during the continental collision (see small black square in Fig. 21.5). This unit exhumes toward the surface during thermal relaxation and detachment of the slab (see small black square in Fig. 21.7). Numbers along the *P-T* path in (a) show time in Myr from the beginning of the collision experiment (Fig. 21.5).

Fig. 21.8). This stage lasts over 12 Myr (from 12.7 Myr to 25.4 Myr, cf. Fig. 21.5b and Fig. 21.7b). After the strength of the slab interior is lowered by the temperature increase, a self-accelerating (due to feedbacks from stress concentration and shear heating, Gerya et al., 2004a), thermomechanical *necking* is activated and leads to rapid (within <1 Myr) detachment of the slab (Fig. 21.7b,c). This necking is driven by thermally activated, stress-sensitive dislocation creep. The depth of breakoff is relatively shallow (around 130 km), but this model feature is sensitive to many model parameters and results may vary widely, ranging from 50 to 500 km (Gerya et al., 2004a; Faccenda et al., 2008b; Mishin et al., 2008; Ueda et al., 2008; Andrews and Billen, 2009; Baumann et al., 2010; Duretz et al., 2011b). Finally, the detached slab rapidly sinks and rotates in a coherent manner (as a rigid body), interacting with the lower model boundary, which is an artefact of the impermeable lower boundary condition. However, rigid slab rotation is a realistic process resulting from slab interaction with the underlying and surrounding mantle (particularly near the 670 km deep spinel-to-perovskite transition), which is confirmed by numerical experiments with larger and deeper models that employ mantle phase transitions (Mishin et al., 2008; Baumann et al., 2010; Duretz et al., 2011b).

Figure 21.8 shows a P-T-time path of a marker representing an upper crustal rock unit deeply subducted during collision and exhumed toward the surface during thermal relaxation and detachment of the slab (see small black square in Figs. 21.5, 21.7). The possibility of computing P-T-time trajectories of various rock units represents a natural very useful feature of the marker-in-cell approach. The synthetic P-T-time paths can be further compared to P-T and P-T-time paths of natural rock complexes derived from petrological data. This comparison poses strong constraints for testing numerical models on the basis of natural data which is now widely used in numerical modelling of various geodynamic processes (e.g. Gerya et al., 2000, 2002, 2008b; Jamieson et al., 2002; Gerya and Maresch, 2004; Stoeckhert and Gerya, 2005; Gerya and Stoeckhert, 2006; Gorczyk et al., 2007a; Faccenda et al., 2008b; Warren et al., 2008; Gerya, 2011 and references therein; Penniston-Dorland et al., 2015).

21.9 Intrusion emplacement into the crust

James Hutton had already conceived the idea of plutonism in the late eighteenth century (e.g. Ellenberger, 1994) but the formation of large magma bodies in the crust, the plutons, still eludes full understanding. Research on the topic has, however, been much focused on the emplacement of granitic magma (e.g. Pitcher, 1979; Ramberg, 1981; Petford et al., 2000). Thermomechanical modelling of magma intrusion is not yet very 'popular' (e.g. Bittner and Schmeling, 1995; Burov et al., 2003; Gerya and Burg, 2007; Burg et al., 2009; Keller et al., 2013; Schubert et al., 2013; Cao et al., 2016; Gorczyk and Vogt, 2018) and is numerically challenging because it involves simultaneous and intense deformation of materials with very contrasting rheological properties. To see the contrast, consider that typical crustal rocks are visco-elasto-plastic whilst the intruding magma is a low viscosity, complex fluid/crystal mixture. Indeed, the staggered finite differences+marker-in-cell (SFD+MIC) numerical methodology discussed in this book is appropriate for such types of experiments and has already been used for thermomechanical modelling of intrusion emplacement (Gerya and Burg, 2007; Burg et al., 2009; Schubert et al., 2013; Gorczyk and Vogt, 2018).

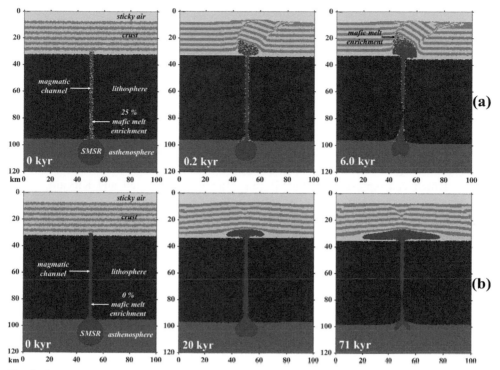

Fig. 21.9. Dynamics of intrusion emplacement into the crust in the case of (a) hot magma ($T = 1700$ K) with mafic melt enrichment (25%) and lowered (10^{14} Pa s) viscosity and (b) colder magma ($T = 1600$ K) with no mafic melt enrichment (0%) and higher (10^{16} Pa s) viscosity. Model resolution is 201×61 nodal points with 192 000 randomly distributed markers. The grid is similar to Fig. 21.1a and has a non-uniform vertical resolution of 1.0–6.4 km and uniform horizontal spacing of 0.5 km. Asthenospheric mantle temperature at the bottom of the model is 1600 K and decreases upward according to an adiabatic gradient of 0.5 K/km. Moho temperature is 900 K for both models. Results are computed with the code **Intrusion_Emplacement.m**.

In the case of trans-lithospheric emplacement (i.e. intrusion of a magmatic body from sub-lithospheric depths into the crust), which is the process assumed for many mafic-ultramafic bodies found in cratons and within magmatic arcs (e.g. Burg et al., 2009), the following model design can be used (Fig. 21.9a,b, 0 kyr). The model domain should obviously include both lithospheric and asthenospheric regions. The envisaged intrusion emplacement area has to be well resolved (grid spacing should be several times smaller than the modelled intrusion size, which typically implies a resolution of 0.5–1 km or better, Gerya and Burg, 2007). Non-uniform grids similar to those used for the slab bending model (Fig. 21.1a) and for the lithospheric extension/collision models (Figs. 21.4, 21.5) are appropriate. The initial thermal structure of the lithosphere, with a 25 km thick crust corresponding to a magmatic arc, is represented by a relatively hot sectioned geotherm

(0°C on the top of the crust, 627°C at the Moho and 1317°C at 93 km depth). An adiabatic gradient of 0.5 K/km is used in the asthenospheric mantle. Dense layering in the crust is used to better visualize the deformation. Boundary conditions are free slip on all boundaries. No far field shortening/extension is introduced to affect the intrusion process.

In this model, we assume that magma is coming from a *sub-lithospheric magmatic source region* (SMSR) (Gerya and Burg, 2007). The SMSR and the magmatic channel across the lithospheric mantle are then prescribed initially. The SMSR and the median channel enriched in mafic melt have an initially uniform magmatic temperature, which can be varied within reasonable limits (e.g. between 1200 and 1500°C). This is the presumed temperature range at the head of a partially molten, hydrous thermal-chemical mantle plume, which can represent the SMSR (e.g. Tamura, 1994; Hall and Kincaid, 2001; Gerya and Yuen, 2003b; Gerya et al., 2004b, Gorczyk et al., 2007b; Castro and Gerya, 2008; Zhu et al., 2009; Behn et al., 2011; Marschall and Schumacher, 2012). We emphasize that the modelled SMSR is only a thermally and chemically distinct region of the hydrated, partially molten mantle rocks and not a chamber of fully molten magma. Partially molten material has a much lower viscosity than the surrounding dry mantle (e.g. Pinkerton and Stevenson, 1992; Caricchi et al., 2007; Costa et al., 2009) and can move through the magmatic channel as a melt/crystal mixture. In that sense, the SMSR is equivalent to a magma reservoir. According to the melting model described below, this hydrated region initially has 10–30% volume of melt (depending on the assumed initial temperature), which varies with depth. Though the bulk composition of the SMSR in the model is ultramafic, the melt composition within the SMSR depends on the degree of melting and thus varies between mafic and ultramafic.

By implementing a pre-existing hot and thus weak (with plastic strength of 1 MPa) magmatic channel, we implicitly accept the general assumption that magma rises from the source area as a result of any lithospheric perturbation. Natural hot channels may originate at an early magmatic stage due to the rapid and localized upward percolation of hot mobile fluids/melts, which are differentiation products at the top of the SMSR. Respectively, enrichment (up to 25%) by mafic melt is prescribed for the channel. Mechanisms of localized upward fluid/melt transport include hydrofracture (e.g. Clemens and Mawer, 1992), diffusion (e.g. Scambelluri and Philippot, 2001), porous flow (e.g. Scott and Stevenson, 1986; Connolly and Podladchikov, 1998; Vasilyev et al., 1998; Ricard et al., 2001) and reactive flow (e.g. Spiegelman and Kelemen, 2003; Keller and Katz, 2016). It is possible to model spontaneous propagation of such channels (Keller et al., 2013) by using the visco-elasto-plastic hydro-thermomechanical codes discussed in Chapter 16. An important point is that fluid/melt percolation diminishes the brittle/plastic strength of rocks through an increase in pore fluid pressure (Eqs. 16.69, 16.70, Chapter 16), which in turn allows massive amounts of partially molten rocks to ascend through the lithosphere. Spontaneous propagation of magmatic rocks from the channel into the crust is mainly controlled by the crustal rheology and density.

According to geological observations, fault tectonics, and hence plastic deformation of the crust, play a significant role during plutonic emplacement resulting from fluid/melt percolation along forming fracture zones (e.g. Clemens and Mawer, 1992). The plastic yield strength of rocks under fluid-present conditions strongly depends on the difference between the total (P^t) and fluid (P^f) pressure (Eqs. 16.69, 16.70). During intrusion, the most intensive percolation of magmatic fluids is expected to follow the pattern of fractured rocks along spontaneously propagating fault zones (Keller et al., 2013). Fluid supply will then increase the pore fluid pressure along the fault zones, thus lowering the plastic yield strength of the fractured rocks. This will in turn further localize deformation along the weakening fault zones. Under such circumstances, the plastic strength of fractured rocks will be inversely correlated with the amount of continuous plastic deformation experienced by the rocks. In order to model this process in a simplified way, we assume that $P^t \geq P^f$ so that the yielding condition is given by Eq. (12.51a) written in the following form:

$$\sigma_{yield} = \sigma_c + \gamma_{int}\left(P^t - P^f\right) = \sigma_c + \gamma_{int}\left(1 - \frac{P^f}{P^t}\right)P^t = \sigma_c + \gamma_{int}(1 - \lambda^f)P^t = \sigma_c + \gamma_{int(eff)}P^t, \quad (21.9)$$

$$\lambda^f = \frac{P^f}{P^t}, \quad (21.10)$$

$$\gamma_{int(eff)} = \gamma_{int}\left(1 - \frac{P^f}{P^t}\right) = \gamma_{int}(1 - \lambda^f), \quad (21.11)$$

where λ^f is the pore fluid pressure ratio and $\gamma_{int(eff)}$ is the effective internal friction coefficient. We then assume that the fluid pressure ratio for a given model rock increases with the plastic strain $\varepsilon_{plastic}$ (Eqs. 14.12, 14.13) experienced by the rock

$$\lambda^f = 1 - \left(1 - \lambda^{fo}\right)\left(1 - \frac{\varepsilon_{plastic}}{\varepsilon_{cr}}\right) \text{ when } \varepsilon_{plastic} < \varepsilon_{cr} \text{ and}$$
$$\lambda^f = 1 \text{ when } \varepsilon_{plastic} > \varepsilon_{cr}, \quad (21.12)$$

or in terms of the effective friction coefficient (as in the sandbox benchmark of Chapter 20),

$$\gamma_{int(eff)} = \gamma^o_{int(eff)}\left(1 - \frac{\varepsilon_{plastic}}{\varepsilon_{cr}}\right) \text{ when } \varepsilon_{plastic} < \varepsilon_{cr} \text{ and}$$
$$\gamma_{int(eff)} = 0 \text{ when } \varepsilon_{plastic} \geq \varepsilon_{cr}, \quad (21.13)$$

$$\gamma^o_{int(eff)} = \gamma_{int}(1 - \lambda^{fo}), \quad (21.14)$$

where λ^{fo} and $\gamma^o_{int(eff)}$ are the initial (before plastic yielding) pore fluid pressure ratio and effective friction coefficient characteristic for the rock, respectively; ε_{cr} is the critical plastic strain required for weakening the rock by percolating fluid (i.e. the strain necessary for reaching condition $P^f = P^t$, $\lambda^f = 1$, $\gamma_{int(eff)} = 0$). In our example, we used $\varepsilon_{cr} = 0.1$ and $\gamma^o_{int(eff)} = \gamma_{int}(1 - \lambda^{fo})$, equal to 0.2 and 0.6 for the crustal and mantle rocks, respectively.

For simplicity, the effective viscosity η of partially molten rocks ($M > 0.1$) is assigned by a low constant value of 10^{14}–10^{16} Pa s. In real melt-crystal aggregates, this viscosity is strongly and non-linearly dependent on the melt fraction (Chapter 16, Eqs. 16.67, 16.68). One can also calculate it using the following relatively simple formula (Pinkerton and Stevenson, 1992; Bittner and Schmeling, 1995)

$$\eta = \eta_o \exp\left[2.5 + (1 - M)\left(\tfrac{1-M}{M}\right)^{0.48}\right], \tag{21.15}$$

where η_o is an empirical parameter depending on rock composition. According to Bittner and Schmeling (1995), $\eta_o = 10^{13}$ Pa s can be taken for partially molten mafic rocks (i.e., $1 \times 10^{14} \leq \eta \leq 2 \times 10^{15}$ Pa s for $0.1 \leq M \leq 1$) and $\eta_o = 5 \times 10^{14}$ Pa s (i.e., $6 \times 10^{15} \leq \eta \leq 8 \times 10^{16}$ Pa s for $0.1 \leq M \leq 1$) can be adopted for felsic rocks.

Compared to previous numerical examples, the new feature that we have to introduce is a treatment of the melting/crystallization processes. Crystallization of the intruding magma and, to some extent, partial melting of host rocks are two important and coeval processes during plutonism (e.g. Marsh, 1982; Best and Christiansen, 2001) because they affect the density and the rheology of both intruding and intruded rocks, respectively. The numerical models presented allow the gradual crystallization of magma and partial melting of the crust (e.g. Bittner and Schmeling, 1995) in the pressure-temperature domain between the wet solidus and dry liquidus of corresponding rocks (Table 21.2). As a first approximation, the volumetric fraction of melt M at constant pressure is assumed to increase linearly with temperature according to the relations (Gerya and Yuen, 2003b; Burg and Gerya, 2005):

$$M = 0 \quad \text{at} \quad T \leq T_{solidus}, \tag{21.16a}$$

$$M = \frac{(T - T_{solidus})}{(T_{liquidus} - T_{solidus})} \quad \text{at} \quad T_{solidus} < T < T_{liquidus}, \tag{21.16b}$$

$$M = 1 \text{ at } T \geq T_{liquidus}, \tag{21.16c}$$

where $T_{solidus}$ and $T_{liquidus}$ are the wet solidus and dry liquidus temperatures of the considered rock, respectively (Table 21.2). It should be noted that more sophisticated melting models for the mantle exist, which take into account mantle composition and water content (e.g. Katz et al., 2003).

The effective density, ρ_{eff}, of partially molten rocks is then calculated from:

$$\rho_{eff} = \rho_{solid}\left(1 - M + M\frac{\rho_{0(molten)}}{\rho_{0(solid)}}\right), \tag{21.17}$$

where $\rho_{0(solid)}$ and $\rho_{0(molten)}$ are the standard densities of solid and molten rock, respectively (Table 21.2) and ρ_{solid} is the density of solid rocks at given P and T computed according to Eq. (2.4b) based on the thermal expansion and compressibility coefficients (Table 21.2).

The effect of latent heating due to equilibrium melting/crystallization is included implicitly by increasing (*in the temperature equation*) the effective heat capacity ($C_{P(eff)}$) and the thermal expansion (α_{eff}) of the partially crystallized/molten rocks ($0 < M < 1$), calculated as (Burg and Gerya, 2005):

$$C_{P(eff)} = C_P + Q_L \left(\frac{\partial M}{\partial T} \right)_{P=\text{const}} \tag{21.18}$$

$$\alpha_{eff} = \alpha + \rho \frac{Q_L}{T} \left(\frac{\partial M}{\partial P} \right)_{T=\text{const}} \tag{21.19}$$

where C_P is the heat capacity of the solid rock and Q_L is the latent heat of melting of the rock (Table 21.2). Pressure and temperature derivatives of the melt fraction M can be obtained numerically by calling an external melt fraction computation routine (see MATLAB function **Melt_Fraction.m** used by the code **Intrusion_emplacement.m**).

Figure 21.9 displays results of an intrusion experiment performed with the code **Intrusion_emplacement.m**. Two emplacement regimes are compared: rapid emplacement of hot, low viscosity magma with higher degree of melting (Fig. 21.9a) and slower emplacement of colder, higher viscosity magma with lower degree of melting (Fig. 21.9b). In the first case, emplacement is mainly controlled by the brittle/plastic deformation (faulting) of the crust. Magma rises rapidly toward the surface (Fig. 21.9a, 6.0 kyr), despite the fact that its density is notably higher (by 150–300 kg/m^3) than that of the crustal host rocks (though it is lower than the density of the non-molten lithospheric mantle). Extrusion of hot, partially molten rocks through the magmatic channel is primarily driven by the density contrast between these partially molten rocks and the mantle lithosphere. To minimize gravitational energy, intrusive rocks of intermediate density should tend to pool along the crust/mantle boundary (i.e. at the neutral buoyancy level). However, this tendency is only realized if the magma viscosity is relatively high and its emplacement, therefore, is relatively slow such that viscous deformation of the lower crust can accommodate the intrusion emplacement along the Moho (Fig. 21.9b). In the first model, however (Fig. 21.9a), the magma viscosity is low and its emplacement is, therefore, fast and can only be accommodated by brittle/plastic deformation. Since the brittle/plastic strength of the crust rapidly decreases with decreasing depth (more precisely with decreasing dynamic pressure), the intrusion propagates upwards to the middle-upper crust where emplacement requires less mechanical work. The restraining gravitational energy produced by the arrival of a dense intrusion in the less dense crust is compensated and overcome by the positive buoyancy of partially molten rocks in the magmatic channel. This phenomenon is comparable to the penetration of a diapir head into lower density rocks (e.g. Ramberg, 1981). This intrusion mechanism is also called trans-lithospheric mantle diapirism (Burg et al., 2009). It is worth emphasizing that the gravitational balance controls the height of the column of molten rock, but not the volume of magmatic rock below and above the Moho. This is expressed in the mechanical equilibrium relation:

$$h_{Channel}(\rho_{Mantle} - \rho_{Magma}) = h_{intrusion}(\rho_{Magma} - \rho_{Crust}), \qquad (21.20)$$

where $h_{Channel}$ is the height of the column of magmatic rock in the channel below the Moho, $h_{Intrusion}$ is the height of magmatic rock above the Moho, ρ_{Mantle}, ρ_{Magma} and ρ_{Crust} are the density of the mantle lithosphere, the intruding magma and the crust, respectively. Since the width of the channel below the Moho is limited by the rigidity of the cold mantle lithosphere, the volume of magma intruding into the crust can be much larger than the volume of rock remaining in the channel (Fig. 21.9a, 6.0 kyr).

21.10 Mantle convection with phase changes

As we discussed in the Introduction, thermomechanical modelling of mantle convection has a rich history dating back to the early 1970s (e.g. Richter, 1978; Schubert, 1992; Bercovici, 2007 and references therein). Not surprisingly, it is one of the most advanced fields of geodynamic modelling in terms of both technical and conceptual progress (e.g. Yuen et al., 2000; Tackley, 2012; Bercovici et al., 2015). Realistic modelling of terrestrial and planetary convection is a challenging topic (e.g. Hansen and Yuen, 1988; Larsen et al., 1995; Yuen et al., 2000; Zhong et al., 2007; Tackley, 2008, 2012; Bercovici et al., 2015 and references therein) and requires the application of sophisticated 3D numerical codes working with spherical geometries at high grid resolution which can almost exclusively only be performed by parallel computing on 'big machines'. Indeed, one important aspect of modelling mantle convection, which is of interest for this chapter, is the incorporation of solid-state phase transitions into such numerical models (e.g. Tackley, 2008).

Solid-state phase transitions are crucial phenomena in the Earth's mantle. Major phase transitions include olivine-spinel at 410 km depth and spinel-perovskite at 670 km depth. These transitions are associated with significant changes in mantle density and seismic wave speeds (Turcotte and Schubert, 2002). It was also suggested that the so-called D″ discontinuity near the core-mantle boundary is related to a perovskite-post-perovskite phase transition (Murakami et al., 2004; Oganov and Ono, 2004). Phase transitions affect the dynamics of mantle convection due to (1) density changes and (2) latent heating (Richter, 1973; Schubert et al., 1975; Christensen and Yuen, 1985; Tackley, 1993; Zhong and Gurnis, 1994).

Phase changes are traditionally included in mantle convection models (e.g. Richter, 1973; Schubert et al., 1975; Christensen and Yuen, 1985; Tackley, 1993; Zhong and Gurnis, 1994) by programming each transition individually (i.e. similarly to what we did with melting reactions in the previous example). However, for realistic mantle compositions, the number of various phase transitions is larger than only three (Fig. 21.10) and these phase transitions involve several minerals of variable composition (so-called *solid solutions*, Table 21.3), which makes the traditional approach rather inconvenient. An alternative method has been developed recently based on Gibbs free energy minimization (Chapter 2).

Table 21.3 *Phase notation and formulas of minerals for the CaO-FeO-MgO-Al₂O₃-SiO₂ pyrolite model (Fig. 21.10)*

Symbol	Phase	Formula*
aki	akimotoite	$Mg_xFe_{1-x-y}Al_{2y}Si_{1-y}O_3$, $x + y \leq 1$
c2c	pyroxene	$[Mg_xFe_{1-x}]_4Si_4O_{12}$
cpv	Ca-perovskite	$CaSiO_3$
cpx	clinopyroxene	$Ca_{2y}Mg_{4-2x-2y}Fe_{2x}Si_4O_{12}$
gt	garnet	$Fe_{3x}Ca_{3y}Mg_{3(1-x+y+z/3)}Al_{2-2z}Si_{3+z}O_{12}$, $x + y \leq 1$
o	olivine	$[Mg_xFe_{1-x}]_2SiO_4$
opx	orthopyroxene	$[Mg_xFe_{1-x}]_{4-2y}Al_{4(1-y)}Si_4O_{12}$
ppv	post-perovskite	$Mg_xFe_{1-x-y}Al_{2y}Si_{1-y}O_3$, $x + y \leq 1$
pv	perovskite	$Mg_xFe_{1-x-y}Al_{2y}Si_{1-y}O_3$, $x + y \leq 1$
rng	ringwoodite	$[Mg_xFe_{1-x}]_2SiO_4$
sp	spinel	$Mg_xFe_{1-x}Al_2O_3$
wad	waddsleyite	$[Mg_xFe_{1-x}]_2SiO_4$
wus	magnesiowuestite	$Mg_xFe_{1-x}O$

* Unless otherwise noted, the compositional variables *x, y,* and *z* can vary between zero and unity and are determined as a function of pressure and temperature by free-energy minimization using the Perple–X program (Connolly, 2005).

This method was initially applied for crustal- and lithospheric-scale thermal (Gerya et al., 2001; Petrini et al., 2001; Kaus et al., 2005) and thermomechanical (Gerya et al., 2004c, 2006; Yamato et al., 2008) models and then expanded to mantle convection models (Tackley, 2008).

The idea of this *petrological-thermomechanical* method is relatively simple (Gerya et al., 2004c, 2006): (i) phase diagrams (*P-T pseudosections*) and related density (ρ) and enthalpy (*H*) maps (see programming exercise 2.3 in Chapter 2) are first computed for the necessary rock compositions in a relevant range of *P-T* conditions and (ii) these maps are then used in thermomechanical experiments for computing density (ρ), effective heat capacity incorporating latent heat ($C_{P(eff)}$) and energetic effects (both adiabatic and latent heating) for isothermal (de)compression (H_P) for material points (markers) based on standard thermodynamic formulas and numerical differentiation in *P-T* space (Fig. 21.11):

$$\rho = \rho_{i,j}\left(1 - \frac{\Delta T_m}{\Delta T}\right)\left(1 - \frac{\Delta P_m}{\Delta P}\right) + \rho_{i+1,j}\left(1 - \frac{\Delta T_m}{\Delta T}\right)\frac{\Delta P_m}{\Delta P}$$

$$+ \rho_{i,j+1}\frac{\Delta T_m}{\Delta T}\left(1 - \frac{\Delta P_m}{\Delta P}\right) + \rho_{i+1,j+1}\frac{\Delta T_m}{\Delta T}\frac{\Delta P_m}{\Delta P},$$

$$(21.21)$$

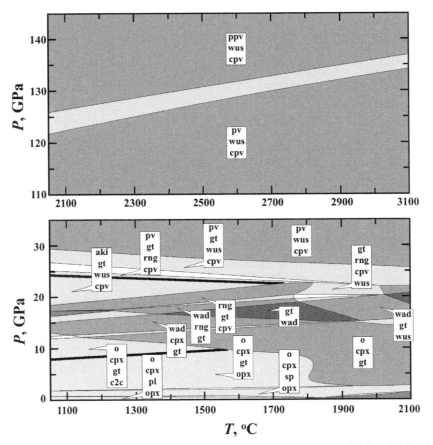

Fig. 21.10. Phase relations for the CaO-FeO-MgO-Al$_2$O$_3$-SiO$_2$ pyrolite model (see Table 21.3 for notation of minerals) computed (Mishin et al., 2008) with the Gibbs free energy minimization program Perple_X (Connolly, 2005). To permit the resolution of phase relations the diagram is split to exclude the large depth interval between the transition zone and core-mantle boundary in which the model does not predict phase transformations. Composition for the pyrolite model is 3.87 wt% CaO, 8.11 wt% FeO, 3.61 wt% Al$_2$O$_3$, 38.59 wt% MgO and 45.82 wt% SiO$_2$.

$$C_{P(eff)} = \left(\frac{\partial H}{\partial T}\right)_{P=\text{const}} = \frac{H_{i,j+1} - H_{i,j}}{\Delta T}\left(1 - \frac{\Delta P_m}{\Delta P}\right) + \frac{H_{i+1,j+1} - H_{i+1,j}}{\Delta T}\left(\frac{\Delta P_m}{\Delta P}\right), \quad (21.22)$$

$$\frac{H_P}{DP/Dt} = \left(1 - \rho\frac{\partial H}{\partial P}\right)_{T=\text{const}}$$

$$= 1 - \frac{H_{i+1,j} - H_{i,j}}{\Delta P}\left(1 - \frac{\Delta T_m}{\Delta T}\right)\frac{\rho_{i+1,j} + \rho_{i,j}}{2} - \frac{H_{i+1,j+1} - H_{i,j+1}}{\Delta P}\left(\frac{\Delta T_m}{\Delta T}\right)\frac{\rho_{i+1,j+1} + \rho_{i,j+1}}{2}.$$

$$(21.23)$$

The temperature equation (9.8) is respectively modified as

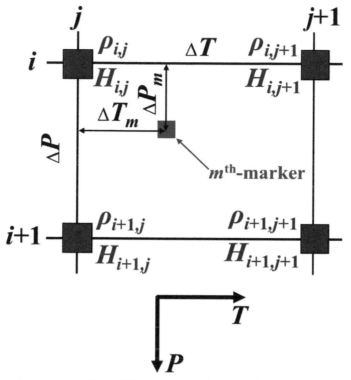

Fig. 21.11. Stencil associated with the *P-T* grid used for the interpolation of physical properties from enthalpy and density look-up tables, to the markers.

$$\rho C_{P(\text{eff})} \frac{DT}{Dt} = -\frac{\partial q_i}{\partial x_i} + H_r + H_s + H_P. \tag{21.24}$$

The stable mineralogy and physical properties for the mantle used in our example are computed (Mishin et al., 2008) with Perple_X (Connolly, 2005) by a free energy minimization approach. For this purpose the Mie–Grueneisen formulation of Stixrude and Bukowinski (1990) was adopted with the parameterization of Stixrude and Lithgow-Bertelloni (2005) augmented for lower mantle phases as described by Khan et al. (2006). This parameterization limits the chemical model to $CaO\text{-}FeO\text{-}MgO\text{-}Al_2O_3\text{-}SiO_2$ (Table 21.3). The mantle rheology is based on the dry olivine flow law (Table 21.1) with an activation volume of 5 cm^3, which allows us to mimic a 1.5–3 order of magnitude increase in viscosity for the lower mantle. This range is often used in mantle convection models (e.g. Tackley, 2000). Olivine is obviously not stable in the deep mantle below the olivine-spinel transition and, therefore, our rheological choice is rather arbitrary. This is related to the limited availability of experimentally calibrated flow laws applicable to the deep mantle such that simplified temperature- and depth-dependent rheological models are

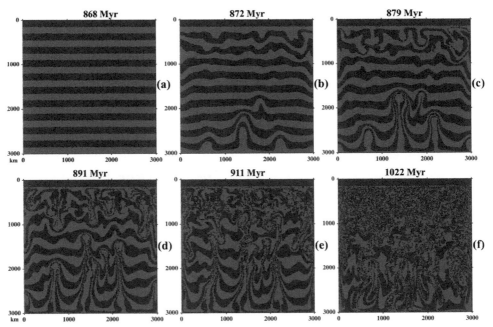

Fig. 21.12. Development of mantle convection processes with phase changes (Fig. 21.10) computed with the petrological-thermomecanical code **Mantle_convection.m** associated with this chapter. Model resolution is 51 × 51 nodal points with 40 000 randomly distributed markers. Grid resolution is uniform in both directions. The model is shown from an arbitrary stage of 868 Myr when non-compositional layering is superimposed for visualizing deformation onto a pre-computed non-steady thermal structure. Note moderate deformation of the lower mantle by localized upwellings and downwellings (thermal plumes), which contrasts with the intense chaotic mixing of the upper mantle.

traditionally used in numerical mantle convection studies (e.g. Richter, 1973; Schubert et al., 1975; Christensen and Yuen, 1985; Zhong and Gurnis, 1994; Tackley, 1993, 2000, 2008).

Figure 21.12 shows several stages of a visco-elasto-plastic mantle convection modelled with the code **Mantle_convection.m** (please note that brittle/plastic deformation plays a negligible role in this relatively simple model with no self-consistent plate generation). The model design is simple: a square 3000 × 3000 km^2 box with free slip boundaries, constant temperature conditions applied at the top and in the bottom and no heat flux condition across the vertical walls. In this model we do not intend to model self-consistent plate generation (e.g. Tackley, 2000, 2008, 2012; Bercovici et al., 2015 and references therein) and therefore use a uniform, relatively coarse grid resolution. Mantle convection is characterized by a semi-layered structure with a strongly convecting upper part (above 670 km depth) and a much more slowly deforming lower mantle with several plumes penetrating

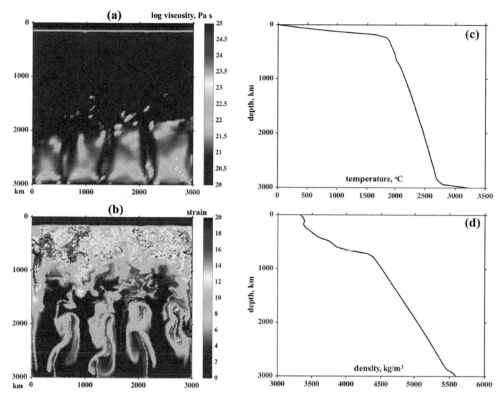

Fig. 21.13. Viscosity (a) and strain (b) maps and horizontally averaged temperature (c) and density (d) profiles for the computed mantle convection model at 1022 Myr (Fig. 21.12f). Note, 1–2 order of magnitude viscosity contrast between the upper and lower mantle obtained with our simplified rheological model based on dry olivine flow law with activation volume of 5 cm^3. Mantle deformation in (b) is semi-layered and is strongly affected by the spinel-perovskite transition at around 670 km depth due to its negative Clapeyron slope (Fig. 21.10), which creates difficulties for both upwellings and downwellings for penetrating this boundary. Cold lithosphere at the top of the model remains undeformed (stagnant lid regime, e.g. Tackley, 2000) due to the prescribed high plastic strength ($\gamma_{int} = 0.6$), which corresponds to a dry mantle. Plate bending processes (Fig. 21.1) in this lithosphere are not modelled due to the imposed free slip upper boundary condition without a weak layer (in contrast to the free surface condition formed by sticky air/water used in all previous examples) as well as due to the very coarse grid resolution employed here (60×60 km^2 in contrast to 2×2 km^2 needed for properly resolving slab bending processes, Fig. 21.1).

from the core-mantle boundary (Figs. 21.12, 21.13b). Density changes (Fig. 21.13d) and thermal effects (Fig. 21.13c) of various phase transitions (Fig. 21.10) are captured with our petrological-thermomechanical numerical approach without programming them individually. This approach makes coding simpler and easily allows changes to be made when testing different models for mantle composition.

21.11 Deformation of a self-gravitating planetary body

Lastly, let us discuss some planetary-scale applications of numerical geodynamic modelling which are becoming more and more widespread in relation to the problem of planetary
accretion and core formation processes (e.g. Stevenson, 1981). One important group of
such applications concerns the internal deformation of a heterogeneous, self-gravitating
body. Numerical modelling of deformation of such a body was discussed in Chapter 11 with
the use of a 'spherical-Cartesian' approach (Honda et al., 1993; Gerya and Yuen, 2007; Lin
et al., 2009). This approach is relatively simple and does not require major rewriting of the
Cartesian code used for the previous examples. All that is required is to add the
numerical solution of the Poisson equation before solving the momentum and continuity
equations (Figs. 11.3, 11.4). Gravitational acceleration components are then
computed from the gravity potential locally (Eqs. 11.12, 11.13) and are used in the
right hand side of the momentum equation (Eqs. 11.4, 11.5). As a boundary condition for
gravity potential, we can use the constant value of gravitational potential (e.g. $\Phi = 0$)
applied at a circular surface located at a distance from the planet (Fig. 18.7, Eqs. 18.12–
18.15). The chosen value of the potential along the surface is arbitrary since it does not
affect the resulting gravitational acceleration field (given by derivatives of the potential).
The use of such a boundary condition is based on the fact that with growing distance from
the planet, both the gravitational acceleration and gravity potential tend to become solely a
functions of the planetary mass (m_p) and the distance to its centre (d)

$$g = G\frac{m_p}{d^2}, \tag{21.25}$$

$$\Phi = \text{const} - G\frac{m_p}{d}. \tag{21.26}$$

However, it should be mentioned that forcing the potential to be uniform outside the
planet at a certain distance from the planetary centre also affects the gravitational field
inside the planet (especially the tangential component). Therefore, the gravity potential
boundary should preferably be located at a *significant distance* from the planetary surface,
which should be comparable to the planetary radius (Lin et al., 2009). This requirement may
warrant using an independent larger grid for the gravity potential and/or an adaptive grid
refinement (Chapter 17, Fig. 17.4).

The initial setup for the numerical experiment on gravitational redistribution of metal and
silicate in a Mars-sized body (3000 km in radius) is shown in Fig. 21.14 (0.03 Myr). This
setup is based on the numerical studies of Golabek et al. (2009, 2011) who in particular
investigated rheological controls on the iron core formation mechanism. The initial temperature is 273 K at the surface and 1200 K at 100 km depth and then rises linearly to
1500 K in the centre of the planet. This initial profile is rather arbitrary and mimics to some
degree the effects of various planetary heating processes: (i) decay of short-lived radioactive isotopes (MacPherson et al., 1995) such as [26]Al (half-life is 0.73 Ma) and [60]Fe (half-
life is 1.5 Ma), (ii) impact heating by accreted planetesimals (Davies, 1985; Melosh, 1990),
(iii) impact-associated gravitational unloading (Asphaug et al., 2006) and (iv) adiabatic

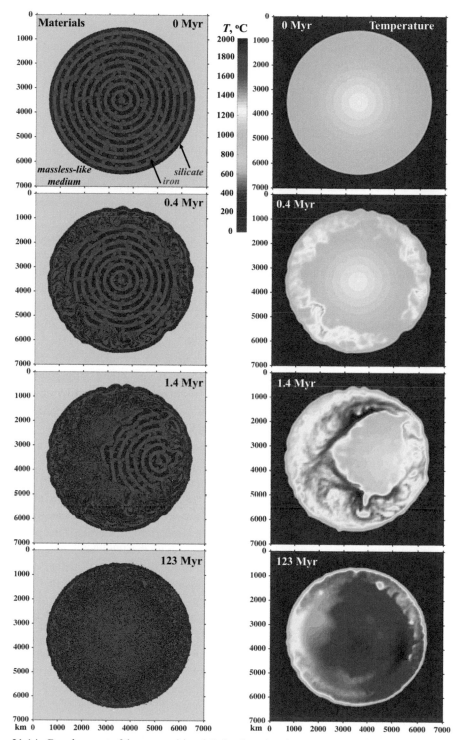

Fig. 21.14. Development of the compositional (left column) and temperature (right column) fields in the numerical experiment on core formation computed with the code **Core_formation.m** associated with this chapter. Model resolution is 101 × 101 nodal points with 160 000 randomly distributed markers.

heating caused by growing pressure in the planetary interior. The planet is heterogeneous and composed of a silicate matrix and randomly distributed iron diapirs with radii varying from 50 to 100 km. Use of a variable sized diapir is related to the fact that the size distribution of the planetesimals and planetary embryos after runaway growth and the formation of Mars-sized bodies should be heterogeneous (Melosh, 1990; Tonks and Melosh, 1992; Stevenson, 2008). As in the previous example, we also apply a coupled petrological-thermomechanical approach for modelling density and thermal properties of the silicate matrix, which is assumed to have pyrolitic composition (Fig. 21.10). The rheology of silicate corresponds to dry olivine (Table 21.1) with an activation volume of 10 cm^3. Constant high density ($10\,000 \text{ kg/m}^3$) and lowered viscosity (10^{20} Pa s) are used for the metal.

Viscosity of the weak massless-like medium (1 kg/m^3) surrounding the planet is also taken to be 10^{20} Pa s, which is 2–5 orders of magnitude lower than that of silicate at the planetary surface. This is sufficient to provide free surface like conditions (Schmeling et al., 2008; Lin et al., 2009). Lower viscosity will require a shorter time step, which will slow down computation. Feedback from shear heating caused by release of gravitational energy is also taken into account since this process may strongly affect temperature distribution inside the differentiating planet (Gerya and Yuen, 2007; Golabek et al., 2009).

Figure 21.14 shows several stages of core formation computed with the code **Core_formation.m** associated with this chapter. Model development corresponds to the so-called 'decomposition mode' (Golabek et al., 2009) which is defined by choosing a relatively large activation volume (10 cm^3) and plastic strength (no Peierls plasticity limit is imposed) of the silicate matrix that effectively makes the pressurized planetary interior rheologically stronger than the outer shell. Therefore, in this model, the iron diapirs near the surface are activated foremost (Fig. 21.14, 0.4 Myr). Even the slightest asymmetries in the initial iron distribution lead to an earlier initiation of diapir sinking in some distinct regions where significantly larger temperatures develop due to strong shear heating processes associated with the diapir sinking. This leads to the activation of neighbouring diapirs and finally to the formation of large iron ponds. The underlying material of the planetary interior is too strong to be deformed by stresses arising from the available iron agglomerations. Therefore, all iron diapirs in the outer region of the planet will be activated finally. This leads to a global temperature rise in the outer layer of the planet (Fig. 21.14, 0.4 Myr). Consequently, a low viscosity shell, basically analogous to a magma ocean, is formed around the remaining highly viscous central region. The iron ponds finally coalesce into an iron-rich ring around the central region (Fig. 21.14, 0.4 Myr), as was suggested by Stevenson (1981) and Ida et al. (1987). Temperature and density dis-equilibrium around the ring leads to the degree-one instability (e.g. Ida et al., 1987) resulting in the formation of advective streams of the iron-rich material around the central region. Iron accumulating on one side of the planet pushes the stiff, non-differentiated planetary interior (including the passive iron diapirs) out of the central region, creating a noticeable asymmetry in the planet (Fig. 21.14, 1.4 Myr). A similar scenario was proposed by Elsasser (1963) and modelled numerically in a simplified way (Honda et al., 1993; Lin et al., 2009), but under the assumption of a cold central region. This translation favours the decompression-related decomposition of the 'exhuming' interiors when the material approaches shallower depths. Decompression causes (via activation volume) a viscosity reduction in the regions of the translated central sphere closer to the planet's surface. The silicates released at the leading side rise, as a result of their low density, as Rayleigh–Taylor instabilities upwards into the high temperature zone (Fig. 21.14, 1.4 Myr). This causes further iron release and makes the process of interior 'decomposition' self-sustaining, which finally results in the formation of an iron core (Fig. 21.14, 123 Myr). Obviously, this idealized core formation scenario is in part artificial (due to the highly idealized initial conditions) and non-unique, and other core formation scenarios may develop instead, depending on variations in planetary accretion history, size, temperature structure and material properties, as well as on the effective silicate rheology (e.g. Golabek et al., 2009 and references therein).

Programming exercise

Exercise 21.1.

Program a model for extension of a continental lithosphere based on the provided oceanic lithosphere extension model (Fig. 21.4). Take the initial compositional and thermal structure of the continental lithosphere to be the same as in the continental collision example (Fig. 21.5a). Use the codes **Extension.m** and **Collision.m** associated with this chapter. Alternatively, you may also want to program a similar model using a uniform grid resolution using your own visco-elasto-plastic codes created during exercises for Chapters 13 and 20.

Outlook

Theory: Where are we now? Where to go next? Current and future directions of numerical geodynamic modelling development: 3D, MPI, OpenMP, PETSc, AMR, FEM, FVM, GPU/cell-based computing, interactive computing, realistic physics, visualization challenges etc.
Exercises: No more exercises!

Where are we now?

Where are we after reading 21 chapters + Introduction and performing (all?) the analytical and programming exercises? We are still only at the door to the wonderful, rapidly developing, world of numerical geodynamic modelling. The method we have learned is based on a finite difference method and marker-in-cell technique (FDM+MIC). Although this holds for many geodynamic situations ('all in one' tool, Gerya and Yuen, 2007), its universality is not absolute and other approaches may be better suited for some numerical problems. The examples provided in this book are written in a very explicit way, with the aim of facilitating learning and understanding, rather than producing massive numerical results (although code examples made for this book run quite fast ☺). The resolution of the numerical examples is also moderate so that corresponding experiments can be completed in a reasonable amount of time (several seconds to several hours) on an ordinary laptop which is, by the way, not exactly the main tool of geodynamic modellers (unless, sometimes, for code development and testing, for looking at results computed on *big machines* and for writing papers and books like this one . . .). Once again, we are at the beginning and if we want to go further, it is worth reading the following.

Where to go next?

Not only modellers ask this question. Some time ago, ten research questions shaping twenty-first-century Earth science were identified by the US National Research Council of the National Academy of Sciences (DePaolo et al., 2008, http://www8 .nationalacademies.org/onpinews/newsitem.aspx?RecordID=12161).

These questions are cited here.

(1) 'How did Earth and other planets form? While scientists generally agree that this solar system's sun and planets came from the same nebular cloud, they do not know enough about how Earth obtained its chemical composition to understand its evolution or why the other planets are different from each other. Although credible models of planet formation now exist, further measurements of solar system bodies and extrasolar objects could offer insight in the origin of the Earth and the solar system.'

(2) 'What happened during Earth's 'dark age' (the first 500 million years)? Scientists believe that another planet collided with Earth during the late stages of its formation, creating debris that became the moon and causing Earth to melt down to its core. This period is critical to understanding planetary evolution, especially how the Earth developed its atmosphere and oceans, but scientists have little information because few rocks from this age are preserved.'

(3) 'How did life begin? The origin of life is one of the most intriguing, difficult, and enduring questions in science. The only remaining evidence of where, when, and in what form life first appeared springs from geological investigations of rocks and minerals. To help answer the question, scientists are also turning toward Mars, where the sedimentary record of early planetary history predates the oldest Earth rocks, and other star systems with planets.'

(4) 'How does Earth's interior work, and how does it affect the surface? Scientists know that the mantle and core are in constant convective motion. Core convection produces the Earth's magnetic field, which may influence surface conditions, and mantle convection causes volcanism, seafloor generation, and mountain building. However, scientists can neither precisely describe these motions, nor calculate how they were different in the past, hindering scientific understanding of the past and prediction of Earth's future surface environment.'

(5) 'Why does Earth have plate tectonics and continents? Although plate tectonics theory is well established, scientists wonder why Earth has plate tectonics and how closely it is related to other aspects of Earth, such as the abundance of water and the existence of the continents, oceans, and life. Moreover, scientists still do not know when continents first formed, how they remained preserved for billions of years, or how they are likely to evolve in the future. These are especially important questions as weathering of the continental crust plays a role in regulating Earth's climate.'

(6) 'How are Earth processes controlled by material properties? Scientists now recognize that macroscale behaviors, such as plate tectonics and mantle convection, arise from microscale properties of Earth materials, including the smallest details of their atomic structures. Understanding materials at the microscale is essential to comprehend the Earth's history and making reasonable predictions about how planetary processes may change in the future.'

(7) 'What causes climate to change – and how much can it change? Earth's surface temperature has remained within a relatively narrow range for most of the last 4 billion years, but how does it stay well-regulated in the long run, even though it can change so abruptly? Study of Earth's climate extremes through history – when climate was extremely cold or hot or changed quickly – may lead to improved climate models that could enable scientists to predict the magnitude and consequences of climate change.'

(8) 'How has life shaped Earth – and how has Earth shaped life? The exact ways in which geology and biology influence each other are still elusive. Scientists are interested in life's role in oxygenating the atmosphere and reshaping the surface through weathering and erosion. They also seek to understand how geological events caused mass extinctions and influenced the course of evolution.'

(9) 'Can earthquakes, volcanic eruptions, and their consequences be predicted? Progress has been made in estimating the probability of future earthquakes, but scientists may never be able to predict the exact time and place an earthquake will strike. Nevertheless, they continue to decipher how fault ruptures start and stop and how much shaking can be expected near large earthquakes. For volcanic eruptions, geologists are moving toward predictive capabilities, but face the challenge of developing a clear picture of the movement of magma, from its sources in the upper mantle, through Earth's crust, to the surface where it erupts.'

(10) 'How do fluid flow and transport affect the human environment? Good management of natural resources and the environment requires knowledge of the behavior of fluids, both below ground and at the surface, and scientists ultimately want to produce mathematical models that can predict the performance of these natural systems. Yet, it remains difficult to determine how subsurface fluids are distributed in heterogeneous rock and soil formations, how fast they flow, how effectively they transport dissolved and suspended materials, and how they are affected by chemical and thermal exchange with the host formations.'

A remarkable fact is that all these questions (especially 1, 2, 4, 5, 6, 9 and 10) are directly related to the topics addressed by numerical geodynamic modelling, and finding answers to these questions is likely to depend on the future efforts of modellers. This will keep us all busy for a while! The general ideas for future technical and conceptual advances in numerical geodynamic modelling are therefore quite clear: (i) *fast computing of high-resolution 3D numerical problems with complex and realistic physics applicable to nature*, and (ii) *obtaining a rigorous understanding of geodynamic and planetary evolution and related processes (such as climate and life) and the key physical parameters controlling them*. This means that modellers have to concentrate on (i) further developing efficient, fast, realistic, high-resolution numerical 3D codes that are now routine tools for geodynamicists, (ii) incorporating new physics (fluid-solid coupling, seismic processes, climate and life evolution etc.) into these codes and (iii) integrating results of numerical models with natural observations and producing comprehensive testable predictions in related fields.

How big are '*big numerical geodynamic problems*'? At the moment, up to *602 billion DOF* (degrees of freedom), which is equivalent to *150 billion nodal points*, run in parallel on up to *1 572 864 processors* (e.g. Cohen, 2005; Burstedde et al., 2008, 2009; Krotkiewski et al., 2008; Tackley, 2008; Rudi et al., 2015; Kaus et al., 2016; Omlin et al., 2017) i.e. close to $5300^3 = 5300 \times 5300 \times 5300$ nodal points which is ca. 1.3 million times larger than the $49^3 = 49 \times 49 \times 49$ 3D problem that we solved with a multigrid in Chapter 19 (Figs. 19.11, 19.12). Also, tens of billions of markers are explored (Gorczyk et al., 2007b) in some models (Fig. Outlook.1), which is tens of thousands of times bigger than the 750 thousand markers explored in our largest experiment, the retreating subduction in Chapter 21 (Figs. 21.2, 21.3). Handling huge models is not a trivial task. Difficulties, as usual, do not come from the hardware side, i.e. from limited availability of corresponding 'big machines' with hundreds of thousands of processors (e.g. Rudi et al., 2015; Kaus et al., 2016) and terabytes of memory – they exist and getting access to them is sufficiently uncomplicated. In addition, recent hardware development trends involve using highly efficient, low-cost *GPU/cell-based* multiprocessor units, which greatly expand the computational capabilities in computational Earth sciences (e.g. Zheng et al., 2013; Omlin, 2016; Omlin et al., 2017, 2018). The key difficulties are (i) to efficiently *parallelize computations* (i.e. to use in parallel *many processors at near peak performance* for the same numerical experiment, e.g. Burstedde et al., 2008, 2009; May et al., 2014; Kaus et al., 2016; Omlin, 2016) and (ii) to program efficient procedures for *data storage and mining*.

For example, the highest resolution FDM+MIC simulations to date use 10 to 40 billion markers (Fig. Outlook.1). These extremely high-resolution simulations result in uncompressed output file sizes of up to over 1 TB for each time step. Similar amounts of memory are required during the runs as well, but a number of supercomputers have ample memory for these large runs (e.g. Burstedde et al., 2008, 2009; Rudi et al., 2015; Kaus et al., 2016). Multiplying to hundreds of time steps, which typically need to be stored for each experiment, and to tens to hundreds of experiments typically needed for every numerical problem studied, implies that enormous storage capabilities are required. If we do not take care of data compression, then supercomputers will spend 99% of their time on *IO operations* (i.e. writing results of our experiment onto hard disks)! This is not exactly what they are made for . . . The problem only appears at a certain resolution size and we would perhaps not even guess about it before. Data compression should also be sufficiently fast so that it does not take more than 10% of CPU time. Efficient and fast data compression algorithms based, for example, on wavelets (e.g. Vasiliev et al., 2004) allow the reduction of storage size by a factor of 100 to 1000 (e.g. Gorczyk et al., 2007b) which brings us back to efficient computing and production of results.

Another data handling option is to perform *fully interactive computing* associated with no/little data storage (e.g. Damon et al., 2008). The idea behind this is the following: multiprocessor hardware is currently so efficient that one can, in principle, repeat almost any kind of numerical experiment within minutes or a few hours. If one spent this time looking at a screen showing the progressing results of an ongoing numerical experiment, and saving a limited amount of characteristic frames as image files, then the regular data

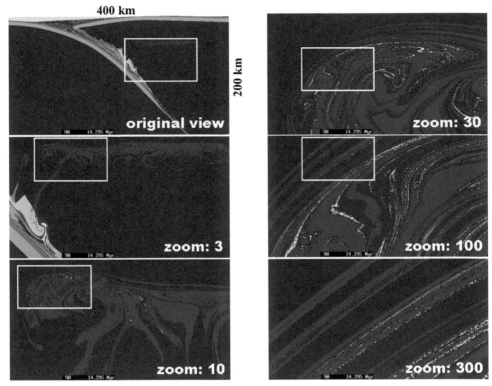

Fig. Outlook.1. High-resolution, multiple scale visualization of mechanical stirring structures related to development of a partially molten mantle wedge plume (Gorczyk et al., 2007b), which spreads underneath the cold and stiff mantle lithosphere of the overriding plate. The 40 billion element (pixels) data set with spatial resolution of around 2 m is based on compressed output from the 2D numerical experiment with 10 billion markers. Each pixel of the original size figure (top left frame) can be resolved as a full-size picture (bottom right frame).

storage would no longer be needed. If you want to learn something more about the same experiment you simply repeat it. Of course, these two strategies of data handling can also be complementary.

Let us now go briefly through a list of things that, from a general perspective, would be useful to read, think of, learn and implement in the future. This list is obviously fragmentary, subjective and non-exhaustive, but will provide some useful hints triggering further thinking.

State of the art overview

A good cross-section covering the current state of the art in numerical geodynamic modelling (who is now doing what and with which numerical technique) can be retrieved from several recent textbooks and special volumes attributed to computational Earth sciences.

- Textbook by Alik Ismail-Zadeh and Paul Tackley: *Computational Methods for Geodynamics*. Cambridge University Press, 2010.
- Textbook by Guy Simpson: *Practical Finite Element Modeling in Earth Science using Matlab*. Wiley-Blackwell, 2017.
- Textbook by Gabriele Morra: *Pythonic Geodynamics: Implementations for Fast Computing*. Lecture Notes in Earth System Sciences. Springer Nature, 2018.
- Special volume: *Mantle Dynamics*, edited by David Bercovici and Gerald Schubert, *Treatise on Geophysics*, Volume 7, Elsevier, 2015.
- Special volume: *200 Years of Geodynamic Modelling*, edited by Vincent Strak, João C. Duarte, Filipe M. Rosas, Louis Moresi and Wouter P. Schellart, *Journal of Geodynamics*, Volume 100, 2016.
- Special volume: *Solid Earth Processes: A memorial volume to Evgenii Burov*, edited by Fernando O. Marques, Taras Gerya, Boris J. P. Kaus, Yuriy Y. Podladchikov and Laurent Jolivet, *Tectonophysics*, Volume 746, 2018.

Efficient direct solvers

Implementation of efficient direct solvers (developed by mathematicians) to numerical codes and tuning them for geodynamic modelling problems can significantly (sometimes up to 100 times) speed up calculation in 2D and can even allow 3D problems to be addressed at sufficiently high resolution (e.g. Braun et al., 2008). The following advanced direct solvers are widely used in geosciences.

(1) WSMP (Watson Sparse Matrix Package, Gupta, 2000, http://www-users.cs.umn.edu/~agupta/wsmp.html).
(2) MUMPS (MUltifrontal Massively Parallel sparse direct Solver, used in MATLAB S=L\R, Amestoy et al., 2001, http://graal.ens-lyon.fr/MUMPS/).
(3) PARDISO (Schenk and Gärtner, 2004, 2006, http://www.pardiso-project.org/).
(4) PaStix (Parallel Sparse matriX package, http://pastix.gforge.inria.fr/).
(5) SuiteSparse (a suite of sparse matrix algorithms, used in MATLAB S=L\R: UMFPACK: multifrontal LU factorization, CHOLMOD: supernodal Cholesky, http://faculty.cse.tamu.edu/davis/suitesparse.html).
(6) SuperLU (a general purpose library for the direct solution of large, sparse, non-symmetric systems of linear equations, http://crd-legacy.lbl.gov/~xiaoye/SuperLU/).

The PARDISO solver is, for example, currently being implemented in the thermomechanical 2D codes originally developed by Gerya and Yuen (2003a, 2007) and resulted in a speedup of 30 times compared to a standard Gaussian solver (Chapter 3), thereby allowing numerical experiments to be performed at high resolution of up to 2000×2000 grid points on a single or multiple processors (which is 400 times more grid points compared to our 101×101 grid in the core formation experiment of Fig. 21.14). Similar performances can also be reached using the MATLAB-based finite element code MILAMIN (Dabrowski et al., 2008, https://sourceforge .net/projects/milamin/). MILAMIN means Million-A-Minute, i.e. one million equations are solved within one minute of time. This corresponds to a resolution of approximately 600×600 nodal points in the case of a 2D Stokes+continuity solver.

Parallelization of numerical codes

Efficient, high-resolution 2D and 3D modelling require the *parallelization of numerical codes*. This allows many processors to be used at the same time, which proportionally speeds up numerical calculations (as long as parallelization is efficiently implemented, see e.g. Moresi et al., 2007; May and Moresi, 2008; Tackley, 2008; Burstedde et al., 2008, 2009; Furuichi et al, 2011; Kronbichler et al., 2012; May et al., 2014, 2015; Rudi et al., 2015; Kaus et al., 2016; Omlin, 2016; Heister et al., 2017). Individual processors need to exchange information during the calculations and their activity is not fully independent and should be thoroughly correlated. This makes parallelization a non-trivial programming task (e.g. Karniadakis and Kirby, 2003). There are several ways to implement code parallelization using open source libraries such as MPI, OpenMP and PETSc etc.

MPI – the Message Passing Interface standard (e.g. http://www-unix.mcs.anl.gov/mpi/). MPI is a library specification for *message-passing* which defines and enables a mechanism of data exchange between different processors that are simultaneously used for the same numerical experiment. MPI was designed for high performance on both massively parallel machines and on workstation clusters. MPI works with both *distributed memory* (different parts of the memory can be accessed by different processors) and *shared memory* (the entire memory can be directly accessed by each processor). Programming MPI requires a significant effort to begin but the resulting codes are very efficient.

OpenMP – Open Multi-Processing (e.g., http://openmp.org/wp/, Chapman et al., 2007). The OpenMP Application Program Interface (API) supports multi-platform shared-memory parallel programming in C/C++ and Fortran on all architectures, including Unix platforms and Windows NT platforms. OpenMP is easy to learn and implement in numerical codes (typically one only has to add a few lines to parallelize loops already present in the code) but its application is limited to shared memory machines.

PETSc – Portable, Extensible Toolkit for Scientific Computation (e.g. https://www .mcs.anl.gov/petsc/). PETSc, pronounced PET-see (the S is silent), is a suite of data structures and routines for the scalable (parallel) solution of scientific applications modelled by partial differential equations. It employs the MPI standard for all message-passing communication and is quite convenient to learn and to use for programming parallel numerical codes (it is a bit similar to MATLAB, actually).

By the way, many available direct solvers already contain parallelization, which makes them even more attractive to employ in our codes.

Adaptive mesh refinement algorithms

Adaptive mesh refinement (AMR, Chapter 17) is a very suitable option when various processes are being modelled on different scales in the same numerical model (e.g. Fig. Outlook.1). In this case, the numerical grid should preferably be able to follow only the area where high numerical resolution is needed. We already discussed 'Swiss cross' and block-structured 2D AMR methods in Chapter 17 and extensively applied the 'Swiss cross' AMR in Chapter 21, where we forced the high resolution part of our grid to follow areas of interest (e.g., Figs. 21.2, 21.4, 21.5). However, the development of block-structured and unstructured AMR methods for 3D thermomechanical models is technically challenging and remains an important modern trend enabling unprecedented high resolution in realistic global and regional geodynamic models with strongly localized deformation along plate boundaries (e.g. Burstedde et al., 2008, 2009; Stadler et al., 2010; Rudi et al., 2015). These methods can be based either on finite differences (e.g. Albers et al., 2000; Gerya et al., 2013) or on finite elements (e.g. Braun et al., 2008; Burstedde et al., 2008, 2009; Stadler et al., 2010; Kronbichler et al., 2012; Rudi et al., 2015; Heister et al., 2017) and require significant work to properly derive a discrete formulation of the governing equations which is conservative, especially in the areas where resolution changes. The *finite element methods (FEM)* (e.g. Zienkiewicz et al., 2005) possess many advantages in this respect (e.g. Moresi et al., 2003, 2007; Simpson, 2017), as they are well suited to dealing with complex, irregular rectangular and triangular meshes and give more accurate results in cases when sharp curved material interfaces need to be followed in numerical models (e.g. Deubelbeiss and Kaus, 2008; Popov and Sobolev, 2008; Simpson, 2017).

An additional method of choice for unstructured meshes is the *finite volume method (FVM)* (Toro, 1999; LeVeque, 2002; Hüttig and Stemmer, 2008). Like the finite difference method, FVM values are calculated at discrete nodes. 'Finite volume' refers to the small volume surrounding each node (e.g. to a cell surrounding a central pressure node in the staggered grid (e.g., Fig. 19.1). In FVM, volume integrals in a partial differential equation that contain a divergence term are converted to surface integrals, using the divergence theorem. These terms are then evaluated as fluxes at the surfaces of each finite volume. Since the flux entering a given volume is identical to that leaving the adjacent volume, these methods are conservative. In fact, the conservative finite differences discussed in Chapters 7 and 10 are equivalent to applying a finite volume method on the relatively simple rectangular grids used in this book.

In the case of various marker-in-cell approaches, which are very common in geodynamic modelling, refinement may (or even should) also be applied to the unstructured grid of markers by using marker reseeding, splitting and merging procedures (Moresi et al., 2003; Keller et al., 2013).

Future numerical geodynamic modelling research directions

Current and future 'hot' conceptual and methodological trends in numerical geodynamic modelling are in good agreement with the 'top ten' research questions listed above.

(1) Using more realistic physical properties of rocks. Composite visco-elasto-plastic rheologies including both diffusion and dislocation creep as well as Mohr–Coulomb, Druker–Prager and Peierls plasticity (Ranalli, 1995; Karato, 2008; Katayama and Karato, 2008; Chapters 6, 12). Including influences of grain size evolution, damage and pinning (e.g. Rozel et al., 2011; Bercovici and Ricard, 2012, 2014). Incorporating realistic rheology for partially molten rocks (see reviews by Caricchi et al., 2007; Costa et al., 2009). Use of compressible time-dependent forms of the continuity equation (Gerya and Yuen, 2007; Tackley, 2008; Chapter 14).

(2) Accounting for phase transformations (including melting). Incorporating both volumetric and thermal effects of various phase transitions in numerical models (Gerya et al., 2004c, 2006; Tackley, 2008; Chapters 16, 21). Adding kinetics of phase transitions.

(3) Hydro-thermomechanical modelling. Understanding of fluid and melt migration in deforming rocks. Programming coupled approaches for modelling fluid/melt generation and transport in actively deforming systems associated with many geodynamic processes in the crust and mantle (e.g. Connolly and Podladchikov, 1998; Schmeling, 2000; Katz, 2008; Dymkova and Gerya, 2013; Keller et al., 2013; Yarushina and Podladchikov, 2015; Keller and Katz, 2016; Omlin et al., 2017, 2018; Chapter 16).

(4) Geochemical-(hydro)-thermomechanical modelling. Including modelling of geochemical processes into (hydro)-thermomechanical experiments (e.g. Xie and Tackley, 2004a,b; Vogt et al., 2013; Baitsch-Ghirardello et al., 2014; Ikemoto and Iwamori, 2014). Development of realistic models including fluid and melt related transport of trace elements (e.g. Baitsch-Ghirardello et al., 2014; Ikemoto and Iwamori, 2014; Fig. Outlook.2) in various geodynamic and planetary environments.

(5) Geomorphological-thermomechanical modelling. Coupling of modelling of surface processes (erosion, sedimentation, sediment transport, fluvial networks, glaciers etc.) with thermomechanical simulations of geodynamic processes (e.g. Kooi and Beaumont, 1994; Willett, 1999; Cloetingh et al., 2007; Braun et al., 2008; Kaus et al., 2008; Ueda et al., 2015; Chapter 21, Fig. 21.6, Fig. Outlook.3). Understanding coupling between crustal, lithospheric and mantle evolution and surface processes.

(6) Numerical modelling of volcanic and plutonic processes. Coupling of modelling of magma conduit physics and volcanic processes (e.g. Melnik and Sparks, 1999; Melnik, 2000; Papale, 1999, 2001) with magma generation and ascent (e.g. Schmelling, 2000; Katz, 2008; Keller et al., 2013), intrusion emplacement (e.g. Burov et al., 2003, Gerya and Burg, 2007; Burg et al., 2009; Keller et al., 2013; Schubert et al., 2013; Colón, et al., 2018; Gorczyk and Vogt, 2018; Chapter 21, Fig. 21.9), magma chamber dynamics (e.g. Oldenburg et al., 1990; Bagdassarov and

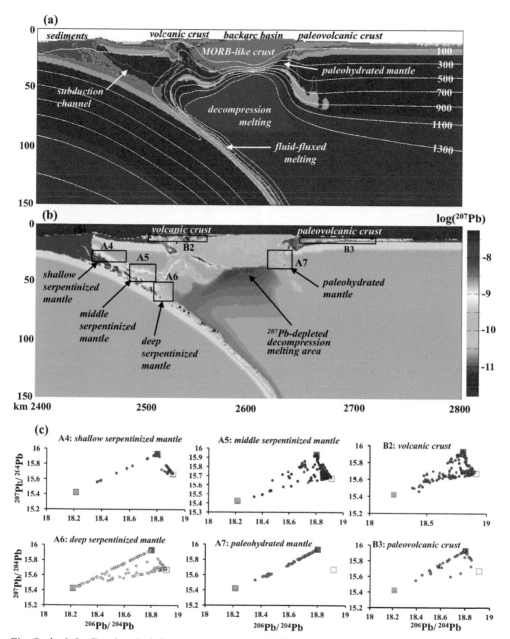

Fig. Outlook.2. Geochemical-thermomechanical modelling of the evolution of lead isotopes during oceanic subduction (Baitsch-Ghirardello et al., 2014). (a) Distribution of different rock types (colour) and isotherms (white lines, in °C). (b) Distribution of ^{207}Pb concentrations; black labelled rectangles A4–A7 and B2, B3, mark respectively hydrated mantle and volcanic rock sampling areas for which isotopic ratios are shown in (c). Time corresponds to 20 Myr since the beginning of subduction.

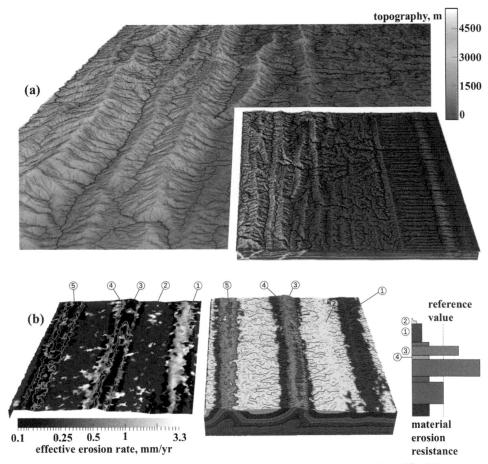

Fig. Outlook.3. High-resolution geomorphological-thermomechanical modelling of fluvial networks (a) and erosion patterns (b) during deformation of accretionary wedges (Ueda et al., 2015).

Fradkov, 1993; Spera et al., 1995; Bergantz, 2000; Simakin and Botcharnikov, 2001; Longo et al., 2006; Ruprecht et al., 2008) and related hydrothermal processes (e.g. Driesner et al., 2006; Driesner and Geiger, 2007; Weis et al., 2012).

(7) Seismo-thermomechanical modelling. Coupling of long-term deformation and short-term (poro)-visco-elasto-plastic deformation processes. Developing numerical approaches for relating long-term geodynamic processes with faulting dynamics, fluid flows, rupture processes and seismicity (e.g. Lapusta et al., 2000; Miller et al., 2004; Wang, 2007; Ben-Zion, 2008; Faccenda et al., 2008a, 2012; Pergler and Matyska, 2008; Regenauer-Lieb and Yuen, 2008; Lapusta and Liu, 2009; Van Dinther et al., 2013a,b, 2014; Herrendörfer et al., 2015, 2018; Dal Zilio et al., 2018; Chapter 15, Fig. Outlook.4).

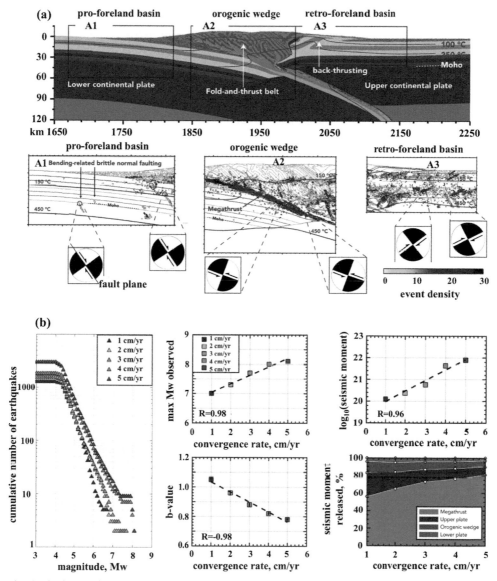

Fig. Outlook.4. Seismo-thermomechanical modelling of continental collision seismicity pattern (a) and Gutenberg–Richter distribution (b) (Dal Zilio et al., 2018).

(8) Realistic modelling of planetary formation, differentiation and early evolution processes: accretion, core formation, magma ocean development, onset of mantle convection. One of the numerical challenges is in coupling of planetary accretion (typically addressed with N-body simulations, e.g. Chambers and Wetherill, 1998; Chambers, 2001), giant impacts and meteorite impacts (modelled with

hydrodynamic codes, e.g. Benz et al. 1986; Canup and Asphaug, 2001; Canup, 2004; Wada et al., 2006; Melosh, 2008; and analytical models, e.g. Senshu et al., 2002), pebble/chondrule accretion (e.g. Bitsch et al., 2015; Bollard et al., 2017) and core formation processes (addressed with continuum mechanics approaches, Honda et al., 1993; Golabek et al., 2008, 2009, 2011; Samuel and Tackley, 2008; Lin et al., 2009; Ricard et al, 2009; also see Fig. 21.14). A possible numerical solution to this multi-disciplinary problem may be based on a combination of various types of codes into hybrid accretion-impact-differentiation tools (Golabek et al., 2011, 2018; O'Neil et al., 2017; Chapters 14, 21, Fig. 21.14, Fig. Outlook.5).

(9) Modelling of planetary surface and interior evolution. Using numerical geodynamic modelling tools for understanding evolution of planets of the Solar system (e.g. Golabek et al., 2011, 2014; Armann and Tackley, 2012; Gerya, 2014a; Lichtenberg et al., 2016, 2018; Fig. Outlook.6) as well as potential evolution scenarios for extrasolar planets (e.g. van Heck and Tackley, 2011; Tackley et al., 2014; Lourenco et al., 2018).

(10) Modelling of modern Earth geodynamics and plate tectonic processes on a global and regional scale (e.g. Gerya et al., 2008a; Stadler et al., 2010; Gerya, 2010b, 2013; Coltice et al., 2012; Crameri et al., 2012b; Li et al, 2013; Burov and Gerya, 2014; Koptev et al., 2015; Mallard et al., 2016; Gerya and Burov, 2018; also see reviews Gerya, 2011, 2016; Brune, 2016; Stern and Gerya, 2018 and references therein, Fig. Outlook.7). Current technical challenges are in providing high numerical resolution in plate boundary regions (e.g. Gerya, 2010b; Stadler et al., 2010; Burov and Gerya, 2014) and using realistic visco-(elasto)-plastic rock rheology and free surface in global models (e.g. Crameri et al., 2012b; Patocka et al., 2017).

(11) Modelling of Precambrian Earth geodynamics, continent formation and initiation of subduction and plate tectonics (e.g. van Thinen et al., 2004; van Hunen and van den Berg, 2008; Sizova et al., 2010, 2014, 2015; van Hunen and Moyen, 2012; Moore and Webb, 2013; Bercovici and Ricard, 2014; Johnson et al., 2014; Rey et al., 2014; Gerya et al., 2015; Fischer and Gerya, 2016a,b; O'Neil et al., 2017; Rozel et al., 2017; also see review Gerya, 2014b and references therein, Fig. Outlook.8). Technical challenges are in coupling thermomechanical evolution of the Earth's crust and mantle with magmatic processes and crustal growth, which critically affect both global (e.g. Rozel et al., 2017; Lourenco et al., 2018) and regional (e.g. Sizova et al., 2015; Fischer and Gerya, 2016b) tectonic regimes.

(12) Coupled modelling of the evolution of Earth's interior, climate, environment and life. Growing evidence suggests that the geodynamic evolution of Earth's interior is intimately interrelated with the evolution of its atmosphere, oceans, landscape and life (e.g. Sobolev et al., 2011; Large et al., 2015; Stern, 2016; Lee et al., 2018). In order to understand this coupling, rigorous conceptual and technical development of new bio-geodynamical numerical modelling tools is needed, which will allow the investigation of multiple feedbacks between the Earth's interior, atmosphere, ocean,

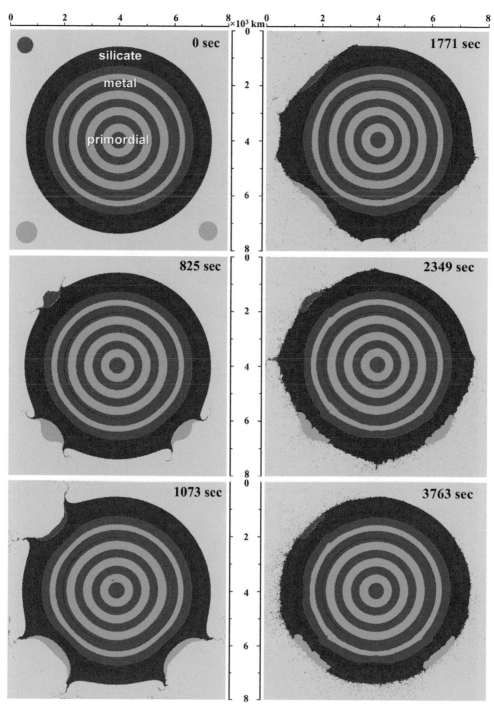

Fig. Outlook.5. Results of a preliminary numerical experiment of a planetary accretion process associated with impacts of meteorites (planetesimals) into a Mars-sized body (Gerya and Yuen, 2007). The experiment is performed using a 2D spherical-Cartesian approach.

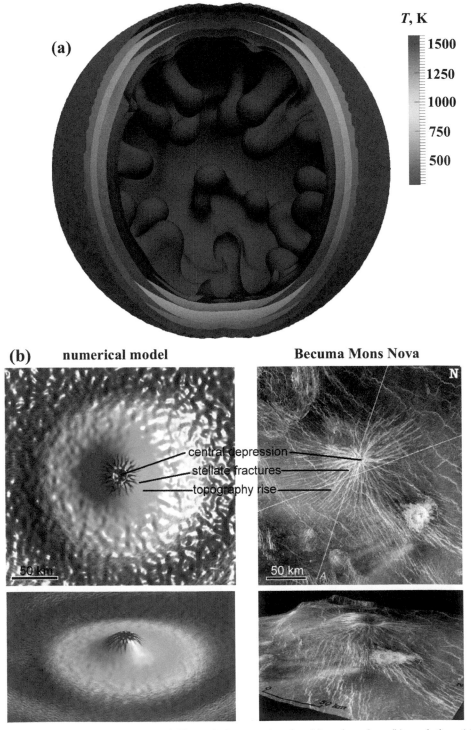

Fig. Outlook.6. 3D numerical modelling of planetary interior (a) and surface (b) evolution. (a) Convection of a partially molten silicate planetesimal without an iron core (Lichtenberg et al., 2016). Experiment is performed using a 3D spherical-Cartesian approach. (b) Development of a nova structure on the surface of Venus (Gerya, 2014a). Experiment is performed using a 3D Cartesian approach.

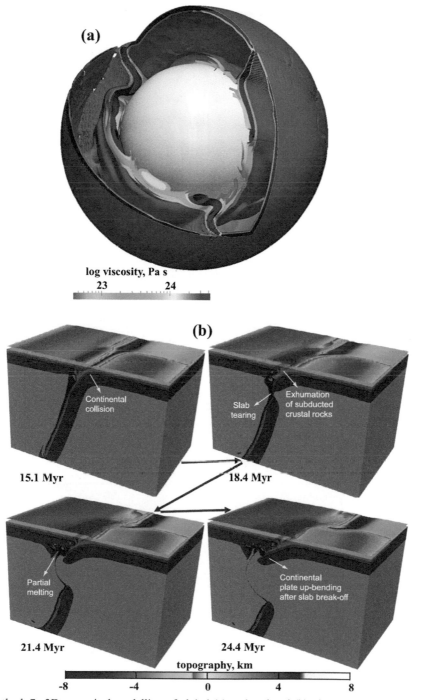

Fig. Outlook.7. 3D numerical modelling of global (a) and regional (b) plate tectonic processes. (a) Mantle convection with free surface, spontaneously forming tectonic plates and terrestrial-style one-sided subduction (Crameri et al., 2012b). (b) Post-subduction continental collision associated with slab tearing (Li et al., 2013).

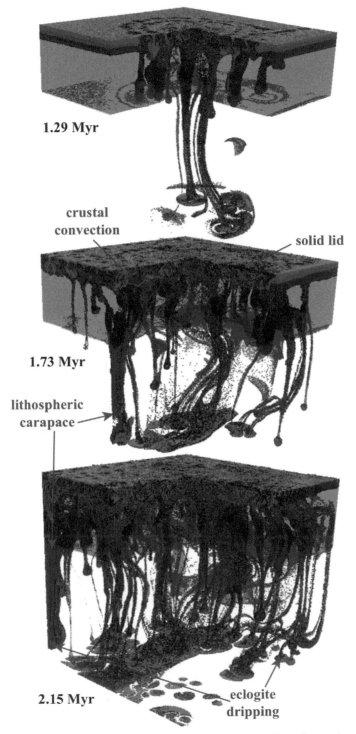

1.29 Myr

crustal convection

solid lid

1.73 Myr

lithospheric carapace

2.15 Myr

eclogite dripping

Fig. Outlook.8. 3D numerical modelling of Archean plume-lid tectonics and crustal growth (Fischer and Gerya, 2016b).

landscape and life. As in the case of planetary formation processes, hybrid modelling approaches are needed, which will couple geodynamic processes with the development of atmosphere, climate (e.g. Gillmann and Tackley, 2014), landscape (e.g. Ueda et al., 2015) and oceans as well as with biological evolution (e.g. Gotelli et al., 2009; Condamine et al., 2013; Descombes et al., 2017).

(13) Using geodynamic inversion and data assimilation approaches for constraining large-scale physical properties (e.g., effective rheology), global and regional physical state (e.g., stress state, mantle convection state) and evolutionary stage (e.g., in the seismic cycle) for the crust, lithosphere and mantle (e.g. Baumann et al., 2014; Baumann and Kaus, 2015; Baumann, 2016; Bocher et al., 2016; van Dinther et al., 2019).

(14) Industrial and environmental applications of numerical geodynamic modelling tools. Technical advances made in numerical geodynamic modelling encourage the application of geodynamic tools for problems of social relevance, such as CO_2 storage, geothermal systems, induced seismicity, hydrocarbon exploration, radioactive waste disposal, deformation of composite materials etc. (e.g. Rass et al., 2014; Hubschwerlen et al., 2017; Pouryazdan et al., 2017; Yarushina et al., 2017).

3D visualization challenges

As discussed in the Introduction, modellers spend much, or even most, of their time on visualizing and understanding numerical models. Since these models are likely to become bigger and more complex in the future, the role of efficient visualization technologies will increase. Visualization of large numerical models is already a non-trivial task in 2D (Rudolph et al., 2004; Gorczyk et al., 2007b) since the amount of graphical information exceeds the resolutions of even the most powerful graphic screens by a factor of thousands (e.g. snapshots from 40 gigapixel database shown in Fig. Outlook.1). It is even more complicated with 3D models (e.g. Billen et al., 2008; Chen et al., 2008, Damon et al., 2008; Kadlec et al., 2008; Jadamec et al., 2018) because, even for a single field the structure cannot be seen at once, and requires processing of many views to gain a proper understanding (Figs. Outlook.7– Outlook.9, also see movie **Cold_Plumes.mpeg** associated with this chapter). Future challenges in this respect are again quite obvious: geology-friendliness, ultrahigh resolution, multiple scales, multiple fields, interactive visualization etc.

Conceptual warning

Discussion of the forthcoming technical advances above may give the impression that the only thing that we have to do is to write larger, more complex 3D codes, which run on parallel supercomputers, and have efficient ways of visualizing results and compressing the data. Obviously, this is only part of what needs to be done. More important is that we obtain an in-depth physical understanding of the dynamics of geological and planetary processes. This understanding should not only be based on the 'powerful numerics', but also on using

Fig. Outlook.9. Different 3D projections of thermal-chemical plumes (Fig. Outlook.1) growing atop of the subducting slab (Zhu et al., 2009). The temperature iso-surface of 1350 K is shown. Rising plumes are colder than the mantle wedge and move upward due to their compositional buoyancy (Gerya and Yuen, 2003b; Zhu et al., 2009).

scaling laws based on simplified theories and on comparison of predictions of such theories with the much more complex numerical simulations and with nature (e.g. Kaus and Podladchikov, 2006; Gerya et al., 2015; Gerya and Burov, 2018). If we find consistency, we have probably learned something essential, and we did not have to perform 2^{35} simulations to understand how nature works.

Conclusion

In conclusion, the future of numerical geodynamic modelling looks bright and there is a lot of exciting work and challenging research to do. Just go on!

It was fun to write all this (for the second time ☺). Thank you for reading and good luck!

Appendix
MATLAB program examples

The following resources can be found here: www.cambridge.org

Introduction

Visualisation_is_important.m (Exercise Introduction.2) – visualization of 'sin' and 'cos' functions with 'plot', 'pcolor', 'contour' and 'surf'.

Chapter 1

Divergence.m (Exercise 1.2) – computation and visualization of velocity, divergence of velocity and time derivatives of density with 'pcolor' and 'quiver'.

Chapter 2

Periclase_EOS.m (Exercise 2.2) – computation and visualization of density, thermal expansion and compressibility for periclase (MgO) using external Gibbs free energy function *G_periclase.m.*

Density_map.m (Exercise 2.3) – loading from data files (*m895_ro, morn_ro*) and visualizing density maps for pyrolite (*m895_ro*) and MORB (*morn_ro*) and density difference between pyrolite and MORB.

Chapter 3

Poisson1D.m (Exercise 3.1) – solution of 1D Poisson equation with finite differences on a regular grid using direct solver '\'.

Poisson2D_direct.m (Exercise 3.2) – solution of 2D Poisson equation with finite differences on a regular grid using direct solver '\'.

Poisson2D_Gauss_Seidel.m (Exercise 3.3) – solution of 2D Poisson equation with finite differences on a regular grid using Gauss–Seidel iteration.

Poisson2D_Jacobi.m (Exercise 3.4) – solution of 2D Poisson equation with finite differences on a regular grid using Jacobi iteration.

Chapter 4

Strain_rate.m (Exercise 4.1) – computation and visualization of velocity field, strain rate, deviatoric strain rate, and second strain rate invariant.

Chapter 5

Streamfunction2D.m (Exercise 5.2) – solution of 2D Stokes and continuity equations with finite differences on a regular grid using stream function – vorticity formulation for a medium with constant viscosity.

Chapter 6

Viscosity_profile.m (Exercise 6.1) – computation and visualization of viscosity profile across the lithosphere.

Viscosity_map.m (Exercise 6.2) – computation and visualization of viscosity map in temperature–log stress coordinates.

Viscosity_comparison.m (Exercise 6.3) – computation and visualization of viscosity maps in temperature–log stress coordinates for a combination of dislocation and diffusion creep; comparison of wet and dry olivine rheology.

Chapter 7

Stokes_continuity_constant_viscosity.m (Exercise 7.1) – solution of 2D Stokes and continuity equations with finite differences on a regular grid using pressure-velocity formulation for a medium with constant viscosity.

Stokes_continuity_variable_viscosity.m (Exercise 7.2) – solution of 2D Stokes and continuity equations with finite differences on a regular grid using pressure-velocity formulation for a medium with variable viscosity.

Chapter 8

Upwind_1D.m (Exercise 8.1) – comparison of upwind, downwind and central differences for 1D advection of a square density wave in a constant velocity field.

FCT_1D.m (Exercise 8.2) – using FCT algorithm for 1D advection of a square density wave in a constant velocity field.

Markers_1D.m (Exercise 8.2) – using marker-in-cell algorithm with regular Eulerian grid for 1D advection of a square density wave in a constant velocity field.

Markers_1Dirregular.m (Exercise 8.3) – using marker-in-cell algorithm with irregular Eulerian grid for 1D advection of a square density wave in a variable velocity field; using bisection algorithm.

Markers_divergence.m (Fig. 8.12) – testing of continuity-based advection in 2D; solution of 2D Stokes continuity and advection equations with finite differences and marker-in-cell technique on a regular grid using pressure-velocity formulation for a medium with variable viscosity; using continuity-based marker velocity calculation combined with the fourth-order accurate in space first-order accurate in time Runge–Kutta marker advection scheme.

Stokes_Continuity_Markers.m (Exercise 8.4) – solution of 2D Stokes continuity and advection equations with finite differences and marker-in-cell technique on a regular grid using pressure-velocity formulation for a medium with variable viscosity; using the first-order accurate in space and time marker advection scheme.

Stokes_Continuity_Markers_Runge_Kutta.m (Exercise 8.5) – solution of 2D Stokes continuity and advection equations with finite differences and marker-in-cell technique on a regular grid using pressure-velocity formulation for a medium with variable viscosity; using the fourth-order accurate in space first-order accurate in time Runge–Kutta marker advection scheme.

Stokes_continuity_based_advection.m (Exercise 8.6) – solution of 2D Stokes continuity and advection equations with finite differences and marker-in-cell technique on a regular grid using pressure-velocity formulation for a medium with variable viscosity; using continuity-based marker velocity calculation combined with the fourth-order accurate in space first-order accurate in time Runge–Kutta marker advection scheme.

Sticky_air.m (Exercise 8.7) – solution of 2D Stokes continuity and advection equations with finite differences and marker-in-cell technique on a regular grid using pressure-velocity formulation for a medium with variable viscosity; using free surface upper boundary condition based on the sticky air approach; using free surface stabilization scheme in the Stokes equation; using continuity-based marker velocity calculation combined with the fourth-order accurate in space first-order accurate in time Runge–Kutta marker advection scheme.

Chapter 9

Shear_adiabatic_heating.m (Exercise 9.3) – solution of 2D Stokes and continuity equations with finite differences on a regular grid using pressure-velocity formulation for a medium with variable viscosity; computation and visualization of shear and adiabatic heating distribution.

Chapter 10

Explicit_implicit_1D.m (Fig. 10.2) – solution of 1D temperature equation on a regular grid for a non-moving medium with constant conductivity; comparison of implicit and explicit methods.

Explicit_Implicit2D.m (Exercise 10.1) – solution of 2D temperature equation on a regular grid for a non-moving medium with constant conductivity; comparison of implicit and explicit methods.

Variable_conductivity.m (Exercise 10.2) – solution of 2D temperature equation on a regular grid for a non-moving medium with variable conductivity; comparison of implicit and explicit methods.

Conduction_advection2D.m (Exercise 10.3) – solution of 2D Eulerian temperature equation with advective terms on a regular grid for a moving medium with constant conductivity; using upwind differences for advection of temperature; comparison of implicit and explicit methods.

Variable_conductivity_advection2D.m (Exercise 10.3) – solution of 2D Eulerian temperature equation with advective terms on a regular grid for a moving medium with variable conductivity; using upwind differences for advection of temperature; comparison of implicit and explicit methods.

Variable_conductivity_markers2D.m (Exercise 10.4) – solution of 2D Lagrangian temperature equation on a regular grid with implicit finite differences for a moving medium with variable conductivity; using marker-in-cell approach for advection of temperature.

Chapter 11

i2vis.m (Exercise 11.1) – 2D thermomechanical viscous Cartesian code; solution of 2D Stokes, continuity, temperature and advection equations with finite differences and marker-in-cell technique on a regular grid using pressure-velocity formulation for a deforming incompressible medium with variable viscosity and thermal conductivity; taking into account radiogenic, shear and adiabatic heating.

i2planet.m (Exercise 11.2, Fig. 11.5) – 2D thermomechanical viscous spherical-Cartesian code; solution of 2D Poisson, Stokes, continuity, temperature and advection equations with finite differences and marker-in-cell technique on a regular grid using pressure-velocity formulation for a deforming self-gravitating incompressible medium with variable viscosity and thermal conductivity; using low viscosity 'sticky space' to enable free-surface condition for a deforming planet; computing local gravitational acceleration components for the Stokes equation based on solution of the Poisson equation; using free surface stabilization in the Stokes equation; taking into account radiogenic, shear and adiabatic heating.

Chapter 12

Viscoelastic_stress.m (Exercise 12.2) – computation of visco-elastic stress buildup/ relaxation with time.

Viscoelastoplastic_strain_rate.m (Exercise 12.3) – computation of visco-elasto-plastic stress buildup and associated viscous, elastic and plastic strain rate evolution with time.

Peierls_creep.m (Exercise 12.4) – computation and visualization of viscosity maps in temperature–log stress coordinates for a combination of dislocation, diffusion and Peierls creep; comparison of wet and dry olivine rheology.

Stress_rotation.m (Exercise 12.5) – comparison of different approaches for the case of 2D stress rotation with constant angular velocity: (A) analytical, (B) Jaumann, (C) Jaumann with effective stress rate computed with fourth-order (in stress space) Runge–Kutta method and (D) 3D finite angle rotation.

Stress_rotation_markers.m (Exercise 12.6) – comparison of different approaches for the case of 2D stress advection and rotation on Lagrangian markers: (A) analytical, (B) Jaumann, (C) Jaumann with effective stress rate computed with fourth-order (in stress space) Runge–Kutta method and (D) 3D finite angle rotation.

Chapter 13

Viscoplastic2D.m – 2D thermomechanical visco-plastic code; solution of 2D Stokes, continuity, temperature and advection equations with finite differences and marker-in-cell technique on a regular grid using pressure-velocity formulation for a deforming incompressible visco-plastic medium with variable viscosity, plastic strength and thermal conductivity; adjusting marker visco-plastic viscosity to satisfy the Drucker–Prager yielding condition; taking into account radiogenic, shear and adiabatic heating.

Viscoelastic2D.m (Exercise 13.1) – 2D thermomechanical visco-elastic code; solution of 2D Stokes, continuity, temperature and advection equations with finite differences and marker-in-cell technique on a regular grid using pressure-velocity formulation for a deforming incompressible medium with variable viscosity, plastic strength, shear modulus and thermal conductivity; taking into account radiogenic, shear and adiabatic heating.

Viscoelastoplastic2D.m (Exercise 13.2) – 2D thermomechanical visco-elasto-plastic code; solution of 2D Stokes, continuity, temperature and advection equations with finite differences and marker-in-cell technique on a regular grid using pressure-velocity formulation for a deforming visco-elasto-plastic incompressible medium with variable viscosity, shear modulus, plastic strength and thermal conductivity; adjusting marker visco-plastic viscosity to satisfy the Drucker–Prager yielding condition; taking into account radiogenic, shear and adiabatic heating.

Viscoelastoplastic2D_iterations.m (Fig.13.3) – same as above but with global Picard iterations on Lagrangian markers.

i2elvis.m (Exercise 13.3) – 2D thermomechanical visco-elasto-plastic code with global Picard iteration; solution of 2D Stokes, continuity, temperature and advection equations with finite differences and marker-in-cell technique on a regular grid using pressure-velocity formulation for a deforming incompressible medium with variable viscosity, shear modulus, plastic strength and thermal conductivity; using global Picard iterations to adjust visco-plastic viscosity on Eulerian nodes; taking into account radiogenic, shear and adiabatic heating.

Chapter 14

Compressible2D.m (Exercise 14.1) – 2D thermomechanical compressible visco-elasto-plastic code with inertia and global Picard iteration; solution of 2D momentum, continuity, temperature and advection equations with finite differences and marker-in-cell technique on a regular grid using pressure-velocity formulation for a deforming compressible medium with inertia and variable viscosity, shear modulus, plastic strength and thermal conductivity; using global Picard thermomechanical iterations to adjust visco-plastic viscosity, temperature and pressure on Eulerian nodes; taking into account radiogenic, shear and adiabatic heating.

Inertia2D.m (Exercise 14.2) – 2D thermomechanical compressible visco-elasto-plastic code with inertia and global Picard iteration for bouncing ball setup; solution of 2D momentum, continuity, temperature and advection equations with finite differences and marker-in-cell technique on a regular grid using pressure-velocity formulation for a deforming compressible medium with inertia and variable viscosity, shear modulus, plastic strength and thermal conductivity; using global Picard thermomechanical iterations to adjust visco-plastic viscosity, temperature and pressure on Eulerian nodes; taking into account radiogenic, shear and adiabatic heating.

i2elvisNS.m (Exercise 14.3) – 2D thermomechanical compressible visco-elasto-plastic code with inertia, tensile plasticity, strain weakening and global Picard iteration for breaking bridge experiment; solution of 2D momentum, continuity, temperature and advection equations with finite differences and marker-in-cell technique on a regular grid using pressure-velocity formulation for a deforming compressible medium with inertia and variable viscosity, shear modulus, plastic strength (both compressive and tensile, subjected to strain weakening), and thermal conductivity; using global Picard thermomechanical iterations to adjust visco-plastic viscosity, temperature and pressure on Eulerian nodes; taking into account radiogenic, shear and adiabatic heating.

Chapter 15

Rate_dependent_friction.m – 2D seismo-thermomechanical compressible visco-elasto-plastic code with rate-dependent friction; solution of 2D momentum, continuity, temperature and advection equations with finite differences and marker-in-cell technique on a regular grid using pressure-velocity formulation for a deforming compressible medium with inertia, rate-dependent friction and variable viscosity, shear modulus, plastic strength and thermal conductivity; using global Picard thermomechanical iterations to adjust visco-plastic viscosity, temperature and pressure on Eulerian nodes; taking into account radiogenic, shear and adiabatic heating.

i2elvisSTM_backtracing.m – 2D seismo-thermomechanical compressible visco-elasto-plastic code with rate- and state-dependent friction and back-tracing of stress, pressure and velocity changes on Eulerian nodes; solution of 2D momentum, continuity, temperature and advection equations with finite differences and marker-in-cell technique on a regular grid using pressure-velocity formulation for a deforming compressible medium with inertia, rate- and state-dependent friction and variable viscosity, shear modulus, rate- and state-dependent friction parameters and thermal conductivity; using global Picard thermomechanical iterations to adjust visco-plastic viscosity, temperature and pressure on Eulerian nodes; taking into account radiogenic, shear and adiabatic heating.

i2elvisSM.m – 2D Eulerian seismo-mechanical compressible visco-elasto-plastic code with rate- and state-dependent friction and variable fault thickness D; solution of 2D momentum, continuity, temperature and advection equations with finite differences on a regular grid using pressure-velocity formulation for a deforming compressible medium with inertia, rate-dependent friction and variable viscosity, shear modulus and rate- and state-dependent friction parameters; using global Picard iterations to adjust visco-plastic viscosity and pressure on Eulerian nodes; neglecting effects of advection for stress, pressure and velocity changes due to small time steps.

i2elvisSTM.m (Exercise 15.1) – 2D seismo-thermomechanical compressible visco-elasto-plastic code with rate- and state-dependent friction; solution of 2D momentum, continuity, temperature and advection equations with finite differences and marker-in-cell technique on a regular grid using pressure-velocity formulation for a deforming compressible medium with inertia, rate- and state-dependent friction and variable viscosity, shear modulus, rate- and state-dependent friction parameters and thermal conductivity; using global Picard thermomechanical iterations to adjust visco-plastic viscosity, temperature and pressure on Eulerian nodes; taking into account radiogenic, shear and adiabatic heating.

Chapter 16

i2visHTM_melting.m – solving of a toy thermodynamic model of mantle melting with 2D hydro-thermomechanical viscous code; solution of 2D momentum, continuity, temperature and advection equations for solid and fluid with finite differences and marker-in-cell technique on a regular grid using pressure-velocity formulation for a deforming viscous compacting/decompacting fluid-bearing medium with variable viscosity and thermal conductivity; using global Picard hydro-thermomechanical iterations to adjust the mass transfer term, temperature and pressure on Eulerian nodes; taking into account radiogenic, shear and adiabatic heating.

i2visHTM_hydration.m – solving of a toy thermodynamic model of rock hydration/ dehydration with 2D hydro-thermomechanical viscous code; solution of 2D momentum, continuity, temperature and advection equations for solid and fluid with finite differences and marker-in-cell technique on a regular grid using pressure-velocity formulation for a deforming viscous compacting/decompacting fluid-bearing medium with variable viscosity and thermal conductivity; using global Picard hydro-thermomechanical iterations to adjust the mass transfer term, temperature and pressure on Eulerian nodes; taking into account radiogenic, shear and adiabatic heating; both solid and fluid are assumed to be incompressible.

i2visHTM.m (Exercise 16.1) – 2D hydro-thermomechanical viscous code; solution of 2D momentum, continuity, temperature and advection equations for solid and fluid with finite differences and marker-in-cell technique on a regular grid using pressure-velocity formulation for a deforming viscous compacting/decompacting fluid-bearing medium with variable viscosity and thermal conductivity; taking into account radiogenic, shear and adiabatic heating; both solid and fluid are assumed to be incompressible.

i2elvisHTM.m (Exercise 16.2) – 2D hydro-thermomechanical visco-elasto-plastic code; solution of 2D momentum, continuity, temperature and advection equations for solid and fluid with finite differences and marker-in-cell technique on a regular grid using pressure-velocity formulation for a deforming visco-elasto-plastic fluid-bearing medium with variable viscosity, shear modulus, plastic strength and thermal conductivity; taking into account the difference between the total and fluid pressure for brittle/plastic yielding (both compressive and tensile); using global Picard iterations to adjust visco-plastic viscosity and pressure (both total and fluid) on Eulerian nodes; taking into account radiogenic, shear and adiabatic heating; both solid and fluid are assumed to be incompressible.

Chapter 17

i2vis_Swiss_cross.m (Exercise 17.1) – 2D thermomechanical viscous code with 'Swiss cross' grid adaptation; solution of 2D Stokes, continuity, temperature and advection equations with finite differences and marker-in-cell technique on an irregular 'Swiss cross' grid using pressure-velocity formulation for a deforming incompressible medium with variable viscosity and thermal conductivity; taking into account radiogenic, shear and adiabatic heating; adaptive changes in the grid structure to follow propagation of a plume.

AMR_mechanical.m (Exercise 17.2) – Eulerian 2D mechanical viscous codes with block-structured grid adaptation and external functions *epsilonxy.m*, *eval_anal_Dani.m*, *eval_- sol_cx.m*, *eval_sol_kz.m*, *gridchange_error_new.m*, *matrixadd_test4_eta_rho.m*, *ronurecalc_test2.m*, *sigmaxx_test.m*, *sigmaxy_test.m*, *sigmaxy_vx_test.m*, *sigmaxy_- vy_test.m*, *sigmayy_test.m*, *splitcelbest.m*, *varindex_test2.m*; solution of 2D Stokes and continuity equations with finite differences on a block-structured adaptive grid using pressure-velocity formulation for a deforming incompressible medium with variable viscosity; iterative adaptive changes in the grid structure are based on local changes in material properties.

AMR_markers.m (Exercise 17.3, Fig. 17.3) – 2D mechanical visco-plastic marker-in-cell code with block-structured grid adaptation and external functions *celinter.m, epsilonxy.m, gridchange_viscoplastic.m, matrixadd_viscoplastic.m, mergecelbest.m, movemark_plastic.m, ronurecalc_markers.m, sigmaxx_test.m, sigmaxxyy.m, sigmaxy_test.m, sigmaxy_vx_test.m, sigmaxy_vy_test.m, sigmaxycalc.m, sigmayy_test.m, splitcelbest.m, varindex_test2.m*; solution of 2D Stokes, continuity and advection equations with finite differences and marker-in-cell technique on a block-structured adaptive grid using pressure-velocity formulation for a deforming incompressible visco-plastic medium with variable viscosity and plastic strength; adjusting marker visco-plastic viscosity to satisfy the Drucker–Prager yielding condition; iterative adaptive changes in the grid structure are based on local changes in visco-plastic viscosity to follow the evolving shear zones pattern.

AMR_planet_T.m (Exercise 17.4, Fig. 17.4) – 2D thermomechanical spherical-Cartesian viscous marker-in-cell code with block-structured grid adaptation and external functions *celinter.m, dfidx.m, dfidy.m, epsilonxy.m, gridchange.m, gxcalc.m, heatsolve.m, matrixadd_planet.m, mergecelbest.m, movemark_T.m, qx.m, qy.m, ronurecalc_T.m, sigmaxx.m, sigmaxy.m, sigmayy.m, splitcelbest.m, varindex_planet.m*; solution of 2D Poisson, Stokes, continuity, temperature and advection equations with finite differences and marker-in-cell technique on a block-structured adaptive grid using pressure-velocity formulation for a deforming incompressible viscous medium with variable viscosity and thermal conductivity; iterative adaptive changes in the grid structure are based on local changes in viscosity to follow the evolving planetary surface and interior.

Chapter 18

Gauss_Seidel_iterations_Poisson_planet.m (Fig. 18.1) – solution of 2D Poisson equation for a circular planetary body with Gauss–Seidel iteration.

Gauss_Seidel_iterations_Poisson.m (Fig. 18.2) – solution of 2D Poisson equation with Gauss–Seidel iteration.

Poisson_Multigrid.m (Fig. 18.5) – solution of 2D Poisson equation with multigrid based on V-cycle and external functions *Poisson_smoother.m, Poisson_restriction.m, Poisson_prolongation.m*; resolution between multigrid levels changes by a factor of two.

Poisson_Multigrid_planet_arbitrary.m – solution of 2D Poisson equation for the case of a circular planetary body embedded in a massless-like medium with multigrid based on V-cycle and external functions ***Poisson_smoother_planet.m***, ***Poisson_restriction _planet.m***, ***Poisson_prolongation_planet.m***; resolution between multigrid levels changes in an arbitrary way.

Stokes_Continuity_Multigrid.m – solution of 2D Stokes and continuity equations for a constant viscosity medium with multigrid based on V-cycle and external functions ***Stokes_Continuity_smoother.m***, ***Stokes_Continuity_restriction.m***, ***Stokes_Continuity_prolongation.m***; resolution between multigrid levels changes by a factor of two.

Variable_viscosity_MultiMultigrid_arbitrary.m (Fig. 18.13) – solution of 2D Stokes and continuity equations for a variable viscosity medium with multi-multigrid based on V-cycle and external functions ***Viscosity_restriction.m***, ***Stokes_Continuity_ viscous_smoother.m***.

Poisson_Multigrid_planet.m (Exercise 18.1) – solution of 2D Poisson equation for the case of a circular planetary body embedded in a massless-like medium with multigrid based on V-cycle and external functions ***Poisson_smoother_planet.m***, ***Poisson_ restriction_planet.m***, ***Poisson_prolongation_planet.m***; resolution between multi-grid levels changes by a factor of two.

Constant_Viscosity_Multigrid_ghost.m (Exercise 18.2, Fig. 18.10) – solution of 2D Stokes and continuity equations for a constant viscosity medium with multigrid based on V-cycle, ghost-node-based smoother ***Stokes_Continuity_smoother_ghost.m*** and external functions ***Stokes_Continuity_restriction.m***, ***Stokes_Continuity_ prolongation.m***; resolution between multigrid levels changes by a factor of two.

Variable_viscosity_Multigrid_arbitrary.m (Exercise 18.3, Fig. 18.12) – solution of 2D Stokes and continuity equations for a variable viscosity medium with multigrid based on V-cycle and external functions ***Viscosity_restriction.m***, ***Stokes_Continuity_viscous_smoother.m***, ***Stokes_Continuity_viscous_restriction. m***, ***Stokes_Continuity_prolongation.m***; resolution between multigrid levels changes in an arbitrary way.

Chapter 19

Temperature3D_Gauss_Seidel.m (Exercise 19.1, Fig. 19.9) – solution of 2D tempera-ture equation on a regular grid for a non-moving medium with variable conductivity; the solution is based on Gauss–Seidel iteration using external function ***Temperature3D_smoother.m***.

Poisson3D_Multigrid_planet_arbitrary.m (Exercise 19.2, Fig. 19.10) – solution of 3D Poisson equation for the case of a spherical planetary body embedded in a massless-like medium with multigrid based on V-cycle and external functions ***Poisson3D_smoother_ planet.m***, ***Poisson3D_restriction_planet.m***, ***Poisson3D_prolongation_planet.m***; resolution between multigrid levels changes in an arbitrary way.

Stokes_Continuity3D_Multigrid.m (Exercise 19.3, Fig. 19.11a) – solution of 3D Stokes and continuity equations for a constant viscosity medium with multigrid based on V-cycle and external functions ***Stokes_Continuity3D_smoother.m***, ***Stokes_Continuity3D_restriction.m***, ***Stokes_Continuity3D_prolongation.m***; resolution between multigrid levels changes by a factor of two.

Variable_viscosity3D_Multigrid.m (Exercise 19.3, Fig. 19.11b) – solution of 3D Stokes and continuity equations for a variable viscosity medium with multigrid based on V-cycle and external functions ***Viscosity_restriction3D.m***, ***Stokes_Continuity3D_viscous_smoother.m***, ***Stokes_Continuity3D_viscous_restriction.m***, ***Stokes_Continuity3D_prolongation.m***; resolution between multigrid levels changes by a factor of two.

Variable_viscosity3D_MultiMultigrid.m (Fig. 19.12) – solution of 3D Stokes and continuity equations for a variable viscosity medium with multi-multigrid based on V-cycle and external functions ***Viscosity_restriction3D.m***, ***Stokes_Continuity3D_viscous_smoother.m***, ***Stokes_Continuity3D_viscous_restriction.m***, ***Stokes_Continuity3D_prolongation.m***; resolution between multigrid levels changes by a factor of two.

Chapter 20

Variable_viscosity_Ramberg.m (Fig. 20.1) – mechanical benchmark for a two-layer Rayleigh-Taylor instability; solution of 2D Stokes, continuity and advection equations with finite differences and marker-in-cell technique using external function ***Stokes_Continuity_solver_ghost.m.***

Variable_viscosity_block.m (Fig. 20.3, Exercise 20.1) – mechanical benchmark for a falling square block; solution of 2D Stokes, continuity and advection equations with finite differences and marker-in-cell technique using external function ***Stokes_Continuity_solver_ghost.m.***

Variable_viscosity_channel.m (Fig. 20.4) – mechanical benchmark for a channel flow with a non-Newtonian rheology; solution of 2D Stokes, continuity and advection equations with finite differences and marker-in-cell technique using external function ***Stokes_Continuity_solver_channel.m.***

Constant_viscosity_channel_T.m (Fig. 20.5) – thermomechanical benchmark for a non-steady temperature distribution in a Newtonian channel; solution of 2D Stokes, continuity, temperature and advection equations with finite differences and marker-in-cell technique using external functions ***Stokes_Continuity_solver_channel.m***, ***Temperature_solver.m.***

Variable_viscosity_Couette_T.m (Fig. 20.6) – thermomechanical benchmark for a steady Couette flow with viscous heating and temperature-dependent viscosity; solution of 2D Stokes, continuity, temperature and advection equations with finite differences and marker-in-cell technique using external functions ***Stokes_Continuity_solver_Couette.m***, ***Temperature_solver.m.***

Solid_Body_Rotation_T.m (Fig. 20.7) – thermal benchmark for advection and diffusion of sharp temperature fronts in a prescribed rigid-body rotation velocity field; solution of 2D temperature and advection equations with finite differences and marker-in-cell technique using external function *Temperature_solver.m*.

Variable_conductivity_channel.m (Fig. 20.8) – thermomechanical benchmark for a steady Newtonian channel flow with variable thermal conductivity; solution of 2D Stokes, continuity, temperature and advection equations with finite differences and marker-in-cell technique using external functions *Stokes_Continuity_solver_Couette. m*, *Temperature_solver.m*.

Variable_viscosity_convection_irregular_grid.m (Figs. 20.9, 20.10) – thermomechanical benchmark for thermal convection with constant temperature- and depth-dependent viscosity; solution of 2D Stokes, continuity, temperature and advection equations with finite differences and marker-in-cell technique on regular/ irregular grid using external functions *Stokes_Continuity_solver_grid.m*, *Temperature_solver_grid.m*; nearly steady-state temperature distribution for cases 1a, 1c and 2a can be loaded from data files *data_1a_regular.txt*, *data_1c_regular.txt*, *data_2a_regular.txt*, *data_1a_irregular.txt*, *data_1c_irregular.txt*, *data_2a_irregular.txt*.

Stress_buildup.m (Fig. 20.11) – mechanical benchmark for stress buildup in a visco-elastic incompressible Maxwell body; solution of 2D Stokes, continuity and advection equations with finite differences and marker-in-cell technique using external function *Stokes_Continuity_solver_grid.m*.

Slab_deformation.m (Fig. 20.12, Exercise 20.2) – mechanical benchmark for recovery of the original shape of an elastic slab; solution of 2D Stokes, continuity and advection equations with finite differences and marker-in-cell technique using external function *Stokes_Continuity_solver_grid.m*.

Sandbox_shortening_ratio.m (Fig. 20.14) – mechanical visco-elasto-plastic benchmark for numerical sandbox shortening experiment; solution of 2D Stokes, continuity and advection equations with finite differences and marker-in-cell technique using external function *Stokes_Continuity_solver_sandbox.m*.

Bouncing_Ball.m (Fig. 20.15) – 2D mechanical visco-elastic benchmark with inertia for the bouncing ball setup; solution of 2D momentum, continuity, temperature and advection equations with finite differences and marker-in-cell technique on a regular grid using pressure-velocity formulation for a deforming compressible medium with inertia; tracing of the movements of the ball centre with time.

Manufactured.m (Exercise 20.3) – mechanical benchmark for a manufactured solution with an incompressible 2D isoviscous flow; solution of 2D Stokes and continuity equations with finite differences; comparison of results with the analytical solution.

Chapter 21

Subducting_slab_bending.m (Fig. 21.1) – thermomechanical visco-elasto-plastic numerical model for spontaneous bending of subducting oceanic slab; the model uses external functions *Stokes_Continuity_solver_sandbox.m*, *Temperature_solver_grid.m*.

Subduction.m (Figs. 21.2, 21.3) – thermomechanical visco-elasto-plastic numerical model for spontaneous retreating oceanic subduction; the model uses external functions *Stokes_Continuity_solver_sandbox.m*, *Temperature_solver_grid.m*.

Extension.m (Fig. 21.4) – thermomechanical visco-elasto-plastic numerical model for oceanic lithosphere extension; the model uses external functions *Stokes_Continuity_solver_sandbox.m*, *Temperature_solver_grid.m*.

Collision.m (Figs. 21.5, 21.6) – thermomechanical visco-elasto-plastic numerical model for post-subduction continental collision; the model accounts for erosion/sedimentation processes and uses external functions *Stokes_Continuity_solver_sandbox.m*, *Temperature_solver_grid.m*.

Collision_and_breakoff.m (Fig. 21.7) – thermomechanical visco-elasto-plastic numerical model for slab breakoff during continental collision; the model accounts for erosion/sedimentation processes and uses external functions *Stokes_Continuity_solver_sandbox.m*, *Temperature_solver_grid.m*.

Intrusion_Emplacement.m (Fig. 21.9) – thermomechanical visco-elasto-plastic numerical model for trans-lithospheric mafic-ultramafic intrusion emplacement into the crust; the model accounts for erosion/sedimentation processes and uses external functions *Stokes_Continuity_solver_sandbox.m*, *Temperature_solver_grid.m*.

Mantle_convection.m (Figs. 21.12, 21.13) – thermomechanical visco-elasto-plastic numerical model for mantle convection with phase changes; the phase changes are treated based on a Gibbs free energy minimization approach with pre-computed density and enthalpy maps in P-T space; these maps are loaded with external function *loading_database.m* from data files *m895_ro*, *m895_hh*, *morn_ro*, *morn_hh*; pre-computed non-steady temperature distribution can be loaded from the data file *convection.txt*; the model also uses external functions *Stokes_Continuity_solver_sandbox.m*, *Temperature_solver_grid.m*.

Core_formation.m (Fig. 21.14) – thermomechanical visco-elasto-plastic numerical model for the deformation of a self-gravitating iron-silicate planetary body; gravity field is computed with external function *Poisson_solver_planet_grid.m*; phase changes in the silicate component are treated based on a Gibbs free energy minimization approach with pre-computed density and enthalpy maps in P-T space; these maps are loaded with the external function *loading_database.m* from data files *m895_ro*, *m895_hh*, *morn_ro*, *morn_hh*; the model also uses external functions *Stokes_Continuity_solver_sandbox.m*, *Temperature_solver_grid.m*.

References

Albers, M. (2000) A local mesh refinement multigrid method for 3D convection problems with strongly variable viscosity. Journal of Computational Physics, **160**, 126–150.

Amestoy, P., Duff, I., Koster, J., L'excellent, A. (2001) A fully asynchronous multifrontal solver using distributed dynamic scheduling. SIAM Journal on Matrix Analysis and Applications, **23** (1), 15–41.

Ampuero, J.-P., Ben-Zion, Y. (2008) Cracks, pulses and macroscopic asymmetry of dynamic rupture on a bimaterial interface with velocity-weakening friction. Geophysical Journal International, **173**, 674–692.

Ampuero, J.-P., Rubin, A. M. (2008) Earthquake nucleation on rate and state faults – aging and slip laws. Journal of Geophysical Research, **113**, B01302.

Anderson, O. L. (1995) Equations of State for Solids in Geophysics and Ceramic Science. Oxford University Press.

Andrews, E. R., Billen, M. I. (2009) Rheologic controls on the dynamics of slab detachment. Tectonophysics, **464**, 60–69.

Armann, M., Tackley, P. J. (2012) Simulating the thermo-chemical magmatic and tectonic evolution of Venus' mantle and lithosphere 1. two-dimensional models. Journal of Geophysical Research, **117**, E12003.

Asphaug, E., Agnor, C. B., Williams, Q. (2006) Hit-and-run planetary collisions. Nature, **439**, 155–160.

Bagdassarov, N. S., Fradkov, A. S. (1993) Evolution of double diffusion convection in a felsic magma chamber. Journal of Volcanology and Geothermal Research, **54** (3–4), 291–308.

Baitsch-Ghirardello, B., Stracke, A., Connolly, J. A. D., Nikolaeva, K. M., Gerya, T. V. (2014) Lead transport in intra-oceanic subduction zones: 2D geochemical–thermo-mechanical modeling of isotopic signatures. Lithos, **208–209**, 265–280.

Barazangi, M., Isacks, B. L., Oliver, J., Dubois, J., Pascal, G. (1973) Descent of lithosphere beneath New Hebrides, Tonga–Fiji and New Zealand: evidence for detached slabs. Nature, **242** (5393), 98–101.

Baumann, T. (2016) Appraisal of geodynamic inversion results: a data mining approach. Geophysical Journal International, **207**, 667–679.

Baumann, T., Kaus, B. J. P. (2015) Geodynamic inversion to constrain the nonlinear rheology of the lithosphere. Geophysical Journal International, **202**, 1289–1316.

Baumann, T., Gerya, T., Connolly, J. A. D. (2010) Numerical modelling of spontaneous slab breakoff dynamics during continental collision. In: Spalla, M. I., Marotta, A. M., Gosso, G. (editors), Advances in Interpretation of Geological Processes. Geological Society, London, Special Publications, **332**, pp. 99–114.

438

Baumann T., Kaus B. J. P., Popov A. (2014) Constraining effective rheology through parallel joint geodynamic inversion. Tectonophysics, **631** (15), 197–211.

Baumgardner, J. R. (1985) Three-dimensional treatment of convective flow in the Earth's mantle. Journal of Statistical Physics, **39**, 501–511.

Beaumont, C., Jamieson, R. A., Nguyen, M. H., Lee, B. (2001) Himalayan tectonics explained by extrusion of a low-viscosity crustal channel coupled to focused surface denudation. Nature, **414**, 738–742.

Behn, M. D., Kelemen, P. B., Hirth, G., Hacker, B. R., Massonne, H. J. (2011) Diapirs as the source of the sediment signature in arc lavas. Nature Geoscience, **4**, 641–646.

Belytschko, T., Liu, W. K., Moran, B. (2000) Nonlinear Finite Elements for Continua and Structures. John Wiley & Sons, Chichester.

Ben-Zion, Y. (2008) Collective behavior of earthquakes and faults: continuum-discrete transitions, evolutionary changes and corresponding dynamic regimes. Reviews of Geophysics, **46**, RG4006.

Ben-Zion, Y., Rice, J. R. (1993) Earthquake failure sequences along a cellular fault zone in a three-dimensional elastic solid containing asperity and nonasperity regions. Journal of Geophysical Research, **98**, 14109–14131.

Ben-Zion, Y., Rice, J. R. (1997) Dynamic simulations of slip on a smooth fault in an elastic solid. Journal of Geophysical Research, **102**, 17771–17784.

Benz,W., Slattery, W., Cameron, A. G. W. (1986) The origin of the moon and the single-impact hypothesis 1. Icarus, **66**, 515.

Bercovici, D. (editor) (2007) Mantle Dynamics. Treatise on Geophysics, Volume **7**. Elsevier.

Bercovici, D., Michaut, C. (2010) Two-phase dynamics of volcanic eruptions: compaction, compression and the conditions for choking. Geophysical Journal International, **182** (2), 843–864.

Bercovici, D., Ricard, Y. (2012) Mechanisms for the generation of plate tectonics by two phase grain-damage and pinning. Physics of the Earth and Planetary Interiors, **202–203**, 27–55.

Bercovici, D., Ricard, Y. (2014) Plate tectonics, damage and inheritance. Nature, **508**, 513–516.

Bercovici, D., Schubert, G. (editors) (2015) Mantle Dynamics. Treatise on Geophysics, 2nd Edition, Volume 7. Elsevier.

Bercovici, D., Ricard, Y., Schubert, G. (2001) A two-phase model for compaction and damage: 1. General theory. Journal of Geophysical Research, **106**, 8887–8906.

Bercovici, D., Tackley, P., Ricard, Y. (2015) The generation of plate tectonics from mantle dynamics. In: Bercovici, D., Schubert, G. (editors), Mantle Dynamics. Treatise on Geophysics, Volume **7**. Elsevier, pp. 271–318.

Bergantz, G. W. (2000) On the dynamics of magma mixing by reintrusion: implications for pluton assembly processes. Journal of Structural Geology, **22**, 1297–1309.

Berman, R. G. (1988) Internally-consistent thermodynamic data for minerals in the system Na_2O-K_2O-CaO-MgO-FeO-Fe_2O_3-Al_2O_3-SiO_2-TiO_2-H_2O-CO_2. Journal of Petrology, **29**, 445–522.

Berner, H., Ramberg, H., Stephanson, O. (1972) Diapirism in theory and experiment. Tectonophysics, **15**, 197–218.

Best, M. G., Christiansen, E. H. (2001) Igneous Petrology. Blackwell Science, Malden.

Bhattacharya, P., Rubin, A. M., Bayart, E., Savage, H. M., Marone, C. (2015) Critical evaluation of state evolution laws in rate and state friction: fitting large velocity steps in simulated fault gouge with time-, slip-, and stress-dependent constitutive laws. Journal of Geophysical Research, **120**, 6365–6385.

Billen, M. I., Kreylos, O., Hamann, B., Jadamec, M. A., Kellogg, L. H., Staadt, O., Sumner, D. Y. (2008) A geoscience perspective on immersive 3D gridded data visualization. Computers & Geosciences, **34**, 1056–1072.

Biot, M. A. (1941) General theory of three-dimensional consolidation. Journal of Applied Physics, **12**, 155–164.

Biot, M. A. (1965) Mechanics of Incremental Deformations: Theory of Elasticity and Viscoelasticity of Initially Stressed Solids and Fluids, Including Thermodynamic Foundations and Applications to Finite Strain. John Wiley & Sons.

Birch, F. (1947) Finite elastic strain of cubic crystals. Physical Review, **71**, 809–824.

Bird, P. (1978) Finite elements modeling of lithosphere deformation: the Zagros collision orogeny. Tectonophysics, **50**, 307–336.

Bitsch, B., Lambrechts, M., Johansen, A. (2015) The growth of planets by pebble accretion in evolving protoplanetary discs. Astronomy & Astrophysics, **582**, A112.

Bittner, D., Schmeling, H. (1995) Numerical modeling of melting processes and induced diapirism in the lower crust. Geophysical Journal International, **123**, 59–70.

Blankenbach, B., Busse, F., Christensen, U., Cserepes, L., Gunkel, D., Hansen, U., Harder, H., Jarvis, G., Koch, M., Marquart, G., Moore, D., Olson, P., Schmeling, H., Schnaubelt, T. (1989) A benchmark comparison for mantle convection codes. Geophysical Journal International, **98** (1), 23–38.

Bocher, M., Coltice, N., Fournier, A., Tackley, P. J. (2016) A sequential data assimilation approach for the joint reconstruction of mantle convection and surface tectonics. Geophysical Journal International, **204** (1), 200–214.

Bollard, J., Connelly, J. N., Whitehouse, M. J., Pringle, E. A., Bonal, L., Jørgensen, J. K., Nordlund, Å., Moynier, F., Bizzarro, M. (2017) Early formation of planetary building blocks inferred from Pb isotopic ages of chondrules. Science Advances, **3**, e1700407.

Boris, J. P., Book, D. L. (1973) Flux-corrected transport. I. SHASTA, a fluid transport algorithm that works. Journal of Computational Physics, **11**, 38–69.

Brace, W. F., Kohlstedt, D. T. (1980) Limits on lithospheric stress imposed by laboratory experiments. Journal of Geophysical Research, **85**, 6248–6252.

Braun, J., Sambridge, M. (1997) Modelling landscape evolution on geological time scales: a new method based on irregular spatial discretization. Basin Research, **9**, 27–52.

Braun, J., Thieulot, C., Fullsack, P., DeKool, M., Beaumont, C., Huismans, R. (2008) DOUAR: a new three-dimensional creeping flow numerical model for the solution of geological problems. Physics of the Earth and Planetary Interiors, **171**, 76–91.

Brosh, E., Shneck, R. Z., Makov, G. (2008) Explicit Gibbs free energy equation of state for solids. Journal of Physics and Chemistry of Solids, **69**, 1912–1922.

Brune, S. (2016) Rifts and rifted margins: a review of geodynamic processes and natural hazards. In: Duarte, J., Schellart, W. (editors), Plate Boundaries and Natural Hazards. American Geophysical Union, Wiley, pp. 39–76.

Budiansky, B. (1970) Thermal and thermoelastic properties of isotropic composites. Journal of Composite Materials, **4**, 286–295.

Buiter, S. J. H., Govers, R., Wortel, M. J. R. (2002) Two-dimensional simulations of surface deformation caused by slab detachment. Tectonophysics, **354**, 195–210.

Buiter, S. J. H., Babeyko, A. Yu., Ellis, S., Gerya, T. V., Kaus, B. J. P., Kellner, A., Schreurs, G., Yamada, Y. (2006) The numerical sandbox: comparison of model results for a shortening and an extension experiment. In: Buiter, S. J. H., Schreurs, G. (editors),

Analogue and Numerical Modelling of Crustal-Scale Processes. Geological Society, London, Special Publications, **253**, pp. 29–64.

Buiter, S. J. H., Schreurs, G., Albertz, M., Gerya, T. V., Kaus, B., Landry, W., le Pourhiet, L., Mishin, Y., Egholm, D. L., Cooke, M., Maillot, B., Thieulot, C., Crook, T., May, D., Souloumiac, P., Beaumont, C. (2016) Benchmarking numerical models of brittle thrust wedges. Journal of Structural Geology, **92**, 140–177.

Burg, J.-P., Gerya, T. V. (2005) The role of viscous heating in Barrovian metamorphism of collisional orogens: thermomechanical models and application to the Lepontine Dome in the Central Alps. Journal of Metamorphic Geology, **23**, 75–95.

Burg, J.-P., Bodinier, J.-L., Gerya, T., Bedini, R.-M., Boudier, F., Dautria, J.-M., Prikhodko, V., Efimov, A., Pupier, E., Balanec, J.-L. (2009) Translithospheric mantle diapirism: geological evidence and numerical modelling of the Kondyor zoned ultramafic complex (Russian Far-East). Journal of Petrology, **50**, 289–321.

Burkett, E. R., Billen, M. I. (2010) Three-dimensionality of slab detachment due to ridge–trench collision: laterally simultaneous boudinage versus tear propagation. Geochemistry, Geophysics, Geosystems, **11**, Q11012.

Burov, E. B., Cloetingh, S. (1997) Erosion and rift dynamics: new thermomechanical aspects of post-rift evolution of extensional basins. Earth and Planetary Science Letters, **150**, 7–26.

Burov, E., Gerya, T. (2014) Asymmetric three-dimensional topography over mantle plumes. Nature, **513**, 85–89.

Burov, E., Poliakov, A. (2001) Erosion and rheology controls on synrift and postrift evolution: verifying old and new ideas using a fully coupled numerical model. Journal of Geophysical Research, **106** (B8), 16461–16481.

Burov, E., Jolivet, L., Le Pourhiet, L., Poliakov, A. (2001) A thermomechanical model of exhumation of high pressure (HP) and ultra-high pressure (UHP) metamorphic rocks in Alpine-type collision belts. Tectonophysics, **342**, 113–136.

Burov, E., Jaupart, C., Guillou-Frottier, L. (2003) Ascent and emplacement of buoyant magma bodies in brittle-ductile upper crust. Journal of Geophysical Research, **108**, 2177.

Burstedde, C., Ghattas, O., Gurnis, M., Stadler, G., Tan, E., Tu, T., Wilcox, L. C., Zhong, S. (2008) Scalable adaptive mantle convection simulation on petascale supercomputers. Proceedings of the 2008 ACM/IEEE conference on Supercomputing, Austin, TX, Article No. 62.

Burstedde, C., Ghattas, O., Stadler, G., Tu, T., Wilcox, L. C. (2009) Parallel scalable adjoint-based adaptive solution of variable-viscosity Stokes flow problems. Computer Methods in Applied Mechanics and Engineering, **198**, 1691–1700.

Busse, F. H., Christensen, U., Clever, R., Cserepes, L., Gable, C., Giannandrea, E., Guillou, L., Houseman, G., Nataf, H.-C., Ogawa, M. (1994) 3-D convection at infinite Prandtl number in Cartesian geometry – a benchmark comparison. Geophysical & Astrophysical Fluid Dynamics, **75**, 39–59.

Byerlee, J. D. (1978) Friction of rocks. Pure and Applied Geophysics, **116**, 615–626.

Cai, M. (2010) Practical estimates of tensile strength and Hoek-Brown strength parameter $m(i)$ of brittle rocks. Rock Mechanics and Rock Engineering, **43**, 167–184.

Canup, R. M. (2004) Simulations of a late lunar-forming impact. Icarus, **168**, 33–456.

Canup, R. M., Asphaug, E. (2001) Origin of the Moon in a giant impact near the end of the Earth's formation. Nature, **412**, 708.

Cao, W., Kaus, B. J. P., Paterson, S. (2016) Intrusion of granitic magma into the continental crust facilitated by multiple pulses and diking: numerical simulations. Tectonics, **35**, 1575–1594.

Caricchi, L., Burlini, L., Ulmer, P., Gerya, T., Vassalli, M. Papale, P. (2007) Non-Newtonian rheology of crystal-bearing magmas and implications for magma ascent dynamics. Earth and Planetary Science Letters, **264**, 402–419.

Castro, A. Gerya, T. V. (2008) Magmatic implications of mantle wedge plumes: experimental study. Lithos, **103**, 138–148.

Chambers, J. E. (2001) Making more terrestrial planets. Icarus, **152**, 205–224.

Chambers, J. E., Wetherill, G. W. (1998) Making the terrestrial planets: N-body integrations of planetary embryos in three dimensions. Icarus, **136**, 304–327.

Chapman, B., Jost, G., van der Pas, R. (2007) Using OpenMP: portable shared memory parallel Programming. MIT Press, Cambridge, MA.

Chemenda, A. I., Burg, J.-P., Mattauer, M. (2000) Evolutionary model of the Himalaya-Tibet system: geopoem based on new modelling, geological and geophysical data. Earth and Planetary Science Letters, **174**, 397–409.

Chen, J., Spiers, C. J. (2016) Rate and state frictional and healing behavior of carbonate fault gouge explained using microphysical model. Journal of Geophysical Research, **121**, 8642–8665.

Chen, S., Zhang, H., Yuen, D., Zhang, S., Zhang, J., Shi, Y. (2008) Volume rendering visualization of 3D spherical mantle convection with an unstructured mesh. Visual Geosciences, **13**, 97–104.

Chester, F. M. (1994) Effects of temperature on friction: constitutive equations and experiments with quartz gouge. Journal of Geophysical Research, **99**, 7247–7261.

Chopin, C. (1984) Coesite and pure pyrope in high-grade blueschists of the Western Alps: a first record and some consequences. Contributions to Mineralogy and Petrology, **86**, 107–118.

Chopin, C. (2003) Ultrahigh-pressure metamorphism: tracing continental crust into mantle. Earth and Planetary Science Letters, **212**, 1–14.

Christensen, U. (1982) Phase boundaries in finite amplitude mantle convection. Geophysical Journal of the Royal Astronomical Society, **68**, 487–497.

Christensen, U. R., Yuen, D. A. (1985) Layered convection induced by phase changes. Journal of Geophysical Research, **90**, 10291–10300.

Chung, W.-Y., Kanamori, H. (1976) Source process and tectonic implications of the Spanish deep-focus earthquake of March 29, 1954. Physics of the Earth and Planetary Interiors, **13** (2), 85–96.

Clauser, C., Huenges, E. (1995) Thermal conductivity of rocks and minerals. In: Ahrens, T. J. (editor), Rock Physics and Phase Relations. AGU Reference Shelf 3. American Geophysical Union, Washington DC, pp. 105–126.

Clemens, J. D., Mawer, C. K. (1992) Granitic magma transport by fracture propagation. Tectonophysics, **204**, 339–360.

Cloetingh, S. A. P. L., Ziegler, P. A., Bogaard, P. J. F., Andriessen, P. A. M., Artemieva, I. M., Bada, G., van Balen, R. T., Beekman, F., Ben-Avraham, Z., Brun, J.-P., Bunge, H. P., Burov, E. B., Carbonell, R., Facenna, C., Friedrich, A., Gallart, J., Green, A. G., Heidbach, O., Jones, A. G., Matenco, L., Mosar, J., Oncken, O., Pascal, C., Peters, G., Sliaupa, S., Soesoo, A., Spakman, W., Stephenson, R. A., Thybo, H., Torsvik, T., de Vicente, G., Wenzel, F., Wortel, M. J. R. (2007) TOPO-EUROPE: the geoscience of coupled deep Earth-surface processes. Global and Planetary Change, **58**, 1–118.

Cloos, M. (1982) Flow melanges – numerical modeling and geologic constraints on their origin in the Franciscan subduction complex, California. Geological Society of America Bulletin, **93**, 330–345.

Cocco, M., Bizzarri, A. (2002) On the slip-weakening behavior of rate- and state dependent constitutive laws. Geophysical Research Letters, **29**, 1516.

Cocco, M., Bizzarri, A., Tinti, E. (2004) Physical interpretation of the breakdown process using a rate- and state-dependent friction law. Tectonophysics, **378**, 241–262.

Cohen, R.E. (editor) (2005) High-Performance Computing Requirements for the Computational Solid Earth Sciences. www.geo-prose.com/computational_SES.html

Colón, D. P., Bindeman, I. N., Gerya, T. V. (2018) Thermomechanical modelling of the formation of a multilevel, crustal-scale magmatic system by the Yellowstone plume. Geophysical Research Letters, **45**, 3873–3879.

Coltice, N., Rolf, T., Tackley, P. J., Labrosse, S. (2012) Dynamic causes of the relation between area and age of the ocean floor. Science, **336** (6079), 335–338.

Condamine, F. L., Rolland, J., Morlon, H. (2013) Macroevolutionary perspectives to environmental change. Ecology Letters, **16**, 72–85.

Connolly, J. A. D. (2005) Computation of phase equilibria by linear programming: a tool for geodynamic modeling and an application to subduction zone decarbonation. Earth and Planetary Science Letters, **236**, 524–541.

Connolly, J. A. D., Kerrick, D. M. (1987) An algorithm and computer program for calculating composition phase diagrams. CALPHAD, **11**, 1–55.

Connolly, J. A. D., Podladchikov, Y. Y. (1998) Compaction-driven fluid flow in viscoelastic rock. Geodinamica Acta, **11**, 55–84.

Connolly, J., Podladchikov, Y. Y. (2000) Temperature-dependent viscoelastic compaction and compartmentalization in sedimentary basins. Tectonophysics, **324**, 137–168.

Costa, A., Caricchi, L., Bagdassarov, N. (2009) A model for the rheology of particle-bearing suspensions and partially molten rocks. Geochemistry, Geophysics, Geosystems, **10**, Q03010.

Crameri, F., Schmeling, H., Golabek, G. J., Duretz, T., Orendt, R., Buiter, S. J. H., May, D. A., Kaus, B. J. P., Gerya, T. V., Tackley, P. J. (2012a) A comparison of numerical surface topography calculations in geodynamic modelling: an evaluation of the 'sticky air' method. Geophysical Journal International, **189**, 38–54.

Crameri, F., Tackley, P. J., Meilick, I., Gerya, T., Kaus, B. J. P. (2012b) A free plate surface and weak oceanic crust produce single-sided subduction on Earth. Geophysical Research Letters, **39**, doi:10.1029/2011GL050046.

Cserepes, L., Rabinowicz, M., Rosemberg-Borot, C. (1988) Three-dimensional infinite Prandtl number convection in one and two layers and implications for the Earth's gravity field. Journal of Geophysical Research, **93**, 12009–12025.

Dabrowski, M., Krotkiewski, M., Schmid, D. W. (2008) MILAMIN: MATLAB-based finite element method solver for large problems. Geochemistry, Geophysics, Geosystems, **9**, Q04030, doi:10.1029/2007GC001719.

Daignières, M., Fremond, M., Friaa, A. (1978) Modèle de type Norton-Hoff généralisé pour l'étude des déformations lithosphériques (exemple: la collision Himalayenne). Comptes Rendus Hebdomadaires des Seances de l'Academie des Sciences, **268B**, 371–374.

Dal Zilio, L., van Dinther, Y., Gerya, T. V., Pranger C. C. (2018) Seismic behaviour of mountain belts controlled by plate convergence rate. Earth and Planetary Science Letters, **482**, 81–92.

Damon, M., Kameyama, M. C., Knox, M., Porter, D. H., Yuen, D., Sevre, E. O. D. (2008) Interactive visualization of 3D mantle convection. Visual Geosciences, **13**, 49–57.

David, C., Wong, T. F., Zhu, W. L., Zhang, J. X. (1994) Laboratory measurement of compaction-induced permeability change in porous rocks – implications for the generation and maintenance of pore pressure excess in the crust. Pure and Applied Geophysics, **143**, 425–456.

Davies, G. F. (1985) Heat deposition and retention in a solid planet growing by impacts. Icarus, **63**, 45–68.

Davies, J. H, Von Blanckenburg, F. (1995) Slab breakoff: a model of lithospheric detachment and its test in the magmatism and deformation of collisional orogens. Earth and Planetary Science Letters, **129**, 85–102.

Day, S. M., Dalguer, L. A., Lapusta, N., Liu, Y. (2005) Comparison of finite difference and boundary integral solutions to three-dimensional spontaneous rupture. Journal of Geophysical Research, **110**, B12307.

de Capitani, C., Brown, H. (1987) The computation of chemical equilibrium in complex systems containing non-ideal solid solutions. Geochimica et Cosmochimica Acta, **51**, 2639–2652.

DePaolo, D. J., Cerling, T. E., Hemming, S. R., Knoll, A. H., Richter, F. M., Royden, L. H., Rudnick, R. L., Stixrude, L., Trefil, J. S. (2008) Origin and Evolution of Earth: Research Questions for a Changing Planet. Committee on Grand Research Questions in the Solid-Earth Sciences, Board on Earth Sciences and Resources, Division on Earth and Life Studies, National Research Council of the National Academies, The National Academies Press, Washington, DC.

Descombes, P., Gaboriau, T., Albouy, C., Heine, C., Leprieur, F., Pellissier, L. (2017) Linking species diversification to palaeo-environmental changes: a process-based modelling approach. Global Ecology and Biogeography, **27**, 233–244.

Deubelbeiss, Y., Kaus, B. J. P. (2008) Comparison of Eulerian and Lagrangian numerical techniques for the Stokes equations in the presence of strongly varying viscosity. Physics of the Earth and Planetary Interiors, **171**, 92–111.

Dieterich, J. H. (1972) Time-dependent friction in rocks. Journal of Geophysical Research, **77**, 3690–3697.

Dieterich, J. H. (1978) Time-dependent friction and the mechanics of stick-slip. Pure and Applied Geophysics, **116**, 790–806.

Dieterich, J. H. (1979) Modeling of rock friction: 1. Experimental results and constitutive equations. Journal of Geophysical Research, **84**, 2161–2168.

Dieterich, J. H. (1981) Constitutive properties of faults with simulated gouge. In: Carter, N. L., Friedman, M., Logan, J. M., Stearns, D. W. (editors), Mechanical Behavior of Crustal Rocks. The Handin volume. Geophysical Monograph Series, volume **24**. Washington, DC, American Geophysical Union, pp. 103–120.

Dieterich, J. H. (1992) Earthquake nucleation on faults with rate- and state-dependent strength. Tectonophysics, **211**, 115–134.

Dieterich, J. H., Kilgore, B. D. (1994) Direct observation of frictional contacts: new insights for state-dependent properties. Pure and Applied Geophysics, **143**, 283–302.

Dobrzhinetskaya, L. F., Eide, E. A., Larsen, R. B., Sturt, B. A., Tronnes, R. G., Smith, D. C., Taylor, W. R., Posukhova, T. V. (1995) Microdiamond in high-grade metamorphic rocks of the Western Gneiss Region, Norway. Geology, **2**, 597–600.

Dohrmann, C., Bochev, P. (2004) A stabilized finite element method for the Stokes problem based on polynomial pressure projections. International Journal for Numerical Methods in Fluids, **46**, 183–201.

Dong, J. J., Hsu, J. Y., Wu, W. J., Shimamoto, T., Hung, J. H., Yeh, E. C., Wu, Y. H., Sone, H. (2010) Stress-dependence of the permeability and porosity of sandstone and shale from TCDP Hole-A. International Journal of Rock Mechanics and Mining Sciences, **47**, 1141–1157.

Dorogokupets, P. I., Karpov, I. K. (1984) Thermodynamics of Minerals and Mineral Equilibria. Nauka, Novosibirsk (in Russian).

Drew, D. (1971) Averaged field equations for two-phase media. Studies in Applied Mathematics, **50**, 133–166.

Drew, D. (1983) Averaged field equations for two-phase flow. Annual Review of Mechanics, **15**, 261–291.

Drew, D., Passman, S. (1999) Theory of Multicomponent Fluids. Applied Mathematics and Science Volume **135**. Springer-Verlag, New York.

Drew, D., Segel, L. (1971) Averaged equations for two-phase flows. Studies in Applied Mathematics, **50**, 205–257.

Driesner, T., Geiger, S. (2007) Numerical simulation of multiphase fluid flow in hydrothermal systems. Fluid–Fluid Interactions, **65**, 187–212.

Driesner, T., Geiger, S., Heinrich, C. A. (2006) Modeling multiphase flow of H_2O-NaCl fluids by combining CSP5.0 with SoWat2.0. Geochimica et Cosmochimica Acta, **70** (18), A147–A147.

Duretz, T., May D. A., Gerya T. V., Tackley, P. J. (2011a) Discretization errors and free surface stabilization in the finite difference and marker-in-cell method for applied geodynamics: a numerical study. Geochemistry, Geophysics, Geosystems, **12**, Q07004.

Duretz, T., Gerya, T. V., May, D. A. (2011b) Numerical modelling of spontaneous slab breakoff and subsequent topographic response. Tectonophysics, **502**, 244–256.

Duretz, T., Gerya, T. V., Spakman, W. (2014) Slab detachment in laterally varying subduction zones: 3-D numerical modeling. Geophysical Research Letters, **41**, 1951–1956.

Dymkova, D., Gerya, T. (2013) Porous fluid flow enables oceanic subduction initiation on Earth. Geophysical Research Letters, **40**, 5671–5676.

Einstein, A. (1906) Eine neue Bestimmung der Moleküldimensionen. Annalen der Physik, **19**, 289–306.

Ellenberger, F. (1994) Histoire de la Géologie. La grande éclosion et ses prémices. Petite Collection d'Histoire des Sciences, 2. Technique et Documentation. Lavoisier, Paris.

Elsasser, W. M. (1963) Early history of the Earth. In: Geiss, J., Goldberg, E. (editors), Earth Science and Meteoritics. North-Holland, Amsterdam, pp. 1–30.

Evans, B., Goetze, C. (1979) The temperature variation of hardness of olivine and its implication for polycrystalline yield stress. Journal of Geophysical Research, **84**, 5505–5524.

Faccenda, M., Burlini, L., Gerya, T. V., Mainprice, D. (2008a) Fault-induced seismic anisotropy by hydration in subducting oceanic plates. Nature, **455**, 1097–1101.

Faccenda, M., Gerya, T. V., Chakraborty, S. (2008b) Styles of post-subduction collisional orogeny: influence of convergence velocity, crustal rheology and radiogenic heat production. Lithos, **103**, 257–287.

Faccenda, M., Gerya, T. V., Burlini, L. (2009) Deep slab hydration induced by bending related variations in tectonic pressure. Nature Geoscience, **2**, 790–793.

Faccenda, M., Gerya, T. V., Mancktelow, N. S., Moresi, L. (2012) Fluid flow during slab unbending and dehydration: implications for intermediate-depth seismicity, slab weakening and deep water recycling. Geochemistry, Geophysics, Geosystems, **13**, Q01010.

Faccenna, C., Bellier, O., Martinod, J., Piromallo, C., Regard, V. (2006) Slab detachment beneath eastern Anatolia: a possible cause for the formation of the North Anatolian fault. Earth and Planetary Science Letters, **242**, 85–97.

Farrington, R. J., Moresi, L.-N., Capitanio, F. A. (2014) The role of viscoelasticity in subducting plates. Geochemistry, Geophysics, Geosystems, **15**, 4291–4304.

Fedorenko, R. P. (1964) The speed of convergence of one iterative process. USSR Computational Mathematics and Mathematical Physics, **4** (3), 227–235.

Fischer, R., Gerya, T. (2016a) Regimes of subduction and lithospheric dynamics in the Precambrian: 3D thermomechanical modeling. Gondwana Research, **37**, 53–70.

Fischer, R., Gerya, T. (2016b) Early Earth plume-lid tectonics: a high-resolution 3D numerical modelling approach. Journal of Geodynamics, **100**, 198–214.

Fornberg, B. (1995) A Practical Guide to Pseudospectral Methods. Cambridge University Press, Cambridge.

Fowler, A. C. (1984) On the transport of moisture in polythermal glaciers. Geophysical and Astrophysical Fluid Dynamics, **29**, 99–140.

Fowler, A. (1985) A mathematical model of magma transport in the asthenosphere. Geophysical and Astrophysical Fluid Dynamics, **33**, 63–96.

Fuchs, K., Bonjer, K.-P., Bock, G., Cornea, I., Radu, C., Enescu, D., Jianu, D., Nourescu, A., Merkler, G., Moldoveanu, T., Tudorache, G. (1979) The Romanian earthquake of March 4, 1977: II, aftershocks and migration of seismic activity. Tectonophysics, **53** (3–4), 225–247.

Furuichi, M., May, D. A., Tackley, P. J. (2011) Development of a Stokes flow solver robust to large viscosity jumps using a Schur complement approach with mixed precision arithmetic. Journal of Computational Physics, **230**, 8835–8851.

Galvan, B., Miller, S. (2013) A full GPU simulation of evolving fracture networks in a heterogeneous poro-elasto-plastic medium with effective-stress-dependent permeability. In: Yuen, D., Wang, L., Chi, X., Johnsson, L., Ge, W., Shi, Y. (editors), GPU Solutions to Multi-scale Problems in Science and Engineering. Lecture Notes in Earth System Sciences. Springer, Berlin, pp. 305–319.

Gassmann, F. (1951) Über die elastizität poröser medien. Vierteljahrsschrift der Naturforschenden Gesellschaft in Zürich, **96**, 1–23.

Geenen, T., Rehman, M., MacLachlan, S. P., Segal, G., Vuik, C., van den Berg, A. P., Spakman, W. (2009) Scalable robust solvers for unstructured FE geodynamic modeling applications: solving the Stokes equation for models with large localized viscosity contrasts. Geochemistry, Geophysics, Geosystems, **10**, Q09002.

Gerya T. V. (2010a) Introduction to Numerical Geodynamic Modelling. Cambridge University Press.

Gerya, T. (2010b) Dynamical instability produces transform faults at mid-ocean ridges. Science, **329**, 1047–1050.

Gerya, T. (2011) Future directions in subduction modeling. Journal of Geodynamics, **52**, 344–378.

Gerya, T. V. (2013) Three-dimensional thermomechanical modeling of oceanic spreading initiation and evolution. Physics of the Earth and Planetary Interiors, **214**, 35–52.

Gerya, T. V. (2014a) Plume-induced crustal convection: 3D thermomechanical model and implications for the origin of novae and coronae on Venus. Earth and Planetary Science Letters, **391**, 183–192.

Gerya, T. V. (2014b) Precambrian geodynamics: concepts and models. Gondwana Research, **25**, 442–463.

Gerya, T. (2016) Origin, evolution, seismicity and models of oceanic and continental transform boundaries. In: Duarte, J., Schellart, W. (editors), Plate Boundaries and Natural Hazards. American Geophysical Union, Wiley, pp. 39–76.

Gerya, T. V., Burg, J.-P. (2007) Intrusion of ultramafic magmatic bodies into the continental crust: numerical simulation. Physics of the Earth and Planetary Interiors, **160**, 124–142.

Gerya, T., Burov., E. (2018) Nucleation and evolution of ridge-ridge-ridge triple junctions: thermomechanical model and geometrical theory. Tectonophysics, **746**, 83–105.

Gerya, T. V., Maresch, W. V. (2004) Metapelites of the Kanskiy granulite complex (Eastern Siberia): kinked P-T paths and geodynamic model. Journal of Petrology, **45**, 1393–1412.

Gerya, T. V., Meilick, F. I. (2011) Geodynamic regimes of subduction under an active margin: effects of rheological weakening by fluids and melts. Journal of Metamorphic Geology, **29**, 7–31.

Gerya, T. V., Stoeckhert, B. (2006) 2-D numerical modeling of tectonic and metamorphic histories at active continental margins. International Journal of Earth Sciences, **95**, 250–274.

Gerya, T. V., Yuen, D. A. (2003a) Characteristics-based marker-in-cell method with conservative finite-differences schemes for modeling geological flows with strongly variable transport properties. Physics of the Earth and Planetary Interiors, **140**, 293–318.

Gerya, T. V., Yuen, D. A. (2003b) Rayleigh–Taylor instabilities from hydration and melting propel cold plumes at subduction zones. Earth and Planetary Science Letters, **212**, 47–62.

Gerya, T. V., Yuen, D. A. (2007) Robust characteristics method for modelling multiphase visco-elasto-plastic thermo-mechanical problems. Physics of the Earth and Planetary Interiors, **163**, 83–105.

Gerya, T. V., Perchuk, L. L., Van Reenen, D. D., Smit, C. A. (2000) Two-dimensional numerical modeling of pressure-temperature-time paths for the exhumation of some granulite facies terrains in the Precambrian. Journal of Geodynamics, **29**, 17–35.

Gerya, T. V., Maresch, W. V., Willner, A. P., Van Reenen, D. D., Smit, C. A. (2001) Inherent gravitational instability of thickened continental crust with regionally developed low- to medium-pressure granulite facies metamorphism. Earth and Planetary Science Letters, **190**, 221–235.

Gerya, T. V., Stoeckhert, B., Perchuk, A. L. (2002) Exhumation of high-pressure metamorphic rocks in a subduction channel – a numerical simulation. Tectonics, **21**, 1056.

Gerya, T. V., Yuen, D. A., Maresch, W. V. (2004a) Thermomechanical modeling of slab detachment. Earth and Planetary Science Letters, **226**, 101–116.

Gerya, T. V., Yuen, D. A., Sevre, E. O. D. (2004b) Dynamical causes for incipient magma chambers above slabs. Geology, **32**, 89–92.

Gerya, T. V., Perchuk, L. L., Maresch, W. V., Willner, A. P. (2004c) Inherent gravitational instability of hot continental crust: implication for doming and diapirism in granulite facies terrains. In: Whitney, D., Teyssier, C., Siddoway, C. S. (editors), Gneiss Domes in Orogeny, GSA Special Paper 380, pp. 97–115.

Gerya, T. V., Podlesskii, K. K., Perchuk, L. L., Maresch, W. V. (2004d) Semi-empirical Gibbs free energy formulations for minerals and fluids. Physics and Chemistry of Minerals, **31** (7), 429–455.

Gerya, T. V., Connolly, J. A. D., Yuen, D. A., Gorczyk, W., Capel, A. M. (2006) Sesmic implications of mantle wedge plumes. Physics of the Earth and Planetary Interiors, **156**, 59–74.

Gerya, T. V., Connolly, J. A. D., Yuen, D. A. (2008a) Why is terrestrial subduction one-sided? Geology, **36**, 43–46.

Gerya, T. V., Perchuk, L. L., Burg, J.-P. (2008b) Transient hot channels: perpetrating and regurgitating ultrahigh-pressure, high-temperature crust-mantle associations in collision belts. Lithos, **103**, 236–256.

Gerya, T. V., May, D. A., Duretz, T. (2013) An adaptive staggered grid finite difference method for modeling geodynamic Stokes flows with strongly variable viscosity. Geochemistry, Geophysics, Geosystems, **14**, 1200–1225.

Gerya, T. V., Stern, R. J., Baes, M., Sobolev, S., Whattam, S. A. (2015) Plate tectonics on the Earth triggered by plume-induced subduction initiation. Nature, **527**, 221–225.

Ghabezloo, S. (2010) Effect of porosity on the thermal expansion coefficient: a discussion of the paper 'Effects of mineral admixtures on the thermal expansion properties of hardened cement paste' by Z.H. Shui, R. Zhang, W. Chen, D. Xuan, Constr. Build. Mater. 24 (9) (2010) 1761–1767. Construction and Building Materials, **24**, 1796–1798.

Ghabezloo, S. (2012) Micromechanical analysis of the effect of porosity on the thermal expansion coefficient of heterogeneous porous materials. International Journal of Rock Mechanics and Mining Sciences, **55**, 97–101.

Gillmann, C., Tackley, P. J. (2014) Atmosphere/mantle coupling and feedbacks on Venus. Journal of Geophysical Research: Planets, **119**, 1189–1217.

Golabek, G. J., Schmeling, H., Tackley, P. J. (2008) Earth's core formation aided by flow channelling instabilities induced by iron diapirs. Earth and Planetary Science Letters, **271**, 24–33.

Golabek, G. J., Gerya, T. V., Kaus, B. J. P., Ziethe, R., Tackley, P. J. (2009) Rheological controls on the terrestrial core formation mechanism. Geochemistry, Geophysics, Geosystems, **10**, Q11007.

Golabek, G. J., Keller, T., Gerya, T. V., Zhu, G., Tackley, P. J., Connolly, J. A. D. (2011) Origin of the martian dichotomy and Tharsis from a giant impact causing massive magmatism. Icarus, **215**, 346–357.

Golabek, G. J., Bourdon, B., Gerya, T. V. (2014) Numerical models of the thermomechanical evolution of planetesimals: application to the acapulcoite-lodranite parent body. Meteoritics & Planetary Science, **49**, 1083–1099.

Golabek, G. J., Emsenhuber, A., Jutzi, M., Asphaug, E. I., Gerya, T. V. (2018) Coupling SPH and thermochemical models of planets: methodology and example of a Mars-sized body. Icarus, **301**, 235–246.

Gorczyk, W., Vogt, K. (2018) Intrusion of magmatic bodies into the continental crust: 3-D numerical models. Tectonics, **37**, 705–723.

Gorczyk, W., Guillot, S., Gerya, T. V., Hattori, K. (2007a) Asthenospheric upwelling, oceanic slab retreat and exhumation of UHP mantle rocks: insights from Greater Antilles. Geophysical Research Letters, **34**, L21309.

Gorczyk, W., Gerya, T. V., Connolly, J. A. D., Yuen, D. A. (2007b) Growth and mixing dynamics of mantle wedge plumes. Geology, **35**, 587–590.

Gotelli, N. J., Anderson, M. J., Arita, H. T., Chao, N., Colwell, R. K., Connolly, S. R., Currie, D. J., Dunn, R. R., Graves, G. R., Green, J. L., Grytnes, J. -A., Jiang, Y. -H., Jetz, W., Lyons, S. K., McCain, C. M., Magurran, A. E., Rahbek, C., Rangel, T. F. L. V.

B., Sobero, J., Webb, C. O., Willig, M. R. (2009) Patterns and causes of species richness: a general simulation model for macroecology. Ecology Letters, **12**, 873–886.

Griffith, A. A. (1924) The theory of rupture. Proceedings of the 1st International Congress on Applied Mechanics, Delft, pp. 54–63.

Gueguen, Y., Dienes, J. (1989) Transport properties of rocks from statistics and percolation. Mathematical Geology, **21**, 1–13.

Gupta, A. (2000) WSMP: Watson sparse matrix package (Part-II: direct solution of general sparse systems). Technical Report RC 21888 (98472), IBM T.J. Watson Research Center, Yorktown Heights, NY.

Gustafsson, B. (2008) High Order Finite-Difference Methods for Time-dependent PDE. Springer Verlag.

Hager, B. H., O'Connell, R. J. (1981) A simple global model of plate dynamics and mantle convection. Journal of Geophysical Research, **86**, 4843–4867.

Hall, C. E., Gurnis, M., Sdrolias, M., Lavier, L. L., Muller, R. D. (2003) Catastrophic initiation of subduction following forced convergence across fractures zones. Earth and Planetary Science Letters, **212**, 15–30.

Hall, C. E., Parmentier, E. M. (2003) Influence of grain size evolution on convective instability. Geochemistry, Geophysics, Geosystems, **4**, 1029.

Hall, P. S., Kincaid, C. (2001) Diapiric flow at subduction zones: a recipe for rapid transport. Science, **292**, 2472–2475.

Hansen, U., Yuen, D. A. (1988) Numerical simulations of thermal-chemical instabilities at the core–mantle boundary. Nature, **334**, 237–240.

Heister, T., Dannberg, J., Gassmöller, R., Bangerth, W. (2017) High accuracy mantle convection simulation through modern numerical methods – II: realistic models and problems. Geophysical Journal International, **210**, 833–851.

Helgeson, H. C., Delany, J. M., Nesbitt, H. W., Bird, D. K. (1978) Summary and critique of the thermodynamic properties of rock-forming minerals. American Journal of Science, **278A**.

Herrendörfer, R., van Dinther, Y., Gerya, T., Dalguer, L. A. (2015) Earthquake supercycle in subduction zones controlled by the width of the seismogenic zone. Nature Geoscience, **8**, 471–474.

Herrendörfer, R., Gerya, T., van Dinther, Y. (2018) An invariant rate- and state-dependent friction formulation for earthquake cycle simulations Part 1: mature fault zone. Journal of Geophysical Research, **123**, 5018–5051.

Hess, P. C. (1989) Origin of Igneous Rocks. Harvard University Press, London.

Hirschmann, M. M. (2000) Mantle solidus: experimental constraints and the effects of peridotite composition. Geochemistry, Geophysics, Geosystems, **1** (10), 1042, doi:10.1029/2000GC000070.

Hofmeister, A. M. (1999) Mantle values of thermal conductivity and the geotherm from phonon lifetimes. Science, **283**, 1699–1706.

Holland, T. J. B., Powell, R. (1990) An enlarged and updated internally consistent thermodynamic data set with uncertainties and correlations: the system K_2O–Na_2O–CaO–MgO–FeO–Fe_2O_3–Al_2O_3–TiO_2–SiO_2–C–H_2O–O_2. Journal of Metamorphic Geology, **8**, 309–343.

Holland, T. J. B., Powell, R. (1998) Internally consistent thermodynamic data set for phases of petrological interest. Journal of Metamorphic Geology, **16**, 309–344.

Holland, T. J. B., Powell, R. (2011) An improved and extended internally consistent thermodynamic dataset for phases of petrological interest, involving a new equation of state for solids. Journal of Metamorphic Geology, **29**, 333–383.

Honda, R., Mizutani, H., Yamamoto, T. (1993) Numerical simulation of earth's core formation. Journal of Geophysical Research, **98**, 2075–2089.

Houseman, G. (1988) The dependence of convection planform on mode of heating. Nature, **332**, 346–349.

Hubschwerlen, N., Zheng, L., Kaempfer, T., Gerya, T. (2017) Thermo-hydro-mechanical effects of a geological repository at macro-scale – a novel modeling approach adapted from plate tectonics. Clay Conference 2017, Abstract Volume, Davos, Switzerland.

Huismans, R., Beaumont, C. (2011) Depth-dependent extension, two-stage breakup and cratonic underplating at rifted margins. Nature, **473**, 74–79.

Hüttig, C., Stemmer, K. (2008) The spiral grid: a new approach to discretize the sphere and its application to mantle convection. Geochemistry, Geophysics, Geosystems, **9**, Q02018.

Ida, S., Nakagawa, Y., Nakazawa, K. (1987) The Earth's core formation due to the Rayleigh-Taylor instability. Icarus, **69**, 239–248.

Ikemoto, A., Iwamori, H. (2014) Numerical modeling of trace element transportation in subduction zones: implications for geofluid processes. Earth, Planets and Space, **66**, 26, doi.org/10.1186/1880–5981-66–26.

Isacks, B., Molnar, P. (1969) Mantle earthquake mechanisms and the sinking of the lithosphere. Nature, **223**, 1121–1124.

Ismail-Zadeh, A., Tackley, P. (2010) Computational Methods for Geodynamics. Cambridge University Press.

Jadamec, M. A., Kreylos, O., Chang, B., Fischer, K. M., Yikilmaz, M. B. (2018) A visual survey of global slab geometries with ShowEarthModel and implications for a three-dimensional subduction paradigm. Earth and Space Science, **5**, doi.org/10.1002/2017EA000349.

Jamieson, R. A., Beaumont, C., Nguyen, M. H., Lee, B. (2002) Interaction of metamorphism, deformation, and exhumation in large convergent orogens. Journal of Metamorphic Geology, **20**, 9–24.

Jenny, P., Pope, S. B., Muradoglu, M., Caughey, D. A. (2001) A hybrid algorithm for the joint PDF equation of turbulent reactive flows. Journal of Computational Physics, **166**, 218–252.

Johannes, W. (1985) The significance of experimental studies for the formation of migmatites. In: Ashworth, V. A. (editor), Migmatites. Blackie, Glasgow, pp. 36–85.

Johnson, T. E., Brown, M., Kaus, B., Van Tongeren, J. A. (2014) Delamination and recycling of Archaean crust caused by gravitational instabilities. Nature Geoscience, **7**, 47–52.

Kadlec, B., Dorn, G., Tufo, H., Yuen, D. (2008) Interactive 3-D computation of fault surfaces using level sets. Visual Geosciences, **13**, 133–138.

Kameyama, M., Yuen, D. A., Karato, S. (1999) Thermal-mechanical effects of low-temperature plasticity (the Peierls mechanism) on the deformation of a viscoelastic shear zone. Earth and Planetary Science Letters, **168**, 159–172.

Kaneko, Y., Lapusta, N., Ampuero, J.-P. (2008) Spectral element modeling of spontaneous earthquake rupture on rate and state faults: effect of velocity-strengthening friction at shallow depths. Journal of Geophysical Research, **113**, B09317.

Karato, S. (2008) Deformation of Earth Materials. Cambridge University Press, New York.

Karato, S., Wu, P. (1993) Rheology of the upper mantle: a synthesis. Science, **260**, 771–778.

Karato, S., Riedel, M. R., Yuen, D. A. (2001) Rheological structure and deformation of subducted slabs in the mantle transition zone: implications for mantle circulation and deep earthquakes. Physics of the Earth and Planetary Interiors, **127**, 83–108.

Karniadakis, G. E., Kirby, R. M. (2003) A Seamless Approach to Parallel Algorithms and their Implementation. Cambridge University Press.

Karpov, I. K., Kiselev, A. I., Letnikov, F. A. (1976) Computer Modeling of Natural Mineral Formation. Nedra Press, Moscow (in Russian).

Katayama, I., Karato, S. (2008) Low-temperature, high-stress deformation of olivine under water-saturated conditions. Physics of the Earth and Planetary Interiors, **168**, 125–133.

Katz, R. F. (2008) Magma dynamics with the enthalpy method: benchmark solutions and magmatic focusing at mid-ocean ridges. Journal of Petrology, **49**, 2099–2121.

Katz, R. F., Spiegelman, M., Langmuir, C. H. (2003) A new parameterization of hydrous mantle melting. Geochemistry, Geophysics, Geosystems, **4**, 1073.

Katz, R. F., Spiegelman, M., Holtzman, B. (2006) The dynamics of melt and shear localization in partially molten aggregates. Nature, **442**, 676–679.

Kaus, B. J. P., Becker, T. W. (2007) Effects of elasticity on the Rayleigh-Taylor instability: implications for large-scale geodynamics. Geophysical Journal International, **168**, 843–862.

Kaus, B. J. P., Podladchikov, Y. Y. (2006) Initiation of localized shear zones in viscoelastoplastic rocks. Journal of Geophysical Research, **111**, B04412.

Kaus, B. J. P., Schmalholz, S. M. (2006) 3D finite amplitude folding: implications for stress evolution during crustal and lithospheric deformation. Geophysical Research Letters, **33**, L14309.

Kaus, B. J. P., Connolly, J. A. D., Podladchikov, Y. Y., Schmalholz, S. M. (2005) Effect of mineral phase transitions on sedimentary basin subsidence and uplift. Earth and Planetary Science Letters, **233**, 213–228.

Kaus, B. J. P., Steedman, C., Becker, T. W. (2008) From passive continental margin to mountain belt: insights from analytical and numerical models and application to Taiwan. Physics of the Earth and Planetary Interiors, **171**, 235–251.

Kaus, B. J. P., Mühlhaus, H., May, D. A. (2010) A stabilization algorithm for geodynamic numerical simulations with a free surface. Physics of the Earth and Planetary Interiors, **181**, 12–20.

Kaus, B. J. P., Popov, A. A., Baumann, T. S., Püsök, A. E., Bauville, A., Fernandez, N., Collignon, M. (2016) Forward and inverse modelling of lithospheric deformation on geological timescales. In: Binder, K., Müller, M., Schnurpfeil, A. (editors), NIC Symposium 2016 – Proceedings. NIC Series, Volume **48**, pp. 299–307.

Kelemen, P., Shimizu, N., Salters, V. (1995) Extraction of mid-ocean-ridge basalt from the upwelling mantle by focused flow of melt in dunite channels. Nature, **375**, 747–753.

Keller, T., Katz, R. F. (2016) The role of volatiles in reactive melt transport in the asthenosphere. Journal of Petrology, **57**, 1073–1108.

Keller, T., May, D. A., Kaus, B. J. P. (2013) Numerical modelling of magma dynamics coupled to tectonic deformation of lithosphere and crust. Geophysical Journal International, **195**, 1406–1442.

Keondzhyan, V. P., Monin, A. S. (1977) Continental drift and large-scale wandering of the Earths' pole. Izvestiya, Physics of the Solid Earth, **13**, 760–772.

Keondzhyan, V. P., Monin, A. S. (1980) Compositional convection in the Earth's mantle. Dokladi Akademii Nauk SSSR, **253**, 78–81.

Khan, A., Connolly, J. A. D., Olsen, N. (2006) Constraining the composition and thermal state of the mantle beneath Europe from inversion of long-period electromagnetic sounding data. Journal of Geophysical Research, **111**, B10102.

Kocks, U. F., Argon, A. S., Ashby, M. F. (1975) Thermodynamics and kinetics of slip. Progress in Materials Science, **19**, 1–291.

Kooi, H., Beaumont, C. (1994) Escarpment evolution on high-elevation rifted margins – insights derived from a surface processes model that combines diffusion, advection, and reaction. Journal of Geophysical Research, **99** (B6), 12191–12209.

Koptev, A., Calais, E., Burov, E., Leroy, S., Gerya, T. (2015) Dual continental rift systems generated by plume–lithosphere interaction. Nature Geoscience, **8**, 388–392.

Kronbichler, M., Heister, T., Bangerth, W. (2012) High accuracy mantle convection simulation through modern numerical methods. Geophysical Journal International, **191**, 12–29.

Krotkiewski, M., Dabrowski, M., Podladchikov, Y. Y. (2008) Fractional steps methods for transient problems on commodity computer architectures. Physics of the Earth and Planetary Interiors, **171**, 122–136.

Kundu, P. K., Cohen, I. M. (2002) Fluid Mechanics. Academic Press.

Landau, L. D., Lifshitz, E. M. (1987) Fluid Mechanics, 2nd English edition. Pergamon Press.

Lapusta, N. (2003) Nucleation and early seismic propagation of small and large events in a crustal earthquake model. Journal of Geophysical Research, **108**, B42205.

Lapusta, N., Barbot, S. (2012) Models of earthquakes and aseismic slip based on laboratory-derived rate and state friction laws. In: Bizzarri, A., Bhat, H. S. (editors), The Mechanics of Faulting: From Laboratory to Real Earthquakes, Volume **661**. Research Signpost, Kerala, pp. 153–207.

Lapusta, N., Liu, Y. (2009) Three-dimensional boundary integral modeling of spontaneous earthquake sequences and aseismic slip. Journal of Geophysical Research, **114**, B09303.

Lapusta, N., Rice, J. R., Ben-Zion, Y., Zheng, G. (2000) Elastodynamic analysis for slow tectonic loading with spontaneous rupture episodes on faults with rate- and state-dependent friction. Journal of Geophysical Research, **105** (23), 23765–23789.

Large, R. R., Halpin, J. A., Lounejeva, E., Danyushevsky, L. A., Maslennikov, V. V., Gregory, D., Sack, P. J., Haines, P. W., Long, J. A., Makoundi, C., Stepanov, A. S. (2015) Cycles of nutrient trace elements in the Phanerozoic ocean. Gondwana Research, **28**, 1282–1293.

Larsen, T. B., Yuen, D. A., Malevsky, A. V. (1995) Dynamical consequences on fast subducting slabs from a self-regulating mechanism due to viscous heating in variable viscosity convection. Geophysical Research Letters, **22**, 1277–1280.

Lee, C.-T. A., Caves, J., Jiang, H., Cao, W., Lenardic, A., McKenzie, N. R., Shorttle, O., Yin, Q., Dyer, B. (2018) Deep mantle roots and continental emergence: implications for whole-Earth elemental cycling, long-term climate, and the Cambrian explosion. International Geology Review, **60** (4), 431–448.

Leeman, J. R., Saffer, D. M., Scuderi, M. M., Marone, C. (2016) Laboratory observations of slow earthquakes and the spectrum of tectonic fault slip modes. Nature Communications, **7**, 11104.

LeVeque, R. (2002) Finite Volume Methods for Hyperbolic Problems. Cambridge University Press.

Levin, V., Shapiro, N., Park, J., Ritzwoller, M. (2002) Seismic evidence for catastrophic slab loss beneath Kamchatka. Nature, **418**, 763–767.

Li, Z., Xu, Z., Gerya, T., Burg, J.-P. (2013) Collision of continental corner from 3-D numerical modeling. Earth and Planetary Science Letters, **380**, 98–111.

Lichtenberg, T., Golabek, G. J., Gerya, T. V., Meyer, M. R. (2016) The effects of short-lived radionuclides and porosity on the early thermo-mechanical evolution of planetesimals. Icarus, **274**, 350–365.

Lichtenberg, T., Golabek, G. J., Dullemond, C. P., Schonbachler, M., Gerya, T. V., Meyer, M. R. (2018) Impact splash chondrule formation during planetesimal recycling. Icarus, **302**, 27–43.

Lin, J.-R., Gerya, T. V., Tackley, P., Yuen, D. (2009) Numerical modeling of protocore destabilization during planetary accretion: methodology and results. Icarus, **204**, 732–748.

Liou, J. G., Tsujimori, T., Zhang, R. Y., Katayama, I., Maruyama, S. (2004) Global UHP metamorphism and continental subduction/collision: the Himalayan model. International Geology Review, **46**, 1–27.

Liu, Y., Rice, J. R. (2007) Spontaneous and triggered aseismic deformation transients in a subduction fault model. Journal of Geophysical Research, **112**, B09404.

Longo, A., Vassalli, M., Papale, P., Barsanti, M. (2006) Numerical simulation of convection and mixing in magma chambers replenished with CO_2-rich magma. Geophysical Research Letters, **33** (21), L21305.

Lourenco, D. L., Rozel, A., Gerya, T., Tackley, P. (2018) Efficient cooling of rocky planets by intrusive magmatism. Nature Geoscience, **11**, 322–327.

Lu, G., Kaus, B. J. P., Zhao, L., Zheng, T. (2015) Self-consistent subduction initiation induced by mantle flow. Terra Nova, **27** (2), 130–138.

Lynch, D. R. (2005) Numerical Partial Differential Equations for Environmental Scientists and Engineers: A Practical First Course. Springer Verlag.

Machetel, P., Rabinowicz, M., Bernardet, P. (1986) Three-dimensional convection in spherical shells. Geophysical & Astrophysical Fluid Dynamics, **37**, 57–84.

MacPherson, G. J., Davis, A. M., Zinner, E. K. (1995) The distribution of aluminium-26 in the early solar system – a reappraisal. Meteoritics, **30**, 365–386.

Mallard, C., Coltice, N., Seton, M., Mueller, R. D., Tackley, P. J. (2016) Subduction drives the organisation of Earth's tectonic plates. Nature, **535**, 140–143.

Marone, C. (1998) Laboratory-derived friction laws and their application to seismic faulting. Annual Review of Earth and Planetary Sciences, **26**, 643–696.

Marone, C., Hobbs, B. E., Ord, A. (1992) Coulomb constitutive laws for friction: contrasts in frictional behavior for distributed and localized shear. Pure and Applied Geophysics, **139**, 195–214.

Marschall, H., Schumacher, J. C. (2012) Arc magmas sourced from mélange diapirs in subduction zones. Nature Geoscience, **5**, 862–867.

Marsh, B. D. (1982) On the mechanics of igneous diapirism, stoping, and zone melting. American Journal of Science, **282**, 808–855.

Massonne, H.-J. (1999) A new occurrence of microdiamonds in quartzofeldspathic rocks of the Saxonian Erzgebirge, Germany, and their metamorphic evolution. Proc. 7th Int. Kimberlite Conf., pp. 533–539.

Matsumoto, T., Tomoda, Y. (1983) Numerical-simulation of the initiation of subduction at the fracture-zone. Journal of Physics of the Earth, **31**, 183–194.

May, D. A., Moresi, L. (2008) Preconditioned iterative methods for Stokes flow problems arising in computational geodynamics. Physics of the Earth and Planetary Interiors, **171**, 33–47.

May, D. A., Schellart, W. P., Moresi, L. (2013) Overview of adaptive finite element analysis in computational geodynamics. Journal of Geodynamics, **70**, 1–20.

May, D. A., Brown, J., Le Pourhiet, L. (2014) pTatin3D: High-performance methods for long-term lithospheric dynamics. Proceedings of the International Conference for High Performance Computing, Networking, Storage and Analysis. IEEE Press, pp. 274–284.

May, D. A., Brown, J., Le Pourhiet, L. (2015) A scalable, matrix-free multigrid preconditioner for finite element discretizations of heterogeneous Stokes flow. Computer Methods in Applied Mechanics and Engineering, **290**, 496–523.

McKenzie, D. (1984) The generation and compaction of partially molten rock. Journal of Petrology, **25**, 713–765.

Mei, S., Bai, W., Hiraga, T., Kohlstedt, D. (2002) Influence of melt on the creep behavior of olivine-basalt aggregates under hydrous conditions. Earth and Planetary Science Letters, **201**, 491–507.

Melnik, O. (2000) Dynamics of two-phase conduit flow of high viscosity gas-saturated magma: large variations of sustained explosive eruption intensity. Bulletin of Volcanology, **62**, 153–170.

Melnik, O., Sparks, R. S. J. (1999) Nonlinear dynamics of lava dome extrusion. Nature, **402**, 37–41.

Melosh, H. J. (1990) Giant impacts and the thermal state of the early Earth. In: Newsom, H. E., Jones, J. H. (editors), Origin of the Earth. Oxford University Press, New York, pp. 69–83.

Melosh, H. J. (2008) Did an impact blast away half of the martian crust? Nature Geoscience, **1**, 412–414.

Meyer, D., Jenny, P. (2004) Conservative velocity interpolation for PDF methods. Proceedings in Applied Mathematics and Mechanics, **4**, 466–467.

Miller, S. A., Collettini, C., Chiaraluce, L., Cocco, M., Barchi, M., Kaus, B. J. P. (2004) Aftershocks driven by a high-pressure CO_2 source at depth. Nature, **427**, 724–727.

Minear, J. W., Toksöz, M. N. (1970) Thermal regime of a downgoing slab and new global tectonics. Journal of Geophysical Research, **75**, 1397–1419.

Mishin, Y. A., Gerya, T. V., Burg, J.-P., Connolly, J. A. D. (2008) Dynamics of double subduction: numerical modeling. Physics of the Earth Planetary Interiors, **171**, 280–295.

Moore, J. D. P, Nielsen, S., Hansen, L. N. (2019) Ductile deformation explains the physics of friction and genesis of earthquakes. Nature (submitted).

Moore, W., Webb, A. (2013) Heat-pipe earth. Nature, **501**, 501–505.

Morency, C., Huismans, R. S., Beaumont, C., Fullsack, P. (2007) A numerical model for coupled fluid flow and matrix deformation with applications to disequilibrium compaction and delta stability. Journal of Geophysical Research, **112**, B10407.

Moresi, L., Zhong, S., Gurnis, M. (1996) The accuracy of finite element solutions of Stokes' flow with strongly varying viscosity. Physics of the Earth and Planetary Interiors, **97**, 83–94.

Moresi, L., Dufour, F., Mühlhaus, H.-B. (2003) A Lagrangian integration point finite element method for large deformation modeling of viscoelastic geomaterials. Journal of Computational Physics, **184**, 476–497.

Moresi, L., Quenette, S., Lemiale, V., Mériaux, C., Appelbe, B., Mühlhaus, H.-B. (2007) Computational approaches to studying non-linear dynamics of the crust and mantle. Physics of the Earth and Planetary Interiors, **163**, 69–82.

Morra, G. (2018) Pythonic Geodynamics: Implementations for Fast Computing. Lecture Notes in Earth System Sciences. Springer Nature.

Murakami, M., Hirose, K., Kawamura, K., Sata, N., Ohishi, Y. (2004) Post-perovskite phase transition in $MgSiO_3$. Science, **304**, 855–858.

Murnaghan, F. D. (1944) The compressibility of media under extreme pressures. Proceedings of the National Academy of Sciences, **30**, 244–247.

Nagata, K., Nakatani, M., Yoshida, S. (2012) A revised rate- and state-dependent friction law obtained by constraining constitutive and evolution laws separately with laboratory data. Journal of Geophysical Research, **117**, B02314.

Nakatani, M. (2001) Conceptual and physical clarification of rate and state friction: frictional sliding as a thermally activated rheology. Journal of Geophysical Research, **106**, 13347–13380.

Naliboff, J. B., Billen, M. I., Gerya, T., Saunders, J. (2013) Dynamics of outer rise faulting in oceanic-continental subduction systems. Geochemistry, Geophysics, Geosystems, **14**, 2310–2327.

Nigmatulin, R. I. (1991) Dynamics of Multiphase Media. Hemisphere, New York.

Nikolaeva, K., Gerya, T. V., Connolly, J. A. D. (2008) Numerical modelling of crustal growth in intraoceanic volcanic arcs. Physics of the Earth and Planetary Interiors, **171**, 336–356.

Noda, H., Shimamoto, T. (2012) Transient behavior and stability analyses of halite shear zones with an empirical rate-and-state friction to flow law. Journal of Structural Geology, **38**, 234–242.

Oganov, A. R., Ono, S. (2004) Theoretical and experimental evidence for a post-perovskite phase of $MgSiO_3$ in Earth's D' layer. Nature, **430**, 445–448.

Oldenburg, C. M., Spera, F. J., Yuen, D. A. (1990) Self-organization in convective magma mixing. Earth-Science Reviews, **29**, 331–348.

Omlin, S. (2016) Development of massively parallel near peak performance solvers for three-dimensional geodynamic modelling. Doctoral Thesis, University of Lausanne, Switzerland.

Omlin, S., Malvoisin, B., Podladchikov, Y. Y. (2017) Pore fluid extraction by reactive solitary waves in 3-D. Geophysical Research Letters, **44**, 9267–9275.

Omlin, S., Räss, L., Podladchikov, Y. Y. (2018) Simulation of three-dimensional viscoelastic deformation coupled to porous fluid flow. Tectonophysics, **746**, 695–701.

O'Neill, C., Marchi, S., Zhang, S., Bottke, W. (2017) Impact-driven subduction on the Hadean Earth. Nature Geoscience, **10**, 793–797.

Papale, P. (1999) Strain-induced magma fragmentation in explosive eruptions. Nature, **397**, 425–428.

Papale, P. (2001) Dynamics of magma flow in volcanic conduits with variable fragmentation efficiency and nonequilibrium pumice degassing. Journal Geophysical Research, **106**, 11043–11065.

Pascal, G., Dubois, J., Barazangi, M., Isacks, B. L., Oliver, J. (1973) Seismic velocity anomalies beneath the New Hebrides island arc: evidence for a detached slab in the upper mantle. Journal of Geophysical Research, **78** (29), 6998–7004.

Patankar, S. V. (1980) Numerical Heat Transfer and Fluid Flow. McGraw-Hill, New York.

Patocka, V., Cadek, O., Tackley, P. J., Cizkova, H. (2017) Stress memory effect in viscoelastic stagnant lid convection. Geophysical Journal International, **209**, 1462–1475.

Peng, Z., Gomberg, J. (2010) An integrated perspective of the continuum between earthquakes and slow-slip phenomena. Nature Geoscience, **3**, 599–607.

Penniston-Dorland, S. C., Kohn, M. J., Manning, C. E. (2015) The global range of subduction zone thermal structures from exhumed blueschists and eclogites: rocks are hotter than models. Earth and Planetary Science Letters, **428**, 243–254.

Pergler, T., Matyska, C. (2008) A hybrid spectral and finite element method for coseismic and postseismic deformation. Physics of the Earth and Planetary Interiors, **163**, 122–148.

Petford, N., Cruden, A. R., McCaffrey, K. J., Vigneresse, J.-L. (2000) Granite magma formation, transport and emplacement in the Earth's crust. Nature, **408**, 669–673.

Petrini, K., Connolly, J. A. D., Podladchikov, Y. Y (2001) A coupled petrological-tectonic model for sedimentary basin evolution: the influence of metamorphic reactions on basin subsidence. Terra Nova, **13**, 354–359.

Pinkerton, H., Stevenson, R. J. (1992) Methods of determining the rheological properties of magmas at subliquidus temperatures. Journal of Volcanology and Geothermal Research, **53**, 47–66.

Pitcher, W. S. (1979) The nature, ascent and emplacement of granitic magma. Journal of the Geological Society London, **136**, 627–662.

Plümper, O., John, T., Podladchikov, Y. Y., Vrijmoed, J. C., Scambelluri, M. (2017) Fluid escape from subduction zones controlled by channel-forming reactive porosity. Nature Geoscience, **10**, 150–156.

Poli, S., Schmidt, M. W. (2002) Petrology of subducted slabs. Annual Review of Earth and Planetary Science, **30**, 207–235.

Popov, A. A., Sobolev, S. V. (2008) SLIM3D: a tool for three-dimensional thermomechanical modeling of lithospheric deformation with elasto-visco-plastic rheology. Physics of the Earth and Planetary Interiors, **171**, 55–75.

Popov, A. A., Lehmann, R., Kaus, B. J. P. (2014a) How to compute stress and effective viscosity for visco-elasto-plastic rheologies in geodynamic codes. German-Swiss Geodynamics Workshop 2014, Abstracts.

Popov, I. Yu., Lobanov, I. S., Popov, S. I.,. Popov, A. I, Gerya, T. V. (2014b) Practical analytical solutions for benchmarking of 2-D and 3-D geodynamic Stokes problems with variable viscosity. Solid Earth, **5**, 461–476.

Pouryazdan, M., Kaus, B. J. P., Rack, A., Ershov, A., Hahn, H. (2017) Mixing instabilities during shearing of metals. Nature Communications, **8**, 1611.

Pusok, A. E., Kaus, B. J. P., Popov, A. A. (2017) On the quality of velocity interpolation schemes for marker-in-cell method and staggered grids. Pure and Applied Geophysics, **174**, 1071–1089.

Pysklywec, R. N. (2006) Surface erosion control on the evolution of the deep lithosphere. Geology, **34**, 225–228.

Ramberg, H. (1968) Instability of layered system in the field of gravity. Physics of the Earth and Planetary Interiors, **1**, 427–474.

Ramberg, H. (1981) The role of gravity in orogenic belts. In: McClay, K. R., Price, N. J. (editors), Thrust and Nappe Tectonics. Geological Society Special Publication, London, pp. 125–140.

Ranalli, G. (1995) Rheology of the Earth. Chapman & Hall, London.

Ranero, C. R., Phipps Morgan, J., Reichert, C. (2003) Bending-related faulting and mantle serpentinization at the Middle America trench. Nature, **425**, 367–373.

Ranero, C. R., Villaseñor, A., Phipps Morgan, J., Weinribe, W. (2005) Relationship between bend-faulting at trenches and intermediate-depth seismicity. Geochemistry, Geophysics, Geosystems, **6**, doi:10.1029/2005GC000997.

Rass, L., Yarushina, V. M., Simon, N. S. C., Podladchikov, Y. Y. (2014) Chimneys, channels, pathway flow or water conducting features – an explanation from numerical modelling and implications for CO_2 storage. In: Dixon, T., Herzog, H., Twinning, S. (editors), 12th International Conference on Greenhouse Gas Control Technologies, GHGT-12, Energy Procedia, **63**, pp. 3761–3774.

Regenauer-Lieb, K., Yuen, D. A. (2008) Multiscale brittle-ductile coupling and genesis of slow earthquakes. Pure and Applied Geophysics, **165**, 523–543.

Revenaugh, J., Parsons, B. (1987) Dynamic topography and gravity anomalies for fluid layers whose viscosity varies exponentially with depth. Geophysical Journal of the Royal Astronomical Society, **90**, 349–368.

Rey, P. F., Coltice, N., Flament, N. (2014) Spreading continents kick-started plate tectonics. Nature, **513**, 405–408.

Ricard, Y., Bercovici, D., Schubert, G. (2001) A two-phase model for compaction and damage 2. Applications to compaction, deformation, and the role of interfacial surface tension. Journal of Geophysical Research, **106**, 8907–8924.

Ricard, Y., Šrámek, O., Dubuffet, F. (2009) Runaway core-mantle segregation of terrestrial planets. Earth and Planetary Science Letters, **284**, 144–150.

Rice, J. R. (1993) Spatio-temporal complexity of slip on a fault. Journal of Geophysical Research, **98** (B6), 9885–9907.

Rice, J., Ruina, A. (1983) Stability of steady frictional slipping. Journal of Applied Mechanics, **50**, 343–349.

Rice, J., Lapusta, N., Ranjith, K. (2001) Rate and state dependent friction and the stability of sliding between elastically deformable solids. Journal of the Mechanics and Physics of Solids, **49**, 1865–1898.

Richter, F. M. (1973) Finite amplitude convection through a phase boundary. Geophysical Journal of the Royal Astronomical Society, **35**, 265–276.

Richter, F. M. (1978) Mantle convection models. Annual Review of Earth and Planetary Science, **6**, 9–19.

Roache, P. J. (1997) Quantification of uncertainty in computational fluid dynamics. Annual Review of Fluid Mechanics, **29**, 123–160.

Roache, P. J. (1998) Verification and Validation in Computational Science and Engineering. Hermosa Publishers, Albuquerque, NM.

Rosen, O. M., Zorin, Y. M., Zayachkovsky, A. A. (1972) A find of a diamond linked with eclogites of the Precambrian Kokchetav massif. Dokladi Akademii Nauk SSSR, **203**, 674–676 (in Russian).

Rozel, A., Ricard, Y., Bercovici, D. (2011) A thermodynamically self-consistent damage equation for grain size evolution during dynamic recrystallization. Geophysical Journal International, **184**, 719–728.

Rozel, A. B., Golabek, G. J., Jain, C., Tackley, P. J., Gerya, T. (2017) Continental crust formation on early Earth controlled by intrusive magmatism. Nature, **545**, 332–335.

Rozhko, A. Y., Podladchikov, Y. Y., Renard, F. (2007) Failure patterns caused by localized rise in pore-fluid overpressure and effective strength of rocks. Geophysical Research Letters, **34**, L22304.

Rubin, A. M., Ampuero, J. (2005) Earthquake nucleation on (aging) rate and state faults. Journal of Geophysical Research, **110**, B11312.

Rubinstein, R., Atluri, S. N. (1983) Objectivity of incremental constitutive relations over finite time steps in computational finite deformation analyses. Computer Methods in Applied Mechanics and Engineering, **36**, 277–290.

Rudge, J. F., Bercovici, D., Spiegelman, M. (2011) Disequilibrium melting of a two phase multicomponent mantle. Geophysical Journal International, **184**, 699–718.

Rudi, J., Malossiy, A. C. I., Isaac, T., Stadler, G., Gurnis, M., Staary, P. W. J, Ineicheny, Y., Bekasy, C., Curioniy, A., Ghattas, O. (2015) An extreme-scale implicit solver for complex PDEs: highly heterogeneous flow in Earth's mantle. SC '15, Proceedings of the International Conference for High Performance Computing, Networking, Storage and Analysis, Article No. 5, doi.org/10.1145/2807591.2807675.

Rudolph, M. L., Gerya, T. V., Yuen, D. A., DeRosier, S. (2004) Visualization of multiscale dynamics of hydrous cold plumes at subduction zones. Visual Geosciences, **9** (1), 59, doi.org/10.1007/s10069-004–0017-2.

Ruina, A. (1983) Slip instability and state variable friction laws. Journal of Geophysical Research, **88**, 10359–10370.

Ruprecht, P., Bergantz, G. W., Dufek, J. (2008) Modeling of gas-driven magmatic overturn: tracking of phenocryst dispersal and gathering during magma mixing. Geochemistry, Geophysics, Geosystems, **9**, Q07017.

Sacks, P. E., Secor, D. T. (1990) Delamination in collisional orogens. Geology, **18**, 999–1002.

Samuel, H., Tackley, P. J. (2008) Dynamics of core formation and equilibration by negative diapirism. Geochemistry, Geophysics, Geosystems, **9**, Q06011, doi:10.1029/2007GC001896.

Scambelluri, M., Philippot, P. (2001) Deep fluids in subduction zones. Lithos, **55**, 213–227.

Schenk, O., Gärtner, K. (2004) Solving unsymmetric sparse systems of linear equations with PARDISO. Journal of Future Generation Computer Systems, **20**, 475–487.

Schenk, O., Gärtner, K. (2006) On fast factorization pivoting methods for symmetric indefinite systems. Electronic Transactions on Numerical Analysis, **23**, 158–179.

Schmeling, H. (1987) On the relation between initial conditions and late stages of Rayleigh-Taylor instabilities. Tectonophysics, **133**, 65–80.

Schmeling, H. (2000). Partial melting and melt segregation in a convecting mantle. In: Bagdassarov, N., Laporte, D., Thompson, A. B. (editors), Physics and Chemistry of Partially Molten Rocks. Kluwer Academic, Dordrecht pp. 141–178.

Schmeling, H., Babeyko, A. Y., Enns, A., Faccenna, C., Funiciello, F., Gerya, T., Golabek, G. J., Grigull, S., Kaus, B. J. P., Morra, G., Schmalholz, S. M., van Hunen, J. (2008) A benchmark comparison of spontaneous subduction models – towards a free surface. Physics of the Earth and Planetary Interiors, **171**, 198–223.

Schmeling, H., Kruse, J. P., Richard, G. (2012) Mineral physics, rheology, heat flow and volcanology. Effective shear and bulk viscosity of partially molten rock based on elastic moduli theory of a fluid filled poroelastic medium. Geophysical Journal International, **190**, 1571–1578.

Schmeling, H., Marquart, G., Grebe, M. (2018) A porous flow approach to model thermal non-equilibrium applicable to melt migration. Geophysical Journal International, **212**, 119–138.

Schmid, D. W., Podladchikov, Y. Y. (2003) Analytical solutions for deformable elliptical inclusions in general shear. Geophysical Journal International, **155**, 269–288.

Schmidt, M. W., Poli, S. (1998) Experimentally based water budgets for dehydrating slabs and consequences for arc magma generation. Earth and Planetary Science Letters, **163**, 361–379.

Schubert, G. (1992) Numerical models of mantle convection. Annual Review of Fluid Mechanics, **24**, 359–394.

Schubert, G., Yuen, D. A., Turcotte, D. L. (1975) Role of phase transitions in a dynamic mantle. Geophysical Journal of the Royal Astronomical Society, **42**, 705–735.

Schubert, M., Driesner, T., Gerya, T. V., Ulmer, P. (2013) Mafic injection as a trigger for felsic magmatism: a numerical study. Geochemistry, Geophysics, Geosystems, **14**, 1910–1928.

Scott, D. R., Stevenson, D. J. (1984) Magma solitons. Geophysical Research Letters, **11**, 61–64.

Scott, D. R., Stevenson, D. J. (1986) Magma ascent by porous flow. Journal of Geophysical Research, **91**, 9283–9296.

Senshu, H., Kuramoto, K., Matsui, T. (2002) Thermal evolution of a growing Mars. Journal of Geophysical Research, **107**, E12, 5118, doi:10.1029/2001JE001819.

Shabana, A. A. (2008) Computational Continuum Mechanics. Cambridge University Press.

Shukla, K. N. (2005) Mathematical Principles of Heat Transfer. Begell House, New York.

Simakin, A., Botcharnikov, R. (2001) Degassing of stratified magma by compositional convection. Journal of Volcanology and Geothermal Research, **105**, 207–224.

Simpson, G. (2017) Practical Finite Element Modeling in Earth Science Using Matlab. Wiley-Blackwell.

Simpson, G., Spiegelman, M. (2011) Solitary wave benchmarks in magma dynamics. Journal of Scientific Computing, **49**, 268–290.

Simpson, G., Spiegelman, M., Weinstein, M. (2010a) A multiscale model of partial melts: 1. Effective equations. Journal of Geophysical Research, **115**, B04410.

Simpson, G., Spiegelman, M., Weinstein, M. (2010b) A multiscale model of partial melts: 2. Numerical results. Journal of Geophysical Research, **115**, B04411.

Sizova, E., Gerya, T., Brown, M., Perchuk, L. L. (2010) Subduction styles in the Precambrian: insight from numerical experiments. Lithos, **116**, 209–229.

Sizova, E., Gerya, T., Brown, M. (2012) Exhumation mechanisms of melt-bearing ultrahigh pressure crustal rocks during collision of spontaneously moving plates. Journal of Metamorphic Geology, **30**, 927–955.

Sizova, E. V., Gerya, T. V., Brown, M. (2014) Contrasting styles of Phanerozoic and Precambrian continental collision. Gondwana Research, **25**, 522–545.

Sizova, E., Gerya, T., Stuewe, K., Brown, M. (2015) Generation of felsic crust in the Archean: a geodynamic modeling perspective. Precambrian Research, **271**, 198–224.

Sleep, N. H. (1974) Segregation of a magma from a mostly crystalline mush. Geological Society of America Bulletin, **85**, 1225–1232.

Sleep, N. H. (1995) Ductile creep, compaction, and rate and state dependent friction within major fault zones. Journal of Geophysical Research, **100**, 13065–13080.

Sleep, N. H. (1997) Application of a unified rate and state friction theory to the mechanics of fault zones with strain localization. Journal of Geophysical Research, **102**, 2875–2895.

Smith, D. C. (1984) Coesite in clinopyroxene in the Caledonides and its implications for geodynamics. Nature, **310**, 641–644.

Sobolev, S. V., Babeyko, A. Y. (1994) Modeling of mineralogical composition, density and elastic-wave velocities in anhydrous magmatic rocks. Surveys in Geophysics, **15**, 515–544.

Sobolev, S. V., Babeyko, A. Y. (2005) What drives orogeny in the Andes? Geology, **33**, 617–620.

Sobolev, S. V., Muldashev, I. A. (2017) Modeling seismic cycles of great megathrust earthquakes across the scales with focus at postseismic phase. Geochemistry, Geophysics, Geosystems, **18**, 4387–4408.

Sobolev, N. V., Shatsky, V. S. (1990) Diamond inclusions in garnets from metamorphic rocks: an environment for diamond formation. Nature, **343**, 742–745.

Sobolev, S. V., Sobolev, A. V., Kuzmin, D. V., Krivolutskaya, N. A., Petrunin, A. G., Arndt, N. T., Radko, V. A., Vasiliev, Y. R. (2011) Linking mantle plumes, large igneous provinces and environmental catastrophes. Nature, **477**, 312–316.

Souza de Neto, E. A., Periæ, D., Owen, D. R. J. (2009) Computational Methods for Plasticity: Theory and Applications. Wiley.

Spakman, W., Wortel, M. J. R., Vlaar, N. J. (1988) The Hellenic subduction zone: a tomographic image and its geodynamic implications. Geophysical Research Letters, **15**, 60–63.

Spera, F. J., Oldenburg, C. M., Christensen, C., Todesco, M. (1995) Simulations of convection with crystallization in the system $KAlSi_2O_6$-$CaMgSi_2O_6$: implications for compositionally zoned magma bodies. American Mineralogist, **80** (11–12), 1188–1207.

Spiegelman, M. (1993) Flow in deformable porous media. Part 1. Simple analysis. Journal of Fluid Mechanics, **247**, 17–38.

Spiegelman, M., Katz, R. F. (2006) A semi-Lagrangian Crank-Nicolson algorithm for the numerical solution of advection-diffusion problems. Geochemistry, Geophysics, Geosystems, **7**, Q04014.

Spiegelman, M., Kelemen, P. B. (2003) Extreme chemical variability as a consequence of channelized melt transport. Geochemistry, Geophysics, Geosystems, **4**, Article No. 1055.

Spiegelman, M., May, D. A., Wilson, C. R. (2016) On the solvability of incompressible Stokes with viscoplastic rheologies in geodynamics. Geochemistry, Geophysics, Geosyststems, **17**, 2213–2238.

Stadler, G., Gurnis, M., Burstedde, C., Wilcox, L. C., Alisic, L., Ghattas, O. (2010) The dynamics of plate tectonics and mantle flow: from local to global scales. Science, **329**, 1033–1038.

Stern, R. J. (2016) Is plate tectonics needed to evolve technological species on exoplanets? Geoscience Frontiers, **7**, 573–580.

Stern, R. J., Gerya, T. (2018) Subduction initiation in nature and models: a review. Tectonophysics, **746**, 173–198.

Stevenson, D. J. (1980) Self regulation and melt migration (can magma oceans exist?). Transactions American Geophysical Union EOS, **61**, 1021.

Stevenson, D. J. (1981) Models of the Earth's core. Science, **214**, 611–619.

Stevenson, D. J. (2008) A planetary perspective on the deep Earth. Nature, **451**, 261–265.

Stevenson, D., Scott, D. (1991) Mechanics of fluid-rock systems. Annual Review of Fluid Mechanics, **23**, 305–339.

Stixrude, L., Bukowinski, M. S. T. (1990) Fundamental thermodynamic relations and silicate melting with implications for the constitution of D". Journal of Geophysical Research, **95**, 19311–19325.

Stixrude, L., Lithgow-Bertelloni, C. (2005) Thermodynamics of mantle minerals – I. Physical properties. Geophysical Journal International, **162**, 610–632.

Stixrude, L., Lithgow-Bertelloni, C. (2011) Thermodynamics of mantle minerals – II. Phase equilibria. Geophysical Journal International, **184**, 1180–1213.

Stoeckhert, B., Gerya, T. V. (2005) Pre-collisional high pressure metamorphism and nappe tectonics at active continental margins: a numerical simulation. Terra Nova, **17**, 102–110.

Tackley, P. J. (1993) Effects of strongly temperature-dependent viscosity on time-dependent, 3-dimensional models of mantle convection. Geophysical Research Letters, **20**, 2187–2190.

Tackley, P. J. (2000) Self-consistent generation of tectonic plates in time-dependent, three-dimensional mantle convection simulations Part 1: Pseudo-plastic yielding. Geochemistry, Geophysics, Geosystems, **1**, 2000GC000036.

Tackley, P. J. (2008) Modelling compressible mantle convection with large viscosity contrasts in a three-dimensional spherical shell using the yin-yang grid. Physics of the Earth and Planetary Interiors, **171**, 7–18.

Tackley, P. J. (2012) Dynamics and evolution of the deep mantle resulting from thermal, chemical, phase and melting effects. Earth Science Reviews, **110**, 1–25.

Tackley, P. J., Ammann, M., Brodholt, J. P., Dobson, D. P., Valencia, D. (2014) Habitable planets: interior dynamics and long-term evolution. In: Haghighipour, N. (editor), Formation, Detection and Characterization of Extrasolar Habitable Planets, Proceedings IAU Symposium No. 293, 2012. Cambridge University Press, pp. 339–349.

Tamura, Y. (1994) Genesis of island arc magmas by mantle derived bimodal magmatism: evidence from the Shirahama Group, Japan. Journal of Petrology, **35**, 619–645.

Thieulot, C. (2014) ELEFANT: a user-friendly multipurpose geodynamics code. Solid Earth Discussion, **6**, 1949–2096.

Thieulot, C. (2017) Analytical solution for viscous incompressible Stokes flow in a spherical shell. Solid Earth, **8**, 1181–1191.

Tikhonov, A. N., Samarsky, A. A. (1972) Equations of Math Physics. Nauka, Moscow (in Russian).

Tonks, W. B., Melosh, H. J. (1992) Core formation by giant impacts. Icarus, **100**, 326–346.

Toro, E. F. (1999) Riemann Solvers and Numerical Methods for Fluid Dynamics. Springer-Verlag.

Torrance, K. E., Turcotte, D. L. (1971) Thermal convection with large viscosity variations. Journal of Fluid Mechanics **47**, 113–125.

Tosi, N., Stein, C., Noack, L., Hüttig, C., Maierová, P., Samuel, H., Davies, D. R., Wilson, C. R., Kramer, S. C., Thieulot, C., Glerum, A., Fraters, M., Spakman, W., Rozel, A., Tackley, P. J. (2015) A community benchmark for viscoplastic thermal convection in a 2-D square box. Geochemistry, Geophysics, Geosystems, **16**, 2175–2196.

Turcotte, D. L., Schubert, G. (2002) Geodynamics. Cambridge University Press, Cambridge.

Turcotte, D. L., Schubert, G. (2014) Geodynamics, 3rd Edition. Cambridge University Press, Cambridge.

Ueda, K., Gerya, T., Sobolev, S. V. (2008) Subduction initiation by thermal-chemical plumes: numerical studies. Physics of the Earth and Planetary Interiors, **171**, 296–312.

Ueda, K., Gerya, T. V., Burg, J.-P. (2012) Delamination in collisional orogens: thermo-mechanical modelling. Journal of Geophysical Research, **117**, B08202.

Ueda, K., Willett, S. D., Gerya, T., Ruh, J. (2015) Geomorphological-thermo-mechanical modeling: application to orogenic wedge dynamics. Tectonophysics, **659**, 12–30.

Vance, D., Bickle, M., Ivy-Ochs, S., Kubik, P. W. (2003) Erosion and exhumation in the Himalaya from cosmogenic isotope inventories of river sediments. Earth and Planetary Science Letters, **206**, 273–288.

van Dinther, Y., Gerya, T. V., Dalguer, L. A., Corbi, F., Funiciello, F., Mai, P. M. (2013a) The seismic cycle at subduction thrusts: 2. Dynamic implications of geodynamic simulations validated with laboratory models. Journal of Geophysical Research, **118**, 1502–1525.

van Dinther, Y., Gerya, T. V., Dalguer, L. A., Mai, P. M., Morra, G., Giardini, D. (2013b) The seismic cycle at subduction thrusts: insights from seismo-thermo-mechanical models. Journal of Geophysical Reasearch, **118**, 6183–6202.

van Dinther, Y., Gerya, T. V., Dalguer, L. A., Corbi, F., Funiciello, F., Mai, P. M. (2014) Modeling the seismic cycle in subduction zones: the role and spatiotemporal occurrence of off-megathrust earthquakes. Geophysical Research Letters, **41**, 1194–1201.

van Dinther, Y., Kuensch, H. R., Fichtner, A. (2019) Ensemble data assimilation for earthquake sequences: probabilistic estimation and forecasting of fault stresses. Geophysical Journal International, DOI: https://doi.org/10.1093/gji/ggz063.

van Heck, H., Tackley, P. J. (2011) Plate tectonics on super-Earths: equally or more likely than on Earth. Earth and Planetary Science Letters, **310**, 252–261.

van Hunen, J., Allen, M. B. (2011) Continental collision and slab break-off: a comparison of 3D numerical models with observations. Earth and Planetary Science Letters, **302**, 27–37.

van Hunen, J., Moyen, J.-F. (2012) Archean subduction: fact or fiction? Annual Review of Earth and Planetary Sciences, **40**, 195–219.

van Hunen, J., van den Berg, A. (2008) Plate tectonics on the early Earth: limitations imposed by strength and buoyancy of subducted lithosphere. Lithos, **103**, 217–235.

van Keken, P., King, S., Schmeling, H., Christensen, U., Neumeister, D., Doin, M.-P. (1997) A comparison of methods for the modeling of thermochemical convection. Journal of Geophysical Research, **102**, 22477–22495.

van Keken, P. E., Currie, C., King, S. D., Behn, M. D., Cagnioncle, A., He, J., Katz, R. F., Lin, S.-C., Parmentier, E. M., Spiegelman, M., Wang, K. (2008) A community benchmark for subduction zone modeling. Physics of the Earth and Planetary Interiors, **171**, 187–197.

van Thienen, P., van den Berg, A. P., Vlaar, N. J. (2004) Production and recycling of oceanic crust in the early Earth. Tectonophysics, **386**, 41–65.

Vasilyev, O. V., Podladchikov, Y. Y., Yuen, D. A. (1998) Modeling of compaction driven flow in poro-viscoelastic medium using adaptive wavelet collocation method. Geophysical Research Letters, **25**, 3239–3242.

Vasilyev, O. V., Gerya, T. V., Yuen, D. A. (2004) The application of multidimensional wavelets to unveiling multi-phase diagrams and in situ physical properties of rocks. Earth and Planetary Science Letters, **223**, 49–64.

Vogt, K., Gerya, T. V., Castro, A. (2012) Crustal growth at active continental margins: numerical modeling. Physics of the Earth and Planetary Interiors, **192–193**, 1–20.

Vogt, K., Castro, A., Gerya, T. (2013) Numerical modeling of geochemical variations caused by crustal relamination. Geochemistry, Geophysics, Geosystems, **14**, 1131–1155.

Von Blanckenburg, F., Davies, J. H. (1995) Slab breakoff: a model for syncollisional magmatism and tectonics in the Alps. Tectonics, **14**, 120–131.

Wada, K., Kokubo, E., Makino, J. (2006) High-resolution simulations of a Moon-forming impact and postimpact evolution. Astrophysical Journal, **638**, 1180–1186.

Wang, K. (2007) Elastic and Viscoelastic Models of Crustal Deformation in Subduction Earthquake Cycles. Columbia University Press, New York.

Wang, H., Agrusta, R., van Hunen, J. (2015) Advantages of a conservative velocity interpolation (CVI) scheme for particle-in-cell methods with application in geodynamic modeling. Geochemistry, Geophysics, Geosystems, **16**, 2015–2023.

Warren, C. J., Beaumont, C., Jamieson, R. A. (2008) Modelling tectonic styles and ultrahigh pressure (UHP) rock exhumation during the transition from oceanic subduction to continental collision. Earth and Planetary Science Letters, **267**, 129–145.

Weinberg, R. B., Schmeling, H. (1992) Polydiapirs: multiwavelength gravity structures. Journal of Structural Geology, **14**, 425–436.

Weis, P., Driesner, T., Heinrich, C. A. (2012) Porphyry-copper ore shells form at stable pressure-temperature fronts within dynamic fluid plumes. Science, **338**, 1613–1616.

Wesseling, P. (1992) An Introduction to Multigrid Methods. John Wiley & Sons, Chichester.

Willett, S. D. (1999) Orogeny and orography: the effects of erosion on the structure of mountain belts. Journal of Geophysical Research, **104** (B12), 28957–28981.

Woidt, W. D. (1978) Finite-element calculations applied to salt dome analysis. Tectonophysics, **50** (2–3), 369–386.

Wong A Ton, S. Y. M., Wortel, M. J. R. (1997) Slab detachment in continental collision zones: an analysis of controlling parameters. Geophysical Research Letters, **24** (16), 2095–2098.

Wortel, M. J. R., Spakman, W. (1992) Structure and dynamics of subducted lithosphere in the Mediterranean region. Proceedings of the Koninklijke Nederlandse Akademie van Wetenschappen, Volume **95**, pp. 325–347.

Wortel, M. J. R., Spakman, W. (2000) Geophysics – subduction and slab detachment in the Mediterranean-Carpathian region. Science, **290**, 1910–1917.

Xie, S., Tackley, P. J. (2004a) Evolution of helium and argon isotopes in a convecting mantle. Physics of the Earth and Planetary Interiors, **146**, 417–439.

Xie, S., Tackley, P. J. (2004b) Evolution of U-Pb and Sm-Nd systems in numerical models of mantle convection. Journal of Geophysics Research, **109**, B11204, doi:10.1029/2004JB003176.

Xu, P. F., Sun, R. M., Liu, F. T., Wang, Q., Cong, B. (2000) Seismic tomography showing, subduction and slab breakoff of the Yangtze block beneath the Dabie–Sulu orogenic belt. Chinese Science Bulletin, **45**, 70–74.

Yamato, P., Burov, E., Agard, P., Le Pourhiet, L., Jolivet, L. (2008) HP-UHP exhumation during slow continental subduction: self-consistent thermodynamically and thermo-mechanically coupled model with application to the Western Alps. Earth and Planetary Science Letters, **271**, 63–74.

Yarushina, V. M., Podladchikov, Y. Y. (2015) (De)compaction of porous viscoelastoplastic media: model formulation. Journal of Geophysical Research, **120**, 4146–4170.

Yarushina, V. M., Bercovici, D., Oristaglio, M. L. (2013) Rock deformation models and fluid leak-off in hydraulic fracturing. Geophysical Journal International, **194**, 1514–1526.

Yarushina, V. M., Bercovici, D., Michaut, C. (2015) Two-phase dynamics of volcanic eruptions: particle size distribution and the conditions for choking. Journal of Geophysical Research, **120**, 1503–1522.

Yarushina, V. M., Podladchikov, Y. Y., Minakov, A., Rass, L. (2017) On the mechanisms of stress-triggered seismic events during fluid injection. In: Vandamme, M., Dangla, P., Pereira, J. M., Ghabezloo, S. (editors), Poromechanics VI: Proceedings of the Sixth Biot Conference on Poromechanics, pp. 795–800.

Yoshioka, S., Wortel, M. J. R. (1995) Three-dimensional numerical modeling of detachment of subducted lithosphere. Journal of Geophysical Research, **100**, 20233–20244.

Yoshioka, S., Yuen, D. A., Larsen, T. B. (1995) Slab weakening: thermal and mechanical consequences for slab detachment. Island Arc, **40**, 89–103.

Yuen, D. A., Balachandar, S., Hansen, U. (2000) Modelling mantle convection: a significant challenge in geophysical fluid dynamics. In: Fox, P. A., Kerr, R. M. (editors), Geophysical and Astrophysical Convection. CRC Press, pp. 257–294.

Zeng, Q., Lia, K., Fen-Chong, T., Dangla, P. (2012) Effect of porosity on thermal expansion coefficient of cement pastes and mortars. Construction and Building Materials, **28**, 468–475.

Zhang, D. Z., Prosperetti, A. (1994) Averaged equations for inviscid disperse two-phase flow. Journal of Fluid Mechanics, **267**, 185–219.

Zheng, L., Zhang, H., Gerya, T., Knepley, M., Yuen, D. A., Shi, Y. (2013) Implementation of a multigrid solver on a GPU for Stokes equations with strongly variable viscosity based on Matlab and CUDA. International Journal of High Performance Computing Applications, doi:10.1177/1094342013478640.

Zhigadla, D. (2015) Numerical modeling of meteorite impacts with continuum-based approach. MSc Thesis, ETH-Zurich.

Zhong, S. (1996) Analytic solutions for Stokes' flow with lateral variations in viscosity. Geophysical Journal International, **124**, 18–28.

Zhong, S., Gurnis, M. (1994) The role of plates and temperature-dependent viscosity in phase change dynamics. Journal of Geophysical Research, **99**, 15903–15917.

Zhong, S. J., Yuen, D. A., Moresi, L. N. (2007) Numerical methods in mantle convection. In: Schubert, G., Bercovici, D. (editors), Treatise in Geophysics, Volume **7**. Elsevier, pp. 227–252.

Zhong, S., McNamara, A., Tan, E., Moresi, L., Gurnis, M. (2008) A benchmark study on mantle convection in a 3-D spherical shell using CitcomS. Geochemistry, Geophysics, Geosystems, **9**, Q10017, doi:10.1029/2008GC002048.

Zhu, W., David, C., Wong, T.-F. (1995) Network modelling of permeability evolution during cementation and hot isostatic pressing. Journal of Geophysical Research, **100**, 15451–15464.

Zhu, G., Gerya, T. V., Yuen, D. A., Honda, S., Yoshida, T., Connolly, J. A. D. (2009) 3-D Dynamics of hydrous thermal-chemical plumes in oceanic subduction zones. Geochemistry, Geophysics, Geosystems, **10**, Q11006.

Zienkiewicz, O. C., Taylor, R. L., Zhu, J. Z. (2005) The Finite Element Method: Its Basis and Fundamentals, 6th Edition. Butterworth and Heinemann Inc.

Zlotnik, S., Fernandez, M., Diez, P., Verges, J. (2008) Modelling gravitational instabilities: slab break-off and Rayleigh–Taylor diapirism. Pure and Applied Geophysics, **165**, 1491–1510.

Index